ENVIRONMENTAL MANAGEMENT

Readings and Case Studies

Edited by
Lewis A. Owen
University of California, Riverside

Tim Unwin
Royal Holloway, University of London

Advisory Editors

Kevin Pickering, University College London
Judith Rees, London School of Economics
Kerry Turner, University of East Anglia
Nigel Woodcock, University of Cambridge

Editorial matter and organization copyright © Lewis A. Owen and Tim Unwin 1997

First published 1997

2 4 6 8 10 9 7 5 3 1

Blackwell Publishers Ltd
108 Cowley Road
Oxford OX4 1JF
UK

Blackwell Publishers Inc.
350 Main Street
Malden, MA 02148
USA

British Library Cataloguing in Publication Data

A CIP catalogue record for this book is available from the British Library.

Library of Congress Cataloging-in-Publication Data

Environmental management: readings and case studies/ edited by Lewis A. Owen and Tim Unwin.
   p.   cm. — (Blackwell readers on the natural environment)
ISBN 0–631–20116–5 — ISBN 0–631–201173 (pbk.)
1.   Environmental management.   I. Owen, Lewis A., 1964–
II. Unwin, P. T. H.   III. Series.
GE300.E59   1997
363.7′05—DC21.   96–47639
                                                                              CIP

Printed in Great Britain by T.J. International, Padstow, Cornwall

CONTENTS

CONTRIBUTORS

D. M. Anderson, Woods Hole Oceanographic Institution, Woods Hole, MA 02543, USA

S. P. Beaton, Department of Chemistry, University of Denver, Denver, CO 80208, USA

E. Blum, Box 192 Vanderbilt Hall, 107 Avenue Louis Pasteur, Boston, MA 02115, USA

T. G. Brydges, Environment Canada, Atmospheric Environment Service, 4905 Dufferin Street, Downsview, Ontario, Canada M3H 5T4

G. C. Daily, Energy and Resources Group, Building T-4, Room 100, University of California, Berkeley, CA 94720, USA

D. Daly, Geological Survey of Ireland, Beggars Bush, Haddington Road, Dublin 4, Ireland

W. Dansgaard, The Niels Bohr Institute, Department of Geophysics, University of Copenhagen, Haraldsgade 6, DK–2200 Copenhagen N, Denmark

P. H. Gleick, Global Environment Program, Pacific Institute for Studies in Development, Environment and Security, Oakland, California, USA

M. Golnarahgi, Division of Applied Sciences, Harvard University, Boston, Massachusetts, USA

M. Grubb, Royal Institute of International Affairs, London, UK

R. W. Hahn, Heinz School, Carnegie Mellen University, Pittsburgh, USA

J. T. Houghton, Meteorological Office, Bracknell, UK

T. P. Hughes, Department of Marine Biology, James Cook University, Townsville, QLD 4811, Australia

M. Hulme, Climatic Research Unit, University of East Anglia, Norwich, NR4 7JJ, UK

M. Koch, Technische Universität, Berlin

M. P. McCormick, Atmospheric Sciences Division, NASA Langley Research Center, Hampton, Virginia 23681–0001, USA

M. F. Myers, Natural Hazards Research and Application Information Center, University of Colorado, Boulder, Colorado, USA

H. Newby, Vice-Chancellor, University of Southampton, Highfield, Southampton SO17 1BJ, UK

D. W. Pearce, Centre for Social and Economic Research on the Global Environment, University College London, Gower Street, London WC1 6BT, UK

F. Pearce, c/o New Scientist Publications, IPC Magazines Ltd., King's Reach Tower, Stamford Street, London SE1 9LS, UK

C. M. Peters, Institute of Economic Botany, New York Botanical Garden, Bronx, New York 10458, USA

D. Pimentel, 5126 Constock Hall, College of Agriculture and Life Sciences, Cornell University, Ithaca, NY 14853–0901, USA

S. L. Pimm, Department of Ecology and Evolutionary Biology, University of Tennessee, Knoxville, TN 87996, USA

M. Redclift, Environmental Section, University of London, Wye, Ashford, Kent TN25 1UR, UK

J. Rees, Department of Geography, London School of Economics, Houghton Street, London, WC2A 2AE, UK

C. Safina, National Audubon Society, Living Oceans Program, Scully Science Center, 550 South Bay Avenue, Islip, NY 11751, USA

D. Skole, Institute for the Study of Earth, Oceans and Space, University of New Hampshire, Durham, NH 03824, USA

R. K. Turner, Executive Director, Centre for Social and Economic Research on the Global Environment, University of East Anglia, Norwich NR4 7JJ, UK

ACKNOWLEDGEMENTS

The editor and publishers wish to thank the following for permission to use copyright material:

American Economic Association for material from Hahn, R.W. (1989) 'Economic prescriptions for environmental problems: how the patient followed the doctor's orders', *Journal of Economic Perspectives*, 3(2) 95–114;

American Association for the Advancement of Science for Pimm, S.L., Russell, G.J., Gittleman, J.L. and Brooks, T.M. (1995) 'The future of biodiversity', *Science*, 269, 347–50. Copyright © 1995 American Association for the Advancement of Science; Daily G.C. (1995) 'Restoring value to the world's degraded lands', *Science*, 269, 351–4. Copyright © 1995 American Association for the Advancement of Science; Beaton, S.P., Bishop, G.A., Yi Zhang, Ashbaugh, L.L., Lawson, D.R. and Stedman, D.H. (1995) 'On-road vehicle emission: regulations, costs and benefits', *Science*, 268, 991–3. Copyright © 1995 American Association for the Advancement of Science; and Pimental, D., Harvey, C., Resosudarmo, P., Sinclair, K., Kurz, D., McNair, M., Crist, S., Shpritz, L., Fitton, L., Saffouri, R. and Blair, R. (1995) 'Environmental and economic costs of soil erosion and conservation benefits', *Science*, 267, 1117–23. Copyright © American Association for the Advancement of Science;

BSA Publications Ltd for Newby, H. (1991) 'One world, two cultures: sociology and the environment', *BSA Bulletin Network*, 50, 1–8;

Cambridge University Press for Warrick, R.A. and Rahman, A.A. (1992) 'Future sea level rise: environmental and socio-political considerations' in *Confronting Climatic Change: Risks, Implications and Responses*, ed. I.M. Mintzer, Stockholm Environment Institute, 97–112;

Earthscan for Koch, M. and Grubb, M. (1993) 'Agenda 21' in *The Earth Summit Agreements: a Guide and Assessment*, ed. Grubb, M., Koch, M., Thomson, K., Munson A., and Sullivan, F., 97–103;

The Geographical Association for Rees, J. (1991) 'Equity and environmental policy', *Geography*, 76(4), 292–303;

Intergovernmental Panel on Climate Change for Houghton, J.T., Jenkins, G.J. and Ephiraums, J.J., eds (1990) *Climate Change: The IPCC Scientific Assessment*, Cambridge University Press, pp. xii–xxxii; and Houghton, J.T. Meira Filho, L.G., Bruce, J., Hoesung Lee, Callander, B.A., Haites, E., Harris, N. and Maskell, K. (1995) *Climate Change 1994: Radiative Forcing of Climate, Executive Summary*, Cambridge University Press, 11–14;

Nature for Dansgaard, W., Johnsen, S.J., Clausen, H.B., Dahl-Jensen, D., Gundestrup, N.S., Hammer, C.U., Hvidberg, C.S., Steffensen, J.P., Sveinbjörnsdottir, A.E., Jouzel, J. and Bond, G. (1993) 'Evidence for general instability of past climate from a 250-kyr ice-core record', *Nature*, 364, 218–20. Copyright © 1993 Macmillan Magazines Ltd; and Peters, C.M., Gentry, A.H. and

Mendelsohn, R.O. (1989) 'Valuation of an Amazonian rainforest', *Nature*, 339, 655–6. Copyright © 1989 Macmillan Magazines Ltd;

IPC Magazines Ltd for Pearce, F. (1995) 'The biggest dam in the world', *New Scientist*, 28 January, 25–9. Copyright © New Scientist; Pearce, F. (1994) 'Rush for rock in the Highlands', *New Scientist*, 8 July, 11–12. Copyright © New Scientist; and Pearce, F. (1995) 'Dead in the water', *New Scientist*, 4 February, 23–31. Copyright © New Scientist;

Oxford University Press for Tietenberg, T.H. (1990) 'Economic instruments for environmental regulation', *Oxford Review of Economic Policy*, 6(1), 17–26, 30–3;

Penguin UK for Harrison, P. (1993) 'Executive summary' in *The Third Revolution: Population, Environment and a Sustainable World*, Penguin Books, 323–30. Copyright © 1992, 1993, Paul Harrison;

Prentice Hall for material from Pearce, D.W. and Turner, R.K. (1990) 'The historical development of environmental economics' in *Economics of Natural Resources and the Environment*, ed. D.W. Pearce and R.K. Turner, Harvester Wheatsheaf, 3–28;

The Royal Society of Edinburgh and the authors for Brydges, T.G. and Wilson, R.B. (1991) 'Acid rain since 1985 – times are changing', *Proceedings of the Royal Society of Edinburgh*, Section B: Biological Sciences, 97, 1–16;

Heldref Publications and the Helen Dwight Reid Educational Foundation for Hulme, M. and Kelly, M. (1993) 'Exploring the links between desertification and climate change', *Environment*, 35(6), 4–11, 39–46. Copyright © 1993; Brundtland, G.H. (1994) 'The solution to a global crisis', *Environment*, 36(10), 16–20. Copyright © 1994; Myers, M.F. and White, G.F. (1993) 'The challenge of the Mississippi floods', *Environment*, 39(10) 5–9, 23–5. Copyright © 1993; Mitchell, R.B. (1995) 'Lessons from intentional oil pollution', *Environment*, 37(4), 10–15, 36–41. Copyright © 1995; Golnaraghi, M. and Kaul, R. (1995) 'Responding to ENSO', *Environment*, 37(1), 16–20, 38–44. Copyright © 1995; and Gleick, P.H. (1994) 'Water, war and peace in the Middle East', *Environment*, 36(3), 6–15, 35–42. Copyright © 1994;

Scientific American, Inc for Safina, C. (1995) 'The world's imperiled fish', *Scientific American*, November, 46–53. Copyright © 1995 by Scientific American, Inc; and Anderson, D.M. (1994) 'Red tides', *Scientific American*, August, 52–58. Copyright © 1994 by Scientific American, Inc;

R.K. Turner for material from Turner, R.K. (1993) 'Sustainability: principles and practice' in R.K. Turner, ed., *Sustainable Environmental Economics and Management: Principles and Practice*, Belhaven, 3, 21;

The University of Wisconsin Press for Daly, H.E. (1991) 'Towards an environmental macroeconomics', *Land Economics*, 67(2), 255–9;

The White Horse Press for Redclift, M. (1993) 'Sustainable development: needs, values, rights', *Environmental Values*, 2, 3–20;

Quaternary Research Association for Daly, D. (1992) 'Quaternary deposits and groundwater pollution', *Quaternary Proceedings*, 2, 79–89;

United Nations Environment Programme and World Health Organization for (1994) 'Air pollution in the world's megacities' from *Urban Air Pollution in Megacities of the World*, as published in *Environment*, 36(2), 4–13, 25–37.

Every effort has been made to trace the copyright holders but if any have been inadvertently overlooked the publishers will be pleased to make the necessary arrangement at the first opportunity.

INTRODUCTION AND GUIDE TO FURTHER READING

This volume has been compiled to present a collection of important and stimulating chapters on environmental management. The book is aimed at geography and environmental undergraduates, as well as teachers, researchers and practitioners involved in environmental management. The selected readings have been chosen with advice from teachers and researchers in environmental science. Care has been taken to provide a range of chapters to help illustrate the variety of sources in which environmental research is presented. The content of the original papers has not been modified, although some have been abbreviated. When papers have been slightly shortened or when plates and/or figures have been omitted there is a short note of the changes in the part introduction.

The volume is divided into seven sections, an introduction and six thematic parts, presenting selected chapters which illustrate key issues and strategies involved in environmental management. Many of the chapters presented are seminal and are of historical importance. Care has been taken, however, to choose chapters which are not too specialized or too long, and that can be clearly followed by an undergraduate student. Each thematic part has a short introduction, and each chapter is briefly explained. These introductions and explanations help place the chapters in context within the volume, as well as to help the student appreciate the chapter's particular value for environmental management. Brief descriptions of key textbooks are also

presented at the end of each part to help direct the reader to more comprehensive texts that are beyond the scope of this book.

It is impossible for a book of this nature to cover every aspect of environmental concern and management. The reader should, therefore, consider using the texts listed in the *suggested further reading* at the end of this Introduction to provide additional background. These texts will help improve and widen the reader's knowledge and understanding of many of the important principles addressed throughout the selected readings. The most comprehensive texts are Pickering and Owen (1997), O'Riordan (1994) and Middleton (1995). These also complement the selected readings.

In recent years, environmental concern has focused on the degradation of the Earth's natural systems mainly due to increased population pressures and reduced resources. Management arguments have been directed towards the actual impacts of human activity on natural environments and the means by which resources can be sustained. The continuing destruction of the biosphere, the pollution of the atmosphere and water resources, and land degradation have attracted much attention and concern over recent years. These sets of issues form the first four thematic parts of this book. Parts I–IV examine the processes and nature of these changes within these realms.

It is important that the reader fully understands these chapters in order to assess the dynamics and nature of environmental change within the Earth's natural systems to provide effective and safe environmental management strategies.

These chapters attempt to provide a basis to aid students of environmental management critically to assess the scientific issues involved in environmental management before addressing the economic, political and social aspects of these problems, which are dealt with in the last two parts of the book.

These chapters highlight the need for technical solutions to environmental problems as well as a consideration of the human context in which they occur.

The environment is continuously being 'managed' or regulated by a wide range of actors, many of whom do not have the conscious realization of the environmental implications of their actions. However, changes to the way such management occurs can have profound implications for the allocation of resources. Environmental management inevitably, therefore, has to confront the critical issues of social equity, economic interest, and the legitimation of power.

Part I, *Managing the Biosphere*, considers the disruption and destruction of the biosphere. It is particularly concerned with ways of preserving biodiversity. It emphasizes the importance of biodiversity for sustaining the world's biogeochemical cycles and preserving the stability of ecosystems, and as an important gene reserve. Part II, *Predicting and Managing Atmospheric Change*, examines the natural- and human-induced atmospheric changes that have the potential to disrupt ecosystems and ultimately lead to economic, social and political problems. This part looks at the mechanics, the evidence and the consequences of climate change. Part III, *Reducing Land Degradation*, examines the natural changes and those induced by human activity which result in the reduction of the quality and value of the land's surface. Part IV, *Managing Water Resources*, examines the pollution of the hydrosphere and considers various issues relating to the reduction of quality and abundance of water, as well as associated resources. Part V, *Economic*

Management, examines the three main themes of environmental economics: the micro-economics of environmental policy, and the economics of sustainable development. Finally, Part VI, *Political and Social Agendas*, considers the ideologies of environmental management, their political context, and the social issues involved in managing the environment.

It is hoped that the reader will find these readings stimulating and useful as a background to developing an interest and/or vocation in environmental management.

Suggested further reading

Botkin, D. and Keller, E. 1995: *Environmental Science: Earth as a Living Planet*. Chichester: Wiley.

This textbook examines the basic principles of environmental science. A large number of case studies provide useful and stimulating reading. Sections include: environment as an idea; the Earth as a system; life and the environment; sustaining living resources; energy; water environment; air pollution; and environment and society. It is a clearly written and useful companion text for students who are beginning to study these issues in environmental management.

Bradshaw, M. and Weaver, R. 1993: *Physical Geography: An Introduction to Earth Environments*. London: Mosby.

Physical Geography is a comprehensive and well illustrated introductory text which outlines the principles of environmental systems. It describes the dynamics of the atmosphere–ocean system, geological and geomorphological processes, soil environments and biomes systems. The human interaction with these natural environments is examined, and the characteristics of ecological systems are outlined.

Jackson, A. R. W. and Jackson, J. M. 1996: *Environmental Science: the Natural Environment and Human Impact*. Harlow: Longman.

This textbook is a good introductory text for undergraduates who are studying environmental science. The book explores the fundamental concepts of the natural environment, the interactions between the lithosphere, hydrosphere, atmosphere and biosphere. It also emphasizes the environmental consequences of human activity as a result of natural resource exploitation.

Lovelock, J. E. 1988: *The Ages of Gaia: A Biography of our Living Earth*. Oxford: Oxford University Press.

The Ages of Gaia is the follow-up book to *Gaia: A new look at life on Earth* (1982), which elaborates on the Gaia view of Earth, examining the interaction between the atmosphere, oceans, the Earth's crust, and the organisms that evolve and live on Earth. Recent scientific developments – including those on global warming, ozone depletion, acid rain and nuclear power – are discussed, providing a thought-provoking look at interdependence, and the role of negative and positive feedbacks in controlling the evolution and adaptability of life.

Middleton, N. 1995: *The Global Casino: An Introduction to Environmental Issues*. London: Edward Arnold.

Middleton's book provides an interesting introductory text for students who are embarking on environmental science courses. It explores a wide range of environmental issues, providing interesting examples from a wide range of environments.

O'Riordan, T. (ed.) 1994: *Environmental Science for Environmental Management*. Harlow: Longman.

This book comprises a compilation of chapters by academics from the University of East Anglia. It provides an interesting and informative introduction to environmental issues, and is an ideal companion for undergraduate students in environmental science and geography. The content ranges from environmental debates, economics and ethics, to climate change, pollution, geomorphological processes, remote sensing, environmental risk assessment, energy and disease prevention.

Pickering, K. T. and Owen, L. A. 1997: *An Introduction to Global Environmental Issues*.

London: Routledge, 2nd edition.

Pickering and Owen's textbook is an essential text for anyone looking for a full introduction to the science behind the world's physical systems and processes. The book explores the world's major environmental concerns including the effects on, and the effect of, human activity on environmental systems. The book includes chapters on: climate change and past climates; global atmospheric change; acid precipitation; water resources and pollution; nuclear issues; energy; natural hazards; the human impact on the Earth's surface; and managing the Earth. There is a section on issues for discussion, a bibliography and glossary. An instructor's manual accompanies the book.

Roberts, N. (eds) 1994: *The Changing Global Environment*. Oxford: Blackwell Publishers.

The Changing Global Environment comprises a collection of chapters on global change in a variety of forms and in a variety of environments. It is an excellent text for both specialists and non-specialist readers who require informative case studies and overviews of a variety of different environmental topics.

Watts, S. (ed.) 1996: *Essential Environmental Science*. London: Routledge.

A vast range of techniques, methods and basic tools necessary for the study of the environment are examined within the manual. It is valuable for practitioners and students at all levels.

Woodcock, N. 1994: *Geology and Environment in Britain and Ireland*. London: University College Press.

This is an excellent introduction to environmental geology, covering a variety of environmental topics from a geological perspective. It is essential reading for undergraduate students in geology, geography and environmental science who may be studying environmental geography/geology.

Part I

Managing the Biosphere

Introduction and Guide to Further Reading

1 Pimm, S. L., Russell, G. J., Gittleman, J. L. and Brooks, T. M. 1995: The future of biodiversity. *Science,* 21 July, 269, 347–50.

This paper discusses the uncertainties of calculating the number of species and the rates at which they are being exterminated. It emphasizes that recent extinction rates are accelerating and that there is a need to obtain more knowledge of areas which are rich in endemics in order to help reduce the rates of extinction.

2 Hughes, T. P. 1994: Catastrophes, phase shifts, and large-scale degradation of a Caribbean coral reef. *Science,* 265, 1547–51.

The nature of coral reef degradation in the Caribbean is described. The chapter provides an excellent example of how natural and human processes can lead to the catastrophic destruction of a sensitive ecosystem and highlights the need for immediate implementation of management procedures.

3 Safina, C. 1995: The world's imperiled fish. *Scientific American,* 273, 46–53.

Overfishing is leading to the accelerated destruction of the world's fisheries. This chapter highlights the major threats to, and the changes in abundance of, important fish stocks. It describes fishing technologies and discusses the economic considerations that are needed to help in the effective management of the world's fisheries.

4 Peters, C. M., Gentry, A. H. and Mendelsohn, R. O. 1989: Valuation of an Amazonian rainforest. *Nature,* 339, 655–6.

This important text attempts to calculate the true value of an area of Amazonian rainforest by considering a whole range of resources within the rainforest. The paper can be used as a basis to highlight the importance of the rainforest as a valuable resource.

5 Blum, E. 1993: Making biodiversity conservation profitable: a case study of the Merck/INbio agreement. *Environment,* 35, 4, 16–20 and 38–45.

This very useful text shows how business and conservation can work together effectively and beneficially for economic gain. It describes how a US pharmaceutical company, Merck, paid a Costa Rican organization INBio to look for species of potential value and to conserve precious habitats. It shows how conventions can be created to safeguard the rights of nations to protect their biogenetic commons.

The biosphere comprises a thin layer of organic matter that covers the Earth. It is generally less than a few metres in thickness over much of the land surface, but may reach several tens of metres in the rainforests. It also extends into the atmosphere, as creatures fly and plant spores are blown by the wind, and it extends into lakes and oceans where creatures live within the water column, and on or within sediments on the sea floor or lake bed. There are approximately 1.4 million formally described species of animals and plants, but there may be as many as 5 to 30 million species in total. Humans are intricately linked to the biosphere, interacting with and affecting nearly all of its component parts. In recent decades there has been a growing concern over the degradation of the biosphere, particularly with regard to the continued extinction of large numbers of animal and plant species. many of these extinctions and much of the degradation of the biosphere is a direct or indirect result of human activity.

Extinctions and biological change have occurred continuously throughout the Earth's history, and are the result of natural changes, induced by continental drift, climate change, volcanic eruptions and meteor impacts. Throughout geological time five major extinction events have been recognized. 439 Ma, at about the boundary of the Ordovician and Silurian periods; 375 Ma, late in the Devonian Period: 240 Ma, at the boundary between the Permian and Triassic periods; 210 Ma in the Triassic Period; and 65 Ma, at the boundary between the Cretaceous and Tertiary periods. The best known are the Cretaceous-Tertiary extinctions when the dinosaurs died out and an estimated 70 per cent of the flora disappeared. Although much debated, this extinction was probably a result of a series of meteorite impacts, which resulted in the ejection of huge volumes of very fine material into the upper atmosphere, which shaded the sun and absorbed the incoming solar radiation, ultimately leading to global cooling (Alvarez and Asaro, 1990). In addition, the impact may have resulted in wildfires which produced pyrotoxins exacerbating the environmental stress. The most dramatic extinction however, occurred at the end of the Permian Period, 240 Ma, when approximately 80–95 per cent of all known marine species became extinct (Erwin, 1994).

Homo sapiens sapiens have only been in existence for about the last 200,000 years, and yet their activities have resulted in the demise of a large number of species. As yet this has not been on the same scale as those described above. If not heeded, however,

human activity may well lead to extinctions on a similar scale. It is possible that human activity may already have been responsible for the extinction of many species of large land mammals (> 40 kg in weight). These became extinct shortly after the last Glacial some 10,000 years ago. More than 90 per cent of such animals became extinct in Australia, 80 per cent became extinct in South America, 73 per cent in North America, and 29 per cent in Europe. It is still controversial, however, as to whether this was directly the result of human impact on the natural environment, as people began to develop farming techniques, and animals were domesticated and hunting techniques improved, or if it was related to climate change (Stuart, 1993). Over the last few centuries, however, there is little doubt that human activities have resulted in the demise of a large number of species, notably tens of species of Moas in New Zealand, the dodo in Mauritius and the Indian cheetah. Unfortunately, most extinctions have occurred unnoticed, because the taxa that were exterminated were not formally described. This is particularly true for much of the life that has become extinct in the tropical rainforests and oceans. The exact nature and extent of biosphere degradation and the reduction in biodiversity is therefore not really known.

Human activity, however, does not only cause extinctions. It commonly leads to major changes in the structure of vegetation and animal populations within particular ecosystems. Particularly notable are agricultural activities, where natural vegetation is replaced by introduced species. Often the new ecosystem is dominated by a few species. Many of these species are of one strain, which reduces the genetic diversity within the region, making the ecosystem less resilient and less able to adapt to environmental changes such as droughts or deluges. Changes in vegetation cover may also have dramatic effects on the soil quality and may lead to disruptions in the nutrient cycles. To improve soil quality and productivity, new management strategies, such as the use of fertilizers, are normally required. Unfortunately, inappropriate application of such techniques can further exacerbate the environmental problems, as is illustrated by the increased concentrations of nitrates in rivers and coastal waters. These nutrients have been washed from fertilized farmland, and the increased nitrate concentrations have led to algal blooms, eutrophication and mass mortalities within aquatic ecosystems.

It must not be forgotten that biological degradation is not limited to the land surface, but has and still is occurring in the oceans and seas. The oceans contain the largest biomass, with phytoplankton its most important component. It is essential for the production of atmospheric O_2 by photosynthesis. Degradation of the marine ecosystem has important implications for cycling of nutrients, particularly carbon, oxygen and phosphorus. These are not only important for biological processes, but also play a critical role in the chemistry and physics of the atmosphere. Changes in the marine ecosystem may be both natural and human-induced. Changes in nutrients, water temperature, water turbidity, the introduction of new species and disease, for example, may alter the abundance of particular organisms and may lead to disruptions within food chains and food webs. Such changes may occur due to the following factors:

1   Changes in climate, natural or human-induced, leading to changes in the temperature of water and/or changes in the amount of ultraviolet radiation penetrating surface waters (Gleason and Wellington, 1993);
2   the addition of nutrients by pollution or changes in the circulation of deep sea currents, changing the productivity of organisms (Milne, 1989);
3   soil erosion due to poor land management and/or heavy rainfall events which increase the sediment loads in the rivers which feed into the sea, thus reducing the penetration of light into the water column, and hence reducing photosynthesis (Kühlmann, 1988); and
4   the migration of new species into regions where new waterways are opened, or as sea levels change due to

tectonic processes and/or climate change, or as new species travel along canals which join different water bodies (Spanier and Galil, 1991).

The realization that the biosphere is important in controlling and sustaining the concentration of atmospheric gases, and regional and local climates has slowly begun to attract the attention of both policy-makers and the public, as well as to dominate much of the scientific literature. In addition, the realization that the preservation of biodiversity is important not only as a gene store, a potentially rich source of yet unrecognized medical products, but as a critical buffer to reduce the effects of environmental change, is attracting much attention. The importance of mitigating the degradation of the biosphere is also essential in helping to sustain hydrological and geomorphological systems within a region. This in turn helps to reduce any increase in the magnitudes and frequencies of hazards that might be associated with environmental change. Nevertheless, increasing human pressures for available land will inevitably lead to changes in the biosphere. An understanding of the nature of the processes and the dynamics within the biosphere is therefore crucial for efficient management.

The need to preserve the biosphere was emphasized at the United Nations Conference on Environment and Development, popularly known as the Earth Summit, in Rio de Janeiro in 1994. At the Earth Summit, the UN Framework Convention on Biodiversity was signed and ratified by 152 countries with the aim of helping to reduce the threat to biodiversity. The convention requests that each country has to prepare a plan for conserving and sustaining biodiversity, and that it must monitor its own genetic stock, as well as providing financial support to aid the implementation of research and protection programmes. Another means of trying to conserve biodiversity was presented in the form of the World Conservation Strategy. This strategy aims to maintain ecological systems, ensure sustainable use of ecosystems and to initiate conservation schemes. These strategies have met with varying degrees of success, and they still require stronger legislation and enforcement to be fully effective. In Part I, a series of chapters is presented to help examine the threats to the biosphere and biodiversity and to investigate the extent of change, as well as providing some insight into possible mechanisms that may be pursued to help conserve biological systems.

The first chapter, by Pimm et al. (1995), discusses the nature of biodiversity and the uncertainties of calculating the number of species and the rates at which they are being exterminated. It examines the nature of the abundance and distribution of species and the key threats to the biosphere.

Not all threats to the biosphere are human-induced. Chapter 2 by Hughes (1994), illustrates this by examining the recent degradation of coral reefs in the Caribbean. In this region, a combination of erosion by hurricanes and disease leading to disruptions within the food chain has resulted in the extensive destruction of vast areas of reef. The chapter also illustrates how natural changes can be exacerbated by human activities. This is an important contribution because it highlights how the marine ecosystem can be degraded by a combination of processes which magnify the resultant effects.

One of the most recent issues regarding biosphere degradation involves the over-exploitation of fish stocks. There is now growing evidence that the world's fisheries have been reduced to such an extent that they may no longer be sustainable. This is particularly evident on the Grand Banks, once one of the world's most prolific fisheries, which has been so drastically depleted that fishing has had to be totally abandoned (MacKenzie, 1995). Management of the world's fisheries involves complex political and economic issues as well as an understanding of the complex biological modelling involved in calculating marine resources. Chapter 3, by Safina (1995), provides an excellent review of the state of the world's fisheries and describes the various kinds of human activity that have been degrading the marine ecosystem. It

also stresses the economic, political and biological considerations involved in managing the marine environment.

It is difficult to assess the 'true' value of specific ecosystems because of the intricate links between atmospheric, hydrological and geomorphological systems. Even an estimation of the commercial value is difficult and is usually under-calculated because not all of the possible biological resources are considered. The chapter by Peters, Gentry and Mendelsohn (1989), describes a comprehensive assessment of the value of an area of Amazonian rainforest. It examines the variety of resources that previously had not been considered in resource value calculations because they had been based primarily on the logging value of the forest. It is also provides a useful argument for the conservation and sustainable development of rainforests. The final chapter in Part I, by Blum (1993), is important because it provides an example of how biodiversity can be valued and how it can be considered a valuable resource. The chapter shows how business and conservation can work together effectively and beneficially to provide economic gain. It describes how a US pharmaceutical company, Merck, paid the organization INBio to examine the potential value of precious habitats in Costa Rica. It also shows how conventions can be created to safeguard the rights of nations to protect their biological resources.

The chapters presented in Part I have provided an illustration of the nature and importance of the biosphere and the need for sustainable management of the biosphere as a valuable resource.

References

Alvarez, W. and Asaro, F. 1990: An extraterrestrial impact. *Scientific American*, October, 44–60.

Erwin, D. H. 1994: The Permo-Triassic extinction. *Nature*, 367, 231–6.

Gleason, D. F. and Wellington, G. M. 1993: Ultraviolet radiation and coral bleaching. *Nature*, 365, 836–8.

Kühlmann, D. H. 1988: The sensitivity of coral reefs to environmental pollution. *Ambio*, 17, 13–21.

MacKenzie, D. 1995: The cod that disappeared. *New Scientist*, 16 September, 24–9.

Milne, R. 1989: North Sea algae threaten British coasts. *New Scientist*, 122, 1663, 37–41.

Spanier, E. and Galil, B. S. 1991: Lessepsian migration: a continuous biogeographical process. *Endeavour*, 15, 102–6.

Stuart, A. J. 1993: Death of the megafauna: mass extinction in the Pleistocene. *Geoscientist*, 2, 17–20.

Other references cited in the text are listed at the beginning of Part I.

Suggested further reading

Eden, M. J. 1989: *Land Management in Amazonia.* London: Belhaven.

In this text, Eden examines the tropical rainforest as a global resource and the needs for its conservation and development. Case studies are presented where, for example, there have been attempts to adapt resource-use systems of native peoples to encourage the more effective and less harmful exploitation of the rainforests.

Gradwohl, J. and Greenberg, R. 1988: *Saving Tropical Forests.* London: Earthscan.

This book comprises a collection of case studies which show how the destruction of the tropical forests might be slowed or even stopped, and how sustainable management could be achieved.

Grumbine, R. E. (ed.) 1994: *Environmental Policy and Biodiversity.* Washington D.C.: Island Press.

This text examines various aspects of biodiversity and environmental policy. It includes papers on: the ethnical and scientific bases of conservation biology; the effectiveness of existing environmental policy in protecting biodiversity; the examination of overall policy goals and processes; and case studies from the United States. It is an essential reference source for students, teachers, researchers and practitioners.

Huggett, R. J. 1995: *Geoecology: An Evolutionary Approach.* London: Routledge.

This text examines the dynamics of geoecosystems. It develops a model for geoecosystems, organized on a hierarchical basis, which continuously respond to changes within themselves, to the near-surface environment (atmosphere, hydrosphere and lithosphere), as well as to geological and cosmic influences.

Park, C. 1992: *Tropical Rainforests.* London: Routledge.

This book is a very readable account of the ecology of tropical rainforests. It examines the rates and patterns of clearance and the possible solutions to local, regional and global changes.

Poore, D. 1989: *No Timber Without Trees: Sustainability in the Tropical Forest.* London: Earthscan.

Based on a study for the International Tropical Timber Organisation, this book assesses the extent to which natural forests are being sustainably managed for timber production. It also discusses how these practices may be improved.

Tivy, J. 1993: *Biogeography: A Study of Plants in the Ecosphere.* Harlow: Longman, 3rd edition.

This is Tivy's third edition of her classic text on biogeography. It explores the variations in forms and functioning of the biosphere at both the regional and global scales. It highlights the interaction between the organic and inorganic components of the ecosphere. Emphasis is placed on the importance of the plant biosphere as the primary biological product which forms the vital food link between organisms. It also emphasizes the role of humans as the dominant ecological factor. It is an essential undergraduate textbook for those studying geography, biology, environmental studies, conservation and ecology.

1

THE FUTURE OF BIODIVERSITY*

Stuart L. Pimm, Gareth J. Russell, John L. Gittleman, and Thomas M. Brooks

Debates about the consequences of human population growth are not new. Our numbers have increased dramatically since Malthus but so has our technology (1). Will technical ingenuity keep pace with increasing population problems? Ingenuity can replace a whale-oil lamp with an electric light bulb, but not the whales we may hunt to extinction. Species matter to us (2). How fast we drive them to extinction is a matter of our future. Critics consider high estimates of current and future extinction rates to be "doomsday myths," contending that it is the "facts, not the species" that are endangered (3). Here, we review these estimates.

Extinctions have always been a part of Earth's history. So what is the background rate of extinction: how fast did species disappear in the absence of humanity (4)? A summary of 11 studies of marine invertebrates suggests that fossil species last from 10^6 to 10^7 years (5). For ease of comparison, we use the number of extinctions (E) per 10^6 species-years (MSY) or E/MSY. If species last from 10^6 to 10^7 years, then their rate of extinction is 1 to 0.1 E/MSY.

These estimates derive from the abundant and widespread species that dominate the fossil record. The species most prone to current extinction are rare and local. Moreover, we emphasize terrestrial verte-

brates in our discussions of current extinctions. There are only two studies of their fossils (5), and these suggest high background rates (1 E/MSY). Interestingly, we can supplement these estimates from our knowledge of speciation rates. These could not be much less than the extinction rates, or the groups would not be here for us to study.

Molecular phylogenies are now produced rapidly and extensively. There is one for 1700 bird species (6). Using the relative time axis of molecular distances, we can elucidate the patterns of species formation. Models in which every lineage has the same, constant probability of giving birth to a new lineage (speciation) or going extinct (death) permit estimation of the rate parameters (7). The rich details of this approach offer hope in testing for important factors controlling the relative rates of background speciation and extinction. Obviously, absolute rates require accurately dated events, such as the first appearance of a species or genus in the fossil record. There are genetic distance and paleontological estimates of divergence times for 72 carnivore and 14 primate species or subspecies (8). Given their importance as a benchmark against which to compare modern extinction rates, we plead for more absolutely timed accounts.

* Originally published in *Science*, 1995, vol. 269, pp. 347–50.

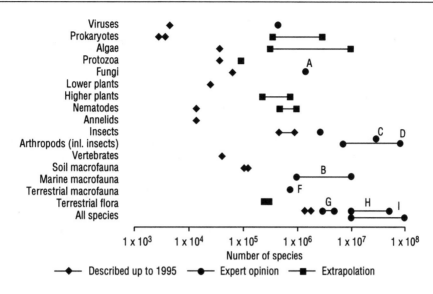

Viruses
Prokaryotes
Algae
Protozoa
Fungi
Lower plants
Higher plants
Nematodes
Annelids
Insects
Arthropods (inl. insects)
Vertebrates
Soil macrofauna
Marine macrofauna
Terrestrial macrofauna
Terrestrial flora
All species

1×10^3 1×10^4 1×10^5 1×10^6 1×10^7 1×10^8

Number of species

◆— Described up to 1995   ●— Expert opinion   ■— Extrapolation

Figure 1.1 [*orig. figure 1*]   Numbers of described species and estimates of species numbers, including expert opinions of taxonomic specialists (12) and various extrapolations (26). (**A**) The British ratio of 6 fungi species:1 plant species and a world total of 2.5×10^5 plant species suggests ~1.5 million species of fungi worldwide. (**B**) A world total of 10^6 to 10^7 species of marine macrofauna comes from the accumulation of new species along sample transects. (**C**) A large sample of canopy-dwelling beetles from one species of tropical trees had 163 species specific to it. There are 5×10^4 tree species, and so $163 \times 5 \times 10^4 \approx 8 \times 10^6$ species of canopy beetles. Because 40% of described insects are beetles, the total number of canopy insects is 2×10^7. Adding half that number for arthropod species on the ground gives a grand total of 3×10^7. (**D**) If only 20% of canopy insects are beetles, but there are at least as many ground as canopy species, then the grand total is 8×10^7. (**E**) Some 63% of the 1690 species on ~500 Indonesian tree species were previously unrecorded. The ~10^6 described insect species thus suggest a total of 2.7×10^6 species. (**F**) Across many food webs, there are roughly three times as many herbivores and carnivores combined as there are plants. This resulting estimate of terrestrial animal species, ~7.5×10^5, is certainly too low, because published food webs omit many species. (**G**) There are about two tropical bird and mammal species for each temperate or boreal species. Yet, of the ~1.5×10^6 described species, about one-third is tropical. The prediction of 3×10^6 species is an underestimate, because not all temperate species are described. (**H**) There is a linear increase in species numbers with decreasing body size. Below a threshold level, however, the numbers drops, perhaps because of sampling bias. If the true pattern remained linear, there would be 1×10^7 to 5×10^7 species. (**I**) We added the more detailed estimates for the numbers of species in the largest groups.

How Many Species Are There?

Any absolute estimate of extinction rate requires that we know how many species there are. In fact, we do not. May (9) shows that the problems of estimating their numbers are formidable. Only 10^6 species are described and ~10^5 – terrestrial vertebrates, some flowering plants, and invertebrates with pretty shells or wings –

are popular enough to be known well. Birds are exceptional in that differences in taxonomic opinion [~8500 to 9500 species (6)] far exceed the annual descriptions of new species (~1). Most species are as yet undescribed in every species-rich group (figure 1.1). Major uncertainties lie in those groups in which we have scant or conflicting evidence of very high diversity. There are ~10^6 described insects, yet estimates range from ~10^7 to nearly 10^8 species. Some potentially rich communities, such as the deep-sea benthos, have been sparsely sampled.

How can we be confident in our extrapolations of extinction rates from the <10^5 well-known species to the ~10^6 described, or to the conservative grand total of ~10^7 (5)? If extinction rates in diverse taxa and regions are broadly similar, then they are likely to be representative. If we understand the underlying mechanisms, we may find they operate universally.

The Past as a Guide to the Future

Unambiguous evidence of human impact on extinction comes from before-and-after comparisons of floras and faunas (10). Polynesians reached the planet's last habitable areas – Pacific islands – within the last 1000 to 4000 years. The bones of many bird species persist into, but not through, archaeological zones that show the presence of humans. No species disappeared in the longer intervals before the first human contact. Adding known and inferred extinctions, it seems that with only Stone Age technology, the Polynesians exterminated >2000 bird species, some ~15% of the world total.

We must infer extinctions, because we will not find the bones of every now-extinct species. From the overlap in species known from bones and those survivors seen by naturalists, sampling theory infers that ~50% of the species are still missing (10). Faunal reconstruction affords a second inference. For example, Steadman (10) contends that every one of ~800 Pacific islands should have had at least one unique species of rail. A few remote islands still have rails. Others lost theirs to introduced rats in the last century. Large volcanic islands typically lost several species of rails. Accessible islands lost their rails earlier, for every survey of bones from islands now rail-free has found species that did not survive human contact.

High extinction rates also followed the Pacific's colonization by Europeans. Since 1778, the Hawaiian islands have lost 18 species of birds; the fate of 12 more is unknown (10). Nor are birds unusual. Of 980 native Hawaiian plants, 84 are extinct and 133 have wild populations of <100 individuals (11). Across the Pacific, a predatory snail introduced to control another introduced snail ate to extinction hundreds of local varieties of land snails (12).

Nor are Pacific islands unusual: of 60 mammalian extinctions worldwide, 19 are from Caribbean islands (12). In the last 300 years, Mauritius, Rodrigues, and Réunion in the Indian Ocean lost 33 species of birds, including the dodo, 30 species of land snails, and 11 reptiles. St. Helena and Madeira in the Atlantic Ocean have lost 36 species of land snails (12).

Importantly, extinction centers are not necessarily on islands nor only in terrestrial environments. The fynbos, a floral region in southern Africa, has lost 36 plant species (of ~8500); 618 more are threatened with extinction (12). Extinctions of 18 (of 282) species of Australian mammals rival those from the Caribbean; 43 more are threatened (12). In the last century, North American freshwater environments lost 21 of 297 mussel and clam species (120 are threatened) and 40 of ~950 fish species (12).

This world tour of extinction centers has remarkable features (12). Recent extinction rates are 20 to 200 E/MSY (figure 1.2) – a small range given, among other things, the uncertainties of whether to average rates over a century or a shorter interval that reflects more recent human impacts. We find high rates in mainlands and islands, in arid lands and rivers, and for both plants and animals. Although we know less about invertebrates, high rates characterize bivalves of continental rivers and island land snails. There is nothing intrinsic to the diverse life histories of these species to predict their being unusually prone to extinction.

What obvious features unite extinction centers? We know the species and places well – as did naturalists a century ago. Importantly, each area holds a high proportion of species restricted to it. Such endemics constituted 90% of Hawaiian plants, 100% of Hawaiian land birds, ~70% of fynbos plants, and 74% of Australian

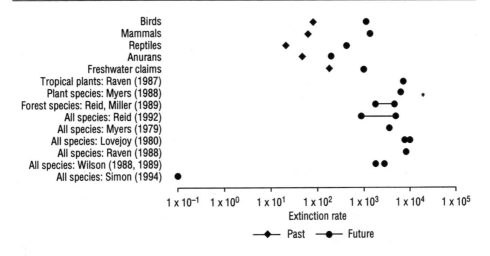

Figure 1.2 [*orig. figure 2*]   Estimates of extinction rates expressed as extinctions per million species-years. For birds through clams, we derive past rates from known extinctions in the last 100 years; we derive future rates by assuming that all currently threatened species will be extinct in 100 years. The latter rates are much higher than the former but are still far too low. The remaining estimates are previously published (1, 16, 27). Myers (1979) (27) assumes an exponential increase in the number of extinctions. Myers (1988) (16) assumes the loss of a small number of areas rich in endemics. With the exception of Simon, the rest are estimates based on the relation between habitat loss and species loss. Simon's claims (1) of one (or a few) species per year (out of a conservative total of 10^7 species) are not scientifically credible.

mammals (12). In contrast, only ~1% of Britain's birds and plants are endemics (12). Remote islands are typically rich in endemics, but so are many areas within continents (13). Past extinctions are so concentrated in small, endemic-rich areas that the analysis of global extinction is effectively the study of extinctions in one or a few extinction centers (12). Why should this be?

Random extinction is the simplest model. Some species groups and some places will suffer more extinctions than others, but generally the more species present, the more there will be to lose. This model does a poor job of predicting global patterns. If island birds were intrinsically vulnerable to extinction, then Hawai'i and Britain with roughly the same number of species of breeding land birds (~135) would have suffered equally. Hawai'i had >100 extinctions, Britain only 3 (12). Nor is the number of species an area houses a good predictor of the total extinctions. Islands house few species and suffer many extinctions.

Imagine a cookie-cutter model where some cause destroys (cuts out) a randomly selected area. Species also found elsewhere survive, for they can recolonize. But some of the endemics go extinct, the proportion depending on the extent of the destruction. We do not assume that island biotas are intrinsically more vulnerable than mainlands. For random species ranges, the number of extinctions correlates weakly with the area's total number of species, but strongly with the number of its endemics. By chance alone, small endemic-rich areas will contribute disproportionately to the total number of extinctions.

This model is consistent with known mechanisms of extinction. Habitat destruction cuts out areas, as the model implies. Introduced species also destroy species regionally. Species need not be entirely within the area destroyed to succumb to extinction: The populations outside may be too small to persist (14). Moreover, across many taxa, range-restricted species have lower local densities than widespread species (15). The former are not only more likely to be cut in the first place, but their surviving populations will have lower densities and thus higher risks of extinction than wide-spread species. This entirely self-evident model emphasizes the localization of endemics – Myers' "hot spots" (16) – as the key variable in understanding global patterns of recent and future extinctions.

Predicting Future Rates of Extinction

Projecting past extinction rates into the future is absurd for no other reason than that the ultimate cause of these extinctions – the human population – is increasing exponentially. For vertebrates, we have worldwide surveys of threatened species (12). It is reasonable to assume that all these species will be extinct in <100 years, thus making future rates 200 to 1500 E/MSY (figure 1.2)?

Some threatened species are declining rapidly and will soon be extinct. Others, not so obviously doomed, have small numbers ($<10^2$). They risk the demographic vagaries of sex (all the young of a generation being of the same sex) and death (all the individuals dying in the same year from independent causes). For these, both models and empirical, long-term studies of island populations suggest times to extinction on the order of decades (17). Population fluctuations, and the environmental vagaries that cause them, drive the extinction of larger populations ($>10^2$) (14). Over 20 years, bird densities can vary 10-fold, and insect densities 10,000-fold (14). Ecologists have been slow to combine models and data. Yet even in the absence of a formal analysis, such fluctuations can obviously doom even quite large populations.

Our predictions may err because some threatened species will survive the century (18). The more serious problem with our predictions is that species not now threatened will become extinct. For birds – the one group for which we have detailed lists of the causes of threats – limited habitat is the most frequently cited factor, implicated in ~75% of threatened species (18). Increasingly well documented studies (19) show that habitat destruction is continuing and perhaps accelerating. Some now-common species will lose their habitats within decades.

Interestingly, accidentally or deliberately introduced species are blamed for only 6% of currently threatened birds (18). Yet introduced species, and the predation, competition, disease, and habitat modification they cause, are the most frequently cited factors in all the extinction centers we discussed above (12). Undoubtedly, many species will be lost to introduced species in ways that we cannot now anticipate. For example, no one considered the birds on the island of Guam to be in danger 30 years ago, but an introduced snake has eliminated all the island's birds since then (14). Were this predator to reach Hawai'i, all its birds would be at risk.

Calibrating Species Loss from Habitat Loss: A Tale of Two Forests

So far, we have sampled well-known, but disparate species whose high extinction rates probably typify the unknown majority. We now consider a typical mechanism of extinction: habitat loss. Can we predict species losses from estimates of habitat losses? The function $S = cA^z$ relates the number of species counted (S) to the area surveyed (A); c and z are constants (20). If the original habitat area, A_o is reduced to An, we expect the original number of species, S_o, to decline eventually to S_n. Now $S_n/S_o = cA_n^z/cA_o^z$ or $(A_n/A_o)z$ – an expression that is independent of c. Across different situations, z varies from 0.1 to 1.0, but it is often taken to be ~¼ (20). This value is typical of islands isolated by sea-level changes, a process that may be the best model for large habitat fragments isolated by deforestation (20, 21).

This recipe forms the basis of the predictions of 1000 to 10,000 E/MSY shown in figure 1.2. To challenge these estimates, critics point to the few bird extinctions after the clearing of North America's eastern forests (3). Is the recipe flawed? Only if interpreted naïvely are these results a poor model for what happens elsewhere.

An extinction "cold spot." European colonists cut >95% of the eastern forests of

North America, but not simultaneously. Locally, forests reclaimed abandoned fields. and regionally forests recovered in the Northeast as settlers moved westward. Of the region's 2.87×10^6 km² area, forests always covered >50% (21). So 16% (=$0.5^{0.25}$) – or 26 of the ~160 forest species – should have gone extinct. Only 4 did so (21). Yet, such predictions are naïve. Not enough time may have elapsed for the extinctions to occur. However, all but 28 of these species occur widely across North America. They would have survived elsewhere even if all the forest had been permanently cleared. The cookie-cutter model restricts the analysis to the region's 28 endemics, whence the predicted and observed number of extinctions correspond ($4 \approx 16\%$ of 28).

Simply, this region has very few endemics and so few species to lose. In contrast, tropical moist forests may hold two-thirds of all species on Earth (22). Despite inevitable differences in their definition, satellite imaging yields detailed and rapidly changing estimates showing their rapid depletion (19). The forests' global extent is variously estimated at 8×10^6 to 12.8×10^6 km² and their rate of clearing as 1.2×10^6 to 1.4×10^6 km² per decade (19).

An extinction "hot spot." The 1.47×10^6 km² of forests in the Philippines and Indonesia (excluding Irian Jaya) hold 545 endemic bird species – 20 times the number in America's eastern forest in half the area (23). Only 0.91 $\times 10^6$ km² of forest remains, and ~10% of the original area is cleared per decade. Using current satellite-based estimates of forest cover, the species-area recipe adequately predicts the number of species endemic to single islands that are currently threatened (figure 1.3A). The recipe, however, overestimates the numbers of currently threatened species that are found on several islands (figure 1.3B) and greatly overestimates the number of currently threatened species that are widespread (figure 1.3C).

Estimates of extinction from habitat losses (figure 1.2) use an area's total number of species, not its smaller number of endemics. Does this reliance on such totals inflate these rates? In general, it does not, because many tropical areas are unusually rich in endemics

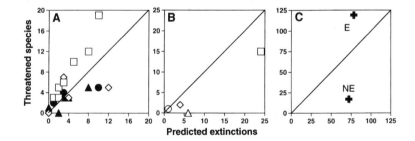

Figure 1.3 [*orig. figure 3*]   The numbers of species currently threatened with extinction plotted against the numbers of species predicted to become extinct. Squares, Philippines; circles, North Wallacea; diamonds, Greater Sundas; triangles, Lesser Sundas. The predictions use satellite estimates of remaining forest cover and the relation between species numbers and area. (**A**) For the subset of species endemic to single islands, these numbers match; they straddle the graph's diagonal. (**B**) For the subset of species that are endemic to several islands within each region, the predicted extinctions consistently exceed the numbers of threatened species. (**C**) When we combine the species found on one or several islands (E) across the entire region, the predictions based on forest losses (78) are slightly smaller than the number of threatened species (119). The remaining subset comprises widely occurring species (NE), Their predicted extinctions (72) far exceed the few (17) actually threatened with extinction. Predictions of extinctions from habitat loss must be scaled to the number of endemics the area contains (23).

(13). For example, 18 areas world-wide are so rich in endemics as to encompass ~20% of the known species of flowering plants in a total area of 0.74×10^6 km² (16). A larger area than this was cleared from the eastern American forests in the 19th century. The fate of these areas obviously dominates the calculations of future extinction rates. Details of land use changes in these areas are critical, but the details are not sufficient in themselves. We also need the detailed patterns of endemism.

Unfortunately, we know the geographical ranges of only a small proportion of the already small proportion of species for which we have names. We do have a comprehensive understanding of the geographical patterns of species richness (20). Its lessons are not encouraging. First, we cannot extrapolate from one species group to the next. For instance, across a continent species richness in frogs may not correlate with the species richness in birds (24). Worse, the direction of the correlation – positive or negative – may differ between continents (24). Second, areas rich in species are not always rich in endemics (24). Simply, our understanding of endemism is insufficient for us to know the future of biodiversity with precision (25).

References and Notes

1   N. Myers and J. L. Simon, *Scarcity or Abundance* (Norton, New York, 1994).

2   P. R. Ehrlich and A. H. Ehrlich, *Extinction: The Causes and Consequences of the Disappearance of Species* (Random House, New York, 1981).

3   J. L. Simon and A. Wildavsky, *New York Times*, 13 May 1993, p. A23; S. Budiansky, *U.S. News World Rep.*, 13 December 1993, p. 81; *Nature* 370, 105 (1994).

4   For extinction to occur, there need not be a background rate – a slow, characteristic winking out of species – punctuated by extreme events such as the one that eliminated the dinosaurs. In the geological record, the number of extinctions per time interval scales continuously as 1/frequency. Thus, there is no mean rate, for the rate depends on the interval over which it is measured. See D. Jablonski, in *Dynamics of Extinction*, D. K. Elliott, Ed. (Wiley, New York, NY, 1986), pp. 193–229.

5   R. M. May, J. H. Lawton, N. E. Stork, in *Extinction Rates*, J. H. Lawton and R. M. May, Eds. (Oxford Univ. Press, Oxford, 1995), pp. 1–24.

6   C. G. Sibley and J. E. Ahlquist, *Phylogeny and Classification of the Birds* (Yale Univ. Press, New Haven, CT, 1990).

7   S. Nee, A. O. Mooers, P. H. Harvey, *Proc. Natl. Acad. Sci. U.S.A.* 89, 8322 (1992); S. Nee, E. C. Holmes, R. M. May, P. H. Harvey, *Proc. R. Soc. London Ser. B* 344, 77 (1994).

8   R. K. Wayne, R. E. Benveniste, D. N. Janczewski, S. J. O'Brien, in *Carnivore Behavior, Ecology, and Evolution*, J. L. Gittleman, Ed. (Cornell Univ. Press, Ithaca, NY, 1989), pp. 465–494; R. K. Wayne, B. Van Valkenburgh, S. J. O'Brien, *Mol. Biol. Evol.* 8, 297 (1991).

9   R. M. May, *Philos. Trans. R. Soc. London Ser. B* 330, 293 (1990).

10 D. W. Steadman, *Science* 267, 1123 (1995); S. L. Pimm, M. P. Moulton, L. J. Justice, *Philos. Trans. R. Soc. London Ser. B* 344, 27 (1994).

11 S. Sohmer, in *Biodiversity and Terrestrial Ecosystems*, C. -I. Peng and C. H. Chou, Eds. (Academia Sinica Monograph Series 14, Taipei, 1994), pp. 43–51.

12 World Conservation Monitoring Centre, *Global Biodiversity: Status of the Earth's Living Resources* (Chapman & Hall, London, 1992); M. P. Nott, E. Rogers, S. L. Pimm, *Curr. Biol.* 5, 14 (1995); J. Parslow, *Breeding Birds of Britain and Ireland: A Historical Survey* (Poyser, Berkhamstead, 1973).

13 International Council for Bird Preservation (ICBP), *Putting Biodiversity on the Map: Priority Areas for Global Conservation* (ICBP, Cambridge, 1992).

14 S. L. Pimm, *The Balance of Nature? Ecological Issues in the Conservation of Species and Communities* (Univ. of Chicago Press, Chicago, IL, 1992).

15 K. J. Gaston, *Rarity* (Chapman & Hall, London, 1994).

16 N. Myers, *Environmentalist* 8, 187 (1988); *ibid.* 10, 243 (1990).

17 S. L. Pimm, J. M. Diamond, T. M. Reed, G. J. Russell, J. M. Verner, *Proc. Natl. Acad. Sci. U.S.A.* 90, 10871 (1993).

18 N. J. Collar, M. J. Crosby, A. J. Stattersfield, *Birds to Watch 2: The World List of Threatened Birds* (BirdLife International, Cambridge, 1994).

19 D. Skole and C. Tucker, *Science* 260, 1905 (1993); Food and Agricultural Organization, *The Forest Resources of the Tropical Zone by Main*

Ecological Regions (Food and Agricultural Organization, Rome, 1992); N. Myers, in *The Causes of Tropical Deforestation*, K. Brown and D. W. Pearce, Eds. (University College Press, London, 1994), pp. 27–40.

20 M. L. Rosenzweig, *Species Diversity in Space and Time* (Cambridge Univ. Press, Cambridge, 1995).

21 S. L. Pimm and R. A. Askins, *Proc. Natl. Acad. Sci. U.S.A.*, in press.

22 P. H. Raven in *Biodiversity*, E. O. Wilson, Ed. (National Academy Press, Washington, DC, 1988), pp. 119–122.

23 T. M. Brooks, S. L. Pimm, N. J. Collar, unpublished results.

24 J. J. Schall and E. R. Pianka, *Science* 201, 679 (1078); J. R. Prendergast, R. M. Quinn, J. H. Lawton, B. C. Eversham, D. W. Gibbons, *Nature* 365, 335 (1993); J. Curnutt, J. Lockwood, H. -K. Luh, P. Nott, G. Russell, *ibid.* 367, 326 (1994).

25 S. L. Pimm and J. L. Gittleman, *Science* 255, 940 (1992).

26 D. L. Hawksworth, *Mycol. Res.* 95, 441 (1991); J. F. Grassle and N. J. Maciolek, *Am. Nat.* 139, 313 (1992); T. L. Erwin, *Coleopt. Bull.* 36, 74 (1982); N. E. Stork, *Biol. J. Linn. Soc.* 35, 321 (1988); I. D. Hodkinson and D. Casson, *ibid.* 43, 101 (1990); S. L. Pimm, J. H. Lawton, J. E. Cohen, *Nature* 350, 669 (1991); P. H. Raven, *Futurist* 19, 8 (1985); R. M. May, *Science* 241, 1441 (1988).

27 M. V. Reid, in *Tropical Deforestation and Species Extinction*, T. C. Whitmore and J. A. Sayer, Eds. (Chapman & Hall, London, 1992), pp. 55–73; P. H. Raven in *Botanic Gardens and the World Conservation Strategy*, D. Bramwell, O. Hamann, V. Heywood, H. Synge, Eds. (Academic Press, London, 1987), pp. 19–29; W. V. Reid and K. R. Miller, *Keeping Options Alive: The Scientific Basis for Conserving Biodiversity* (World Resources Institute, Washington, DC, 1989); N. Myers, *The Sinking Ark: A New Look at the Problem of Disappearing Species* (Pergamon Press, Oxford, 1979); T. E. Lovejoy, *The Global 2000 Report to the President*, vol. 2 (Council on Environmental Quality, Washington, DC, 1980), p. 328; P. H. Raven, in *Biodiversity*, E. O. Wilson and F. M. Peter, Eds. (National Academy Press, Washington, DC, 1988), pp. 1190–1222; E. O. Wilson, in *Biodiversity*, E. O. Wilson and F. M. Peter, Eds. (National Academy Press, Washington, DC, 1988), pp. 3–18; *Sci. Am.* 261, 108 (September 1989).

28 We thank H. -K. Luh, K. Norris, P. Nott, J. Tobias, and three anonymous reviewers for help and comments. S.L.P. is supported by a Pew Scholarship in Conservation and the Environment.

2

Catastrophes, Phase Shifts, and Large-Scale Degradation of a Caribbean Coral Reef*

Terence P. Hughes

Coral reefs are renowned for their spectacular diversity and have significant aesthetic and commercial value, particularly in relation to fisheries and tourism. However, many reefs around the world are increasingly threatened, principally from overfishing and from human activities causing excess inputs of sediment and nutrients such as pollution, deforestation, reef mining, and dredging (1). There is a pressing need to monitor coral reefs to assess the spatial and temporal scale of any damage that may be occurring and to conduct research to understand the mechanisms involved.

Here I describe dramatic shifts in reef community structure that have largely destroyed coral reefs around Jamaica. The results presented here summarize the most comprehensive reef monitoring program yet conducted in the Caribbean, in which annual censusing has been carried out for 17 years at multiple sites and depths along 300 km of coastline. In addition, Jamaican reefs are among the best studied in the world, with a wealth of information available on marine ecology and reef status since the 1950s (2). These long-term observations provide a basis for evaluating the role of rare events such as hurricanes and for quantifying gradual trends in coral cover and diversity over a decadal time scale.

Jamaica (18°N, 77°W) is the third largest island in the Caribbean and lies at the center of coral diversity in the Atlantic Ocean (2). Over 60 species of reef-building corals occur there, four of which are spatial dominants: branching elkhorn and staghorn corals, *Acropora palmata* and *Acropora cervicornis*, which form two distinctive zones on the shallow fore-reef; massive or platelike *Montastrea annularis*, the most important framework coral; and encrusting or foliose *Agaricia agaricites* (3). Reefs fringe most of the north Jamaican coast along a narrow (<1 to 2 km) belt and occur sporadically on the south coast on a much broader (>20 km) shelf. Sea-grass beds and mangrove are often closely associated with reefal areas and provide significant nurseries for commer-

* Originally published in *Science*, 1994, vol. 265, pp. 1547–51.

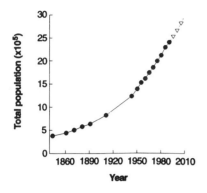

Figure 2.1 [*orig. figure 1*] Population growth of Jamaica, based on numerous sources (5).

cially important reef fisheries (4). Similar ecosystems, with minor variations in community composition, occur throughout the Caribbean (2).

Jamaica's population growth trajectory is typical of most Third World countries (figure 2.1). The population was less than half a million before 1870, then doubled by 1925 and again by 1975, rising to 2.5 million today. Exponential growth continues, with a further 20% increase expected in the next 15 years (5). Environmental changes on land are conspicuous, with virtually all of the native vegetation having been cleared for agriculture and urban development. Major transformations are also occurring on Jamaica's coral reefs.

Overfishing (1960s to Present)

Chronic overfishing is an ever increasing threat to coral reefs worldwide as coastal populations continue to grow (for example, figure 2.1) and exploit natural resources (6). Extensive studies in Jamaica by Munro (7) showed that by the late 1960s fish biomass had already been reduced in preceding decades by up to 80% on the extensive (but narrow) fringing reefs of the north coast, mainly a result of intensive artisanal fish-trapping. By 1973, the number of fishing canoes deploying traps on the north coast was approximately 1800 (or 3.5 canoes per square kilometer of coastal shelf), which was

two to three times above sustainable levels (7). The taxonomic composition of fish has changed markedly over the past 30 to 40 years. Large predatory species, such as sharks, lutjanids (snappers), carangids (jacks), ballistids (triggerfish), and serranids (groupers) have virtually disappeared, while turtles and manatees are also extremely rare. The remaining fish, including herbivores such as scarids (parrotfish) and acanthurids (surgeonfish), are small, so that fully half of the species caught in traps recruit to the fishery below the minimum reproductive size. Indeed, because adult stocks on the northern coast of Jamaica have been sharply reduced for several decades, populations today may rely heavily on larval recruitment from elsewhere in the Caribbean (7). This sequence of changes was repeated more recently along the southern coast of Jamaica. There, the broader coastal shelf has become increasingly accessible to a modernizing fishing fleet, with the number of motorized canoes almost doubling from the 1970s to the mid-1980s (8). Despite this increased fishing effort, the catch from the south coast remained the same over this 15-year period (that is, the catch per unit effort declined by half). The species composition of the fishery has also changed markedly, indicative of severe overfishing nationwide (6–8).

The ecological effects of the drastic reduction in fish stocks on Jamaica's coral reef as a whole were not immediately obvious. Throughout the 1950s to the 1970s the reefs appeared to be healthy; coral cover and benthic diversity were high (3) (figure 2.2). There were relatively few macroalgae throughout this period despite the paucity of large herbivorous fish as a result mainly of grazing by huge numbers of the echinoid *Diadema antillarum* (9, 10). The major predators of adult *Diadema* are fish [for example, ballistids, sparids (porgies), and batrachoidids (toadfish) (11)] that are now rare in Jamaica. Other fish (such as scarids and acanthurids) compete strongly with *Diadema* for algal resources, as evidenced by competitor removal experiments (12). Therefore, the unusually high abundance of *D. antillarum* on overfished reefs such as

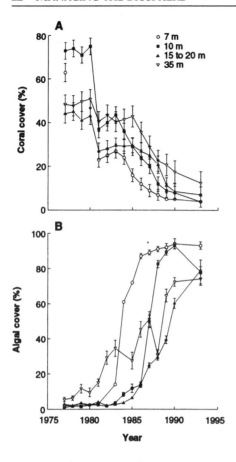

Figure 2.2 [*orig. figure 3*]   Degradation of Jamaican coral reefs over the past two decades. Small-scale changes in (**A**) coral cover and (**B**) macroalgal cover over time at four depths near Discovery Bay (32)

Jamaica's was almost certainly a result of the over-exploitation of reef fisheries. Hay (13) investigated this hypothesis on a geographic scale and found that densities of echinoids were much greater on overfished than on pristine reefs throughout the Caribbean. A mass mortality of *Diadema* in 1983 had far-reaching consequences, in part because of the prior reduction (for several decades) of stocks of herbivorous and predatory fish.

Hurricane Damage (1980)

Hurricanes, typhoons, or cyclones are predictable, recurrent events and an integral part of the natural dynamics of a coral reef (14). The regeneration of a healthy reef system is facilitated by rapid colonization of larval recruits, but in Jamaica this crucial recovery mechanism has been hindered by human influences (that is, by overfishing, which contributed to a prolonged macroalgal bloom causing recruitment failure in corals).

Extensive damage was inflicted on Jamaican coral reefs by Hurricane Allen, a category 5 hurricane that struck in 1980, following a period of almost four decades without a major storm (15). Damage was the greatest at shallow sites (figure 2.2A). The hurricane smashed shallow-water branching species, most notably the elkhorn and staghorn corals (*Acropora palmata* and *A. cervicornis*). In addition, beds of the soft coral *Zoanthus*, which occupied large areas of the inner reef flat, were damaged by *A. Palmata* rubble pushed shoreward by storm waves. Corals with more robust morphologies or living in deeper water (>10 to 15 m, figure 2.2A) were much less susceptible to physical destruction, so the hurricane increased the relative abundances of species with encrusting or massive-shaped colonies (15, 16). Immediately following Hurricane Allen, there was a short-lived algal bloom (primarily composed of the ephemeral Rhodophyte *Liagora*) probably caused by a pulse of nutrient release from terrestrial runoff and suspended reef sediments and from a temporary depression of herbivory by *Diadema* and other herbivores. Within a few months, however, the algae disappeared and substantial coral recruitment began (16). Recruitment by *Acropora* was minimal and broken fragments survived poorly (17), but other corals, notably brooding agaricids and *Porites*, settled in large numbers onto free space generated by the hurricane (18). For the next 3 years up to 1983, cover increased slowly as the reef began to recover (figure 2.2A). However, recovery from Hurricane Allen was short-lived and was soon reversed by biological events that were less selective and ultimately more destructive and wide-spread than even this powerful hurricane.

Disease and Algal Blooms (1983 to Present)

The echinoid species *Diadema antillarum* suffered mass mortality from a species-specific pathogen throughout its entire geographic range from 1982 to 1984 (18). In Jamaica, densities of *Diadema* were reduced by 99% from pre-die-off estimates of close to 10 per square meter on shallow fore-reefs,

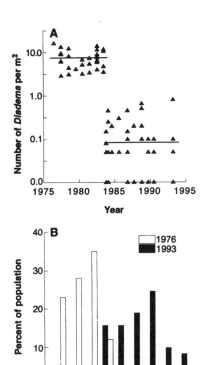

Figure 2.3 [*orig. figure* 4] Long-term dynamics of the echinoid *Diadema antillarum* on Jamaican reefs. (**A**) Abundances over time based on estimates at 14 sites along >100 km of coastline over nearly two decades. Note the 99% drop in 1983 (from a mean of 9 to 0.09 per square meter), with no recovery after 10 years. (**B**) Population structure of *Diadema* (33) before and after the 1983 die-off.

and there has been no significant recovery in the subsequent 10 years (figure 2.3A). Before 1983, *Diadema* were small (19–21), presumably because of food limitation caused by the prevailing high densities of this species (20). Following the die-off, the mean and maximum size of individuals increased greatly, whereas individuals in smaller size classes became uncommon, indicative of low rates of recruitment (figure 2.3B). Individuals today are large, well fed, and have well-developed gonads. However, densities may be too low for effective spawning success because fertilization in *Diadema* is strongly density-dependent (21).

Without *Diadema*, and with the continued depression of herbivorous fish from trapping, the entire reef system of Jamaica has undergone a spectacular and protracted benthic algal bloom that began in 1983 and continues today at all depths (up to 40 m or deeper) (figure 2.2B). Before the echinoid die-off, cover of fleshy macroalgae was typically less than 5% except intertidally, within damsel fish territories, or in very deep water (>25 m) where *Diadema* were scarce (9, 10, 13) (figure 2.2B). In the initial stages of the bloom, algae were small and ephemeral, but within 2 to 3 years weedy species were replaced by longer lived, late successional taxa (notably *Sargassum, Lobophora, Dictyota, and Halimeda*) that formed extensive mats up to 10 to 15 cm deep (10, 22). As a result of this preemption of space, larval recruitment by all species of corals has failed for the past decade (16). Most adult colonies that survived Hurricane Allen have been killed by algal overgrowth, especially low-lying species with encrusting or platelike morphologies (16). Additional mortality occurred following bleaching events in 1987, 1989, and 1990 (23). The most abundant coral on the fore-reef today is mound-shaped *Montastrea annularis*, but even this robust, dominant species has declined to 0 to 2% cover at a depth of 10 m in 1993 (24). This decline in a long-lived coral such as *Montastrea* is particularly significant because it is resistant to hurricanes and is the chief frame-builder of Jamaican reefs. Its slow recruitment and growth rate (25) ensure that the decline of the past 10

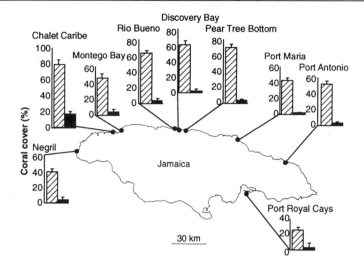

Figure 2.4 [*orig. figure 5*]   Large-scale changes in community structure at fore-reef sites along >300 km of the Jamaican coastline, surveyed in the late 1970s (1977, hatched bars) and the early 1990s (1993, solid bars) (34).

years will not be reversed for many decades.

The scale of damage to Jamaican reefs is enormous. Censuses at sites 5 to 30 km apart along >300 km of coastline in 1977 to 1980 and again in 1990 to 1993 show a decline in coral cover from a mean of 52 to 3% and an increase in cover by fleshy macroalgae from 4 to 92% (figure 2.4). Indeed, the classic zonation patterns of Jamaican reefs, described by Goreau and colleagues just two to three decades ago (3), no longer exist. A striking phase shift has occurred from a coral-dominated to an algal-dominated system (figure 2.5).

Implications and Prospects for the Future

This spectacular sequence of events highlights the dynamic and complex nature of coral reefs; points to the fundamental importance of fish, herbivory, and recovery of the reefs from physical disturbance to their functioning; and provides a clear demonstration of how quickly (one to two decades) a seemingly healthy coral reef can be severely damaged on a spatial scale similar to the size of most tropical island-nations (hundreds of kilometers). Although it was not widely recognized at the time, Jamaica's

reefs were already extensively damaged by the late 1970s (from direct and indirect effects of overfishing) to the extent that the synergistic effects of two subsequent hurricanes and the *Diadema* die-off were sufficient to cause a radical phase shift to algae (figure 2.5). Paradoxically, the changes have occurred although reef systems have demonstrable robustness on a geological time scale. For example, coral reefs have continued to flourish despite major fluctuations in sea level occurring on a time scale of 10^3 to 10^5 years (26). However, the ability of coral reefs to cope with such disturbances in the past is no guarantee of continued resilience in the face of unprecedented and much more rapid anthropogenic stresses. It is highly probable that global reef growth is currently being outpaced by reef degradation (1), with unknown consequences for the future.

A great deal has been learned about the functioning of coral reefs from the litany of disasters described here, and the opportunity should be seized to implement scientifically based environmental management procedures that would facilitate processes of recovery. Clearly, the Jamaican reef system needs more herbivory to allow coral recruitment to resume (27).

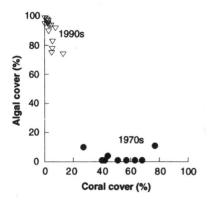

Figure 2.5 [*orig. figure 6*] Large-scale community phase shifts on Jamaican reefs, from coral- to algal-dominated systems (34).

Herbivorous fish (mostly juvenile scarids) responded immediately to the *Diadema* die-off by changing their spatial distribution and increasing their grazing rates in shallow water (28). However, this behavioral response is unlikely to be reflected later in increased fish abundance because of continued overfishing. Clearly, current stocks of herbivorous fish are not capable of reducing algal abundance in the absence of *Diadema* (figure 2.2B). Similarly, other echinoids have not increased in abundance to compensate for the loss of *Diadema* (10, 29). Recovery of *Diadema* has not yet taken place and is likely to be slow if densities have fallen below some threshold level required for successful spawning (21). Even a full recovery of *Diadema* would leave the reef reliant once more on a single dominant herbivore and vulnerable to a recurrence of disease. Future hurricanes will reinforce rather than reverse the phase shift, as illustrated by the more recent impact of Hurricane Gilbert in 1988. Also a category 5 hurricane, it swept much of the algal covering off the reef and caused further damage to corals. However, the algae recovered fully within a few weeks of Hurricane Gilbert (figure 2.2B), mainly from regenerating filaments and holdfasts, long before successful recruitment of corals could resume. Thus, further hurricanes are likely to act in a ratchet fashion, further depressing

coral abundances and favoring the phase shift to algae (figure 2.5).

There is an urgent need, therefore, to control overfishing, a call that had first been made by Munro 20 years ago (7), before more recent events demonstrated the key role of fish and echinoid herbivores in the overall functioning of Jamaica's coral reefs. On the basis of our knowledge of the demography and life histories of fish (7, 8, 30) and corals (25, 31), it will take far longer to rebuild stocks than the two to three decades it has taken to destroy them. Severe, long-term damage has already occurred, and the trajectories of coral and algal abundance (figures 2.2 and 2.5) predict a gloomy future unless action is taken immediately.

References and Notes

1   H. A. Lessios, P. W. Glynn, D. R. Robertson, *Science* 222, 715 (1983); C. S. Rogers, *Proc. 5th Int. Coral Reef Symp.* 6, 491 (1985); B. E. Brown, *Mar. Polut. Bull.* 18, 9 (1987); B. Salvat, Ed., *Human Impacts on Coral Reefs: Facts and Recommendations* (Antenne Museum Ecole Pratique des Hautes Etudes, French Polynesia, 1987); C. F. d'Elia, R. W. Buddemeier, S. V. Smith, Eds., *Workshop on Coral Bleaching, Coral Reef Ecosystems and Global Change: Report of Proceedings* (Maryland Sea Grant College, College Park, 1991); T. J. Done, *Hydrobiologia* 247, 121 (1992); *Global Aspects of Coral Reefs: Health, Hazards and History* (University of Miami, Miami, FL, 1993).

2   S. M. Wells, Ed., *Coral Reefs of the World*, vol. 1 of *United Nations Environment Program Regional Seas Directories and Bibliographies* (International Union for the Conservation of Nature, Cambridge, 1988). There are over 500 refereed publications since the 1950s based on coral reef research conducted at the Discovery Bay and Port Royal Marine Laboratories, which are on the north and south Jamaican coasts, respectively.

3   T. F. Goreau, *Ecology* 40, 67 (1959); J. Lang, *Am. Sci.* 62, 272 (1973); W. D. Liddell and S. L. Ohlhorst, *Bull. Mar. Sci.* 40, 311 (1987).

4   J. D. Parrish, *Mar. Ecol. Prog. Ser.* 58, 143 (1989).

5   D. Hall, *Free Jamaica, 1838–1865: An Economic History* (Yale Univ. Press, New Haven, CT, 1959); B. T. Walsh, *Economic Development and Population Control: A Fifty-Year Projection for Jamaica* (Praeger, New York, 1970); D. Watts, *The West Indies: Patterns of Development,*

Culture and Environmental Change Since 1492 (Cambridge Univ. Press, Cambridge, 1987); B. T. Walsh, *The Sex and Age Distribution of the World Populations* (United Nations, Department of Economic and Social Development, New York, 1993).

6   G. R. Russ, in *The Ecology of Coral Reef Fishes*, P. F. Sale, Ed. (Academic Press, New York, 1991), chap. 20.

7   J. L. Munro, *ICLARM Stud. Rev.* 7, 1 (1983); *Jam. J.* 3, 16 (1969).

8   J. A. Koslow, F. Hanley, R. Wicklund, *Mar. Ecol. Prog. Ser.* 43, 201 (1988).

9   J. C. Ogden, R. A. Brown, N. Salesky, *Science* 182, 715 (1973); P. W. Sammarco, *J. Exp. Mar. Biol. Ecol.* 45, 245 (1980); R. C. Carpenter, *J. Mar. Res.* 39, 749 (1981); P. W. Sammarco, *J. Exp. Mar. Biol. Ecol.* 65, 83 (1982).

10 T. P. Hughes, D. C. Reed, M. J. Boyle, *J. Exp. Mar. Biol. Ecol.* 113, 39 (1987).

11 J. E. Randall, *Caribb. J. Sci.* 4, 421 (1964); *Stud. Trop. Oceanogr.* 5, 665 (1967); D. R. Robertson, *Copeia* 1987, 637 (1987).

12 A. H. Williams, *Ecology* 62, 1107 (1981); M. E. May and P. R. Taylor *Oecologia* 65, 591 (1985).

13 M. E. Hay, *Ecology* 65, 446 (1984).

14 J. H. Connell, *Science* 199, 1302 (1978); T. P. Hughes, Ed., "Disturbance: Effects on Coral Reef Dynamics," Special Issue of *Coral Reefs* 12 (no. 3 and 4), 115 (1993).

15 J. D. Woodley et al., *Science* 214, 749 (1981); J. Porter et al., *Nature* 294, 249 (1981).

16 T. P. Hughes, *Ecology* 70, 275 (1989).

17 N. Knowlton, J. C. Lang, M. C. Rooney, P. Clifford, *Nature* 294, 251 (1981); N. Knowlton, J. C. Lang, B. D. Keller, *Smithson. Contrib. Mar. Sci.* 31, 1 (1990).

18 R. P. M. Bak, M. J. E. Carpay, E. D. De Ruyter Van Steveninck, *Mar. Ecol. Prog. Ser.* 17, 105 (1984); H. A. Lessios, D. R. Robertson, J. D. Cubit, *Science* 226, 335 (1984); T. P. Hughes, B. D. Keller, J. B. C. Jackson, M. J. Boyle, *Bull. Mar. Sci.* 36, 377 (1985); W. Hunte, I. Cote, T. Tomascik, *Coral Reefs* 4, 135 (1986); H. Lessios, *Annu. Rev. Ecol. Syst.* 19, 371 (1988); R. C. Carpenter, *Mar. Biol.* 104, 67 (1990).

19 A. H. Williams, *J. Exp. Mar. Biol. Ecol.* 75, 233 (1984).

20 D. R. Levitan, *Ecology* 70, 1419 (1989).

21 R. H. Karlson and D. R. Levitan, *Oecologia* 82, 44 (1990); D. R. Levitan, *Biol. Bull.* 181, 261 (1991).

22 W. D. Liddell and S. L. Ohlhorst, *J. Exp. Mar. Biol. Ecol.* 95, 271 (1986); E. D. de Ruyter Van Steveninck and R. P. M. Bak, *Mar. Ecol. Prog. Ser.* 34, 87 (1986); E. D. de Ruyter Van Steveninck and A. M. Breeman, *ibid.* 36, 81 (1987); D. R. Levitan, *J. Exp. Mar. Ecol.* 119, 167 (1988).

23 R. D. Gates, *Coral Reefs* 8, 193 (1990); T. J. Goreau and A. H. Macfarlane, *ibid.*, p. 211; J. D. Woodley, unpublished data.

24 Based on estimates of coral cover in 1993 at Rio Bueno, Discovery Bay, Pear Tree Bottom, and Ocho Rios (spanning 40 km of the north Jamaican coast). Twenty replicate 10-m line-intercept transects were run at a depth of 10 m at each site.

25 P. Dustan, *Mar. Biol.* 33, 101 (1975); R. P. M. Bak and M. S. Engel, *ibid.* 54, 341 (1979); R. P. M. Bak and B. E. Luckhurst, *Oecologia* 47, 145 (1980); K. W. Rylaarsdam, *Mar. Ecol. Prog. Ser.* 13, 249 (1983); C. S. Rogers, H. C. Fitz III, M. Gilnack, J. Beets, J. Hardin, *Coral Reefs* 3, 69 (1984); T. P. Hughes and J. B. C. Jackson, *Ecol. Monogr.* 55, 141 (1985); T. P. Hughes, *6th Int. Coral Reef Symp.* 2, 721 (1988).

26 K. J. Mesolella, *Science* 156, 638 (1967); N. D. Newell, *Sci. Am.* 226, 54 (June 1971); R. W. Buddemeier and D. Hopley, *Proc. 5th Int. Coral Reef Symp.* 1, 253 (1988); J. B. C. Jackson, *Am. Zool.* 32, 719 (1992).

27 There is no evidence that the nationwide algal bloom in Jamaica was caused by increased nutrients, because it occurred throughout the Caribbean immediately following the *Diadema* die-off (16, 20), usually far from sources of pollution. Some groundwater input does occur into the shallow margins of the back-reef at Discovery Bay, which enhances nitrates and reduces salinity close to the shore [C. F. D'Elia, K. L. Webb, J. W. Porter, *Bull. Mar. Sci.* 31, 903 (1981)]. These conditions produce localized areas around submarine springs, typically 2 to 3 m in diameter, which contain characteristic brackish-water algal assemblages (dominated by *Chaetomorpha, Enteromorpha,* and *Ulva)* that are quite unlike those occurring on the reef further offshore. None of the sites in figures 2.2 to 2.5 are located close to urban areas or point sources of pollution, with the exception of the Port Royal cays on the south coast near Kingston.

28 R. C. Carpenter, *Proc. Natl. Acad. Sci. U.S.A.* 85, 511 (1988); *Mar. Biol.* 104, 79 (1990); D. Morrison, *Ecology* 69, 1367 (1988).

29 Densities of *Echinometra viridis, Eucidaris tribuloides, Lytochinus williamsi,* and *Trypneustes ventricosus* in 1973 were reported for two Jamaican patch reefs by P. W. Sammarco [*J. Exp. Mar. Biol. Ecol.* 61, 31 (1982)]. The combined total then was 27.5 and 54.0 per

square meter, respectively. By 1986, the combined total had fallen two- to threefold (10). In 1993, mean densities (number per square meter ± SE) on these same reefs were 14.0 ± 1.5 and 14.4 ± 1.2.

30 P. F. Sale, Ed., *The Ecology of Coral Reef Fishes* (Academic Press, New York, 1991).

31 J. H. Connell, in *Biology and Geology of Coral Reefs*, O. A. Jones and R. Endean, Eds. (Academic Press, New York, 1973), chap. 7; J. B. C. Jackson, *Bioscience* 41, 475 (1991); T. P. Hughes, D. J. Ayre, J. H. Connell, *Trends Ecol. Evol.* 7, 292 (1992).

32 Coral and algal abundance (percent cover) shown here were measured from annual photographs of 10 to 20 permanent 1-m² plots at each depth (7, 10, and 15 to 20 m at Rio Bueno; 35 m at Pinnacle 1). All corals (approximately 38,000 records over 17 years) were traced and digitized to obtain relative abundances, while algal cover was estimated by super-imposing a grid of dots on each image (100 per square meter) and counting those covering algae. The small-scale trends reported here for permanent plots mirror almost exactly the results from a larger scale program that was based on replicate 10-m line-intercept transects. For example, in 1993 mean coral cover (±SE) estimated from 20 random transects at each of the 7-, 10-, 15- to 20-, and 35-m stations in figure 2.2A was 5.0 ± 0.8, 5.4 ± 1.2, 5.6 ± 0.9, and 12.8 ± 2.4, respectively. Reef degradation at an even larger scale is shown in figure 2.4.

33 Data for 1976 are from (19), based on a random collection of 97 *Diadema antillarum* from the East Back Reef at Discovery Bay, Jamaica. Data for 1993 are based on 207 individuals from the same site.

34 Coral and macroalgal cover in figures 2.4 and 2.5 is based on 5 to 10 10-m line-intercept transects run at 10 m from 1976 to 1980 (mostly in 1977 and 1978) on fore-reefs at Negril, Chalet Caribe, Rio Bueno, Discovery Bay (two locations), Pear Tree Bottom, Port Maria, Port Antonio (on the north coast), and Port Royal (on the south coast). These measurements were repeated in 1990 to 1993 with 20 transects, with the addition of five more north coast sites.

35 I thank J. H. Connell, F. Jeal, and J. B. C. Jackson for providing encouragement over 20 years; J. D. Woodley and the staff of Discovery Bay Marine Laboratory for excellent logistic support; M. J. Boyle, G. Bruno, M. Carr, L. Dinsdale, F. Jeal, M. Gleason, S. Pennings, D. Reed, L. Sides, L. Smith, J. Tanner, C. Tyler, and many others for field and lab assistance; and D. Bellwood, H. Choat, T. Done, B. Willis, and the Coral Group at James Cook University (JCU), whose comments improved an early draft of the manuscript. Supported by the National Science Foundation, the National Geographic Society, the Whitehall Foundation, the Australian Research Council, and JCU. This is contribution no. 133 of Coral Group at JCU.

3

THE WORLD'S IMPERILED FISH*

Carl Safina

The 19th-century naturalist Jean-Baptiste de Lamarck is well known for his theory of the inheritance of acquired characteristics, but he is less remembered for his views on marine fisheries. In pondering the subject, he wrote, "Animals living in . . . the sea waters . . . are protected from the destruction of their species by man. Their multiplication is so rapid and their means of evading pursuit or traps are so great, that there is no likelihood of his being able to destroy the entire species of any of these animals." Lamarck was also wrong about evolution.

One can forgive Lamarck for his inability to imagine that humans might catch fish faster than these creatures could reproduce. But many people – including those in professions focused entirely on fisheries – have committed the same error of thinking. Their mistakes have reduced numerous fish populations to extremely low levels, destabilized marine ecosystems and impoverished many coastal communities. Ironically, the drive for short-term profits has cost billions of dollars to businesses and taxpayers, and it has threatened the food security of developing countries around the world. The fundamental folly underlying the current decline has been a wide-spread failure to recognize that fish are wildlife – the only wildlife still hunted on a large scale.

Because wild fish regenerate at rates determined by nature, attempts to increase their supply to the marketplace must eventually run into limits. That threshold seems to have been passed in all parts of the Atlantic, Mediterranean and Pacific: these regions each show dwindling catches. Worldwide, the extraction of wild fish peaked at 82 million metric tons in 1989. Since then, the long-term growth trend has been replaced by stagnation or decline.

In some areas where catches peaked as long ago as the early 1970s, current landings have decreased by more than 50 percent. Even more disturbingly, some of the world's greatest fishing grounds, including the Grand Banks and Georges Bank of eastern North America, are now essentially closed following their collapse – the formerly dominant fauna have been reduced to a tiny fraction of their previous abundance and are considered commercially extinct.

Recognizing that a basic shift has occurred, the members of the United Nation's Food and Agriculture Organization (a body that encouraged the expansion of large-scale industrial fishing only a decade ago) recently concluded that the operation of the world's fisheries cannot be sustained. They now acknowledge that substantial damage has already been done to the marine environment and to the many economies that depend on this natural resource.

Such sobering assessments are echoed in

* Originally published in *Scientific American,* 1995, November, pp. 46–53.

the U.S. by the National Academy of Sciences. It reported this past April that human actions have caused drastic reductions in many of the preferred species of edible fish and that changes induced in composition and abundance of marine animals and plants are extensive enough to endanger the functioning of marine ecosystems. Although the scientists involved in that study noted that fishing constitutes just one of the many human activities that threaten the oceans, they ranked it as the most serious.

Indeed, the environmental problems facing the seas are in some ways more pressing than those on land. Daniel Pauly of the Fisheries Center at the University of British Columbia and Villy Christensen of the International Center for Living Aquatic Resources Management in Manila have pointed out that the vast majority of shallow continental shelves have been scarred by fishing, whereas large untouched tracts of rain forest still exist. For those who work with living marine resources, the damage is not at all subtle. Vaughn C. Anthony, a scientist formerly with the National Marine Fisheries Service, has said simply: "Any dumb fool knows there's no fish around."

A War on Fishes

How did this collapse happen? An explosion of fishing technologies occurred during the 1950s and 1960s. During that time, fishers adapted various military technologies to hunting on the high seas. Radar allowed boats to navigate in total fog, and sonar made it possible to detect schools of fish deep under the oceans' opaque blanket. Electronic navigation aids such as LORAN (Long-Range Navigation) and satellite positioning systems turned the trackless sea into a grid so that vessels could return to within 50 feet of a chosen location, such as sites where fish gathered and bred. Ships can now receive satellite weather maps of water-temperature fronts, indicating where fish will be traveling. Some vessels work in concert with aircraft used to spot fish.

Many industrial fishing vessels are floating factories deploying gear of enormous proportions: 80 miles of submerged longlines with thousands of baited hooks, bag-shaped trawl nets large enough to engulf 12 jumbo jetliners and 40-mile-long drift nets (still in use by some countries). Pressure from industrial fishing is so intense that 80 to 90 percent of the fish in some populations are removed every year.

For the past two decades, the fishing industry has had increasingly to face the result of extracting fish faster than these populations could reproduce. Fishers have countered loss of preferred fish by switching to species of lesser value, usually those positioned lower in the food web – a practice that robs larger fishes, marine mammals and seabirds of food. During the 1980s, five of the less desirable species made up nearly 30 percent of the world fish catch but accounted for only 6 percent of its monetary value. Now there are virtually no other marine fish that can be exploited economically.

With the decline of so many species, some people have turned to raising fish to make up for the shortfall. Aquaculture has doubled its output in the past decade, increasing by about 10 million metric tons since 1985. The practice now provides more freshwater fish than do wild fisheries. Saltwater salmon farming also rivals the wild catch, and about half the shrimp now sold are raised in ponds. Overall, aquaculture supplies one fifth of the fish eaten by people.

Unfortunately, the development of aquaculture has not reduced the pressure on wild populations. Strangely, it may do the opposite. Shrimp farming has created a demand for otherwise worthless catch because it can be used as feed. In some countries, shrimp farmers are now investing in trawl nets with fine mesh to catch everything they can for shrimp food, a practice known as biomass fishing. Much of the catch are juveniles of valuable species, and so these fish never have the opportunity to reproduce.

Fish farms can hurt wild populations because the construction of pens along the coast often requires cutting down mangroves – the submerged roots of these salt-tolerant trees provide a natural nursery for shrimp and fish. Peter Weber of the Worldwatch Institute reports that aquaculture is one of the major reasons that half the world's mangroves have been destroyed. Aquaculture also threatens marine fish because some of its most valuable products, such as groupers, milkfish and eels, cannot be bred in captivity and are raised from newly hatched fish caught in the wild: the constant loss of young fry then leads these species even further into decline.

Aquaculture also proves a poor replacement for fishing because it requires substantial investment, land ownership and large amounts of clean water. Most of the people living on the crowded coasts of the world lack all these resources. Acquaculture as carried out in many undeveloped nations often produces only shrimp and expensive types of fish for export to richer countries, leaving most of the locals to struggle for their own needs with the oceans' declining resources.

Madhouse Economics

If the situation is so dire, why are fish so available and, in most developed nations, affordable? Seafood prices have, in fact, risen faster than those for chicken, pork or beef, and the lower cost of these foods tends to constrain the price of fish – people would turn to other meats if the expense of seafood far surpassed them.

Further price increases will also be slowed by imports, by overfishing to keep supplies high (until they crash) and by aquaculture. For instance, the construction of shrimp farms that followed the decline of many wild populations has kept prices in check.

So to some extent, the economic law of supply and demand controls the cost of fish. But no law says fisheries need to be profitable. To catch $70-billion worth of fish, the fishing industry recently incurred costs totaling $124 billion annually. Subsidies fill much of the $54 billion in deficits. These artificial supports include fuel-tax exemptions, price controls, low-interest loans and outright grants for gear or infrastructure. Such massive subsidies arise from the efforts of many governments to preserve employ-

ment despite the self-destruction of so many fisheries.

These incentives have for many years enticed investors to finance more fishing ships than the seas' resources could possibly support. Between 1970 and 1990, the world's industrial fishing fleet grew at twice the rate of the global catch, fully doubling in the total tonnage of vessels and in number. This armada finally achieved twice the capacity needed to extract what the oceans could sustainably produce. Economists and managers refer to this situation as overcapitalization. Curiously, fishers would have been able to catch as much with no new vessels at all. One study in the U.S. found that the annual profits of the yellowtail flounder fishery could increase from zero to $6 million by removing more than 100 boats.

Because this excessive capacity rapidly depletes the amount of fish available, profitability often plummets, reducing the value of ships on the market. Unable to sell their chief asset without major financial loss, owners of these vessels are forced to keep fishing to repay their loans and are caught in an economic trap. They often exercise substantial political pressure so that government regulators will not reduce allowable takes. This common pattern has become widely recognized. Even the U.N. now acknowledges that by enticing too many participants, high levels of subsidy ultimately generate severe economic and environmental hardship.

A World Growing Hungrier

While the catch of wild marine fish declines, the number of people in the world increases every year by about 100 million, an amount equal to the current population of Mexico. Maintaining the present rate of consumption in the face of such growth will require that by 2010 approximately 19 million additional metric tons of seafood become available every year. To achieve this level, aquaculture would have to double in the next 15 years, and wild fish populations would have to be restored to allow higher sustainable catches.

Technical innovations may also help

produce human food from species currently used to feed livestock. But even if all the fish that now go to these animals – a third of the world catch – were eaten by people, today's average consumption could hold for only about 20 years. Beyond that time, even improved conservation of wild fish would not be able to keep pace with human population growth. The next century will therefore witness the heretofore unthinkable exhaustion of the oceans' natural ability to satisfy humanity's demand for food from the seas.

To manage this limited resource in the best way possible will clearly require a solid understanding of marine biology and ecology. But substantial difficulties will undoubtedly arise in fashioning scientific information into intelligent policies and in translating these regulations into practice. Managers of fisheries as well as policy-makers have for the most part ignored the numerous national and international stock assessments done in past years.

Where regulators have set limits, some fishers have not adhered to them. From 1986 to 1992, distant water fleets fishing on the international part of the Grand Banks off the coast of Canada removed 16 times the quotas for cod, flounder and redfish set by the North-west Atlantic Fisheries Organization. When Canadian officials seized a Spanish fishing boat near the Grand Banks early this year, they found two sets of logbooks – one recording true operations and one faked for the authorities. They also discovered nets with illegally small mesh and 350 metric tons of juvenile Greenland halibut. None of the fish on board were mature enough to have reproduced. Such selfish disregard for regulations helped to destroy the Grand Banks fishery.

Although the U.N. reports that about 70 percent of the world's edible fish, crustaceans and mollusks are in urgent need of managed conservation, no country can be viewed as generally successful in fisheries management. International cooperation has been even harder to come by. If a country objects to the restrictions of a particular agreement, it just ignores them.

In 1991, for instance, several countries arranged to reduce their catches of swordfish from the Atlantic; Spain and the U.S. complied with the limitations (set at 15 percent less than 1988 levels), but Japan's catch rose 70 percent, Portugal's landings increased by 120 percent and Canada's take nearly tripled. Norway has decided unilaterally to resume hunting minke whales despite an international moratorium. Japan's hunting of minke whales, ostensibly for scientific purposes, supplies meat that is sold for food and maintains a market that supports illegal whaling worldwide.

Innocent Bystanders

In virtually every kind of fishery, people inadvertently capture forms of marine life that, collectively, are known as "bycatch" or "bykill." In the world's commercial fisheries, one of every four animals taken from the sea is unwanted. Fishers simply discard the remains of these numerous creatures overboard.

Bycatch involves a variety of marine life, such as species without commercial value and young fish too small to sell. In 1990 high-seas drift nets tangled 42 million animals that were not targeted, including diving seabirds and marine mammals. Such massive losses prompted the U.N. to enact a global ban on large-scale drift nets (those longer than 2.5 kilometers) – although Italy, France and Ireland, among other countries, continue to deploy them.

In some coastal areas, fishing nets set near the sea bottom routinely ensnare small dolphins. Losses to fisheries of several marine mammals – the baiji of eastern Asia, the Mexican vaquita (the smallest type of dolphin), Hector's dolphins in the New Zealand region and the Mediterranean monk seal – put those species' survival at risk. Seabirds are also killed when they try to eat the bait attached to fishing lines as these are played out from ships. Rosemary Gales, a research scientist at the Parks and Wildlife Service in Hobart, Tasmania, estimates that in the Southern Hemisphere more than 40,000 albatross are hooked and drowned every year after grabbing at squid used as

> **Box 3.1   No Place Like Home**
> Although much of my work has been focused on overfishing, I have also come to see that marine habitats are being destroyed or degraded in numerous ways. In many temperate regions the larger, bottom-dwelling animals and plants – which feed and shelter fish – have been heavily damaged by trawling, a form of fishing that rakes nets over the shallow continental shelves. In the tropical Indo-Pacific, many people catch fish by stunning them with cyanide – a poison that kills the coral that makes up the fishes' habitat. Some fishers herd their prey into nets by pounding the corals with stones; a boat fishing in this way can destroy up to a square kilometer of living reef every day.
>
> Marine habitats also suffer assaults from aquaculture, agriculture and clear-cutting for logging. In the Pacific Northwest of the U.S. and Canada, intensive deforestation, hydroelectric dams and water diversion have destroyed thousands of miles of salmon habitat. Most species of sturgeon are also becoming endangered in this way throughout the Northern Hemisphere. Profuse sedimentation following deforestation degrades habitats in many parts of the tropics as well. Sediments can kill coral reefs by clogging them, blocking sunlight and preventing settlement of larvae.
>
> In 1989 the tropical marine ecologist Robert Johannes helped to select the tiny Pacific island country of Palau as one of the world's seven undersea wonders – akin to the seven wonders of the ancient world – because of its spectacular and largely unspoiled coral reefs. When I visited him in Palau early this year, I frequently witnessed long plumes of red sediment bleeding off new, poorly made roads into coral lagoons after every heavy rain. Runoff from intact jungle was, in contrast, as clear as the rain itself. Untreated sewage was also flowing into reefs near the capital's harbor. Such nutrient-rich pollution allows algae to grow at unnatural rates, killing corals by altering this delicate balance with internal symbiotic algae.
>
> —C. S.

bait on longlines being set for bluefin tuna. This level of mortality endangers six of the 14 species of these majestic wandering seabirds.

In some fisheries, bykill exceeds target catch. In 1992 in the Bering Sea, fishers discarded 16 million red king crabs, keeping only about three million. Trawling for shrimp produces more bykill than any other type of fishing and accounts for more than a third of the global total. Discarded creatures outnumber shrimp taken by anywhere from 125 to 830 percent. In the Gulf of Mexico shrimp fishery 12 million juvenile snappers and 2,800 metric tons of sharks are discarded annually. Worldwide, fishers dispose of about six million sharks every year – half of those caught. And these statistics probably underestimate the magnitude of the waste: much bycatch goes unreported.

There remain, however, some glimmers of hope. The bykill of sea turtles in shrimp trawls had been a constant plague on these creatures in U.S. waters (the National Research Council estimated that up to 55,000 adult turtles died this way every year). But these deaths are being reduced by recently mandated "excluder devices" that shunt the animals out a trap door in the nets.

Perhaps the best-publicized example of bycatch involved up to 400,000 dolphins killed annually by fishers netting Pacific yellowfin tuna. Over three decades since the tuna industry began using huge nets, the eastern spinner dolphin population fell 80 percent, and the numbers of offshore spotted dolphin plummeted by more than 50 percent. These declines led to the use of so-called dolphin-safe methods (begun in 1990) whereby fishers shifted from netting around dolphin schools to netting around logs and other floating objects.

This approach has been highly successful: dolphin kills went down to 4,000 in 1993. Unfortunately, dolphin-safe netting methods are not safe for immature tuna, billfish, turtle or shark.

On average, for every 1,000 nets set around dolphin herds, fishers inadvertently capture 500 dolphins, 52 billfish, 10 sea turtles and no sharks. In contrast, typical bycatch from the same number of sets

---

Box 3.2   Economies of Scales

Fishing adds only about 1 percent to the global economy, but on a regional basis it can contribute enormously to human survival. Marine fisheries contribute more to the world's supply of protein than beef, poultry or any other animal source.

Fishing typically does not require land ownership, and because it remains, in general, open to all, it is often the employer of last resort in the developing world – an occupation when there are no other options. Worldwide, about 200 million people depend on fishing for their livelihoods. within Southeast Asia alone, more than five million people fish full-time. In northern Chile 40 percent of the population lives off the ocean. In Newfoundland most employment came from fishing or servicing that industry – until the collapse of the cod fisheries in the early 1990s left tens of thousands of people out of work.

Although debates over the conservation of natural resources are often cast as a conflict between jobs and the environment, the restoration of fish populations would in fact boost employment. Michael P. Sissenwine and Andrew A. Rosenberg of the U.S. National Marine Fisheries Service have estimated that if depleted species were allowed to rebuild to their long-term potential, their sustainable use would add about $8 billion to the U.S. gross domestic product – and provide some 300,000 jobs. If fish populations were restored and properly managed, about 20 million metric tons could be added to the world's annual catch. But reinstatement of ecological balance, fiscal profitability and economic security will require a substantial reduction in the capacity of the commercial fishing industry so that wild populations can recover.

The necessary reductions in fishing power need not come at the expense of jobs. Governments could increase employment and reduce the pressure on fish populations by directing subsidies away from highly mechanized ships. For each $1 million of investment, industrial-scale fishing operations require only one to five people, whereas small-scale fisheries would employ between 60 and 3,000. Industrial fishing itself threatens tens of millions of fishers working on a small scale by depleting the fish on which they depend for subsistence.

For some fisheries, regulators have purposefully promoted inefficiency as a way to limit excessive catches and maintain the living resource. For example, in the Chesapeake Bay, law requires oyster-dredging boats to be powered by sail (left), a restriction on technology that has helped this fishery survive. In New England, regulators outlawed the use of nets pulled between two boats ("pair trawls") because this technique was too effective at catching cod. Managers of the U.S. bluefin-tuna fishery allocate 52 percent of the catch to commercial boats that deploy the least capable gear – handlines or rod and reel – even though the entire allowed amount could easily be extracted with purse-seine nets. In this instance, vessels with the more labor-intensive tackle account for nearly 80 percent of direct employment; those that have large nets provide only 2 percent. Numerous other regulations on sizes and total amount of the catch, as well as allocation and allowable equipment, can be viewed as acknowledgments of the need to curb efficiency in order to achieve wider social and ecological benefits.

—C. S.

around floating objects includes only two dolphins but also 654 billfish, 102 sea turtles and 13,958 sharks. In addition, many juvenile tuna are caught under floating objects.

One solution to the bycatch from nets would be to fish for tuna with poles and lines, as was practiced commercially in the 1950s. That switch would entail hiring back bigger crews, such as those laid off when the fishery first mechanized its operations.

The recent reductions in the bycatch of dolphins and turtles provide a reminder that although the state of the world's fisheries is precarious, there are also reasons for optimism. Scientific grasp of the problems is still developing, yet sufficient knowledge has been amassed to understand how the difficulties can be rectified. Clearly, one of the most important steps that could be taken to prevent overfishing and excessive bycatch is to remove the subsidies for fisheries that would otherwise be financially incapable of existing off the oceans' wildlife – but are now quite capable of depleting it.

Where fishes have been protected, they have rebounded – along with the social and economic activities they supported. The resurgence of striped bass along the eastern coast of the U.S. is probably the best example in the world of a species that was allowed to recoup through tough management and an intelligent rebuilding plan.

During the past year, the U.N. has been making historic progress in forging new conservation agreements dealing with high-seas fishing. Such measures, along with regional and local efforts to protect the marine environment, should help guide the world toward a sane and sustainable future for life in the oceans.

Further Reading

Bluefin Tuna in the West Atlantic: Negligent Management and the Making of an Endangered Species. Carl Safina in *Conservation Biology*, Vol. 7, No. 2, pages 229–234; June 1993.

Global Marine Biological Diversity: A Strategy for Building Conservation into Decision Making. Edited by E. Norse. Island Press, 1993.

Where Have All The Fishes Gone? Carl Safina in *Issues in Science and Technology*, Vol. 10, pages 37–43; Spring 1994.

The State of World Fisheries and Aquaculture. United Nations Food and Agriculture Organization, Rome, 1995.

Understanding Marine Biodiversity. Report of the National Research Council's Committee on Biological Diversity in Marine Systems. National Academy Press, 1995.

4

VALUATION OF AN AMAZONIAN RAINFOREST*

Charles M. Peters, Alwyn H. Gentry, and Robert O. Mendelsohn

Tropical forest resources have traditionally been divided into two main groups: timber resources, which include sawlogs and pulp-wood; and non-wood or 'minor' forest products, which include edible fruits, oils, latex, fibre and medicines. Most financial appraisals of tropical forests have focused exclusively on timber resources and have ignored the market benefits of non-wood products. The results from these appraisals have usually demonstrated that the net revenue obtainable from a particular tract of forest is relatively small, and that alternative uses of the land are more desirable from a purely financial standpoint. Thus there has been a strong market incentive for destructive logging and widespread forest clearing.

We contend that a detailed accounting of non-wood resources is required before concluding *a priori* that tropical deforestation makes financial sense. To illustrate our point, we present data concerning inventory, production and current market value for all the commercial tree species occurring in one hectare of species-rich Amazonian forest. These data indicate that tropical

forests are worth considerably more than has been previously assumed, and that the actual market benefits of timber are very small relative to those of non-wood resources. Moreover, the total net revenues generated by the sustainable exploitation of 'minor' forest products are two to three times higher than those resulting from forest conversion.

Our findings are based on an appraisal of an area along the Rio Nanay near to the small village of Mishana (3°47′S, 73°30′W), 30 km south-west of the city of Iquitos, Peru. Annual precipitation in the region averages 3,700 mm; soils are predominantly infertile white sands. The inhabitants of Mishana are detribalized indigenous people called *ribereños* who make their living practising shifting cultivation, fishing and collecting a wide variety of forest products to sell in the Iquitos market.

A systematic botanical inventory of 1.0 ha of forest at Mishana showed 50 families, 275 species and 842 trees ≥10.0 cm in diameter (1). Of the total number of trees on the site, 72 species (26.2%) and 350 individuals (41.6%) yield products with an actual market

* Originally published in *Nature*, 1989, vol. 339, pp. 655–6.

Table 4.1 [*orig. table 1*]   Annual yield and market value of fruit and latex produced in one hectare of forest at Mishana, Rio Nanay, Peru.

Common name	Species	No. trees	Annual production per tree	Unit price (US$)	Total value (US$)
Aguaje	*Mauritia flexuosa* L.	8	88.8 kg	10.00/40 kg	177.60
Aguajillo	*Mauritiella peruviana* (Becc.) Burrett	25	3.0 kg	4.00/40 kg	75.00
Charichuelo	*Rheedia* spp.	2	100 fruits	0.15/20 fruits	1.50
Leche huayo	*Couma macrocarpa* Barb. Rodr.	2	1,060 fruits	0.10/3 fruits	70.67
Masaranduba	*Manilkara quianensis* Aubl.	1	500 fruits	0.15/20 fruits	3.75
Naranjo podrido	*Parahancornia peruviana* Monach.	3	150 fruits	0.25/fruit	112.50
Sacha cacao	*Theobroma subincanum* Mart.	3	50 fruits	0.15/fruit	22.50
Shimbillo	*Inga* spp.	9	200 fruits	1.50/100 fruits	27.00
Shiringa	*Hevea quianensis* Aubl.	24	2.0 kg	1.20/kg	57.60
Sinamillo	*Oenocarpus mapora* Karst.	1	3,000 fruits	0.15/20 fruits	22.50
Tamamuri	*Brosimum rubescens* Taub.	3	500 fruits	0.15/20 fruits	11.25
Ungurahui	*Jessenia bataua* (Mart.) Burret	36	36.8 kg	3.50/40 kg	115.92
Totals		117			697.79

Fruit yields measured for *M flexuosa, J. bataua, P. peruviana* and *C. macrocarpa*; estimated yields for other fruit trees based on interviews with local collectors.

value in Iquitos. Edible fruits are produced by seven dicotyledonous and four palm species, sixty species produce commercial timber and one species, *Hevea guianensis* Aubl., produces rubber. The forest also contains medicinal plants, lianas and several understorey palms of commercial importance in the region, yet these species were too small to be included in our sample.

Annual production rates for all the fruit trees and palms in our sample area were either measured by counting and weighing all the fruits produced by a sub-sample of adult trees (4 species), or estimated from interviews with local collectors (7 species). Latex yields for wild *Hevea* trees were taken from the literature (2). The merchantable volume of each timber tree was calculated using published regression equations

relating diameter at breast height (137 cm) to commercial height (3).

We collected average retail prices for different forest fruits in 1987 by making monthly surveys of the Iquitos produce market. We obtained rubber prices, which are officially controlled by the Peruvian government, from the agrarian bank office. Four independent sawmill operators in Iquitos were interviewed to determine the mill price of each timber species. All market data were converted using Peruvian intis to 1987 US dollars using an exchange rate of 20 intis to the dollar.

The labour investment associated with fruit collection and latex tapping was estimated in man days per year based on interviews and direct observation of local collecting techniques. We then calculated

Table 4.2 [orig. table 2] Merchantable volume and stumpage value of the commercial timber tree in one hectare of forest at Mishana, Rio Nanay, Peru.

Commercial name	Genera included	No. trees	Wood volume (m³)	Mill price (per m³) (US$)	Stumpage value (US$)
Aguano masha	Trichilia	4	0.55	14.80	4.88
Almendro	Caryocar	1	0.08	14.80	0.71
Azucar huayo	Hymenaea	1	0.10	14.80	0.89
Cumala	Iryanthera, Virola	83	19.77	19.00	225.38
Espintana	Guatteria, Xylopia	7	1.47	21.00	18.52
Favorito	Osteophloeum	2	3.90	14.80	34.63
Ishpingo	Endlicheria	4	0.82	14.80	7.28
Itauba	Mezilaurus	3	0.29	14.80	2.57
Lagarto caspi	Calophyllum	2	0.25	40.30	6.04
Loro micuna	Macoubea	1	1.37	14.80	12.17
Machimango	Eschweilera	5	0.76	20.15	9.19
Machinga	Brosimum	10	24.61	14.80	218.53
Moena	Aniba, Ocotea	6	0.75	42.00	18.90
Palisangre	Dialium	1	0.27	14.80	2.39
Papelillo	Cariniana	1	1.19	14.80	10.57
Pashaco	Parkia	19	4.19	14.80	37.21
Pumaquiro	Aspidosperma	12	10.22	14.80	90.75
Quinilla	Chrysophyllum, Pouteria, Manilkara	34	9.18	31.80	175.15
Remo caspi	Swartzia, Aspidosperma	28	11.65	14.80	103.45
Requia	Guarea	4	1.06	14.80	9.41
Tortuga caspi	Duquetia	1	0.13	14.80	1.15
Yacushapana	Terminalia	2	0.71	14.80	6.31
Yutubanco	Heisteria	2	0.53	14.80	4.70
Totals		233	93.85		1,000.78

Twenty-three commercial names represent 28 genera and 60 tree species.

cumulative harvest costs using a wage rate of US$2.50 per man day, the minimum wage in Peru during 1987. Based on previous studies conducted in Mishana (4), we assumed that transport costs for fruit and latex are 30% of the total market value of these products. Studies by the Food and Agriculture Organization (5) suggest that logging and transport costs in the Peruvian lowlands equal 30–50% of the total value of the timber harvested; we used a 40% extraction expense in our study.

Based on our estimates of the density, productivity and market price of each fruit tree and palm, one hectare of forest at

Mishana produces fruit worth almost $650 each year (table 4.1). Annual rubber yields amount to about $50. After deducting the costs associated with collecting and transporting this material to market, net annual revenues from fruit and latex are $400 and $22, respectively.

Given that both fruit and latex can be collected every year, the total financial value of these resources is considerably greater than the current market value of only one year's harvest. Clearly, the net revenue generated by all future harvests must also be estimated and included in the analysis. We used a simple discounting model to calculate the net present value (NPV) as V/r, where V is the net revenue produced each year and r is a 5% inflation-free discount rate, of these annuities. Assuming that 25% of the fruit crop is left in the forest for regeneration, the NPV of sustainable fruit and latex harvests is estimated at $6,330 per hectare.

The Mishana forest also contains 93.8 m^3ha^{-1} of merchantable timber (table 4.2). If liquidated in one felling, this sawtimber would generate a net revenue of £1,000 on delivery to the sawmill. A logging operation of this intensity, however, would damage much of the residual stand and greatly reduce, if not eliminate, future revenues from fruit and latex trees. The net financial gains from timber extraction would be reduced to zero if as few as 18 fruit trees were damaged by logging.

Periodic selective cutting presumably would be more compatible with annual fruit and latex collection. Yield functions derived for the 60 commercial timber species at Mishana using stand-table projection and a mean diameter increment of 0.3 cm yr^{-1} indicate a maximum sustainable harvest of about 30 m^3ha^{-1} every 20 years. Multiplying this volume by a weighted average market price of $17.21 m^{-3} and deducting harvest and transport costs gives a net revenue of about $310 at each cutting cycle. Discounting a perpetual series of these periodic revenues back to the present yields a net present value (NPV = $V/(1 - e^{-11})$, where t is the number of years between harvests), of $490 for timber.

Based on the assumption of sustainable timber harvests and annual fruit and latex collection for perpetuity, we estimate that the tree resources growing in one hectare of forest at Mishana possess a combined financial worth of $6,820. Fruits and latex represent more than 90% of the total market value of the forest, and the relative importance of non-wood products would increase even further if it were possible to include the revenues generated by the sale of medicinal plants, lianas and small palms. In view of the disproportionately low NPV of wood resources and the impact of logging on fruit and latex trees, timber management is a marginal financial option in this forest.

We acknowledge that these projections are subject to temporal changes in market prices, production levels and harvest intensities that could either increase or decrease the actual market benefits obtainable from the forest. Moreover, we realize that not every hectare of tropical forest will have the same market value as our plot in Mishana. Further studies are needed to determine how floristic composition, productivity and distance to markets influence the financial worth of a forest. yet we believe that the NPVs calculated in this study provide a useful economic benchmark for comparing alternative land-use practices and management options for Amazonian forests.

Our results indicate that the financial benefits generated by sustainable forest use tend to exceed those that result from forest conversion. Using identical investment criteria, that is discounting a perpetual series of net revenues at a 5% interest rate, the NPV of the timber and pulpwood obtained from a 1.0-ha plantation of *Gmelina arborea* in Brazilian Amazonia is estimated at $3,184 (ref. 6), or less than half that of the forest. Similarly, gross revenues from fully-stocked cattle pastures in Brazil are reported (7) to be $148 $ha^{-1}yr^{-1}$. The present value of a perpetual series of such pastures discounted back to the present is only $2,960, and deducting the costs of weeding, fencing and animal care would lower this figure significantly. Both these estimates are based on the optimistic assumption that plantation forestry and grazing lands are sustainable land-use practices in the tropics.

The results from our study clearly demonstrate the importance of non-wood forest products. These resources not only yield higher net revenues per hectare than timber, but they can also be harvested with considerably less damage to the forest. Without question, the sustainable exploitation of non-wood forest resources represents the most immediate and profitable method for integrating the use and conservation of Amazonian forests. Why has so little been done to promote the marketing, processing and development of these valuable resources?

We believe that the problem lies not in the actual value of these resources, but in the failure of public policy to recognize it. Tropical timber is sold in international markets and generates substantial amounts of foreign exchange; it is a highly visible export commodity controlled by the government and supported by large federal expenditures. Non-wood resources, on the other hand, are collected and sold in local markets by an incalculable number of subsistence farmers, forest collectors, middlemen and shop-owners. These decentralized trade networks are extremely hard to monitor and easy to ignore in national accounting schemes.

The non-market benefits of tropical forests have been discussed recently (8, 9). Tropical forests perform vital ecological services, they are the repository for an incredible diversity of germplasm, and their scientific value is immeasurable. The results from this study indicate that tropical forests can also generate substantial market benefits if the appropriate resources are exploited and properly managed. We suggest that comparative economics may provide the most convincing justification for conservation and use of these important ecosystems.

Acknowledgements
We thank P. Ashton, M. Balick, K. Clark, D. Nepstad, C. Padoch, T. Panayotou and G. Prance for helpful comments.

References
1   Gentry, A. H. *Proc. natn. Acad. Sci, U.S.A.* 85, 156 (1988).
2   Polhamus, L. G. *Rubber: Botany, Production and Utilization* (Hill, London, 1962).
3   Villanueva-Agustin, A. *Correlaciones entre los Valores Dimensionales de los Arboles de los Bosques de Puerto Almedra, Iquitos* (Instituto de Investigaciones de la Amazonia Peruana, Iquitos, 1985).
4   Padoch, C. *Adv. econ. Bot.* 5, 74–89 (1988).
5   *Estructura de los Costos de Extraccion y Transporte de Madera Rolliza en la Selva Baja* (PNUD/FAO/PER/80/006A, Lima, 1980).
6   Sedjo, R. A. *The Comparative Economics of Plantation Forestry* (Resources for the Future, Washington, DC, 1983).
7   Buschbacher, R. J. *Biotropica* 19, 200–207 (1987).
8   Poore, D. *Unasylva* 28, 127–143 (1976).
9   *Research Priorities in Tropical Biology* (National Academy of Sciences, Washington, DC, 1980).

5

MAKING BIODIVERSITY CONSERVATION PROFITABLE: A CASE STUDY OF THE MERCK/INBIO AGREEMENT*

Elissa Blum

This land is a place where we know where to find all that it provides for us – food from hunting and fishing, and farms, building and tool materials, medicines. This land keeps us together within its mountains; we come to understand that we are not just a few people or separate villages, but one people belonging to a homeland.

The Akawaio Indians Upper Mazaruni District, Guyana[1]

Like the Akawaio Indians of Guyana, all of the world's human population depends on biological diversity for sustained development. Natural resources provide agricultural products for food, fiber for clothes, timber for fuel and construction, and chemicals for medicinal uses. Forests protect soils from decay and absorb carbon dioxide from the environment. Insects and microorganisms, many too small to be seen by the naked eye, process wastes, renew soils, and provide nutrients for plants.[2]

All of these resources are threatened, however, because biological diversity is decreasing at an ever-increasing rate – even faster than rates of evolutionary replacement.[3] Resources and species are being exploited and destroyed without regard to their long-term value. If people continue to exploit biological diversity as they have in the past, humans in the future will not be able to depend on natural resources. Clearly, methods to conserve biological diversity must be devised and implemented. All

* Originally published in *Environment*, 1993, vol. 35, no. 4, pp. 16–45.
[1] World Resources Institute, World Conservation Union, and United Nations Environment Programme, *Global Biodiversity Strategy* (Washington, D.C.: WRI, 1992).
[2] Council on Environmental Quality, *United States of America National Report* (Washington, D.C.: U.S. Government Printing Office, 1992), 292.
[3] T. Eisner, "Chemical Prospecting: A Proposal for Action," in F. H. Bormann and S. R. Kellert, eds., *Ecology, Economics, Ethics: The Broken Circle* (New Haven, Conn.: Yale University Press, 1991), 196.

populations must find ways to ensure the sustainable use of biological resources by balancing present developmental progress with the needs of future generations.

One method by which this balance could be achieved is through the proliferation of agreements such as the one between Merck & Company and the Instituto Nacional de Biodiversidad in Costa Rica (INBio). According to the terms of a September 1991 agreement, INBio collects and processes plant, insect, and soil samples in Costa Rica and gives them to Merck for evaluation as prospective medicines. The contract provides Costa Ricans with an economically beneficial alternative to deforestation and concurrently advances the research efforts of Merck. Are agreements like the one between Merck and INBio a solution to the perceived incompatibility of economic growth and biodiversity preservation within developing nations? Before considering the collaborative agreement, it is first necessary to understand why biodiversity preservation is a problem.

Preserving Biodiversity

Biological diversity, "the variety among living organisms and the ecological communities they inhabit," consists of three categories, or levels: genetic diversity, species diversity, and ecosystem diversity.[4] Species diversity, the level most often studied, is defined as the extant number and variety of species of plants, animals, fungi, microorganisms, and other living beings.

It has been estimated that between 5 million and 30 million species exist; most biologists regard 10 million as the best approximation. Only 1.4 million of these species have been named by taxonomists. Tropical forests, predominantly in Central and South America and Southeast Asia, contain from 50 to 90 percent of all species, including two-thirds of all vascular plant

species and up to 96 percent of insect species. At current deforestation rates, it is estimated that between 4 and 8 percent of all rain forest species would be in danger of extinction by 2015, and from 17 to 35 percent would be in danger of extinction by 2040.[5]

Why is so much of the world's biodiversity in danger of destruction? One of the prime reasons is the incompatibility of short-term economic growth with the sustainable development of natural resources. When an economically struggling country has a choice between logging a forest to sell timber for high profits and leaving the forest intact without monetary compensation, the nation usually chooses the profitable alternative. Because immediate economic gains to the nation are more important than the future environmental costs, deforestation occurs without regard to its long-term effects on biodiversity preservation.

This type of destructive activity occurs because the environment is a market externality; because no economic gain or loss is recorded in a basic accounting network when biological resources are used, the resources are considered to be free.[6] Externalities occur when there are no formal ownership rights to a good. Because it is difficult for a country to realize profitable returns from the environment, many nations are averse to expending large sums of money to preserve it.

Despite previous failures of the market to preserve biodiversity, market reform offers nations the best incentive to sustain their natural resources. According to Jessica T. Mathews, a vice president of the World Resources Institute (WRI) in Washington, D.C., the "marketplace is the most efficient way to achieve [positive] environmental ends . . . if you can make the marketplace work."[7] To make it work, countries must find ways to make resource conservation more profitable than resource exploitation. They can only do so by expressing

---

[4] World Resources Institute, *World Resources 1992–93: A Guide to the Global Environment* (New York: Oxford University Press, 1992), 127–28.

[5] Ibid., 127–31.

[6] R. Repetto, Presentation at Princeton University, New Jersey, 26 October 1992.

[7] J. T. Mathews, Presentation at Princeton University, New Jersey, 26 October 1992.

sovereignty over their natural resources, accepting the responsibility to conserve them, and limiting access to those who abide by certain regulations.[8]

The first step toward this type of natural resource control was taken at a summit in Managua, Nicaragua, on 6 June 1992. At this meeting, the presidents of Belize, Costa Rica, El Salvador, Guatemala, Honduras, Nicaragua, and Panama signed a non-binding resolution that encouraged the passage of laws to regulate and restrict the extraction of natural resources from their countries.[9] The goal of the resolution was to prevent foreigners from invading their nations' wildland and extracting valuable resources without compensating the host nations.

The presidents were motivated to sign the agreement by the realization that foreign companies especially biotechnology and pharmaceutical firms, were using Central American natural resources to develop drugs from which Central Americans realized no profits. The case of the Madagascar periwinkle is often cited as the prime example of this problem. The multi-million dollar cancer drug vincristine was developed from natural resources extracted from the Madagascar periwinkle, and yet Madagascar received no payment or share of the royalties.[10] Other countries have been exploited as well; WRI recently estimated that Costa Rica has lost more than $4 billion during the past 20 years from unrealized returns on natural resources.[11]

Prior to this resolution, no Central American country had ever passed laws regulating the extraction of biological resources. Now, many of these countries have made a commitment to do so. Accord-ing to the Central American Commission on the Environment and Development, laws enacted by individual nations under this resolution should include guidelines for extraction of resources, payment of royalties from products developed from biological resources, future access to patents on the resources, rules ensuring the transfer of technology to the host country, access to advanced research materials, and rules for sharing the expertise necessary to market the developed products.[12]

Exploiting Natural Resources

The need for developing countries to assert sovereignty over their natural resources arose in part because of the extraction of resources by biotechnology and phar-maceutical companies for commercial development. Biotechnology, "The science of recombining the genes of plants and organisms to derive improved plants, drugs, and foods," has revolutionized ways in which natural resources are used in product development.[13] Prior to the 1950s, drugs were often developed from natural resources. However, they were discovered in a haphazard and time-intensive manner. From the 1950s to the 1980s, pharmaceutical companies found that synthetic chemicals and genetic engineering were more prof-itable than screening plants. Therefore, natural resource research slowed. Recently, however, the drug discovery process has reverted to more traditional methods. The usefulness of natural resources in drugs is derived in part from the fact that tropical plants often have strong chemical defenses to repel predators.[14] Although the more simple of these compounds have been

---

[8] World Bank, *World Development Report 1992: Development and the Environment* (New York: Oxford University Press, 1992), 64.

[9] *BNA International Environment Daily*, "Central American Presidents Resolve to Pass Laws Restricting Use of Resources," 11 June 1992.

[10] C. Joyce, "Prospectors for Tropical Medicines," *New Scientist*, 19 October 1991, 36–40.

[11] *Inter Press Service*, "Costa Rica: Plant Life Exploitation Rights Scheme 'A Bad Deal,'" 15 January 1992.

[12] *BNA International Environment Daily*, "Central American Presidents to Sign Separate Accord on Biological Resources," 4 June 1992.

[13] G. Browning, "Biodiversity Battle," *National Journal*, 8 August 1992, 1827.

[14] Joyce, note 10 above, pages 37–38.

synthesized in the laboratory, pharmaceutical companies are realizing that it would be impossible to replicate all of nature's more complex compounds. By using new bioassays and technology to test natural resources, the identification of certain natural resources as prospective drugs has become easier and more accurate. Thus, the drug discovery process has moved back into the rain forest and so increased the number of reasons to exploit the environment.[15]

Biotechnology and pharmaceutical companies also have an incentive not to exploit the environment, however, because doing so would eliminate one of their prime research bases. Traditional medicines, which are derived from natural resources, form the basis of health care for approximately 80 percent of the populations of developing nations.[16] Chances are good that, among the 10 million or so unidentified species in the rain forests, "there's something in there that can do everything we want done."[17] The destruction of biodiversity would mean the destruction of unique chemicals that may not be replicable in the laboratory.[18] Although the number of drugs developed directly from natural resources may be small, it has been estimated that one-quarter of all medicinal prescriptions in the United States are based in some way upon plants or microbes or are synthetic chemicals derived from them.[19] Examples of drugs derived from natural resources include penicillin, developed from a mold; taxol, a cancer drug from a yew tree bark; and the antibiotic streptomycin, developed from a soil sample.[20]

The Biodiversity Convention

The seeming incompatibility of extracting natural resources for profits and protecting biological diversity within the tropics was one of the prime concerns of the representatives negotiating the Biodiversity Convention at the Earth Summit in Rio de Janerio last June. The treaty developed at the conference was designed to ensure the use of biological diversity in a sustainable way and to put guidelines into place that would make economic use of natural resources more compatible with their conservation. The United States was one of the few nations that participated in the summit but did not sign the treaty.[21]

One of the main reasons why the United States refused to sign the treaty was because it did not seem to protect intellectual property rights to the extent that biotechnology and pharmaceutical companies desired. The United States – the world-leader in the commercial development of biotechnology products, whose annual sales are expected to reach $50 billion by 2000 – feared that the vague language of the treaty would be interpreted in such a way that the United States would lose its comparative advantage in the field.[22] The treaty sections most highly contested by biotechnology companies include paragraphs 15 and 16, which they felt would let developing countries make wide exclusions from intellectual property protection and would permit them to enact laws with overly broad compulsory licensing clauses.[23] The drug firms feared that these laws could force them to transfer patent rights to a new drug to a developing country.

[15] Ibid.; and L. Roberts, "Chemical Prospecting: Hope for Vanishing Ecosystems?" *Science* 256 (22 May 1992): 1142–43.
[16] World Resources Institute et al., note 1 above, page 4.
[17] M. Mellon, Personal communication with the author, 20 October 1992.
[18] Eisner, note 3 above, pages 196 and 199.
[19] Joyce, note 10 above, pages 37–38.
[20] G. Bylinsky, "The Race for a Rare Cancer Drug," *Fortune*, 13 July 1992, 100–02; Eisner, note 3 above, pages 197–98; and *Merck World*, "A Modern-Day Noah's Ark," no. 3 (1992).
[21] K. Miller, Personal communication with the author and Peter Bartolino, 26 October 1992.
[22] Browning, note 13 above.
[23] *International Trade Reporter*, "Biodiversity Treaty Risks Interfering with Patent Protections, Official Says," 17 June 1992.

The Merck/INBio Agreement

The type of contractual agreement that many hope would be encouraged by the biodiversity treaty is exemplified by the recent agreement between Merck & company, the world's largest pharmaceutical company, and the Instituto Nacional de Biodiversidad in Costa Rica. The agreement, announced on 19 September 1991, is a two-year "collaborative research agreement" under which Merck agreed to pay INBio a sum of $1 million for all of the plant, insect, and soil samples the institute could collect in addition to a percentage of the royalties from any drugs that Merck develops from samples provided by INBio.[24] The royalties, which were estimated to be in the range of one to three percent but were not made known to the public, will be split equally between INBio and the Costa Rican Ministry of Natural Resources.[25]

Under the agreement, Merck gets exclusive rights, called "right of first refusal," to evaluate the approximately 10,000 samples that INBio agreed to supply to Merck.[26] INBio may enter into similar agreements with other parties, but it must not supply the same species samples that it supplies to Merck unless given explicit permission to do so.[27]

According to Thomas Eisner, an entomologist at Cornell University, "The agreement is a win-win situation. It protects the proprietary rights of the industry, while at the same time recognizing that it is to the advantage of industrial nations to help with the custodianship of natural resources."[28] Merck benefits from the contract by receiving a limited number of plant, insect, and soil samples, already identified and classified. Because the samples are extracted and processed before they reach the Merck laboratories, Merck can immediately focus upon testing the samples for chemical activity.[29] The agreement specifies that, if Merck discovers any active ingredients from which it develops commercial products, the company will retain all patent rights to the developed product.[30] According to Georg Albers-Schonberg, executive director of Merck's Natural Products Division, Merck is skeptical about achieving immediate results from chemical prospecting. If not for the agreement, however, regions containing potential drugs would be destroyed, and the prospect of discovering an active ingredient for a drug would decrease even further.[31]

INBio and Costa Rica benefit from the agreement as well. One of the most important yet least tangible of the benefits incurred by Costa Rica is the ability to use its natural resources in a sustainable way while concurrently strengthening the Costa Rican economy. If INBio can create more jobs, profits, and a better educated constituency by cataloging and selling rights to the country's natural resources than by destroying its resources, it makes economic sense to keep the resources intact.[32] On the more tangible side, in addition to the $1 million from Merck, INBio has benefited from technology transfer in the form of equipment donations worth $135,000 to

[24] Merck & Co., "Summary of the INBio–Merck Agreement" (Unpublished memorandum, Rahway, N. J., September 1992); and idem, "INBio of Costa Rica and Merck Enter into Innovative Agreement to Collect Biological Samples While Protecting Rain Forest" (Press release, Rahway, N.J., 19 September 1991).

[25] Roberts, note 15 above.

[26] Ibid.; and J. Preston, "A Biodiversity Pact with a Premium" *Washington Post*, 9 June 1992, A16.

[27] Merck & Co., "Summary," note 24 above.

[28] T. Reynolds, "Drug Firms, Countries Hope to Cash in on Natural Products," *Journal of the National Cancer Institute* 84, no. 15 (5 August 1992): 1147–48.

[29] Merck & Co., "Summary," note 24 above.

[30] *Pharmaceutical Business News*, "Pharmaceutical Companies Go 'Chemical Prospecting' for Medicines," 21 August 1992.

[31] Joyce, note 10 above, page 39.

[32] Merck & Co., "Searching for Medicines While Preserving the Rain Forest: The INBio/Merck Agreement" (Unpublished memorandum, Rahway, N.J., 12 May 1992).

carry out the chemical extraction process. Merck supplied INBio with two natural products chemists to set up the extraction laboratories and to train scientists in the purification techniques. Costa Rican scientists are also permitted to visit certain Merck laboratories if they wish to learn more. Eventually, INBio hopes that the training will aid Costa Rica in developing its own biotechnology industry so that drug discovery can take place entirely within national boundaries.[33]

The royalties that INBio receives from Merck will be used to support the conservation of Costa Rica's biodiversity. Ten percent of the initial $1 million plus 50 percent of the royalties go directly to Costa Rica's Ministry of Natural Resources; the rest of the profits are used to preserve the environment at the discretion of INBio and its board of directors.[34] The prospect of royalties from commercially developed products could have a drastic effect on Costa Rica's economy. According to the World Resources Institute, if INBio receives 2 percent of the royalties from the sale of 20 products based on its samples, INBio would receive more money than Costa Rica does from the sale of coffee and bananas, two prime exports.[35] This money could be used to build a Costa Rican biotechnology industry and to strengthen Costa Rica's economy.

Developing the Agreement

The idea for the agreement developed from the technique of chemical prospecting, "the exploratory process by which new, useful natural products are discovered."[36] Chemical prospecting, a term coined by Thomas Eisner, involves three processes: first, natural resources are screened for their chemical or biochemical activity; second, the

active components within the natural resources are isolated and characterized; and third, the active components are screened for certain activity.[37] Eisner believes that, by shifting chemical screening from industry to the nation in which the resource is located, an incentive could be created to preserve the natural environment.[38] Linking the biological inventorying that already occurs in tropical nations by institutes such as INBio to chemical prospecting would also decrease the randomness of the screening process. A person searching for a plant with a certain type of chemical repellent could look in the inventory catalog for plants that are free of certain kinds of insects and screen those plants rather than plants randomly selected from the rain forest. This link would hasten the identification of potentially useful medicines and other plant- and insect-based products and would not hinder efforts at conserving the environment; industries that find a useful chemical within a natural resource would try to replicate synthetically the chemical structure of the active component of the resource within the laboratory rather than continue to isolate it from the actual resource.[39]

Eisner's plan makes money for the countries in which the resources are located and motivates them to preserve rather than destroy their environment. The resource-rich nations could form alliances with universities and industries for certain upfront or royalty-based fees to insure that some of the profits from commercialization are captured by the country that initially located, cataloged, and screened the natural resource. Local residents could be trained to carry out the chemical screening to give them an alternative to environmentally destructive sources of employment and to

[33] Roberts, note 15 above.
[34] Merck & Co., note 32 above.
[35] World Resources Institute et al., note 1 above, page 152.
[36] Eisner, note 3 above, page 196.
[37] Ibid., page 199. The term *chemical prospecting* was first introduced in Reynolds, note 28 above.
[38] T. Eisner, "Prospecting for Nature's Chemical Riches," *Issues in Science and Technology*, Winter 1989–90, 31–34.
[39] Ibid.; and Eisner, note 3 above.

strengthen the science base of their country.[40]

Eisner has suggested that, to finance the prospecting, foundations, universities, government laboratories, and industry should all support the developing nations with money, materials, and scientific training in varying degrees. Because it would be laboratories within these institutions that would benefit from the more accurate and faster screening process, these institutions should support developing nations in their efforts.[41] In addition, debt-for-nature swaps, by which indebted countries can convert some of their debt into local investments to preserve the environment, could help finance the screening process.[42]

Because of his idea for institutional support of chemical prospecting, Eisner called Paul Anderson, vice president of medicinal chemistry at Merck Sharp & Dohme Research Laboratories, to suggest that Merck support INBio with a grant. Anderson met with Eisner and realized that there was potential for more direct cooperation. Lynn Caporale, of Merck's Academic & Industry Relations Department, was familiar with Costa Rica and felt that a more direct agreement would be in accord with Merck's high level of research and previous success with finding drugs from natural resources.[43] Merck had already developed many products based upon environmental samples. In the 1940s, Merck helped develop the antibiotic streptomycin from a soil sample. In the 1940s and 1950s, Merck made two major natural resource discoveries: Vitamin B-12, useful for treatment of pernicious anemia, was found in a fermentation broth and mevalonic acid, a link in the body's synthesis of cholesterol, was dis-

covered in a yeast extract. Based upon these successes, Merck developed natural resource screening techniques and discovered the drugs Mefoxin, Primaxin, and Mevacor.[44]

During a conference in October 1990 with Eisner, Anderson, and Rodrigo Gamez of the University of Costa Rica, the idea for the cooperative, profit-sharing agreement between Merck and INBio was solidified.[45] Others instrumental to the success of the agreement were Daniel Janzen, of the University of Pennsylvania, and Bob Bisset, senior director of corporate licensing at Merck and Merck's lead negotiator of the deal.[46] The specifics of the agreement are that the soil samples collected by INBio are sent directly to Merck's lab in Madrid, Spain for analysis. The plant and insect samples are frozen and clinically extracted in a partnership with the University of Costa Rica. INBio supplies Merck with a list of the cataloged and classified plant and insect samples, from which Merck may choose a limited number to receive and test. These specific extracts are sent to Merck laboratories, where they are run through various bioassays to test for activity against certain receptors or enzymes known to be associated with specific diseases.[47] Paul Anderson, Georg Albers-Schonberg, and Keith Bostian, Merck's executive director of microbiology and molecular genetics, head the screening effort once the chemicals are sent to Merck. The laboratories of senior research microbiologist Gerald Bills, senior research fellow George Garrity, research fellow Robert B. Borris, and staff chemist Keith Witherup work on the identification of potential drugs from the natural resource samples.[48]

Although it may take a few years before Merck finds any commercially marketable

---

[40] Eisner, note 3 above.
[41] Eisner, note 38 above.
[42] Eisner, note 3 above.
[43] *Merck World*, note 20 above.
[44] Ibid.
[45] Roberts, note 15 above.
[46] Joyce, note 10 above; and *Merck World*, note 20 above.
[47] Roberts, note 15 above.
[48] *Merck World*, note 20 above.

chemicals within the natural resources supplied by INBio, the company does hope to make more discoveries through the collaborative agreement than through the more random screening processes, which currently are estimated to yield just one drug discovery from every 10,000 samples. Even if Merck does not find any active chemicals in usable form in INBio's natural resource samples, the company may find an extract that can be chemically modified to form a marketable product.[49]

A Unique Situation?

The agreement between Merck and INBio would not have been possible if INBio and Costa Rica had not had certain distinct characteristics, such as the technical expertise in Costa Rica to process samples, a transportation system to speed them to Merck Sharp & Dohme Research Laboratories in the United States and Spain, a scientific collection system already in place, and an established Merck network through the facilities of the Merck manufacturing and Merck Human Health Divisions in Costa Rica.[50]

Costa Rica's high level of dedication to preserving its biodiversity and vast supply of natural resources put it in a unique position to negotiate agreements with foreign firms. More than one quarter of Costa Rica's land is protected in some type of national park or preserve.[51] It is estimated that Costa Rica contains 12,000 plant species, 80 percent of which have been described, and 300,000 insect species, only 20 percent of which have been described.[52] Scientists estimate that Costa Rica is home to between 5 and 7 percent of the world's species and has more biological diversity per acre than any other nation. This vast species variability is due to the diverse climates within Costa Rica; the

country is mountainous, touches both the Atlantic and Pacific oceans, and has ecosystems representative of South America, the West Indies, and tropical North America.[53] Costa Rica's government is stable and democratic and its population is extremely well educated (the government estimates a 98-percent adult literacy), and yet it still faces many of the same economic pressures that often force developing nations to succumb to sacrificing environmental protection for more profitable endeavors.[54]

To prevent environmental deterioration, Hernan Bravo, Costa Rica's minister of natural resources, says, the National Assembly is working on developing a Wildlife Conservation Law to give the government the ability to negotiate with foreign firms that desire to use materials from Costa Rica's environment. This would be the first law of its kind in Central America and is expected to be passed and enacted by the end of the year. Other Central American countries plan to follow suit in accordance with the resolution that they signed in Nicaragua in June 1992. According to Bravo, the law will guarantee that companies that invest their technology and expertise in Costa Rica in exchange for natural resources will not be forced to turn over product patents and abundant profits far into the future but, rather, will be bound to abide by mutually agreed-upon contracts similar to the one signed by Merck and INBio.[55]

Another unique characteristic of Costa Rica that makes it appealing as a partner in agreements with foreign firms is the stature of INBio. INBio is a nonprofit, private, scientific organization founded in 1989 according to recommendations by the Costa Rican government. Initially designed by Daniel Janzen and Rodrigo Gamez, INBio was instituted to study biodiversity and its

[49] Roberts, note 15 above.
[50] *Merck World*, note 20 above.
[51] Joyce, note 10 above.
[52] L. Tangley, "Cataloging Costa Rica's Diversity," *BioScience* 40, no. 9 (October 1990): 633–36.
[53] *Merck World*, note 20 above.
[54] Ibid.; and Joyce, note 10 above.
[55] *BNA International Environment Daily*. "New Measure Would Cover Extraction of Genetic Resources from Rain Forest," 21 July 1992.

socioeconomic value as a way to demonstrate the economic benefits of natural resource preservation.[56] The institute, now directed by Gamez, inventories and catalogs Costa Rica's biological diversity and prepares a computer database of species names, conservation status, distribution, abundance, way of life, and potential uses in medicine, agriculture, and industry. The cost of inventorying all of Costa Rica's species is estimated at $50 million, which will be spent over the 10 years that, INBio approximates, it will take to complete the inventory.[57] INBio hopes that some of this money will come from agreements, such as the one with Merck, and eventually from royalties. Additional funding for the inventory has been supplied by a debt-for-nature swap, local and international grants, development assistance, and private foundations.[58]

To raise funds with its biodiversity catalog, INBio has been developing various arrangements that, it hopes, will lead to cooperative agreements with foreign firms, organizations, or governments. One arrangement, called "the lottery" by Ana Sittenfeld, director of research and development for INBio, consists of INBio providing the foreign firm with a predetermined number of coded compounds. If the company finds a commercial use for the compound and desires more of it, the firm must agree to share a certain percentage of the royalties with INBio before the source of the compound is revealed. This scheme will work because of a bar-code system; because the company does not know the name of the species it is testing until INBio agrees to reveal this information, it cannot circumvent INBio and go directly to the environment to collect more of the beneficial sample. A second possible type of arrangement for cooperation would follow a request by a

foreign firm for samples of compounds that exhibit certain characteristics, such as species that exhibit a resistance to certain types of insects or fungal infections. In each type of agreement, INBio contends that it will permit the company to keep the patent rights to any commercially developed products as long as INBio receives royalties that it can use to conserve Costa Rica's environment.[59] All funds received by INBio, above operating costs, are placed in a special fund managed in cooperation with the Costa Rican government to ensure that the money is directed to conservation activities.[60]

In addition to cataloging Costa Rica's biodiversity, INBio has attempted to improve environmental education and expand the scientific work force among Costa Ricans. According to Gamez, economic measures to improve the environment, such as the Merck/INBio agreement, are not enough to preserve the biodiversity; cultural and intellectual measures must be adopted as well so that the people understand the importance of the environment.[61] To train workers to identify and collect samples for its database, INBio administers a six-month course in parataxonomy, including classes in botany, entomology, and ecology. More than 30 lay-people interested in environmental conservation have been trained so far to collect, dry, pin, and box Costa Rica's species.[62] During the first six months after the first class of parataxonomists was trained, more than four times as many insect samples were collected as had been placed in Costa Rica's insect collection during the previous 100 years.[63]

Response to the Agreement

Acclaim for the Merck/INBio agreement has been widespread, with representatives of environmental groups, developing

[56] Merck & Co., note 32 above; and Roberts, note 15 above.
[57] Tangley, note 52 above.
[58] World Resources Institute et al., note 1 above.
[59] Joyce, note 10 above.
[60] World Resources Institute et al. note 1 above.
[61] Tangley, note 52 above.
[62] Roberts, note 15 above; and Joyce, note 10 above.
[63] World Resources Institute, note 4 above.

nations, biotechnology and pharmaceutical firms, and the U.S. government supporting the unique endeavor. Roberto Rojas, Costa Rica's minister of foreign trade, contends that the deal with Merck was "very good. . . We do think we need protection because we know we can serve as a vehicle for providing profits to companies, but that deal was satisfactory for us."[64] Lee Claro, from Upjohn Pharmaceutical's Washington, D.C., office, thinks that the deal was admirable, as well.[65]

There are some people, however, who do not think that the agreement was the best one that could have been made. Various people have said that they are skeptical about the deal because the actual agreement has not been released.[66] Rodrigo Gamez says that certain parts of the deal, such as the exact percentage of the royalties, were not made public to prevent other inventorying organizations from underpricing INBio in an effort to steal their business.[67] Costa Rican Congresswoman Brigitte Adler feels that INBio may have sold off Costa Rica's natural resources too cheaply.[68] But INBio is selling a product whose returns are uncertain; Merck may discover no prospective products from the agreement and may have paid $1 million for nothing, or it may discover 10 prospective drugs that could yield the company millions. INBio has responded to some people's expressed fear that it would gain in some way from the agreements rather than use all the profits to preserve biodiversity, by saying that most such fears have been caused by a misunderstanding of INBio's intent. Costa Rica's Ministry of Natural Resources has already received the first installment of Merck's fee to INBio, which will be used to support a marine park

on Coco Island, off Costa Rica.[69] In response to the fear that the agreement will fail to protect the environment because it depends on the work of INBio rather than on the work of the local population, Thomas Eisner contends that the deal does involve the local population, it trains parataxonomists, increases the environmental education efforts, and encourages the seven regional national park management bodies to increase local participation in decision-making.[70]

The Impact of the Treaty

Although most scientists, environmentalists, and government officials feel that the Merck/INBio agreement is the type of contract that they would like to see come out of the biodiversity treaty, differing opinions exist regarding the feasibility of such agreements under the treaty in its current form. Eisner thinks that deals like the Merck/INBio agreement would occur anyway but would be facilitated by both countries being party to the treaty. The United States' refusal to sign the treaty is detrimental, Eisner believes, "We lose by not signing that treaty, although as in the first throes of any agreement, some wording needs to be refined. [International collaborations] are going to happen whether we sign or not, but the goodwill attendant to having signed would have put the United States in a much better position to approach these developing countries."[71] Kenton Miller of WRI feels that certain countries, such as Venezuela, might refuse to work with the United States because, by not signing, it did not state its intentions to preserve its own biodiversity. These nations will not cooperate with

---

[64] *BNA International Environment Daily,* note 55 above.

[65] L. Claro, Personal communication with the author, 6 November 1992.

[66] P. Jutro, Presentation at Princeton University, New Jersey, 26 October 1992.

[67] N. Adams, "Merck Drug to Pay Royalties on Costa Rican Forest,"*National Public Radio Show: All Things Considered,* interview of Rodrigo Gamez, 10 June 1992.

[68] *BNA International Environment Daily,* note 12 above.

[69] Roberts, note 15 above.

[70] P. Aldhous, "'Hunting Licence' for Drugs," *Nature* 353 (26 September 1991): 290.

[71] T. Eisner as quoted in T. Reynolds, "Biodiversity Treaty Still Unsigned by U.S.," *Journal of the National Cancer Institute* 84, no. 15 (5 August 1992): 1148.

countries they think might take advantage of them.[72]

Others feel that the biodiversity treaty as it exists would preclude agreements like that which was signed by Merck and INBio. According to Richard Godown, president of the International Biotechnology Association, "The convention would tie the hands of negotiators: When they sat down to make a deal there would be an enormous slug of mandatory contract language . . . It would result in a bum deal."[73] Godown fears that the treaty is too specific regarding the terms of cooperation and technology transfer and that it would prevent contracting parties from negotiating the best possible terms based upon the specific nature of each situation.

However, Richard Wilder, an intellectual property rights lawyer in Washington, D.C., and a member of the Association of Biotechnology Companies' patent committee, says that, from a legal standpoint, there is nothing in the treaty that would prevent the Merck/INBio agreement.[74] The treaty does not require any certain type of contract. Rather, its vague language can be interpreted by different parties to support different agreements with varying levels of intellectual property rights. What the biotechnology and pharmaceutical companies fear is that developing nations will interpret the vague language as supporting weak intellectual property protection, while industry will interpret the vague language as supporting the opposite position.

According to Eisner, "The INBio deal should be a blueprint for other countries to follow. It is a way to provide economic growth along with protecting our precious biological diversity."[75] And, indeed, while the U.S. government is waiting to solve the dilemma over the signing of the treaty, other cooperative research agreements similar to the one between Merck and INBio are being set up.[76] Other pharmaceutical companies are studying the agreement, and other developing countries, including China, Chile, Mexico, India, Indonesia, Nepal, and Nicaragua, are investigating INBio in an effort to set up similar institutions in their own nations.[77] However, Walter Reid at WRI cautions that INBio as it exists in Costa Rica may not be an appropriate model for other countries. Costa Rica is unique in that it has a stable democratic government and a well-known commitment to preserving biodiversity. Other nations may abandon other efforts at conservation in hope of large economic profits from prospecting agreements or may not really use the profits derived from their inventories for resource conservation. In an effort to prevent this type of exploitation of an otherwise positive conservation mechanism, Reid is working with INBio, the Rainforest Alliance, and the African Center for Technology Studies to create guidelines for future prospecting agreements and institutions.[78]

Each new INBio-like institution would have to be adapted to the unique culture, environment, and political climate of its nation. For example, because India is composed largely of community-owned land and vast state-owned forests, such an institution within India would probably have to be decentralized. Indonesia, which already has a national genetic resources institute, may only need an information management system and a group of parataxonomists to establish a prospecting institute. Because Indonesia, India, Brazil, and Mexico each have their own pharmaceutical industries, cooperative agreements could take

[72] Miller, note 21 above.
[73] S. Usdin, "Biotech Industry Played Key Role in U.S. Refusal to Sign BioConvention," *Diversity* 8, no. 2 (1992): 8–9.
[74] R. Wilder, Personal communication with the author, 30 October 1992.
[75] *BNA International Environment Daily*, "Deal Between Drug Firm, Costa Rica Called Example of What Treaty Would Do." 17 June 1992.
[76] Joyce, note 10 above; and Miller, note 21 above.
[77] Miller, note 21 above; Roberts, note 15 above; and Tangley, note 52 above.
[78] Roberts, note 15 above.

place nationally as well as internationally.[79] In nations without a government stable enough to support a prospecting and screening institute, previously established institutions in politically stable countries, such as INBio in Costa Rica, could expand their scope and catalog other nations' resources.[80]

Other types of market-based natural resource exchange programs have been taking place alongside the Merck/INBio agreement, and it is likely that characteristics of some of these agreements would be useful in designing a unique prospecting establishment for each resource-rich nation. Shaman Pharmaceuticals Inc., a small California pharmaceutical firm founded in 1989 by Lisa Conte, is bypassing institutional mechanisms and going directly to native shamans (medicine men) in Brazil and Argentina to isolate traditional plant-derived medicines. According to Conte, "We are using the traditional use of plants as a prioritisation tool to have a more efficient process of drug discovery. By using traditional knowledge there is a greater likelihood of yielding an active compound or a pharmaceutical."[81] Shaman compensates the "ethnobotanists" who help Conte discover drugs by helping them to preserve the rain forests; in addition, if the company needs very large quantities of a plant, it pays people from the local communities to harvest it. A percentage of the company's profits are devoted to the Healing Forest Conservancy, a nonprofit organization founded by Conte to preserve the rain forests and assist indigenous peoples.[82] Conte contends that, by using the native shamans in the search for drugs, "We are creating an economical alternative to the destruction of the rain forest. If they can make a living by collecting our products, it gives them an incentive for leaving the forests intact."[83] Her company has already had a high level of success with its drug discovery process; two of its drugs are now undergoing clinical tests. One is the antiviral agent Provir, useful for treating respiratory infections and herpes 1 and 2, and the other is an antifungal agent useful for treating certain fungal infections.[84]

In another cataloging effort, the U.S. National Cancer Institute in Bethesda, Maryland, is at work creating a repository of natural resource specimens. It has just renewed two five-year contracts with the Missouri and New York botanical gardens, whose botanists collect samples that may be viable treatments for cancer or AIDS. The institute is currently working on a "material transfer agreement" under which it will share some of its profits with governments or departments in the countries in which it prospects.[85]

Policy Recommendations

Over the past 20 years, the United States has enacted much state and federal legislation addressing specific aspects of biodiversity. It never, until the Earth Summit, addressed the issue of comprehensive biodiversity legislation.[86] To rectify this situation, the government should undertake a number of initiatives. First, the United States should sign the biodiversity treaty to demonstrate its positive intent to preserve the environment.

Not all interested parties feel that signing the treaty in its current form is in the best interest of the United States. Both the Industrial Biotechnology Association and the Association of Biotechnology Companies pressured President Bush to withhold his signature from the treaty. The

[79] Miller, note 21 above.
[80] Ibid.
[81] *Pharmaceutical Business News*, note 30 above.
[82] *Industrial Bioprocessing*, "Back to Nature for Chemicals and Drugs," October 1991, 3.
[83] *Pharmaceutical Business News*, note 30 above.
[84] *Industrial Bioprocessing*, note 82 above; and *Pharmaceutical Business News*, note 30 above.
[85] Joyce, note 10 above.
[86] Council on Environmental Quality, note 2 above.

trade associations reaffirmed their support of biodiversity conservation but expressed dismay at the process and outcome of the convention. According to Walter Reid, "When you talk to people who were on the U.S. delegation [in Rio], by and large what they say is [that] nobody was talking to them at all about the [biodiversity] treaty, from the environmental community to the biotech industry. There was virtually no attention being given to [the treaty], much to our frustration."[87]

Richard Wilder says that biotechnology companies have rational reasons for opposing the treaty: "I think that the language the way it is now provides the possibility . . . for countries either to decide not to provide intellectual property protection for biotechnology or to dilute it through very broad compulsory licensing provisions, either possibility of which would be damaging to the biotech industry, which depends upon having strong intellectual property protection to develop new technologies . . . and, once it's developed, to transfer it or to otherwise make use of it."[88]

Others, however, feel that the United States should have signed the treaty to show its good intentions and then worked with the flawed language to make it more appealing to U.S. industry. Although article 37 of the treaty states that no country can sign it with reservations to certain clauses, many of the signatories, including Italy, France, and England, signed the treaty and stated the way in which they will interpret its vague clauses.[89] Kenton Miller thinks that the United States should "join with other countries after we sign [the treaty] in sponsoring continuing negotiating meetings to look at those aspects that are indeed far from clear . . . From a Supreme Court point of view, we ought to have it all clear before we sign it, but in most countries they don't hold to that rigor . . . So [not signing because of the vague language] made us sort of stand out."[90]

Although certain aspects of the treaty do merit clarification, this is not reason enough for the United States to withhold its signature from the treaty. The U.S. government must prove its determination to work with other nations to preserve biodiversity. Therefore, it should sign the treaty and issue a statement delineating how it interprets the vague articles, especially articles 15 and 16. The government should then meet with representatives from environmental groups, biotechnology and pharmaceutical firms, and nongovernmental organizations to determine the best possible ways to modify the treaty. Alterations to the treaty should focus upon using the market, rather than legislation, to promote biodiversity preservation. Changes should be made to permit and encourage market agreements like the Merck/INBio agreement without infringing upon the right of individual parties to negotiate contracts and intellectual property rights on a case-by-case basis. These improvements should be proposed to all signatories at the general meeting to be held in Nairobi in August 1993. Because this meeting will be held prior to the treaty ratification deadline, the United States may be able to renegotiate the treaty to suit its interests before the issue is brought up before congress.

In addition to signing the biodiversity treaty, the United States should encourage the legal adoption of the guidelines currently in development for designing INBio-like institutions internationally. Such guidelines could be attached as an appendix to the biodiversity treaty or could constitute a separate agreement. The United States should take a leading role in promoting such institutions by becoming one of the first countries to institute a national biodiversity prospecting center to collect and catalog its own biodiversity. Such a center has already been proposed as part of the National Biodiversity Conservation Act drafted by David Blockstein during his fellowship in

[87] Browning, note 13 above.
[88] Wilder, note 74 above.
[89] Ibid.
[90] Miller, note 21 above.

Representative James H. Scheuer's (D–N.Y.) office in 1987 and 1988.[91]

Each national biodiversity prospecting center should catalog its inventory on a database. A brief summary of the content of each of these databases should be compiled by the U.N. Environment Programme and made available to industrial groups internationally. Each party should be able to negotiate its own agreements based upon certain requirements, including the fair transfer of natural resources, the commitment to use the resources in a non-destructive way, and the equitable payment of a certain level of start-up expenses and royalties. The existence of a number of these prospecting centers would provide enough incentive to abide by the guidelines; any establishment that tried to subvert the regulations would lose its customers to another nation's biodiversity center.

If these policy recommendations were instituted, developing nations would be able to resolve the seeming incompatibility of sustained economic growth and environmental preservation. Merck/INBio-like agreements represent a profitable alternative to deforestation and provide nations with a greater incentive to preserve their biodiversity than is provided by any type of legislative action or regulation. Thus, the enactment and enforcement of these policy recommendations would help preserve the world's biodiversity. They cannot stand alone, however; environmental education must be improved both in the classroom and in society at large so that populations learn to value biodiversity as they value other national assets. The United States should foster scientific training and exchange programs with other nations to promote the flow of information and to heighten awareness of environmental issues.

[91] Tangley, note 52 above.

Part II

Predicting and Managing Atmospheric Change

Introduction and Guide to Further Reading

6 Dansgaard, W., Johnsen, S. J., Clausen, H. B., Dahl-Jensen, D., Gundestrup, N. S., Hammer, C. U., Hvidberg, C. S., Steffensen, J. P., Sveinbjörnsdottir, A. E., Jouzel, J. and Bond, G. 1993: Evidence for general instability of past climate from a 250-kyr ice-core record. *Nature,* 364, 218–20.

This is a seminal paper that summarizes the important results from the joint European Greenland Ice-core Project (GRIP). Analyses of the oxygen isotopes within the GRIP Summit ice core provide evidence for cycles of abrupt and rapid climate change in Greenland over the past 250,000 years.

7 McCormick, M. P., Thomason, L. W. and Trepte, C. R. 1995: Atmospheric effects of the Mt Pinatubo eruption. *Nature,* 373, 399–404.

In this paper, McCormick et al. provide a comprehensive summary of the effects of the 1991 Mt Pinatubo eruption. It is important because it provides an example of how a volcanic eruption can lead to atmospheric pollution and climate change. It also illustrates the complexities and uncertainties of attributing past volcanic to climate change.

8 United Nations Environment Programme and the World Health Organisation 1994: Air pollution in the world's megacities. *Environment,* 36, 2, 4–13 and 25–37.

The Global Environment Monitoring System (GEMS), under the auspices of the United Nations Environment Programme and the World Health Organisation assess the status of air pollution in the world's largest cities in this excellent report. It is useful because it discusses the sources, the dispersal and impacts of air pollution, as well as suggesting control strategies and recommendations for the reduction of the pollutants.

9 Beaton, S. P., Bishop, G. A., Yi Zhang, Ashbaugh, L. L., Lawson, D. R. and Stedman, D. H. 1995: On-road vehicle emissions: regulations, costs, and benefits. *Science,* 268, 991–3.

In this paper, Beaton et al. present useful information on the US regulations regarding exhaust emissions from vehicles. The paper is important because it discusses the value and the effectiveness of the models that are used to calculate vehicle emissions at monitoring sites in California. The paper continues by discussing and examining the implications for emission control policy and cost effectiveness.

10 Houghton, J. T., Jenkins, G. J. and Ephiraums, J. J. (eds) 1990: *Climate Change: The IPCC Scientific Assessment.* Cambridge: Cambridge University Press, pp. xii–xxxii; and Houghton, J. T., Meira Filho, L. G., Bruce, J., Hoesung Lee, Callander, B. A., Haites, E., Harris, N. and Maskell, K. 1995: *Climate Change 1994: Radiative Forcing of Climate, Executive Summary.* Cambridge: Cambridge University Press, pp. 11–14.

The first part of this selected reading

reproduces the *Policymakers Summary* of the Intergovernmental Panel on Climate Change (IPCC). It is the result of the work of more than 1000 experts and is used as a basis for policy decisions related to climate change. It presents a summary of the best estimates on the nature of natural- and human-induced climate change, as well as the likely effects of atmospheric change. The second part of this selected reading is the executive summary for the revised IPCC report, that considers the models for radiative forcing based on new scenarios for future energy needs and uses.

11 Warrick, R. A. and Rahmam, A. A. 1992: Future sea level rise: environmental and socio-political considerations. In: Mintzer I. M. (ed.) *Confronting Climatic Change: Risks, Implications and Responses.* Cambridge: Cambridge University Press, pp. 97–112.

This paper considers the effects of climate change on sea level and is useful because it summarizes the nature of sea level rise related to global warming and its possible consequences. It discusses both the environmental and socio-political consequences of the likely changes and considers the nature of the uncertainties.

12 Golnaraghi, M. and Kaul, R. 1995: Responding to ENSO. *Environment,* 37, 1, 16–20 and 38–44.

The mechanics of the El Niño Southern Oscillation (ENSO) and a consideration of the various models which can be used to help explain its dynamics are outlined in this paper. The paper is also useful because it discusses the environmental and social impacts of the ENSO, the history of events and management strategies.

During the last few decades there has been growing interest in the nature of climate change. Throughout the 1970s, the development of oxygen isotope analysis of foraminifera in deep sea sediment cores provided a new proxy for reconstructing the details of climate change throughout late Tertiary and Quaternary times (Shackleton and Opdyke, 1973). From these studies it was recognized that there had been rapid fluctuations from times of warmth, with high sea-levels, to times of extreme cold when glaciers and ice sheets expanded and covered large areas of the globe, with lower sea-levels, throughout the last few million years. The warm periods, known as interglacials, lasted from about 10,000 to 20,000 years, while the cold periods, glacials, lasted about 100,000 to 200,000 years. This pattern of change is believed to reflect variations in the amount of solar radiation distributed over different regions of the globe at different times of the year, and is the result of cyclic variations in the angle and rotation of the Earth's axis, as well as the variations in the Earth's orbit around the Sun. These orbital variations have become known as Milankovitch cycles, named after the

astronomer Milutin Milankovitch who first suggested, long before the palaeoclimatic pattern change had been determined by proxy data, that such a mechanism could help force climate change (Milankovitch, 1941). Furthermore, the data showed that these climate changes were very abrupt and rapid. These and other data show that the duration of the present interglacial so far is approximately 10,000 years. This suggests that we might expect a return to glacial conditions sometime in the next 10,000 years. Any concern over this, however, is overshadowed by the likelihood that human activity has the potential drastically to alter the atmosphere and cause climate change. As long ago as 1896, the Swedish chemist Svante Arrhenius suggested that a doubling of atmospheric CO_2 could cause an average temperature rise of 5–6°C. It was not, however, until the early 1970s that increased CO_2 production and rising atmospheric turbidity were recognized as important factors capable of changing climate. The cooling effect of increased turbidity was initially thought to outweigh the warming effect of greenhouse gases (Schneider and Mesirow, 1976). But by the early 1980s research showed that the warming effect of greenhouse gases was likely to be greater (Manabe, Wetherald and Stouffen, 1981; Mitchell, 1983; National Research Council, 1982 and 1983). Awareness increased particularly because the 1980s included the six warmest years in the past century (Kemp, 1994). Furthermore the last few years of the 1980s and first few years of 1990s have experienced increasingly severe weather conditions and several El Niño events (Wuethrich, 1995). Whether these climatic conditions are related to 'global warming', however, has yet to be determined.

Such interest and concern has stimulated the growth of palaeoclimatology and the study of atmospheric science. In 1985, a conference on the 'International Assessment of the Role of Carbon Dioxide and other Greenhouse gases in Climate variations and Associated Impacts' was convened at Villach in Austria. This resulted in the establishment in the following year of the Advisory Group on Greenhouse Gases (AGGG) under the auspices of the International Council of Scientific Unions (ICSU), the United Nations Environment Program (UNEP) and the World Meteorological Organisation (WMO). The AGGG undertook reviews and assessments related to greenhouse gases and published a series of reports (e.g., AGGG, 1986 and 1987). Other studies of the impact of global warming involved the Department of Energy in the United States (e.g., Mac-Cracken and Luther, 1985a and b) and the Commission of European Communities (CEC). With increased interest from government agencies, the WMO and the UNEP established the Intergovernmental Panel on Climate Change (IPCC) in 1988 to assess the evidence for climate change and to model the impacts of future change. This involved several hundred scientists from around the world and it resulted in a series of reports. The first report summarized the scientific assessment (Houghton, Jenkins and Ephiraums, 1990), followed by reports on the impact assessment and response strategies (e.g. Tegart, Sheldon and Griffith, 1990). Supplementary reports were produced in 1992 and 1994, summarizing the current knowledge on global warming and provided predictions for future climate trends (Houghton, Callander and Varney, 1992; Houghton et al. 1995). The IPCC assessments formed a critical component of the United Nations Climate Change Convention at the UN Conference on Environment and Development (UNCED) in Rio de Janeiro in 1992. Strategies were suggested for the reduction of emissions to such levels so as to allow ecosystems to adapt naturally to climate change, to ensure that food production is not threatened, and to allow economic development to proceed in a sustainable manner. The nations which ratified the Framework Convention on Climate Change (FCCC) at Rio de Janeiro met in Berlin during April 1995 to evaluate levels of reduction and the implemented strategies. This resulted in the Berlin Mandate, which outlined a strategy to return greenhouse emissions to 1990 levels by the year 2000. In 1996, the IPCC published the Second Assessment Report (SAR), which includes three volumes of the IPCC Working Groups (I – The Science of Climate Change; II – Scientific–Technical Analysis of Impacts, Adaptations, and Mitigation of Climate Change; III – The Economic and Social Dimensions of Climate Change) and the Synthesis Report are known under the common title, Climate Change 1995. These documents provide a comprehensive assessment of new and recent literature on climate change. The Synthesis Report brings together material contained in the three volumes of the full report that is relevant to governments' interpretation of the UN Framework Convention on Climate Change (UNFCCC) objective. The Synthesis Report is the consequence of IPCC decision, following a resolution of the Executive Council of the WMO (July 1992), to include an examination of approaches to Article 2, the Objective of the UN Framework Convention on Climate Change (FCCC), in its work programme. Although the IPCC was due to be disbanded shortly after this meeting, the parties involved in the FCCC approved its continuation and appointed it as the advisory body for the scientific aspects of climate change. Its other duties will be to consider the likely impact of such change, and the economic implications including an outline of future scenarios of energy use and social patterns to help design a protocol for the reduction of emissions by the year 2000.

Since the mid-1980s, there has also been growing concern over the depletion of stratospheric ozone (O_3) concentrations in the summers over the Antarctic and Arctic. Stratospheric O_3 depletion was first observed and recorded by the British Antarctic Survey in 1977, but it was not until the publication of Farman, Gardiner and Shanklin (1985) that concern arose about the possible implications of the observations. Stratospheric ozone, which is concentrated in a layer at 20–30 km above the Earth's surface, is important in helping to reduce the amount of ultraviolet (uv) radiation reaching the Earth, essentially providing a protective layer shielding life from the adverse effects of too much uv radiation. In addition, O_3 plays an important part in the heat balance within the Earth's atmosphere. Farman, Gardiner and Shanklin (1985) suggested that the recorded levels of depletion were the result of the reactions of certain ozone-destroying anthropogenic emissions of chlorofluorocarbons (CFCs) and other chemicals. The result of this and other research papers during the late 1980s and early 1990s led to international concern and the ratification of the Montreal Protocol in September 1987. In this, leading industrial nations agreed to reduce the production and use of the five most harmful CFC ozone-depleting gases over the 1990s. The target levels set by the Montreal Protocol, however, were modified at the London Ministerial Conference on Ozone in 1990 so that by 1997 there would be an 85 per cent reduction in the production of CFCs and their total ban by the year 2000. The parties to the Montreal Protocol met again in 1992 in Copenhagen where the controls on ozone-depleting substances were once more revised. Despite this international cooperation, levels of ozone-depleting substances continue to increase in the atmosphere and the full consequences of their effects are still to be assessed fully.

Increased atmospheric pollution in urban areas has become another major concern over recent years. Until the 1960s, much of the atmospheric pollution in urbanized regions was the result of industrial emissions and domestic burning of fossil fuels. This resulted in severe smogs, such as the infamous 'pea-souper' smog of December 1952 in London, when nearly 4,000 people died after inhaling acidic water droplets. Strict legislation, such as the 1963 Clean Air Act in the United States, and its amendments in 1970, 1977 and 1990, define tight controls on the amounts of emission that are acceptable. In developed countries, legislation has reduced much of this pollution problem. However, in many of the newly industrialized countries, such as China, the problem still persists because legislation has not been implemented or enforced. Urban pollution over the last few decades is now primarily the result of vehicle emissions. In Athens, for example, the problem of pollution has become so acute that strict controls on vehicles entering the city area has to be enforced during periods of increased atmospheric pollution. In the United Kingdom, the number of people affected from asthma has increased considerably over the last few decades, and many believe this may be the result of atmospheric pollution produced by vehicle emissions (Lee and Manning, 1995). Governments are beginning to realize the dangers from vehicle pollution, as has become evident from the encouragement to phase out lead in petroleum. The full effects of other pollutants, such as tropospheric ozone and subparticulates from diesel emissions, have still not been assessed. Recently in the United Kingdom, the Royal Commission on Environmental Pollution concluded after a very detailed study that transport levels should be drastically reduced not only to reduce local and regional pollution, but also to reduce emissions of greenhouse gases which may contribute to global warming (Royal Commission on Environmental Pollution, 1994).

Policy-makers are now beginning to appreciate that the management of global environments, resources and energy must take into account the effects of their actions on the atmosphere and the consequences of potential future climate change. These may include the migration of climatic belts which will alter the distribution of biomes and change the characteristics of growing seasons, thus altering food production. Sea-levels may change, flooding low lying coastal regions and affecting groundwater resources. In addition, the frequency and magnitude of severe storms may increase, leading to disastrous effects for human populations living in coastal areas.

To evaluate the effects of human activity on climate, managers must be able to assess these changes and to resolve the various components of change which relate to the natural processes that force climate change. Palaeoclimatic reconstructions are used to help understand these processes and the nature of climate change. They are also important in helping to formulate theories and models of change. The most recent and exciting palaeoclimatic reconstructions have been the result of ice core studies in Greenland. The analysis of gases trapped in ancient glacial ice has provided an almost annual record of climate over the last 250,000 years. This has shown that climatic changes throughout the last glacial-interglacial cycle have occurred very rapidly over periods of a few decades (Lorius et al. 1988; Dansgaard et al. 1993; GRIP Members, 1993). This work has stimulated considerable research to help explain such variations.

The first chapter, by Dansgaard et al. (1993), is one of several recent papers that presents the work of the European Greenland Ice-core Project (GRIP). It provides evidence for general instability of past climate over the last 250,000 years. The data emphasize the abruptness and rapidity of climate change.

The next group of chapters concentrates on atmospheric pollution. Pollution can result from both natural and human factors, and can be both a local and a global problem, resulting in atmospheric change, affecting the biosphere and the quality of life for inhabitants of polluted regions. McCormick, Thomason and Trepte (1995) examine the effects of the eruption of Mt Pinatubo in June

1991. This natural process released large quantities of gases, such as SO_2, and dust into the atmosphere. These aerosols were transported around the globe, causing a variety of different effects, such as altering insolation and O_3 levels throughout different regions of the world. The chapter by the United Nations Environmental Programme (UNEP) and the World Health Organisation (WHO) (1994) examines the effects of atmospheric pollution caused by human activity. Such pollution is of growing concern particularly because of the rising incidence of pulmonary diseases. It examines the nature of air pollution in the world's polluted megacities. The chapter is significant because it suggests control strategies and recommendations for the reduction of pollutants.

The next reading in Part II is by Beaton et al. (1995), who discuss the increasing problem of air pollution caused by vehicle emissions. The chapter provides a North American perspective on vehicle emissions controls. It considers the US regulations regarding the exhaust emissions from vehicles and discusses the value and effectiveness of models used to calculate emissions. In addition, the chapter also discusses the implications for emission control policy and cost effectiveness.

Modelling climate change is particularly important to manage the influence of the climate change effectively and efficiently, as well as for providing information for sustainable development. The next chapter presents the summary of the most comprehensive study on climate change and modelling of the human impact on the atmosphere. This study was undertaken during the later part of the 1980s and early 1990s by the Intergovernmental Panel on Climate Change (IPCC) and was published in one of a series of reports (Houghton et al., 1990, 1992 and 1995). These studies made use of General Circulation Models (GCMs) to simulate the resultant climate change that might occur if the composition of the atmosphere changed, notably by increasing the levels of atmospheric CO_2. GCMs are essentially mathematical models of the Earth's surface and atmosphere, where the atmospheric processes are defined by physical equations. By varying the inputs, for example the levels of atmospheric CO_2, the resulting climate change can be examined. Although the models are extremely sophisticated and have to be run on supercomputers, the simulation of the physical boundaries and processes are still quite crude. Models are continually being modified as new knowledge and data are being obtained to provide estimate of change. The summary of the most recent knowledge of climate change resulting from such studies and observations is presented, providing the most recent best estimates on the nature of natural- and human-induced climate change and the likely effects of atmospheric change.

Lastly, the physical consequences of climate change are discussed. The chapter by Warrick and Rahman (1992) examines the environmental and socio-political consequences and considerations of sea level rise induced by global warming. It summarizes the nature of sea level rise and the possible consequences, as well as considering the nature of the uncertainties. The chapter by Golnaraghi and Kaul (1995) describes the dynamics of the El Niño Southern Oscillation (ENSO) and discusses the environmental and social impacts of the increased frequency of El Niño events.

The chapters presented in Part II have examined the complex issues surrounding atmospheric change by providing examples of papers from a variety of sources. Environmental managers must be aware of the implications of their actions on climate and also the likely consequences of future climate change on projects they are undertaking and policies they initiate.

References

AGGG (Advisory Group on Greenhouse Gases) 1986: Understanding CO_2 and Climate.

AGGG (Advisory Group on Greenhouse Gases) 1987: Understanding CO_2 and Climate.

Farman, J. C., Gardiner, B. G. and Shanklin, J. D. 1985: Large losses of total ozone in Antarctica reveal seasonal $C10_x/NO_x$ interaction. *Nature*, 315, 207–10.

GRIP (Greenland Ice-core Project) Members, 1993: Climate instability during the last interglacial period recorded in the GRIP ice core. *Nature*, 364, 203–7.

Houghton, J. T., Callander, B. A. and Varney, S. K. (eds) 1992: *Climate Change 1992: The Supplementary Report to the IPCC Scientific Assessment*. Cambridge: Cambridge University Press, 200 pp.

Houghton, J. T., Jenkins, G. J. and Ephiraums, J. J. (eds) 1990: *Climate Change: The IPCC Scientific Assessment*. Cambridge: Cambridge University Press, 365 pp.

Houghton, J. T., Meira Filho, L. G., Bruce, J., Hoesung Lee, Callander, B. A., Haites, E., Harris, N. and Maskell, K. (eds) 1995: *Climate Change 1994: Radiative Forcing of Climate Change and an Evaluation of the IPCC IS92 Emission Scenarios*. Cambridge: Cambridge University Press, 339 pp.

Kemp, D. D. 1994: *Global Environmental Issues: A Climatological Approach*. London: Routledge, 224 pp.

Lee, J. and Manning, L. 1995: Environmental lung disease. *New Scientist*, 16 September, Inside Science No. 84.

Lorius, C., Barkov, N. I., Jouzel, J., Korotkevich, Y. S., Kotylakov, V. M. and Raynaud, D. 1988: Antarctic ice core: CO_2 and change over the last climatic cycle. *EOS*, 68, 681–4.

Manabe, S., Wetherald, R. T. and Stouffer, R. J. 1981: Summer dryness due to an increase of atmospheric CO_2 concentration. *Climatic Change*, 3, 347–86.

Milankovitch, M. M. 1941: *Canon of Insolation and the Ice-age Problem*. Koniglich Serbische Akademie, Belgrade. English translation by the Israel Program for Scientific Translations, published for the US Department of Commerce, and the National Science Foundation, Washington, D.C. (1969).

Mitchell, J. F. B.. 1983: The seasonal response of a general circulation model to changes in CO_2 and sea temperatures. *Quarterly Journal of the Royal Meteorological Society*, 109, 113–52.

National Research Council 1982: *Carbon Dioxide and Climate: A Second Assessment*. Washington, D.C.: National Academy Press.

National Research Council 1983: *Changing Climate*. Washington, D.C.: National Academy Press.

Royal Commission on Environmental Pollution 1994: *Transport and the Environment*. Her Majesty's Stationery Office: London.

Schneider, S. H. and Mesirow, L. 1976: *The Genesis Strategy*. New York: Plenum Press.

Shackleton, N. J. and Opdyke, N.D. 1973: Oxygen-isotope and palaeomagnetic stratigraphy of Pacific cores V28–239, late Pliocene to latest

Pleistocene. In: Cline, R. M. and Hays, J. D, (eds), *Investigations of Late Quaternary Paleogeography and Palaeoecology.* Boulder, Col.: Geological Society of America memoir, 45, 449–64.

Tegart, W. J., Sheldon, G. W. and Griffith, D. C. 1990: *Climate Change: the IPCC Impact Assessment.* Canberra. Australian Government Publishing Service.

Wuethrich, B. 1995: El Niño goes critical. *New Scientist,* 4 February, 33–5.

Other references cited in text are listed in the selected readings below.

Suggested further reading

Barry, R. G. and Chorely, R. J. 1992: *Atmosphere, Weather and Climate.* London: Routledge, 6th edition.

This is an essential text that describes the Earth's atmospheric processes and climatic conditions. It provides an introduction to the basic principles and chemical characteristics of the atmosphere, and a comprehensive description of regional climates, leading to a discussion of the processes of climate change.

Blunden, J. and Reddish, A. (eds) 1991: *Energy, Resources and Environment.* London: Hodder and Stoughton.

Energy, Resources and Environment is an accessible text that deals with the nature of energy and mineral resources, their extraction, refining and disposal, and the environmental problems associated with obtaining these resources. The possible alternatives to conventional energy resources are explored. These include recycling, energy conservation, solar energy, wind and water energy. In addition, the political implications of using such alternative energy resources and the politics associated with the disposal of radioactive waste are examined in considerable detail.

Bryant, E. A. 1991: *Natural Hazards.* Cambridge: Cambridge University Press.

Natural Hazards is a good introductory text for undergraduates providing an interdisciplinary treatment of a variety of natural hazards and their management including oceanographical, climatological, geological and geomorphological.

Elsom, D. M. 1992: *Atmospheric Pollution: A Global Problem:* Oxford: Blackwell Publishers, 2nd edition.

This excellent text describes the major types of atmospheric pollution containing useful sections on photochemical ozone pollution, traffic pollution and the pollution control strategies.

Gates, D. M. 1993: *Climate Change and its Biological Consequences.* Sunderland, Massachusetts: Sinauer Associates, Inc.

The causes and evidence for climate and vegetational changes, and possible future climate change and the resultant biological consequences are outlined in this useful text.

Houghton, J. T., Jenkins, G. J. and Ephiraums, J. J. (eds) 1990: *Climate Change: The IPCC Scientific Assessment.* Cambridge: Cambridge University Press. Houghton, J. T., Callander, B. A. and Varney, S. K. (eds) 1992: *Climate Change 1992: The Supplementary Report to the IPCC Scientific Assessment.* Cambridge: Cambridge University Press. Houghton, J. T., Meira Filho, L. G., Bruce, J., Hoesung Lee, Callander, B. A., Haites, E., Harris, N. and Maskell, K. (eds) 1995: *Climate Change 1994: Radiative Forcing of Climate Change and an Evaluation of the IPCC IS92 Emission Scenarios.* Cambridge: Cambridge University Press.

These three reports are the result of the Intergovernmental Panel on Climatic Change, set up by the World Meteorological Organisation and the United Nations Environment Programme. The reports assess the potential effects of human activity on the Earth's climate, and they provide the best estimates and summary of current knowledge on climate change.

Kemp, D. D. 1994: *Global Environmental Issues: A Climatological Approach.* London: Routledge, 2nd edition.

Kemp provides an interesting climatological approach to global problems. His textbook includes good sections on the greenhouse effect, acid rain, ozone depletion, drought, and the possible effects of a nuclear winter. It is appropriate for students who are studying environmental science and/or geography and who wish to understand more about the role of climate in environmental change.

Maunder, W. J. 1994: *Dictionary of Global Climate Change.* London: UCL Press, 2nd edition.

All students with interests in climate change will find this dictionary very useful. It clearly explains meteorological and climatological terms, outlining important conventions and protocols, and summarizing the activities of research groups, and political and non-governmental organizations.

Nieuwenhuis, P. and Wells, P. (eds) 1994: *Motor Vehicles in the Environment: Principles and Practice*. Chichester: John Wiley and Sons.
This is an excellent text, highlighting the various issues involved in the greening of the motor industry, emissions legislation, and the planning of green integrated traffic systems. Examples are presented from the USA and Europe, providing a range of perspectives on the implications of the green revolution on the production and use of motor vehicles.

Nilsson, S. and Pitt, D. 1991: *Mountain World in Danger: Climate Change in the Forests and Mountains of Europe*. London: Earthscan.
Mountain World in Danger provides an account of the possible changes in forest and mountain environments throughout Europe which may occur as a result of global warming and acid rain. Possible strategies in response to these changes are presented in the hope that steps may be undertaken by policy makers to retard the potentially harmful impact.

O'Riordan, T. and Jager, J. (eds) 1995: *Politics of Climate Change*. London: Routledge.
This edited volume provides a critical analysis of the political, moral and legal responses to climate change in the midst of significant socio-economic policy shifts. It also examines how climate change was placed on the policy agenda of the EU, and the role it played in the evolution of the United Nations Framework Convention and the subsequent Conference of Parties.

Parry, M. 1990: *Climatic Change and World Agriculture*. London: Earthscan.
As the chief scientist on the Intergovernmental Panel on Climate Change (IPCC), Parry provides an excellent account of the likely patterns of change in climate and world agriculture that may occur as a consequence of global warming. The uncertainties associated with this issue, the sensitivity of the world food system, vegetational patterns and animal life, to global climate change, the geographical limits to different types of farming, and the range of possible ways to adapt agriculture to mitigate any potential hazards caused by global warming are examined in detail.

Paterson, M, 1996: *Global Warming and Global Politics*. London: Routledge.
Global Warming and Global Politics examines the major theories within the discipline of international relations, and considers the emergence of global warming as a political issue.

Smith, K. 1992: *Environmental Hazards: Assessing Risk and Reducing Disaster*. London: Routledge.
This is a useful text describing most of the major environmental hazards, integrating both the earth and social sciences.

Whyte, I. 1995: *Climatic Change and Human Society*. London: Edward Arnold.
This book examines the various ways that climatic change can interact with society. It is useful for students who are studying geography, politics and/or economics.

Williams, M. A. J., Dunkerley, D. L., Deckker, P. De, Kershaw, A. P. and Stokes, T. 1993: *Quaternary Environments*. London: Edward Arnold.
Quaternary Environments is a comprehensive text that examines the environmental changes that have taken place throughout Quaternary time. It emphasizes the interactions between geological, biological and hydrological processes that have caused environmental change throughout this period and that have resulted in the evolution of contemporary environments.

6

Evidence for General Instability of Past Climate from a 250-kyr Ice-core Record*

W. Dansgaard, S. J. Johnsen,
H. B. Clausen, D. Dahl-Jensen,
N. S. Gundestrup, C. U. Hammer,
C. S. Hvidberg, J. P. Steffensen,
A. E. Sveinbjörnsdottir, J. Jouzel, and
G. Bond

Recent results (1, 2) from two ice cores drilled in central Greenland have revealed large, abrupt climate changes of at least regional extent during the late stages of the last glaciation, suggesting that climate in the North Atlantic region is able to reorganize itself rapidly, perhaps even within a few decades. Here we present a detailed stable-isotope record for the full length of the Greenland Ice-core Project Summit ice core, extending over the past 250 kyr according to a calculated timescale. We find that climate instability was not confined to the last glaciation, but appears also to have been marked during the last interglacial (as explored more fully in a companion paper (3)) and during the previous Saale–Holstein glacial cycle. This is in contrast with the extreme stability of the Holocene, suggesting that recent climate stability may be the exception rather than the rule. The last interglacial seems to have lasted longer than is implied by the deep-sea SPECMAP record (4), in agreement with other land-based observations (5, 6). We suggest that climate instability in the early part of the last interglacial may have delayed the melting of the Saalean ice sheets in America and Eurasia, perhaps accounting for this discrepancy.

* Originally published in *Nature*, 1993, vol. 364, pp. 218–22.

In 1990–92, the joint European Greenland Ice-core Project (GRIP) drilled an ice core to near the bedrock at the very top of the Greenland ice sheet (72.58°N, 37.64°W; 3,238 m above sea level (7); annual mean air temperature –32°C). The 3,028.8-m-long core was recovered by an electromechanical drill, ISTUK (8). Less than 1 m of core in total was lost in the drilling process. The deepest 6 m is composed of silty ice with pebbles. The core quality is excellent, except for a 'brittle zone' in the depth interval between 800 and 1,300 m.

Here we discuss a timescale for the entire core and present a continuous profile of $\delta^{18}O$ (hereafter denoted by δ, the relative deviation of the $^{18}O/^{16}O$ ratio in a sample from that in standard mean ocean water). In polar glacier ice, δ is mainly determined by its temperature of formation (9). Profiles of several climate-related parameters spanning the last interglacial period (known as the Eemian period, or Eem) are being investigated by GRIP participants (3).

The timescale back to 14.5 kyr BP (thousands of years before present) is derived by counting annual layers downward from the surface (1). Beyond 14.5 kyr BP a timescale is calculated as

$$t = \int_0^z dz/\lambda_z$$

t being the age of the ice and λ_z the annual layer thickness at depth z. A steady-state ice flow model (10) is modified by introducing a sliding layer at bedrock (11), and using a δ-dependent accumulation rate (λ_H) derived from all available λ_z data,

$$\lambda_H = \lambda_{H0} \exp [0.144(\delta + 34.8)] \text{ m of ice}$$

equivalent per year,

λ_{H0} being the present value, 0.23 m yr^{-1} (ref. 1).

The other flow model parameters are as follows: The total thickness of the ice sheet $H = 3,003.8$ m of ice equivalent; the thickness of the intermediate shear layer $h = 1,200$ m; the ratio between the strain rates at the top of the silty ice and at the surface, $f_b = 0,15$; and the thickness of the silty ice layer, $dh = 6$ m. The latter value may be too low, but higher values would significantly influence only the calculated ages of the deepest 50 m (ice older than 250 kyr) which will not be discussed here.

The h and f_b values are chosen so as to assign well-established ages to two characteristic features in the δ record: 11.5 kyr for the end of the Younger Dryas event (1, 12) and 110 kyr for the marine isotope stage (MIS) 5d (4), which appear at depths of 1,624 m and 2,788 m, respectively, in the δ record. These are points a and b in Figure 6.1B. Back to 35 kyr BP the calculated timescale agrees essentially with that presented in ref. 1.

One of the assumptions behind the timescale calculation is that the stratigraphy has remained undisturbed, so that all annual layers are represented in a continuous sequence and thinned according to the depth-dependent vertical strain described by the flow model. This may fail at great depths if there is folding close to a hilly bedrock, and/or random thinning of layers of different rigidity (boudinage effect (13)).

Large-scale folding caused by bedrock obstacles hardly exists in the Summit area, however, because the bedrock is gently sloping ($\leqslant 40$ m per km) in a large area around the drill site (14). Furthermore, according to radio-echo sounding records (L. Hempel, personal communication) the shape of internal reflection layers suggests that the long-term position of the ice divide was only 5 km west of the present Summit. Consequently, the ice movement at Summit has been essentially vertical in the past, confirming that the ice cannot have travelled long distances over hilly bedrock. Nonetheless, at great depths the boudinage effect may have caused small-scale disturbances. Some layers may have thickened at the expense of others now missing in the core. If so, the layer sequence may not be strictly continuous, but even then the broad outline of the timescale may still be valid.

Visible cloudy bands, probably indicative of former surfaces (15), lie almost perpendicular to the core axis down to ~ 2,900 m depth, corresponding to 160 kyr BP. Then follows a 54-m increment of apparently disturbed stratigraphy (S. Kipfstuhl, personal communication), possibly caused by the boudinage effect. Finally, the regular layer sequence is re-established from 2,954 m (210 kyr BP), but special caution

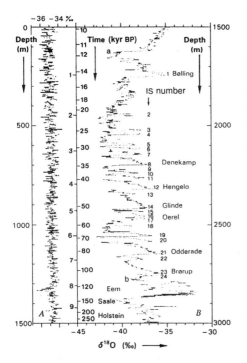

Figure 6.1 [*orig. figure 1*]   The continuous GRIP Summit δ¹⁸O record plotted in two sections on a linear depth scale: *A*, from surface to 1,500 m; *B*, from 1,500 to 3,000 m depth. Each point represents 2.2 m of core increment. Glacial interstadials are numbered to the right of the *B* curve. The timescale in the middle is obtained by counting annual layers back to 14.5 kyr BP, and beyond that by ice flow modelling. The glacial interstadials of longest duration are reconciled with European pollen horizons. (16)

should be applied beyond 160 kyr BP. The American GISP2 deep ice core being drilled 30 km west of Summit (2), may provide verification of the timescale.

A continuous δ record along the upper 3,000 m of the GRIP core is plotted on a linear depth scale in two sections of figure 6.1. The upper half of the record (figure 6.1A, δ scale on top) spans nearly the past 10 kyr, and each δ value represents the snow deposition through a few years near surface, increasing to 20 years at 10 kyr BP. Apart from the δ minimum at 8,210 ± 30 yr BP (1), the record

indicates a remarkably stable climate during the past 10 kyr.

In the contrast, the rest of the record shown in figure 6.1B is dominated by large and abrupt δ shifts. Because of plastic thinning of the layers as they approach the bedrock, these 1,500 m of ice represent a much longer period of time than the 1,500 m above. On the right side of the record is an extension of the numbering of glacial interstadials (IS) introduced previously (1). Furthermore, a series of European pollen horizons (16) (¹⁴C dated back to 60 kyr BP) is reconciled with the longest lasting δ-based interstadials. The ¹⁴C datings are all in essential agreement with our timescale once recent corrections to the ¹⁴C scale (12) have been considered.

Figure 6.2 is a composite of five different chronological records. Figure 6.2D shows the upper 2,982 m of the Summit δ record plotted on the linear timescale in 200 yr increments. The vertical line is drawn for comparison with the Holocene mean δ value, −35‰. On the right-hand side of this line is a division of the last glacial cycles in European terminology. The adopted timescale is further supported, first by the numerous common features (particularly during the last glaciation) between the smoothed version and the other four records in figure 6.2 (exemplified by dashed arrows); second, because a maximum entropy spectrum (17) comprising the total Summit record contains significant signals on the two Milankovitch cycles 41 kyr (obliquity) and, considerably weaker, 24/18 kyr (precession).

The violent δ shifts observed in Greenland cores are less pronounced in the δ record along the Vostok, East Antarctica, ice core (5) (figure 6.2E) probably because the shifts in Greenland are connected to rapid ocean/atmosphere circulation changes in the North Atlantic region (18, 19). Both records, as well as the δ record from Devils Hole (6) (figure 6.2A), and the records in figure 6.2C introduced below, imply MIS 5e (Eem) as an interglacial of considerably longer duration than estimated from sea sediment δ records, such as the SPECMAP record (4) in figure 6.2B. The disagreement may be explained, in

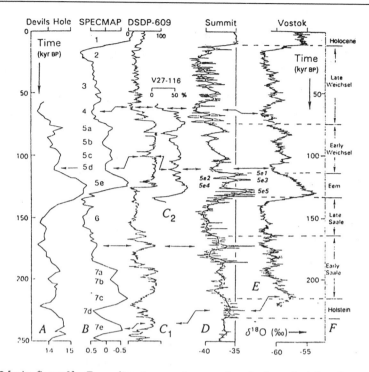

Figure 6.2 [*orig. figure 2*] Four climate records spanning the last glacial cycles and plotted on a common linear timescale. *A*, $\delta^{18}O$ variations in vein calcite from Devils Hole, Nevada (6). Dating by U/Th. *B*, The SPECMAP standard isotope curve (4) with conventional marine isotope stages and sub-stages. Dating by orbital tuning. *C*, part 1, grey-scale measurements along 14.3 m of ocean sediment cores DSDP site 609. Part 2, per cent $CaCO_3$ in V27–116 (through stage 5) from locations west-southwest and west of Ireland. Scale in arbitrary units on top (24). Dating by orbital tuning. *D*, $\delta^{18}O$ record along the upper 2,982 m of the GRIP Summit ice core. Each point represents a 200-yr mean value. The heavy curve is smoothed by a 5-kyr gaussian low-pass filter. Dating by counting annual layers back to 14.5 kyr BP, and beyond that by ice flow modelling. Along the vertical line, which indicates the Holocene mean δ value, is added an interpretation in European terminology. *E*, δD record from Vostok. East Antarctica (5), converted into a $\delta^{18}O$ record by the equation $\delta D = 8\delta^{18}O + 10‰$. Dating by ice flow modelling.

part, by the climate instability recorded (figure 6.2D) in the early stages of Eem, which must have slowed the melting of the Saalean ice sheets in America and Eurasia. Further evidence of delayed sea level rise is found in records of the isotopic composition of atmospheric oxygen in the Vostok core (5). If MIS 5e is defined as the period between the first and last years of higher than Holocene δ, it lasted nearly 20 kyr, from 133 to 114 kyr BP, according to the Summit record.

In figure 6.2D (and in other Summit records (3)), 5e stands out as an interglacial abruptly interrupted several times by periods as cool as 5a and 5c. A similar episode in 5e was previously demonstrated in the Camp Century and Devon Island ice cores (20, 21), and given the new timescale, it could well be identical with one of the 5e cool spells indicated in figure 6.2D. The duration of 5e2 and 5e4 was apparently 2 and 6 kyr, and they define a tripartition of 5e. Carbon dioxide analyses along the Vostok ice core also suggest a partition of 5e (22), but figure 6.2D differs remarkably from most pollen and deep sea records, which show the Eem as a generally warm and stable period (see for example, ref. 16 and figure 6.2B).

The apparent disagreement may not be serious. Substantial global climate changes are generally more pronounced at higher latitudes. The cooling in western Europe corresponding to the long-lasting δ minimum 5e4 in figure 6.2D may therefore not have been deep enough to cause any marked change in the vegetation, and thereby in the pollen records. Sea sediment δ records are primarily indicative of the continental ice volume, which does not necessarily vary with temperature during times of no ice in North America and Eurasia. Furthermore, the resolution is limited by bioturbation and coarse sampling. Significant fluctuations of 5e have been recorded, however, in North Atlantic areas of high sedimentation rate (≥5 cm kyr^{-1}, for example in core V28-56 from the Norwegian sea (23), and as colour and $CaCO_3$ concentration changes in central North Atlantic sediment cores (24).

The colour record from Deep Sea Drilling Project site 609 is plotted in figure 6.2C, part 1, on a timescale tuned (25) to orbital variations. This record has a resolution comparable to that of figure 6.2D. Disregarding MIS 2 and 3, where detrital grains of carbonate corrupt the grey scale, nearly all of the δ shifts can be recognized in the colour record, including some of the δ shifts in 5e. The colour scale reaches 'saturation' (note that 5a, and 5c and the warm phases of 5e appear equally dark), which may be why only one of the cool stages in 5e looks significantly different in colour. Several fluctuations in 5e are registered in cores recovered farther north (24); for example see the inset of per cent $CaCO_3$ (indicative of biological activity) through MIS 5 from V27-116 in figure 6.2C, part 2.

The glacial cycles spanned by the Summit ice core appear different in figure 6.2D. For example, MIS 6 (late Saale) was apparently less cold and variable than late Weichsel, and a few spells of extreme warmth occurred in early Saale. Furthermore, the Holstein interglacial (26) (possibly MIS 7e) seems more stable than Eem, but less stable than Holocene.

In conclusion high resolution records suggest that, apart from the Holocene, insta-bility has dominated the North Atlantic climate over the last 230,000 years. This applies to the Weichsel glaciation (MIS 2 to 5d), to the Eem interglacial (MIS 5e) whose progress was very different from the Holocene, the Saale glaciation (MIS 6 to 7d), and Holstein, the preceding interglacial (MIS 7e), according to the interpretation given in figure 6.2D. This emphasizes the question of whether the Holocene will remain stable in spite of the growing atmospheric pollution.

Acknowledgements

We thank C. C. Langway and H. Tauber for reading the manuscript and making corrections and additions. This work is a contribution of the international Greenland Ice-core Project (GRIP) organized by the European Science Foundation. We thank the GRIP participants and supporters for their cooperative effort. We also thank the funding agencies in Belgium, Denmark, France, Germany, Iceland, Italy, Switzerland and the United Kingdom, as well as the XII Directorate of CEC, the University of Iceland Research fund, the Carlsberg Foundation, and the Commission for Scientific Research in Greenland, for financial support.

References

1 Johnsen, S. J. et al. *Nature* 359, 311–313 (1992).
2 Taylor, K. C. et al. *Nature* 361, 432–436 (1993).
3 GRIP Members *Nature* 364, 203–207 (1993).
4 Martinson, D. G. et al. *Quat. Res.* 27, 1–29 (1987).
5 Jouzel, J. et al. *Nature* (in the press).
6 Winograd, I. J. et al. *Science* 258, 255–260 (1992).
7 Elkholm, S. & Keller, K. *Nat. Survey and Cadastre, Goedetic Div., Tech. Rep, No. 6* (Copenhagen, Denmark, 1993).
8 Gundestrup, N. S. & Johnsen, S. J. in *Greenland Ice Cores: Geophysics, Geochemistry and Environment* (eds Langway, C. C. Jr. Oeschger, H. & Dansgaard, W.) 19–22 (Am. Geophys. Un. Geophys. Monogr. 33, Washington DC, 1985).
9 Johnsen, S. J., Dansgaard, W. & White J. W. C. *Tellus* B41, 452–468, (1992).
10 Dansgaard, W. & Johnsen, S. J. *J. Glaciol.* 8, 215–223 (1969).
11 Johnsen, S. J. & Dansgaard, W. in *The Last Deglaciation: Absolute and Radiocarbon Chronologies* (eds Bard, E. & Broecker, W. S.) (NATO ASI Series 2, 1992).

12 Bard, E. et al. *Radiocarbon* (in the press).
13 Staffelback, T., Stauffer, B. & Oeschger, H. *Ann. Glaciol.* 10, 167–179 (1988).
14 Hodge, S. M. et al. *J. Glaciol.* 36, 17–27 (1990).
15 Hammer, C. U. et al. *J. Glaciol.* 20, 3–26 (1978).
16 Behre, K. -E. & van der Plicht, J. *J. Veg. Hist. Archaeobot.* 1, 111–117 (1992).
17 Ulrych, T. J. & Bishop, T. N. *Rev. Geophys. Space Phys.* 13, 183–200 (1975).
18 Oeschger, H. et al. in *Climate Processes and Climate Sensitivity* (ed. Hansen, F.) 299–306 (Am. Geophys. Un. Geophys. Monogr. 29, Washington DC, 1984).
19 Broeker, W. S., Bond G., Klas, M., Bonani, G. & Wolfli, W. *Palaeoceanography* 5, 469–477 (1990).
20 Dansgaard, W., Clausen, H. B., Johnsen, S. J. & Langway, C. C. *Quat. Res.* 2, 396–398 (1972).
21 Paterson, W. S. B. et al. *Nature* B266, 508–511 (1977).
22 Raynaud, D. et al. *Science* 259, 926–934 (1993).
23 Kellog, T. B. *Marine Micropaleontol.* 2, 235–249 (1977).
24 Bond, G., Broecker, W., Lotti, R. S. & McManus, J. in *Start of a Glacial* (eds Kukla, G. J. & Went, E.) 185–205 (NATO ASI Series I 3 Springer, Heidelberg, 1992).
25 Ruddiman, W. F., Raymo, M. E., Matinson, D. G., Clement, B. M. & Bachman, J. *Palaeoceanogr.* 4, 353–412 (1989).
26 Linke, E., Katzenberg, O. & Grün. R. *Quat. Sci. Rev.* 4, 319–331 (1985).

7

ATMOSPHERIC EFFECTS OF THE Mt PINATUBO ERUPTION*

M. Patrick McCormick, Larry W. Thomason, and Charles R. Trepte

It is well known that human activity is perturbing the chemical composition and radiative balance of the Earth's atmosphere. Studies (1) of the sensitivity of our climate to increasing concentrations of greenhouse gases, so named for their ability to retain heat in the atmosphere, predict that the increase of CO_2 concentration from a pre-industrial value of ~270 p.p.m. to 600 p.p.m. by the middle of the next century, along with expected increases in other greenhouse gases, will increase global surface temperature by 2–5°C. This picture is complicated by the increasing concentration of anthropogenic aerosols in the lower troposphere, which act to mitigate greenhouse warming (2). In the stratosophere, chlorine concentrations have increased because of anthropogenic chlorofluorocarbon production (3). Reactive chlorine compounds play an important role in the chemical processes that give rise to the Antarctic ozone hole, and the weight of evidence suggests that they also contribute to global losses of ozone, with possible concomitant increases in the intensity of biologically harmful ultraviolet-B radiation reaching the Earth's surface (4).

The 15 June 1991 eruption of Mt Pinatubo (15° N, 121° E), on the island of Luzon in the Republic of the Philippines, showed that there is also significant natural, though episodic, variability in the composition of the atmosphere, particularly that of the stratosphere. The Mt Pinatubo eruption forced the evacuation of more than 200,000 people and caused the immediate deaths of more than 300, many from the collapse of homes due to the combination of heavy ashfall and rain from the nearly simultaneous passage of Typhoon Yunya. The threat to life and property from remobilization of tephra ashfall and pyroclastic material persisted through the remainder of 1991 and required careful monitoring by local authorities (5).

The volcanic plume associated with the 15 June eruption was observed to have reached an altitude of more than 30 km. In addition to particulate matter, the eruption also injected gaseous SO_2 into the stratosphere (6), which in turn was transformed into H_2SO_4/H_2O aerosol (particles suspended in air). With an estimated (7) peak aerosol mass loading of 30×10^{12} g (30 Tg), the eruption of Mt Pinatubo caused what is

* Originally published in *Nature*, 1995, vol. 373, pp. 399–404.

believed to be the largest aerosol perturbation to the stratosphere this century (table 7.1 updated from ref. 8), but it is still smaller than the estimated aerosol perturbations from the eruptions of Tambora in 1815 (>100 Tg) and Krakatau in 1883 (~50 Tg) (9). Given both recent volcanic history and inferences from ancient eruptions such as that of Toba (~1,000 Tg of SO_2) ~ 75,000 yr ago (9), the 1991 Mt Pinatubo eruption is clearly not exceptional in its modification of the stratospheric aerosol loading. On the other hand, the Mt Pinatubo eruption is unique in the sense that it has been the most intensely observed eruption on record, as its volcanic cloud has been monitored by ground-based and aircraft lidars and solar photometers, various balloon-borne and airborne aerosol counters and other instruments, and many satellite-borne instruments. From a radiative perspective, the aerosol contributed to the end of several years of globally warm surface temperatures experienced in the late 1980s and early 1990s. The aerosol loading has also been associated with chemical and dynamical perturbations affecting the concentration of NO_2, reactive chlorine and ozone. Attempts to model the dispersion of the aerosol and the radiative and chemical impact of the eruption are 'acid tests' of our understanding of atmospheric processes (figure 7.1).

Table 7.1 [orig. table 1]   Major twentieth-century eruptions.

Volcano	Date	Estimated aerosol loading (Tg)
Stratospheric background	Possibly 1979	<1
Katmai	June 1912	20
Agung	March 1963	16–30
Fuego	October 1974	3–6
El Chichon	April 1982	12
Mt Pinatubo	**June 1991**	**30**
Cerro Hudson	August 1991	3

Mt Pinatubo Aerosol and Dynamical Processes

Following the 15 June eruption, the evolving cloud of water vapour, sulphurous gases and aerosol moved westwards and circled the globe in approximately 22 days (6). According to aircraft, ship, and ground-based lidar measurements, the bulk of Mt Pinatubo aerosol was initially concentrated between 20 and 27 km in altitude, although some was observed as low as the tropopause and as high as 30 km. The total mass of SO_2 injected by the eruption was measured (10) by the total ozone mapping spectrometer (TOMS) to be approximately 20 Tg. This SO_2 was eventually converted into H_2SO_4/H_2O aerosol, beginning with an oxidative process involving $OH^·$ in a characteristic time (6) of 30 days. The SO_2 mass estimate is consistent with the 30-Tg peak in total stratospheric aerosol mass loading estimated by the solar occultation instrument SAGE II (the stratospheric aerosol and gas experiment-II) in the last few months of 1991.

A notable feature of the Mt Pinatubo eruption was the rapid movement of a substantial fraction of volcanic material across the Equator to about 10° S during the first two-week period following the eruption. Within the next few weeks, the cloud continued to disperse longitudinally, but more slowly in the meridional direction, and occupied the latitude band of approximately 20° S to 30° N (11–14). The degree of early cross-equatorial transport was unusual because the early dispersion of volcanic material from other eruptions near the subtropics, such as El Chichon (17° N, 93° W), remained mostly in their hemisphere of origin. The amount of Mt Pinatubo volcanic material transported south at altitudes above 24 km was highlighted by aerosol observations by SAGE II in July and September and by the improved stratospheric and mesospheric sounder (ISAMS) (15) in October which showed that maximum aerosol opacity resided near 10° S.

Recent numerical transport simulations suggest that the southward displacement of the aerosol plume within the topics was the

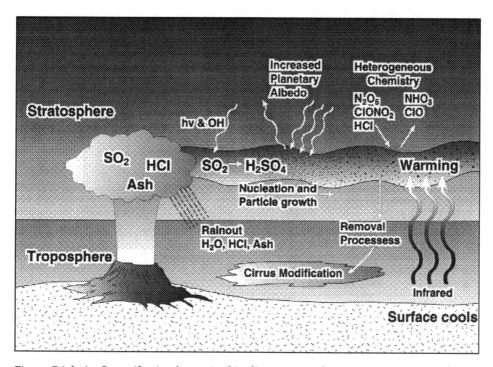

Figure 7.1 [*orig. figure 1*]   As shown in this diagram, a volcanic eruption can produce a significant perturbation to the Earth–atmosphere system by injecting material into the stratosphere where, depending on the magnitude and altitude of the injection, it may persist for several years. The injected material may include ash, which typically does not remain for more than a few months, and gaseous components including water vapour, sulphur dioxide and hydrochloric acid. Most hydrochloric acid is dissolved into condensing water vapour and quickly rains out of the original cloud. Aerosols are produced when the sulphur dioxide (SO_2) is chemically transformed into sulphuric acid (H_2SO_4) which rapidly condenses into aerosols because it has a very low saturation vapour pressure. The new aerosol increases the Earth's albedo by reflecting solar radiation back into space, and can warm the stratosphere by absorbing upwelling infrared radiation. Sedimentation and atmospheric circulation eventually transport the aerosol into the troposphere where it may modify cloud optical properties (particularly cirrus) and further modify the Earth's radiative processes. An additional effect of an eruption is the increased efficiency of heterogeneous chemical processes (that is, processes that take place on the surface of aerosols). This effect, coupled with anthropogenically increasing stratospheric chlorine levels, leads to ozone destruction by modifying the chemistry of reactive chlorine and nitrogen. This ozone removal process is similar to that which produces the Antarctic ozone hole, except that the surface in the latter case is provided by polar stratospheric clouds (4).

result of meridional circulations induced by local heating, rather than planetary-scale transport patterns (16). The heating took place through the absorption of upwelling infrared radiation by the newly nucleated aerosol, and was responsible for warming the lower tropical stratosphere through the remainder of 1991. Peak temperatures at the 30-hPa level in the atmosphere (~24 km) were found in September 1991 to be as much as three standard deviations above the 26-year monthly mean (increases of approximately 3.5 K) (17). The positive temperature anomalies gradually decreased during 1992 as the tropical aerosol dispersed. Strong local heating not only affected horizontal transport, but also enhanced upward motions in the tropics. Satellite observations detected the lifting of aerosol to altitudes exceeding 35 km by October 1991. Because ozone concentration peaks near 25 km in the tropics, a more vigorous circulation at low latitudes, moreover, would have carried ozone-poor air to higher altitudes. At Latitudes below 25 km, ozone is long-lived and a gradual photochemical relaxation to pre-eruption levels would have occurred in about 1–3 months. This mechanism could be responsible for a reduction of columnar ozone of about 6–8% observed over the Equator during the first several months after the eruption (18–21).

One feature that seems to be characteristic of large eruptions at low latitudes is the relatively slow transport timescales of volcanic material out of the tropics. Observations following the eruptions of El Chichon in 1982 and Nevada del Ruiz in 1985 revealed the presence of enhanced reservoirs of volcanic aerosol at low latitudes which persisted for several years (22). Poleward transport of aerosol into the winter hemisphere tends to be suppressed when easterly winds lie over the Equator (23). During these conditions, strong horizontal wind shear (velocity gradient) lies in the subtropics, separating the tropical easterlies and extratropical westerlies in the winter hemisphere. Wind shear inhibits the horizontal mixing of air. When westerly winds lie over the Equator, however, horizontal wind shear is weaker in the sub-tropics and meridional mixing occurs more readily. The reversal between easterly and westerly winds in the lower tropical stratosphere is known as quasi-biennial oscillation (QBO). It dominates the circulation in the lower tropical stratosphere and has a period of about 28 months.

A similar signature of relatively slow tropical aerosol transport to high latitudes was evident after the eruption of Mt Pinatubo. Easterly winds prevailed above about 23 km in the lower tropical stratosphere for the first 6 months after the eruption. Above this height, airborne lidar (24, 25) and satellite (26) observations of the volcanic plume found steep meridional

gradients of aerosol near 20° N and 20° S. At lower altitudes, however, volcanic aerosol dispersed more quickly towards the poles, and it was primarily this low-altitude transport that was responsible for the initial lidar observations of the Pinatubo aerosol (and the spectacular sunsets) in northern mid-latitudes shortly after the eruption (27–29).

Over the course of time, volcanic aerosol is transported from the stratosphere into the troposphere where a variety of mechanisms (mostly associated with clouds and precipitation) effect the deposition of aerosol to the Earth's surface. Gravitational sedimentation of the aerosol, subsidence (downward air motions) in the winter hemisphere, and episodic transport of stratospheric air into the troposphere through tropopause folds contribute to the removal of aerosol from the stratosphere. By the end of 1993, about 2½ years after the eruption, the global stratospheric aerosol mass loading had decreased to approximately 5 Tg. Since the middle of 1992, the total stratospheric aerosol mass has decreased with a 1/e-folding time of approximately 1 year. Measurements by lidars, Sun photometers, and satellite instruments (for example, SAGE II), showed that the longitudinally averaged stratospheric opacity at a wavelength of 1μm had declined from a maximum in excess of 0.2 in the tropics in late 1991 to a range over the globe of 0.006–0.03 by mid-1993. These values still represent enhancement by a factor of 5–10 of values observed in 1989 and 1990, when the global average stratospheric opacity at 1 μm was between 0.001 and 0.003.

Effect on the Earth's Radiative Processes

The introduction of large amounts of sulphuric acid aerosol into the stratosphere increases the planetary albedo (essentially the Earth's reflectivity of solar radiation) because these aerosol particles are efficient scatterers but only weak absorbers at solar wavelength. An analysis of data from the Earth radiation budget experiment (ERBE) aboard the Earth radiation budget satellite (ERBS) found that the albedo in July, August and September 1991 increased compared to a five-year mean (30). The global albedo in August 1991, for example, was 0.250, or more than five standard deviations greater than the five-year mean of 0.236. The greatest effect occurred over cloud-free regions, where the normally low albedo increased by more than 20%. An increase in albedo was also noted in the normally high-albedo regions associated with deep convective cloud systems. This may represent an indirect effect of the volcanic aerosols in the sense that aerosol particles transported into the upper troposphere are incorporated into deeper convective clouds and alter their microphysical structure (31). Increasing the number of aerosol particles available to act as cloud condensation and ice nuclei tends to reduce the mean size of the resulting cloud particles. Because smaller particles are more efficient scatterers of visible radiation, this process increases the albedo of the cloud (32). It has also been suggested that optically thin (but radiatively important) cirrus may be subject to similar microphysical changes (32, 33), although it is not at present clear if any changes in the optical properties of cirrus have occurred. The changes in the Earth's albedo observed by ERBE resulted in a net cooling of approximately 8 W m^{-2} between 5° S and 5° N, with a net cooling of 4.3 Wm^{-2} between 40° S and 40° N. The meridional distribution of the cooling and its magnitude in the tropics reflects the confinement of the aerosol as discussed earlier.

To a first approximation, an increase in stratospheric opacity, such as that following the eruption of Mt Pinatubo, should cool the Earth because the aerosol increases the albedo of the Earth–atmosphere system and therefore increases the amount of solar radiation scattered back into space. But microphysical properties of the aerosol (in particular, aerosol size) determine whether the infrared absorptivity of stratospheric aerosols, which acts to warm the Earth (that is, the greenhouse effect), dominates the scattering of incoming solar radiation back into space (that is, the albedo effect), which acts to cool the Earth. It has been shown (34) that if the aerosol effective radius (the area-weighted mean radius of the aerosol size

distribution) exceeds ~2 μm, the warming effect overrides the cooling effect. Estimates of effective aerosol radius following the Pinatubo eruption indicate that an increase from an average of 0.3 μm before the eruption to about 1 μm afterwards occurred, and thus cooling would be expected (35–37).

As previously noted, substantial heating in the stratosphere was observed immediately after the eruption. This heating was sufficient to cause tropical stratospheric temperatures to increase as much as three standard deviations above the 26-year mean at 30 hPa (~24 km). Analyses by Christy and Spencer (38) using the satellite-borne microwave sounding unit (MSU) also indicate warmer than average stratospheric temperatures in 1991 and 1992. By the end of 1993, however, stratospheric temperatures had decreased to the lowest values ever observed by MSU. This cooling may be related to the ozone loss observed in the stratosphere (to be discussed later), as ozone is an effective absorber of solar radiation and is responsible for substantial heating in the stratosphere.

The negative radiative forcing associated with the eruption of Mt Pinatubo exceeded the magnitude of the positive forcing associated with the greenhouse gases during the second half of 1991 and much of 1992, and remained significant through 1993 (figure 7.2). In response to this forcing, the mean tropospheric temperature was 0.2°C below normal in 1992 compared with a base period of 1958–91 (39). The 1992 El Niño–Southern Oscillation (ENSO, a feature of Pacific Ocean circulation with important couplings to tropospheric weather patterns) would have caused a warmer than average troposphere. As a result, the 1992 tropospheric temperature anomaly, adjusted for ENSO (40), was −0.4°C, a decrease of more than 0.7°C from 1991. Data from MSU yield a similar global decline in lower tropospheric temperatures of 0.5°C in mid-1992, with much of the decrease occurring in the Northern Hemisphere (0.7°C) (41). In a global average context, the measured temperature anomaly is consistent with the value of −0.5°C estimated from a global climate model (42). It should be noted that the small mean changes

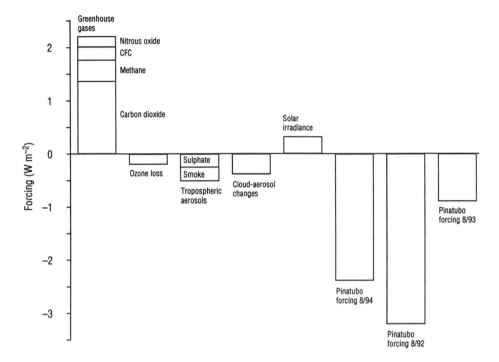

Figure 7.2 [*orig. figure 3*]   Pre-industrial-to-modern global climate forcing, 1850–1990 (histogram adapted from Hansen et al. *(79)*). Climate forcing is the change in the Earth's radiative balance resulting from either natural or anthropogenic sources. The largest forcing (~2 Wm^{-2}) is due to the increase in greenhouse gases associated with industrialization, and the net forcing in this period remains positive when several smaller (and negative) effects, such as increasing tropospheric opacity due to human activities and global ozone loss, are included. Mt Pinatubo is estimated to have produced a transient negative forcing which exceeded greenhouse-gas forcing between mid-1991 and the end of 1992. It also made 1992 one of the coolest of the past 30 years. As the effect of Mt Pinatubo continues to decrease, the increasing magnitude of greenhouse-gas forcing should reveal itself in a return to the globally warm years observed in much of the 1980s.

in global temperature are not evenly distributed, but rather consist of regions of both relative cooling and warming whose locations and magnitudes are dependent on season (42). The temperature anomaly in 1993 was smaller than in 1992 with a value of −0.1°C, which, adjusted for the phase of the ENSO, is equivalent to an anomaly of −0.2°C (J. Angell, personal communication).

This drop in global mean temperatures is similar to that observed in 1964 which had the largest anomaly in the 1958–92 analysis period. Significantly, 1964 is the year following the eruption of Agung, which appears to be the second largest volcanic

eruption (after Pinatubo) to have occurred since 1958. The departure is also comparable to the estimated impact of Tambora (−0.4 to −0.7°C) and greater than that estimated for El Chichon (−0.2°C) and Krakatau (−0.3°C). (9) That the Pinatubo cooling exceeds that of Krakatau is surprising because the latter is believed to have put more SO$_2$ into the stratosphere. However, this anomaly may simply reflect uncertainties in the estimates of both stratospheric aerosol loading and global surface temperature, particularly for the earlier eruption.

Effect on Stratospheric Chemical Processes

Observations of the input of hydrochloric acid (HCl) soon after the eruption of Mt Pinatubo showed little or no increase compared to pre-eruption levels (43, 44). The measurements are consistent with modelling efforts by Tabazadeh and Turco (45) that suggest that the bulk of erupted chlorine is removed within the volcanic plume by condensation into supercooled water drops which are then rained out. These findings argue strongly that stratospheric chlorine is primarily a product of anthropogenic chlorofluorocarbons rather than volcanic eruptions (46).

The concentrations of chlorine species, and particularly reactive chlorine species such as Cl within the stratosphere are crucial in determining the rate of ozone destruction. Ozone loss associated with reactive chlorine figures prominently in the formation of the Antarctic ozone hole. In the Antarctic, the extreme low temperatures of the winter and early spring permit the formation of polar stratospheric clouds (PSCs) (47). The cloud particles provide sites for heterogeneous chemical reactions (that occur on the surface of an aerosol or cloud particle) which transform relatively inert forms of chlorine such as HCl and $ClONO_2$ into more reactive forms such as Cl_2. The rates of such reactions are strongly dependent on the total particulate surface-area density. When the chemically modified air is transported into sunlit regions (usually in late August and September at high southern latitudes) Cl_2 disassociates into highly reactive Cl which in turn is responsible for ozone destruction via gas-phase chemical reactions (4).

An additional factor governing the depletion in the ozone hole is that the initial stage of PSC formation involves sequestration of a significant fraction of stratospheric nitric acid (HNO_3) in condensed forms. Because HNO_3 is unreactive, PSCs act as a sink for more reactive forms of nitrogen such as NO_2 and N_2O_5. Reactive nitrogen is a buffer to ozone loss because it is involved in reactions which convert reactive chlorine into non-reactive forms. There is abundant evidence that further growth of PSC particles by the deposition of ice followed by sedimentation of the larger particles leads to an irreversible denitrification (and dehydration) of the polar stratosphere, further enhancing ozone loss (48–50). The lack of an ozone hole in the Arctic (so far) can be attributed at least in part to generally warmer winter temperatures and, as a result, fewer PSC occurrences (4), less dentrification and a shorter season.

After a large eruption such as that of Mt Pinatubo, stratospheric aerosol surface-area density may be large enough for heterogeneous reactions to perturb the concentrations of relevant chemical species not only in polar latitudes but also in middle and tropical latitudes. For instance, ozone loss (although much smaller than that reported below associated with the Mt Pinatubo eruption) attributed to increased aerosol loading was reported following the eruption of El Chichon in 1982 (51). Because elevated chlorine levels in the stratosphere are the result of chlorofluorocarbon emissions, this is an interesting example of the coupling of human and natural perturbations of stratospheric composition that lead to additional perturbations, in this case ozone destruction (52). It has also been noted that the increase in important greenhouse gases like CO_2, CH_4 and N_2O (as well as CO) have slowed or stopped in the aftermath of the Pinatubo eruption (53). The cause for these changes is still under investigation.

An early indicator during the post-Pinatubo period of a connection between aerosols and ozone loss was found in association with aerosol formed by the eruption of Cerro Hudson (46° S; August 1991) rather than Mt Pinatubo. Stratospheric aerosol from Cerro Hudson was concentrated at low altitude (<16 km) and was able to penetrate deeply into high southern latitudes soon after that eruption. At that time, Pinatubo aerosol in the Southern Hemisphere was primarily residing above 16 km, with the impedance to meridional transport associated with the polar vortex boundary effectively blocking the aerosol from high latitudes until the vortex dissipated in November 1991 (54). Before 1991, Antarctic ozone loss had not been observed below the base of the vortex (~14 km). but in the densest portions of the Cerro Hudson aerosol layer (11–13 km), ozone decreased by 50% in the 30 days following the arrival of the aerosol over Antarctica (55). The aerosol surface-area density in the layer was as high as 100 $\mu m^2 cm^{-3}$, or 20–30 times higher than observed in 1990. Aerosol levels at these altitudes remained unusually high during the austral springs of 1992 and 1993 because of the downward transport of Pinatubo aerosol within the Antarctic vortex (50, 56). Ozone levels in the 12–14 km range were extremely low in 1992 and 1993 with near-zero values in mid-October (57) in both years. Record low column ozone was also observed in these years (for example, 105 Dobson units (DU) on 11 October 1992 and 91 DU on 12 October 1993; column ozone amounts of less than 100 DU were noted on several occasions in 1993 (57)).

In the austral autumn of 1992, more evidence of unusual heterogeneous processing was observed over Antarctica. Elevated levels of chlorine dioxide (OClO) were detected over McMurdo station as early as 13 April (58). Before 1992, elevated OClO was observed only during winter and spring (June–September) as a by-product of heterogeneous chemical reactions on the surface of PSCs. The OClO detected early in 1992 and the ozone losses in 1991 may be the by-products of heterogeneous reactions on the surface of volcanic aerosols:

$$N_2O_5 + H_2O \rightarrow 2HNO_3 \qquad (1)$$
$$ClONO_2 + H_2O \rightarrow HNO_3 + HOCl \qquad (2)$$

where the moderating influence represented by the photolysis of HNO_3 back to NO_2 (allowing benign $ClONO_2$ to form) (59) was suppressed by the low solar illumination of Antarctica in the autumn. These findings suggest that there was considerable poten-tial for ozone destruction by reactive chlorine in the Antarctic during times when it was too warm for PSCs to occur (58).

On the other hand, evidence of global-scale heterogeneous processing following the eruption of Mt Pinatubo was noted as early as August 1991, when measurements in New Zealand (60) showed a reduction in the column NO_2 that reached as much as 40% in October 1991. Significant column NO_2 decreases have also been reported in northern mid-latitudes (61). Similar, although smaller, decreases in the column NO_2 were also reported after the eruption of El Chichon in 1982 (51, 62). SAGE II measurements showed that the loss of NO_2 occurred globally, largely at altitudes below 25 km and exceeded 50% at some altitudes. Stratospheric NO_2 concentrations remained suppressed throughout 1992 before re-turning to normal ranges in late 1993. The decreases in NO_2 suggest a repartitioning of reactive nitrogen species in favour of nitric acid (HNO_3) via the heterogeneous reaction (1), (63) a conclusion supported by recent reports of elevated gaseous HNO_3 at low latitudes (64) and mid-latitudes (65, 67).

As previously mentioned, column ozone in the tropics decreased by 6–8% in the months following the Pinatubo eruption. The bulk of the loss was observed below 28 km and was as large as 20% near 24–25 km (refs 18, 68, 69). Also, small increases were noted above 30 km. Ozone losses in the lower tropical stratosphere were probably produced by increased lofting associated with stratospheric heating, associated with the new aerosol. In this scenario, ozone is lofted to altitudes at which photochemical processes are more effective in destroying it (19, 20). Other processes may also have contributed to the ozone perturbations in the tropics, including those involving SO_2. Before conversion to sulphuric acid aerosol, SO_2 absorbs ultraviolet radiation thereby reducing the photolysis of O_2 and thus the production of ozone, particularly below 25 km. Above 25 km, photolysis of SO_2 contributes to chemical processes which produce (70) O_3. It has also been suggested that increased scattering of solar radiation by volcanic aerosol may perturb the ozone balance (71).

Six months after the eruption of Mt Pinatubo, global mean column ozone, as measure by TOMS and shown in figure 7.3, began to show a significant downward trend with respect to the mean 1979–90 annual

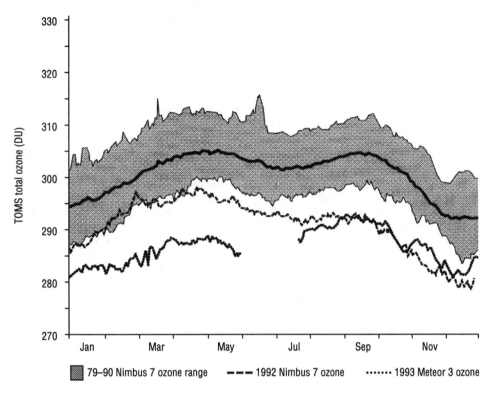

Figure 7.3 [*orig. figure* 4]   Global mean ozone in Dobson units (DU) from the total ozone mapping spectrometer (TOMS) measurements as a function of time. Dashed line, TOMS/NIMBUS 7 measurements for 1992; dotted line, TOMS/METEOR 3 measurements for 1993. These two lines are super-imposed on the range of TOMS-observed global (65° N–65° S) mean ozone between 1979 and 1990 (solid line). The small dotted line is the average total ozone for this period. TOMS data show a notable decrease in column ozone amounts beginning in early 1992 and persisting to the end of 1993. The loss reached as much as 6% of the column mean in April 1992 (72, 73).

ozone cycle. By mid-1992, the global mean column ozone amount was significantly less than the minimum values measured between 1979 and 1990 (72). This downward trend continued well into 1993; in April of that year a deficit of approximately 6% (~18DU) was observed, nearly 4% less than ever observed by TOMS before the eruption (72, 73). A slow recovery toward pre-eruption values began in the second half of 1993. It should be noted, however, that global values are somewhat deceptive because TOMS observed much larger decreases in the Northern Hemisphere than in the Southern Hemisphere (73). Dobson spectrophotometer observations from

several sites in the United States showed a downward trend beginning in May 1992 which accelerated in the first half of 1993, establishing record low values between January and April of that year (74). On average, these sites measured a decrease of more than 10% compared to the 1979–91 monthly means. Departures were noted throughout 1992 and 1993, although the decrease was smaller in magnitude during May–August. Ozone-sonde data from Boulder, Colorado indicated (75) that this loss was greatest between 13 and 22 km altitude and averaged about 20%. Small increases in ozone concentration were noted above 25 km, though it should be noted that the total column amount is small above that altitude. The altitude of the most significant losses was highly correlated with the altitude of highest aerosol loading.

The observation of highly perturbed values of NO_2 and HNO_3, and the general correlation of the perturbations with enhanced aerosol, strongly indicates the importance of heterogeneous chemical processes in the destruction of ozone. Many of the likely processes are similar to those associated with ozone loss in the Antarctic, in which a loss of reactive nitrogen to non-reactive nitric acid led to enhanced levels of reactive chlorine. The reactions that produce increased reactive chlorine and thus increase ozone destruction include (76) reactions (1) and (2), and other reactions involving $ClONO_2$, HCl and HOCl. Ozone destruction reaches a maximum in the winter and early spring because the direct chlorine reactions are strongly temperature dependent (progressing more rapidly with decreasing temperature), and the photolysis of HNO_3 back to reactive nitrogen (which moderates ozone destruction by reactive chlorine in the lower stratosphere) is inhibited by the low solar illumination. Hanson et al. (76) also pointed out that these reactions occur slowly, and that long periods under favourable conditions were required for significant ozone destruction. In the tropics, where solar illumination is high, little ozone loss is noted beyond mid-1992 despite the presence of very high aerosol loading. Modelling efforts by Hanson et al. (76) and Rodriguez et al. (77) corroborate the importance of heterogeneous reactions in the observed ozone loss.

Looking to the Future

The eruption of Mt Pinatubo was a disaster for those living nearby, but for those trying to understand the natural and anthropogenic processes for global change, the eruption represented perhaps a once in a lifetime opportunity. Observations of the Mt Pinatubo aerosol, various chemical species, and the radiative consequences during both the most heavily loaded period and the subsequent recovery provide data needed to test dynamic, chemical and radiative modelling of the atmosphere. Verification of our understanding of the Earth-atmosphere climate system is crucial because future forecasts by such models affect national and international policies and regulations that will in turn directly affect our economic livelihood and quality of life. The Montreal Protocol and its subsequent modifications governing the production of chlorofluorocarbons are good examples of such actions. Of course, much more work is in progress about the effects of Mt Pinatubo than is discussed here. It is also clear that much more work is needed to exploit fully the opportunity provided by nature to enhance our knowledge of how this complex atmosphere and climate system works, and to provide us with a glimpse into the future.

Acknowledgements
We thank J. Herman for providing the TOMS ozone trends, and L. Poole, G. Kent, J. Hansen and S. Solomon for helpful comments and suggestions.

References
1 Cess, R. D. et al. *Science* 245, 513–516 (1989).
2 Charlson, R. J. et al. *Science* 255, 423–430 (1992).
3 Gunson, M. R. et al. *Geophys. Res. Lett.* 21, 2223–2226 (1994).
4 Solomon, S. *Nature* 347, 347–354 (1990).
5 Pinatubo Volcano Observatory *Earth in Space* 4, 5–10 (1992).
6 Bluth, G. J. S., Doiron, S. D., Schnetzler, C. C., Krueger, A. J. & Walter, L. S. *Geophys. Res. Lett.* 19, 151–154 (1992).
7 McCormick, M. P. & Veiga, R. E. *Geophys. Res. Lett.* 19, 155–158 (1992).
8 Kent, G. S. & McCormick, M P. *Optics News* 14, 11–19 (1988).
9 Pinto, J. R., Turco, R. P. & Toon, O. B. *J. Geophys. Res.* 94, 11165–11174 (1989).
10 Bluth, G. J. S. *Nature* 327, 386 (1993).
11 Stowe, L. L., Carey, R. M. & Pellegrino, P. P. *Geophys. Res. Lett.* 19, 159–162 (1992).
12 DeFoor, T. E., Robinson, E. & Ryan, S. *Geophys. Res. Lett.* 19, 187–190 (1992).
13 Winker, D. M. & Osborn, M. T. *Geophys. Res. Lett.* 19, 167–170 (1992).
14 Long, C. S. & Stowe, L. L. *Geophys. Res. Lett.* 21, 2215–2218 (1994).

15 Grainger, R. G. et al. *Geophys. Res. Lett.* 20, 1283–1286 (1993).

16 Young, R. E., Houben, H. & Toon, O. B. *Geophys. Res. Lett.* 21, 369–372 (1994).

17 Labitzke, K. & McCormick, M. P. *Geophys. Res. Lett.* 19, 207–210 (1992).

18 Grant, W. B. et al. *Geophys. Res. Lett.* 19, 1109–1112 (1992).

19 Kinne, S. et al. *Geophys. Res. Lett.* 19, 1927–1930 (1992).

20 Schoeberl, M. R. et al. *Geophys. Res. Lett.* 20, 29–32 (1993).

21 Brasseur, G. & Granier, C. *Science* 257, 1239–1242 (1992).

22 Yue, G. K., McCormick, M. P. & Chiou, E. -W. *J. Geophys. Res.* 96, 5209–5219 (1991).

23 Trepte, C. R. & Hitchman, M. H. *Nature* 355, 626–628 (1992).

24 Browell, E. V. et al. *Science* 261, 1155–1158 (1993).

25 Winker, D. M. & Osborn, M. T. *EOS* (abstr.) 73, 625 (1992).

26 Trepte, C. R., Veiga, R. E. & McCormick, M. P. *J. Geophys. Res.* 98, 18563–18573 (1993).

27 Osborn, M. T., Winker, D. M., Woods, D. D., DeCoursey, R. J. & Trepte, C. R. *Proc. 17th int. Laser Radar Conf.*, 446–449 (Sendai, Japan, July 25–29, 1994).

28 Hayashida, S. *Global Volcanism Network Bull.* Vol. 6, sectn. 16, 14–16 (Smithsonial Institution, Washington DC, 1991).

29 Jager, H. *Geophys. Res. Lett.* 19, 191–194 (1992).

30 Minnis P. et al. *Science* 259, 1411–1415 (1993).

31 Twomey, S. A., Piepgrass, M. & Wolfe, T. *Tellus* 36B, 356–366 (1984).

32 Sassen, K. *Science* 257, 516–519 (1992).

33 Jensen, E. J. & Toon, O. B. *Geophys. Res. Lett.* 19, 1759–1762 (1992).

34 Lacis, A., Hansen, J. E. & Sato, M. *Geophys. Res. Lett.* 19, 1607–1610 (1992).

35 Russell, P. et al. *J. geophys. Res.* 98, 22969–22987 (1993).

36 Valero, F. P. J. & Pilewskie, P. *Geophys. Res. Lett.* 19, 163–166 (1992).

37 Brock, C. A. et al. *Geophys. Res. Lett.* 20, 2555–2558 (1993).

38 Monastersky, R. *Science News* 145, 70 (1994).

39 Halpert, M. S. et al. *Eos* 74, 433 (September 21, 1993 American Geophysical Union EOS Transactions).

40 Angell, J. *Geophys. Res. Lett.* 17, 1093–1096 (1990).

41 Dutton, E. G. & Christy, J. R. *Geophys. Res. Lett.* 19, 2313–2316 (1992).

42 Hansen, J. E., Lacis, A., Ruedy, R. Sato, M. *Geophys. Res. Lett.* 19, 215–218 (1992).

43 Mankin, W. G., Coffey, M. T. & Goldman, A. *Geophys. Res. Lett.* 19, 179–182 (1992).

44 Wallace, L. & Livingston, W. *Geophys. Res. Lett.* 19, 1209 (1992).

45 Tabazadeh, A. & Turco, R. P. *Science* 260, 1082–1086 (1993).

46 Taubes, G. *Science* 260, 1580–1583 (1993).

47 McCormick, M. P., Steele, H. M., Hamill, P., Chu, W. P. & Swissler, T. J. *J. atmos. Sci.* 39, 1387–1397 (1982).

48 Fahey, D. H. et al. *J. geophys. Res.* 94, 11299–11316 (1989).

49 Kelly, K. K. et al. *J. geophys. Res.* 94, 11317–11358 (1989).

50 Thomason, L. W. & Poole, L. R. *J. geophys. Res.* 98, 23003–23012 (1993).

51 Hoffmann, D. J. & Solomon, S. *J. geophys. Res.* 94, 5029–5041 (1989).

52 Rodriguez, J. M., Ko, M. K. W. & Sze, N. D. *Nature* 352, 134–137 (1991).

53 Kerr, R. A. *Science* 263, 1562 (1994).

54 Pitts, M. C. & Thomason, L. W. *Geophys. Res. Lett.* 20, 2451–2454 (1993).

55 Deshier, T. et al. *Geophys. Res. Lett.* 19, 1819–1822 (1992).

56 Kent, G. S., Trepte, C. R., Farrukh, U. O. & McCormick, M. P. *J. atmos. Sci.* 14, 1536–1551 (1985).

57 Hofmann, D. J., Oitsman, S. J., Lathrop, J. A., Harris, J. M. & Vömel, H. *Geophys. Res. Lett.* 21, 421–424 (1994).

58 Solomon, S., Sanders, R. W., Garcia, R. R. & Keys, J. G. *Nature* 363, 245–248 (1993).

59 Fahey, D. W. et al. *Nature* 363, 509–514 (1993).

60 Johnston, P. V., McKenzie, R. L., Keys, J. G. & Matthews, W. A. *Geophys. Res. Lett.* 19, 211–213 (1992).

61 Koike, M., Kondo, Y., Matthews, W. A., Johnston, P. V. & Yamazaki, K. *Geophys. Res. Lett.* 20, 1975–1978 (1993).

62 Johnston, P. V. & McKenzie R. L. *J. geophys. Res.* 94, 15499–15511 (1989).

63 Kawa, S. R. et al. *Geophys. Res. Lett.* 20, 2507–2510 (1993).

64 Rinsland, C. P. et al. *J. geophys. Res.* 99, 8213–8219 (1994).

65 Koike, M. et al. *Geophys. Res. Lett.* 21, 397–400 (1994).

66 Mills, M. J. et al. *Geophys. Res. Lett.* 20, 1187–1190 (1993).

67 Webster, C. R., May, R. D., Allen, M., Jaeglé, L. & McCormick, M. P. *Geophys. Res. Lett.* 21, 53–56 (1994).

68 Hofmann, D. J. et al. *Geophys. Res. Lett.* 20, 1555–1558 (1993).

69 Hofmann, D. J. et al. *Geophys. Res. Lett.* 21, 65–68 (1993).

70 Bekki, S., Toumi, R. & Pyle, J. A. *Nature* 362, 331–333 (1993).

71 Michelangeli, D. V., Allen, M. & Yung, Y. L. *J. geophys. Res.* 94, 18429–18443 (1989).

72 Gleason, J. F. et al. *Science* 260, 523–526 (1993).

73 Herman, J. R. & Larko, D. *J. geophys. Res.* 99, 3483–3496 (1994).

74 Komhyr, W. D. et al. *Geophys. Res. Lett.* 21, 201–204 (1993).

75 Hofmann, D. J. et al. *Geophys. Res. Lett.* 21, 65–68 (1994).

76 Hanson, D. R., Ravishankara, A. R. & Solomon, S. *J. geophys. Res.* 99, 3615–3629 (1994).

77 Rodriguez, J. M. et al. *Geophys. Res. Lett.* 21, 209–212 (1994).

78 Hervig, M. E., Russell, J. M., Gordley, L. R., Park, J. H. & Drayson, S. R. *Geophys. Res. Lett.* 20, 1291–1294 (1993).

79 Hansen, J. E., Lacis, A., Ruedy, R., Sato, M. & Wilson, H. *Res. Explor.* 9, 143–158 (1993).

8

AIR POLLUTION IN THE WORLD'S MEGACITIES*

United Nations Environment Programme and the World Health Organization

Many of the major cities of the world are beset by environmental problems, not the least of which is deteriorating air quality. Exposure to air pollution is now an almost inescapable part of urban life throughout the world. The available information shows that the air quality guidelines of the World Health Organization (WHO) are regularly being exceeded in many cities – in some cases, to a great extent. Given the rate which these cities are growing and the general absence of pollution control measures in many of them, air pollution will probably worsen, and the quality of life of many urban residents will continue to deteriorate. Although some progress has been made in controlling air pollution in many industrialized countries over the last two decades, air quality – particularly in the larger cities of the developing countries – is worsening. The WHO Commission on Health and Environment, which recently concluded its work, identified urban air pollution as a major environmental health problem deserving high priority for action.[1]

To assess the problems of urban air pollution in the world's largest cities, WHO and the United Nations Environment Programme (UNEP) initiated a detailed study of air quality in 20 "megacities."[2] The study was carried out within the framework of the WHO/UNEP urban air quality monitoring and assessment program known as GEMS/Air, which is a component of the Global Environment Monitoring System. For the purposes of this study, megacities were defined as urban agglomerations with current or projected populations of 10

* Originally published in *Environment*, 1994, vol. 36, no. 2, pp. 4–13, 25–37.
[1] World Health Organization, *Our Planet Our Health: Report of the WHO Commission on Health and Environment* (Geneva, 1992).
[2] World Health Organization and United Nations Environment Programme. *Urban Air Pollution in Megacities of the World* (Oxford, England: Blackwell Publishers, 1992). The draft report was reviewed by an expert group in November 1991. See World Health Organization, *Urban Air Pollution Monitoring. Report of a Meeting of UNEP/WHO Government-Designated Experts, Geneva, 5–8 November 1991.* WHO/PEP/92.2, UNEP/GEMS/92.A.1 (Geneva, 1992).

million or more by the year 2000 (see figure 8.1).[3] Although there are 24 such megacities, a lack of resources and time required that only 20 be studied. Dacca, Lagos, and Tehran were excluded because of a general lack of information and Osaka because of its similarity to Tokyo. The urban agglomerations chosen include three in North America, three in South America, one in Africa, eleven in Asia, and two in Europe: the Metropolitan Area of Buenos Aires in Argentina; Greater São Paulo Area and the Metropolitan Area of Rio de Janeiro in Brazil; Beijing Municipality and Metropolitan Shanghai in China; Greater Cairo in Egypt; the Calcutta Metropolitan District, Delhi, and Greater Bombay in India; Metropolitan Jakarta in Indonesia; Metropolitan Tokyo in Japan; Greater Seoul in Korea; the Metropolitan Area of Mexico City in Mexico; Karachi in Pakistan; Metropolitan Manila in the Philippines; Moscow in Russia; Metropolitan Bangkok in Thailand; Greater London in the United Kingdom; and the Los Angeles Metropolitan Area and New York City Metropolitan Area in the United States.

Between 1970 and 2000, the combined populations of these 20 cities will double, with Manila exhibiting the most rapid growth (4.65 average annual rate of growth from 1970 to 1985) and New York the lowest (−0.24 average annual rate of growth from 1970 to 1985). Currently, the most rapidly growing megacities are Lagos and Dacca, whose populations from 1990 to 2000 are expected to grow by 64 and 76 percent, respectively.[4] These megacities are not necessarily the world's most polluted cities. The primary reasons for singling out the megacities are that they all have serious air pollution problems; they encompass large land areas and many people (the total population of the 20 megacities in 1990 was estimated to be 234 million); and many other cities are heading for megacity status. This

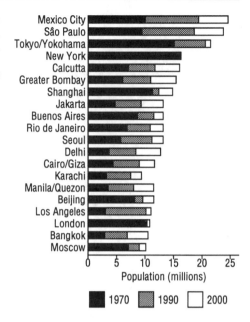

Figure 8.1 [*orig. figure 1*] Estimated populations of 20 megacities in 1970 and 1990 and their projected populations in 2000.

last point is of particular importance. A review of air pollution problems in the megacities and identification of their difficulties in finding solutions can serve as a warning of the problems facing other rapidly growing urban areas and as a guide for solving and preventing some of them.

The compilation of a global assessment of air pollution in megacities is difficult. Data sets on the pollutants and their health and environmental impacts are often non-existent, incomplete or out of date. There are also differences in methodology and reporting between countries and even within countries and cities. Shortcomings in the data used here, including problems of representativeness and comparability, are noted where appropriate. Nevertheless, these data and analyses provide the first valid and comprehensive overview of the air

[3] United Nations, *Prospects of World Urbanization 1988.* Population Studies no. 112 (New York, 1989). The cities designated as megacities here do not include some of the world's largest cities, such as Paris and Milan, whose low birth rates will probably prevent their total populations from surpassing 10 million by 2000.
[4] Ibid.

pollution situation and trends in the megacities.

Understanding Urban Air Pollution

The air pollution problems of the megacities differ greatly and are influenced by a number of factors, including topography, demography, meteorology, and the level and rate of industrialization and socioeconomic development. These problems are also increasingly important because, as the populations of urban areas increase, the number of people exposed to urban air pollution will also increase. United Nations' estimates indicate that by the year 2000, 47 percent of the global population will be living in urban areas.[5] In 1990, 69 cities had populations of 3 million or more, and by 2000, 85 cities will probably be in this category.[6]

Sources of urban air pollution

Growing urban populations and levels of industrialization inevitably lead to greater energy demand, which usually causes more pollutant emissions. The combustion of fossil fuels for domestic heating, for power generation, in motor vehicles, in industrial processes, and in disposal of solid wastes by incineration is the principal source of air pollutant emissions in urban areas. The most common air pollutants in urban environments are sulfur dioxide (SO_2), the nitrogen oxides (NO and NO_2, collectively represented as NO_x), carbon monoxide (CO), ozone (O_3), suspended particulate matter (SPM), and lead.

Sulfur dioxide contributes to acid rain and sulfate aerosol particulate pollution. Oxides of nitrogen contribute to particulate pollution and acid deposition and are precursors of ozone, which is the main constituent of photochemical smog. Particulate pollution also includes smoke and dust. Ozone, CO, SPM, and lead all have proven adverse health effects.

Combustion of fossil fuels in stationary sources leads to the production of SO_2, NO_x,

and particulates – both primary particulates in the form of fly ash and soot and secondary particulates, such as sulfate and nitrate aerosols formed in the atmosphere following gas-to-particle conversion. The domestic combustion of solid fuels, mainly coal and wood, also represents a significant source of these pollutants in some cities, particularly those in the developing world. Gasoline-powered motor vehicles are the principal sources of NO_x, CO, and lead, whereas diesel-fueled engines emit significant quantities of particulates and SO_2 in addition to NO_x.

Ozone, a photochemical oxidant, is not emitted directly from combustion sources but is formed in the lower atmosphere from NO_x and volatile organic compounds (VOCs), or reactive hydrocarbons, in the presence of sunlight. VOCs are emitted from a variety of manmade sources, including road traffic; production and use of organic chemicals such as solvents; transport and use of crude oil; use and distribution of natural gas; and, to a lesser extent, waste disposal sites and wastewater treatment plants.

Although detailed emissions inventories are not widely available for individual cities, on the basis of observed trends in national emissions and recent increases in vehicle registrations, it may be concluded that motor vehicles now constitute the main source of air pollutants in the majority of cities in industrialized countries. This is particularly true of the pollutants CO, NO_x, and, to a lesser extent, SPM.

The cities of developing countries, in contrast, exhibit a greater variety of air pollution sources. The relative contributions of mobile and stationary sources to air pollutant emissions differ markedly among cities and depend on the level of motorization and the level, density, and type of industry present. Cities in Latin America, for example, tend to have higher vehicle densities than those in other developing regions and therefore are likely to experience a higher contribution from motor vehicles to

[5] Ibid.
[6] Ibid.

the total urban pollution load. Motor vehicle contributions are less important in cities with lower levels of motorization, such as those in Africa, and in cities located in temperate regions that are dependent on coal or biomass fuels for space heating and other domestic purposes, such as cities in China and parts of Eastern Europe. It is also worth noting that, in many developing countries, vehicle fleets tend to be older and poorly maintained – a factor that increases the significance of motor vehicles as a pollutant source.

At present, the vehicle fleet of the world is concentrated in the high-income economies of the world. In 1988, the Organization for Economic Cooperation and Development (OECD) countries alone accounted for 80 percent of the world's cars, 70 percent of trucks and buses, and more than 50 percent of two- and three-wheeled vehicles. Since 1950, the global vehicle fleet has grown tenfold and is expected to double within the next 20 or 30 years from the present total of 630 million vehicles.[7] The rate of growth of the world's vehicle fleet is projected to surpass that of both the total and urban population (see figure 8.2).

Much of the expected growth in vehicle numbers is likely to occur in developing countries and in Eastern Europe. In contrast, much of the demand for motor vehicles in the developed countries will be for vehicle replacement. The contribution of motor vehicles to the pollution load is thus set to increase in the developing countries. In the absence of stringent control measures for traffic-related pollutants the air quality will undoubtedly deteriorate in these regions.

In addition to the more common, or "traditional," air pollutants, a large number of toxic and carcinogenic chemicals are increasingly being detected in urban air, albeit at low concentrations. Examples include selected heavy metals, such as beryllium, cadmium, and mercury; trace organics, such as benzene, polychlorodibenzo-dioxins and -furans, for-

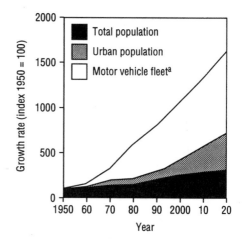

Figure 8.2 [*orig. figure* 2] Estimated and projected rates of increase of the total population, the urban population, and motor vehicle numbers from 1950 to 2020.

maldehyde, vinylchloride, and polyaromatic hydrocarbons; radionuclides, such as radon; and fibers, such as asbestos. Such chemicals are emitted from a wide range of sources, including waste incinerators, sewage-treatment plants, industrial and manufacturing processes, dry-cleaning establishments, building materials, and motor vehicles. Although emissions of these chemicals are generally low compared with those of the traditional pollutants, these pollutants may pose significant risks to health in view of their extremely high toxicity or carcinogenicity or both. Because the measurement of low concentrations of toxic chemical contaminants in air presents analytical difficulties, very little monitoring, let alone routine monitoring, is currently conducted.

Dispersion and transport

The two major influences on the transport and dispersion of air pollutant emissions are the meteorology of the city and the topography of the area in relation to the population.

The climates of the megacities range from

---

[7] A. Faiz, K. Sinha, M. Walsh, and A. Valma, *Automotive Air Pollution: Issues and Options for Developing Countries*, WPS 492 (Washington, D.C.: World Bank, 1990).

cooler humid continental (with severe cold seasons, as in Beijing) to desert (Cairo) or tropical (with high temperatures and humidity and no cold season, as in Bangkok). The severity of the cold season determines the amount of residential heating that is required and thus the increased emissions of, for example, SO_2 in winter (see box 8.1). In cities with a moderate climate, the pollution load tends to be distributed more evenly over the year.

Box 8.1   Beijing: Winter Blues

Beijing, the capital city of China, sits between two rivers on the northwestern border of the Greater North China Plain. Its population was 8.3 million in 1970 and 9.3 million in 1985. In 1990, the population density of the central area was thought to be 27,600 people per square kilometer. By 2000, Beijing is projected to have 11.47 million inhabitants.(1) Winters in Beijing are cold: The mean monthly temperature for January is 4.6°C, and minimum temperatures below −20°C are not uncommon. Summers are hot and humid. The mean monthly maximum of 26.1°C is reached in July. A maximum hourly temperature of 40°C has been recorded. These temperature extremes, combined with the city's heavy reliance on coal for heat and power, produce a large difference between Beijing's winter and summer air pollution levels – particularly levels of sulfur dioxide (SO_2) and suspended particulate matter (SPM), as shown in figure 8.3.

More than 5,700 industrial enterprises operate in Beijing, including 24 power plants, 28 non-ferrous and 25 ferrous metal smelters, 18 coking plants, 194 chemical plants, and 483 metal products factories. Virtually all of the city's electrical power generation is based on coal combustion,(2) and coal is the source of 70 percent of all energy used in Beijing.(3) The municipal government has attempted to control industrial pollution by relocating certain industries, such as electroplating, away from the central area, and new enterprises are actively discouraged from entering the city. The major iron and steel complex of Shijing, 24 kilometers west of the city, has both blast and electric furnaces and is a major source of industrial emissions.

Annual coal consumption in Beijing is 21 million tonnes. About 30 percent of the coal is used in industry, 27 percent for coking, 23 percent for residential heating, 14 percent for electricity generation, and 6 percent for other uses. Beijing coal has a fairly low sulfur content. Coal for domestic combustion is 0.3 percent sulfur, and coal for small-scale industry and commercial establishments is from 0.5 to 1.0 percent sulfur.

Since 1981, SPM and SO_2 measurements have been made daily at four Global Environment Monitoring Systems (GEMS) stations operated by the Ministry of Health. In 1990, an emissions estimate for SO_2 was prepared for Beijing by the World Bank.(4) Despite the coal's low sulfur content, SO_2 emissions totalled about 526,000 tonnes in 1985. However, the estimates are not comprehensive as they do not consider some emission sources, such as motor vehicles, heating boilers, and commercial stoves. Since 1985, growing industrialization has certainly caused increased emissions.

Industrial sources account for 87 percent of SO_2 emissions. Of these, more than half come from just 24 power plants and 18 coking facilities. Household stoves account for yearly emissions of 73,000 tonnes of SO_2 (13 percent of the total), but their relative contribution is certainly much higher in the cold winters. Measures to reduce further the sulfur content of coal and to promote the production of briquettes for more efficient burning of domestic coal have been initiated.

Annual mean SO_2 concentrations remained relatively constant from 1985 to 1989. Monitoring stations in the city center measure annual means of more than 100 micrograms per cubic meter of air ($\mu g/m^3$) until 1989 – far above the annual mean guideline values of 40 to 60 µg/m³ established by the World Health Organization (WHO). Annual means measured at the suburban stations meet the WHO annual guidelines. The WHO 98-percentile guideline of 150µg/m³, not to be exceeded on more than 2 percent of the days, is exceeded especially at the central stations, which have measured daily means above 500µg/m³ on 2 percent of the days. Figure 8.3 shows the typical seasonal variation of SO_2. The highest SO_2 pollution – up to 250µg/m³ – occurs between November and March, when large amounts of coal are burned for domestic heating. Despite nearby industrial sources, summer SO_2 levels are very low. Maximum hourly concentrations of SO_2 occur between 7:00 and 8:00 a.m. and between 6:30 and 8:00 p.m., the domestic cooking periods.

Levels of SPM also vary between the seasons. Although Beijing is subject throughout the year to severe dust fallout from dust storms originating in the western plains, these storms tend to be especially severe in the late winter and early spring. Yet, whereas natural sources account for 60 percent of SPM emissions during summer, they account for only 40 percent during winter.(5) This smaller percentage share is due to the overwhelming amount of anthropogenic particulates emitted during the winter.

Anthropogenic SPM emissions have been estimated by the World Bank to be about 116,000 tonnes per year.(6) However, this inventory did not consider emissions from motor vehicles, power plants, or cooking. Thus, the figures must be considered to be lower than the actual situation. The inventory shows that industrial stationary sources account for 59 percent of anthropogenic SPM emissions. Heating boilers and household stoves, which contribute 34 percent of all SPM emissions, are especially important sources of emissions during the winter. The contribution of these sources to human exposure is particularly great because of the low emission heights (chimneys) and high emission densities (the local coal has a very high ash content, which leads to inefficient burning and dense SPM emissions).

Annual mean SPM levels measured at the GEMS monitoring sites exceed the WHO annual mean guideline of 60 to 90 µg/m³. Even if one considered one-half of the total to be wind-blown natural dust, the particulates from coal combustion alone would be above the WHO guideline value. The SPM problem of Beijing is also reflected in the daily values. The WHO 98-percentile guideline of 230 µg/m³ (not to be exceeded

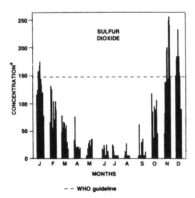
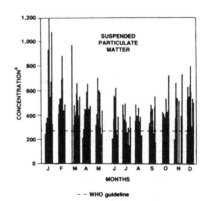

[a] Micrograms per cubic meter of air.
Note: Sixteen days per month sampling frame. Zero levels indicate days when no measurements were taken.
Source: Data from the Global Environment Monitoring Sytem.

Figure 8.3   Daily concentrations of sulfur dioxide and suspended particulate matter in Beijing in 1989.

on 2 percent of the days) was exceeded on almost all of the days in 1989 at one site. The highest daily measurements were above 1,000 µg/m³.

The seasonal variation in SPM concentrations is not quite as evident as that of SO₂ because of the natural dust fraction. However, daily mean concentrations are significantly higher during the winter. It is clear that the SPM pollution loads in winter represent a severe health risk for susceptible people, especially when combined with high SO₂ concentrations.

Because motor vehicles are relatively few in Beijing, carbon monoxide (CO) should be a minor problem.(7) However, domestic coal combustion could endanger quite high ambient CO levels in winter because of the low emission heights and high emission densities. A study of personal CO exposures of non-smokers in Beijing reported that seasonal mean outdoor exposures were of the order of 1.9 milligrams per cubic meter of air (mg/m³) in summer and 5 mg/m³ in winter.(8) Personal indoor exposures, however, averaged 2.8 mg/m³ in summer and 17.9 mg/m³ in winter. These data reflect the relative absence of traffic-generated CO in Beijing and the high indoor exposures from domestic cooking with coal briquettes and other fuels. In fact, indoor CO levels as high as 104 mg/m³ have been measured during cooking periods. The WHO guidelines for CO exposure are just 10 mg/m³ over 8 hours and 30 mg/m³ over any 1 hour.

Clearly, many Beijing residents are exposed daily to high concentrations of SO₂, SPM and CO during the long, cold winters. Respirable acid aerosols from domestic coal combustion in winter also constitute a risk to human health. However, because from 40 to 60 percent of SPM is wind-blown dust, measures to control anthropogenic SPM emissions have only limited effects.

A specific pollution problem results from the many small domestic combustion sources. Although the total amount of emissions from these small sources is moderate, the sources contribute a high proportion of the SO₂, SPM and CO in ambient air because household stacks are low and pollutant dispersion is limited. For instance, although household stoves contribute just 13 percent of total SO₂ emissions, these emissions represent about 38 percent of the ambient SO₂ concentrations.(9)

In 1988, the Beijing Municipal Government passed regulations to implement the 1987 Air Pollution Prevention Act, which requires industries to monitor and report their own emissions. The Beijing government has also set stricter emission standards than the national standards. Emission sources are given deadlines to meet these standards. If they are unable to do so, they must either shut down or move away from the area. As of 1988, only a few emission sources had problems with compliance.

Emission control programs recently put into effect:

- increase the supply of coal gas and natural gas for industrial use;
- convert urban residential fuel from coal to liquified petroleum gas and natural gas;
- require residential sources still using coal to burn briquettes and shaped coal to reduce emissions;
- develop central heating plants to replace smaller boilers with a single large installation with emission controls;
- modify existing boilers (where the location of industry makes central heating impractical) to reduce emissions by automatic feeding and ash removal (all boilers with a capacity greater than one tonne must install scrubbers); and
- pave dirt roads and plant trees, flowers, and grass to reduce wind-blown dust.

Efforts to control Beijing's air pollution are limited by the huge capital investments necessary for control technologies. In spite of the use of environmental impact assessments for new projects, it is anticipated that emissions will continue to increase and that there will be difficulty in meeting the air quality standards.

1 United Nations, *Prospects of World Urbanization 1988,* Population Studies no. 112 (New York, 1989).
2 A. Krupnick and I. Sebastian, *Issues in Urban Air Pollution: Review of the Beijing Case,* Environment Working Paper no. 31 (Washington D.C.: World Bank, 1985).
3 Ibid.
4 Krupnick and Sebastian, note 2 above.
5 D. Zhao and B. Sun, "Atmospheric Pollution from Coal Combustion in China," *Journal of the Air Pollution Control Association* 36 (1986): 371–74.
6 Krupnick and Sebastian, note 2 above.
7 Ibid.
8 World Health Organization, *Human Exposure to Carbon Monoxide and Suspended Particulate Matter in Beijing, People's Republic of China,* GEMS: PEP/85.11 (Geneva, 1985).
9 Krupnick and Sebastian, note 2 above.

Thermal inversions are another particular problem for cities in temperate and cold climates. Under normal dispersive conditions, hot pollutant gases rise as they come into contact with the colder air masses at high altitudes. However, under certain circumstances, the air temperature may increase with altitude, and an inversion layer forms a few tens or hundreds of meters above the ground. This inversion layer may then trap pollutants close to the emission sources and act as a heat cover, prolonging the inversion. These conditions are of greatest concern when wind speeds are low.

Isothermal conditions – that is, when there is no change in temperature with altitude – may have a similar effect.

Another meteorological phenomenon that greatly influences air quality is the "urban heat island." The heat generated by a city causes the air to rise, drawing in colder and possibly more polluted air from surrounding industrial areas. On the other hand, cities in warm, sunny locations with high traffic densities tend to be especially prone to the net formation of O_3 and other photochemical oxidants from precursor emissions (see box 8.2).

Box 8.2   Los Angeles: Traffic and Smog
Los Angeles, California, and its environs constitute the second largest population center in the United States. From just over 4 million in 1950, the city's population grew to an estimated 10.47 million in 1990, and the projected population in 2000 is 10.91 million.(1) By 2010, the population could reach 15.7 million.(2)

The Los Angeles basin has a Mediterranean climate. The surrounding mountains, subsidence inversions, and high solar intensity produce ideal conditions for the atmospheric stagnation conducive to pollutant reactions and buildup. During periods of stagnating high pressure, the air circulation pattern, which takes pollutants out to sea at night and returns them to land during the day, allows air pollutants to build up in the airshed until the passage of a new weather front.

There is little primary heavy industry remaining in Los Angeles because steel, tire, and car assembly plants have left the area. A refinery area, an iron and steel plant, and secondary lead smelters inland have been strictly controlled since the 1970s by the Los Angeles Air Pollution Control District (LAAPCD) and its successor, the South Coast Air Quality Management District (SCAQMD). Although some power plants in the basin burn natural gas, most of the electric power is provided by hydroelectric plants. Nevertheless,

Los Angeles suffers the worst air pollution in the United States, and the primary source is cars (see figure 8.4).

Los Angeles developed with almost no public transportation network. Consequently, the residents must rely on motor vehicles for almost all transportation. With 8 million vehicles, the basin has possibly the greatest number of vehicles per person (0.67) in the world. If the population increases by 5 million by 2010, vehicle-miles traveled will rise 68 percent and 40 percent more trips will be taken.(3)

Cars and other motor vehicles are the primary source of the city's infamous smog – primarily ozone (O_3) formed by the photochemical reaction of two motor vehicle exhaust emissions: oxides of nitrogen (NO_x) and volatile organic compounds (VOCs). When LAAPCD began enforcing air pollutant emission controls in 1947, hourly O_3 concentrations exceeding 1,200 micrograms per cubic meter of air ($\mu g/m^3$) were reported. In the 1960s, O_3 concentrations frequently exceeded 1,000 $\mu g/m^3$. In the 1970s, the state government began instituting vehicle emission standards that were stricter than those promulgated by the U.S. Environmental Protection Agency (EPA) for the nation at large. Despite the strict controls, the maximum O_3 concentration from 1986 to 1991 decreased only to 700 $\mu g/m^3$, mostly because of an 81-percent population increase from 1960 to 1990 and allied increases in vehicular traffic.(4) This enormous population increase resulted in many motorists commuting from 95 to 130 kilometers each way between their work places and secure, affordable, single-family housing outside the central area.

SCAQMD was formed in 1976 when it was recognized that the smog extended well beyond the borders of Los Angeles into neighboring counties. It now monitors air quality at 37 stations distributed throughout the basin. SCAQMD's data show that nitrogen dioxide (NO_2) emissions have been significantly reduced.(5) However, estimated total emissions of NO_x in 1987 were more than 440,000 tonnes. Seventy-six percent of this total is attributable to mobile sources, and 55 percent to "on-road" mobile sources.(6)

Nitrogen dioxide is the reddish-brown gas that produces the distinctive color of the smog haze over Los Angeles. According to EPA, Los Angeles is the only city in the United States that has failed to attain the federal air quality standards for NO_2.(7) The one-hour guideline of 400 $\mu g/m^3$ established by the World Health Organization was exceeded in 1990 at 8 of the 24 stations reporting NO_2 data. The highest reported hourly value was 526 $\mu g/m^3$. Annual mean concentrations in 1990 ranged from 39 $\mu g/m^3$ to 104 $\mu g/m^3$.(8) Projected estimates for the basin show that emissions should decrease until 2000, but, after that, they will stabilize or even increase, depending on the level of population growth.

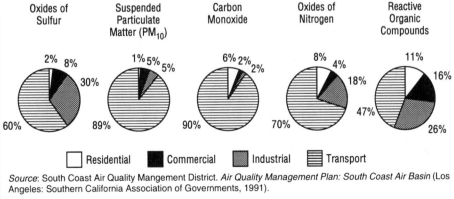

Source: South Coast Air Quality Mangement District. *Air Quality Management Plan: South Coast Air Basin* (Los Angeles: Southern California Association of Governments, 1991).

Figure 8.4   Percentage contribution of anthropogenic emissions sources in 1987.

The NO_x emission reductions proposed for O_3 control in SCAQMD's three-tier pollution reduction plan were designed to decrease NO_2 concentrations, and it is estimated that the air quality standards for NO_x will be met by the year 2000.(9) However, because a U.S. National Research Council report has concluded that reading NO_x emissions is critical to reading O_3,(10) the proposed emission controls for NO_x will probably be tightened in the future.

Los Angeles has the most serious O_3 problem in the United States.(11) In 1990, the maximum hourly O_3 concentration was 660 µg/m³. At one site, the federal standard for O_3 was exceeded on 103 days. The state standard was exceeded at all monitoring stations in 1990.(12)

Because ozone is a secondary pollutant, it can only be reduced by controlling its precursors: NO_x and VOCs. Thus, increasingly strict emissions controls have been enforced on NO_x and hydrocarbons.(13) In the past, emissions of VOCs (more than 500,000 tonnes in 1987) and NO_x have been controlled in the basin by strict industrial and motor vehicle standards set primarily to control O_3. SCAQMD's three-tier plan for organic compounds and NO_x is designed to meet the California and federal air quality standards for O_3 by 2010. These emissions controls are to be applied to industry, motor vehicles, vehicle fuel and consumer product formulations, and consumer life-style activities. Some amendments may be necessary to comply with the new requirements of the federal Clean Air Act of 1990. The emission controls of the three-tier plan represent the most severe air quality management requirements ever proposed for any city. Los Angeles is also seriously pursuing an economic incentive program, which uses marketable permits to supplement or substitute for some of the more stringent control requirements.

Of all the megacities, only Mexico City has higher O_3 levels. The stringent emission control actions taken by SCAQMD are necessary to prevent the continuation of the air pollution's deleterious health effects, which have been documented in several studies. These control activities will continue to place great constraints on the living patterns of Los Angeles residents, and the even stricter controls planned for the coming decade will affect everyone living there.

1 United Nations, *Prospects of World Urbanization 1988,* Population Studies no. 112 (New York, 1989).
2 Southern California Association of Governments, South Coast Air Quality Management District, *Air Quality Management Plan: South Coast Air Basin* (Los Angeles, 1990).
3 Southern California Association of Governments, *Regional Mobility Plan* (Los Angeles, 1989).
4 Southern California Association of Governments, South Coast Air Quality Management District. *Air Quality Management Plan: South Coast Air Basin* (Los Angeles, 1991).
5 U.S. Environmental Protection Agency, Office of Air Quality Planning and Standards, *National Air Quality and Emissions Trends Report* (Research Triangle Park, N.C., 1991).
6 Southern California Association of Governments, note 4 above.
7 U.S. Environmental Protection Agency, note 5 above.
8 Southern California Association of Governments, note 4 above.
9 Southern California Association of Governments, note 2 above.
10 U.S. National Research Council, *Rethinking the Ozone Problem* (Washington, D.C.: National Academy Press, 1991).
11 U.S. Environmental Protection Agency, note 5 above.
12 Southern California Association of Governments, note 4 above.
13 Southern California Association of Governments, note 2 above; and A. C. Lloyd, J. M. Lents, C. Green, and P. Nemeth, "Air Quality Management in Los Angeles: Perspectives on Past and Future Emission Control," *Journal of Air Pollution Control Association* 39 (1989): 696–703.

The topography of the megacity also influences the manner in which pollutants are transported and dispersed. The megacities fall into the following three groupings:

- Beijing, Cairo, Delhi, and Moscow have relatively level topography and their climates are not influenced by a body of water.
- Bangkok, Bombay, Buenos Aires, Calcutta, Jakarta, Karachi, London, Manila, New York, Shanghai, and Tokyo have relatively level topography and their climates are influenced by a body of water.
- Los Angeles, Mexico City, Rio de Janeiro, São Paulo, and Seoul have variable topography and their climates are influenced by surrounding mountains.

The presence of a significant body of water can lead to microclimatological effects and to on-shore and off-shore diurnal wind patterns. Hills that surround cities often act as a downwind barrier, trapping pollution close to the city. When the city is surrounded by high mountains, like Los Angeles and Mexico City, the pollutants may be trapped within the airshed for several days (see box 8.3). Internal topographical features, such as mountain ridges, can also serve as barriers to air pollutant transport within the megacity.

Box 8.3   Mexico City: A Topographical Error

The population of Mexico City was estimated to be 19.37 million in 1990, more than one-fifth of Mexico's total population.(1) Population densities in the city range from almost 7,000 persons per square kilometer in the center to 500 persons per square kilometer in the outskirts. Projections indicate that the population will grow at a rate of 1.4 percent per year to 24.44 million by 2000.

The city sits in the Mexican Basin at a mean altitude of 2,240 meters. The basin is surrounded by mountains, two of which are more than 5,000 meters high. Two valley channels, located to the northeast and northwest, funnel air to the center and to the southwest of the city. Because of this topography and the light winds, ventilation is poor, and surface as well as upper air temperature inversions occur frequently. During winter, inversions occur up to 25 days per month. The expansion of the urban area and consumption of energy have markedly modified the valley's microclimate. The heat island effect is also pronounced in Mexico City, and hot spots can be up to 12°C above the temperature in suburban and rural areas.(2) These climatological and topographical factors trap pollutant emissions close to the city.

More than 30,000 industries – of all types and sizes – and 12,000 service facilities are located in the valley. Among the industries, 4,000 use combustion or transformation processes generating major atmospheric emissions. Some 2.5 million motor vehicles – buses, minibuses, taxis, trucks, vans, and private cars – are responsible for 44 percent of the total energy consumption in the city.(3) Motor vehicles are by far the main pollution source, as they burn 40 thousand barrels of diesel fuel and 1 million barrels of leaded gasoline each day.

Table 8.1 shows the pollutant emissions of various sources in the valley in 1989. Sulfur emissions have probably declined since then, however, because in 1991, the two power plants in the area switched to natural gas, and the PEMEX refinery was closed on environmental grounds.

Air pollution monitoring in Mexico city started in the 1950s and became more systematic in the late 1960s.(4) In 1966, with the support of the Pan American Health Organization, 14 monitoring stations were installed to measure smoke, suspended particulate matter (SPM), and sulfur dioxide (SO_2). In the early 1970s, the Mexican authorities and the United Nations Environment Programme developed a program to improve environmental quality in several cities. One component of the program was to install a manual network of 22 stations for monitoring SO_2 and SPM, which was completed by mid-1976.(5) Finally, with technical assistance from the U.S. Environmental Protection Agency, an automatic monitoring network became operational in 1985. The network, known as the Red Automatica, covers the greater part of the valley, and its stations measure SO_2, carbon monoxide (CO), ozone (O_3), oxides of nitrogen (NO_x), and non-methane hydrocarbons (HCNM). However, only 5 stations – one in each zone in addition to one at the center – are equipped with a complete set of monitors to measure the five pollutants. A manual network of 16 stations for SO_2 and SPM is still fully operational as well.

In the late 1960s and early 1970s, annual mean SO_2 levels were in the range of 40 to 190 micrograms per cubic meter of air ($\mu g/m^3$), and daily maxima occasionally went as high as 900 $\mu g/m^3$.(6) From 1986 to 1991, five sites of the Red Automatica measured SO_2 levels from 50 to 160 $\mu g/m^3$ – well above the guideline values set by the World Health Organization (WHO). Mean levels of SO_2 in Mexico City range from 80 to 200 $\mu g/m^3$, and daily maxima are between 200 and 550 $\mu g/m^3$. However, no clearcut trends in the data are visible from 1986 to 1991. For these five years, SO_2 levels were still comparable to those of the previous 20 years.

In the late 1960s and early 1970s, mean SPM levels were generally in the 60 to 150 $\mu g/m^3$ range.(7) The corresponding maximum daily values varied between 200 and 1,000 $\mu g/m^3$. From 1976 to 1991, average SPM levels increased slightly. The current concentration range is between 100 and 500 $\mu g/m^3$. The national air quality standards and the WHO guidelines for SPM are frequently exceeded in Mexico City.

Lead emissions are not quantified in table 8.2; however, it can be assumed that gasoline-powered motor vehicles are the major source of ambient lead in Mexico City. The lowering of the maximum lead content of gasoline from 3 milliliters of tetraethyl-lead per gallon to less than 1 milliliter per gallon has undoubtedly decreased lead emissions in recent years. After sharply rising during the 1970s, ambient lead concentrations have generally decreased since 1982.(8) Although lead levels frequently meet the national quarterly evaluation standard of 1.5 $\mu g/m^3$, they generally fail to meet the WHO guidelines.

WHO guidelines and national air quality standards for CO are also exceeded at several monitoring sites.(9) A recent exposure assessment study confirmed that CO levels measured at fixed-site stations even underestimate the actual exposures of pedestrians, street sellers, and commuters.(10) Carbon monoxide levels peak on weekdays between 7:00 and 9:00 a.m., when low temperatures, atmospheric stability (inversions), and heavy vehicular traffic occur simultaneously. There is another peak in the evening, but it is usually lower than the morning one.

Because of its altitude, the city has a relatively low oxygen content in its atmosphere. This phenomenon causes CO emissions to increase because of incomplete fuel combustion, and it exacerbates CO's health effects, especially among children, pregnant women, and asthmatics.

The maximum hourly nitrogen dioxide (NO_2) levels seem to have been more or less stationary from 1986 to 1991, and there is no clearcut trend for either annual mean levels or maximum hourly levels.(11) Mean NO_2 range from 113 to 207 $\mu g/m^3$, while hourly extremes range from 301 to 714 $\mu g/m^3$. The latter are often above the WHO guideline of 400 $\mu g/m^3$ and the national air quality standard of 395 $\mu g/m^3$. Nevertheless, safe levels of NO_2 were not exceeded on more than 5 percent of the days in any one year from 1986 to 1991.

Given its topography, climate, and relatively high NO_x and hydrocarbon emissions, Mexico City is almost the ideal place to generate O_3. Ozone levels in the city are exceptionally high. From 1986 to 1990, annual mean O_3 fluctuated around 200 $\mu g/m^3$, with lows around 100 $\mu g/m^3$ and highs between 300 and 400 $\mu g/m^3$.(12) The worst situation occurs in the southwest sector. Hourly O_3 levels often reach 600 $\mu g/m^3$ there, and extreme values have been measured at up to 900 $\mu g/m^3$. The national air quality standard is often exceeded from 80 to 100 hours per month.

Table 8.1   Emissions inventory for the metropolitan area of Mexico City.

Sector	Sulfur dioxide	Suspended particulate matter	Carbon monoxide	Oxides of nitrogen	Non-methane hydrocarbons
	(thousand tonnes per year)				
Energy					
PEMEX[a]	14.7	1.1	52.6	3.2	31.7
Power plants[b]	58.2	3.5	0.5	6.6	0.1
Industry					
Industry	65.7	10.2	15.8	28.8	39.9
Services	22.0	2.4	0.4	3.9	0.1
Transport					
Private cars	3.5	4.4	1,328.1	41.9	141.0
Taxis	0.8	1.0	301.1	9.5	31.9
Combis and minibuses	0.8	1.0	404.4	10.0	42.7
Urban buses	5.2	0.2	6.2	8.0	2.4
Suburban buses	13.0	0.6	12.6	18.2	5.3
Gasoline trucks	0.9	1.1	779.5	16.9	67.8
Diesel trucks	20.0	0.9	16.5	26.1	7.2
Other	0.2	0.1	5.0	2.7	1.6
Environmental degradation					
Erosion	0.0	419.4	0.0	0.0	0.0
Forest fires, etc.	0.1	4.2	27.3	0.9	199.7
Total[c]	**205.7**	**450.6**	**2,950.6**	**177.3**	**572.1**

[a] Closed in 1991.
[b] Switched to natural gas in 1991.
[c] Because emission values are rounded to the nearest 100 tonnes, the sums of columns may not be the same as the total.
Source: Gobierno de la Republica, *Programa Integral Contra in Contaminacion Atmosferica* (Mexico DF, 1990).

Mexico City has a contingency program to cope with prolonged episodes of high pollution levels. During an episode, certain actions are taken to try to reduce emission of pollutants. These measures include reducing the activity of highly polluting industries and restricting vehicle circulation. In extreme cases, primary schools are temporarily closed to prevent potential damage to children's health.

In addition to switching its power plants to run on natural gas and closing the PEMEX refinery, the city has lowered the sulfur content of its fuel oil and diesel fuel; retrofitted buses, trucks, and vans to burn natural gas; limited car CO emissions to 2.11 grams per kilometer; and restricted commuter traffic.(13) Nevertheless, the air quality is deteriorating so rapidly that the city must act urgently on two fronts. First, the economic foundation of the city must be transformed – with nonpolluting activities replacing the old industries – following a comprehensive strategy that halts pollution growth through the development of better technologies. Second, the city must mandate emissions controls, protection of wooded areas, and the use of better fuels.

In his inauguration speech in December 1988, President Carlos Salinas de Gortari said: "I am giving precise, urgent and imperative instructions to the Mayor of the Federal District to act immediately and efficiently to promote community participation in the fight against pollution." In response to that instruction, a Comprehensive Pollution Control Programme for the Mexico City Metropolitan Zone was developed.(14) The priority areas for action in this strategic plan are:

- the oil industry's refinery activities and the distribution and reformulation of fuels;
- transportation, especially energy efficiency and pollution emission control;
- private industry and service facilities, which need improved technology as well as energy efficiency and pollution emission control;
- thermoelectric power plants, which, as the largest consumers of fuel in the city, must continue to use clean fuels;
- reforestation and ecological restoration of the deforested soils, improvement of areas without sewerage, creation of ecological reserves, and transformation of open air garbage dumps; and
- research, ecological education, and social communication by the responsible agencies.

These actions will not enable the city to recover the air quality it had half a century ago, when just 1.5 million people lived there, but it should help. The program will require a great effort from the government and society. It is a medium-term program – just for the 1990s – but it requires a strategic, common effort that must be sustained for decades.

1 United Nations, *Prospects of World Urbanization 1988*. Population Studies no. 112 (New York, 1989).
2 Departamento del Distrito Federal, *Comprehensive Pollution Control Programme for the Mexico City Metropolitan Zone* (Mexico DF, 1991).
3 Departamento del Distrito Federal and EDOMEX, *Programa Integral del Transporte* (Mexico DF: Departamento del Distrito Federal and Gobierno del Estado de Mexico, 1989); and Departamento del Distrito Federal, Programa de Emergencia, Proyectos, *Programa Integral Contra la Contaminacion Atmosferica* (Mexico DF, 1989).
4 Pan American Health Organization, *Red Pan Americana de Muestro de la Contaminacion del Aire: Informe Final 1967–1980*, CEPIS 23 (Lima, 1982).
5 M. E. Marquez, *Informe Sobre la Calidad del Aire en Algunas Ciudades del Pais* (Mexico DF: Proyecto Mexico PNUD, Subsecretaria de Mejoramiento del Ambiente, 1977).
6 Pan American Health Organization, note 4 above.
7 V. Fuentes-Gea and A. Garcia-Gutierrez, "Air Pollution Trends in Mexico City: TSP (1976–1984)" (paper presented at the 83rd Annual Meeting of the Air and Waste Management Association, Pittsburgh, Pa., 1990).
8 R. Gonzalez-Garcia, Secretaria de Desanrollo Urbano y Ecologia, Mexico, DF, personal communication, 1991.
9 Ibid.
10 X. Fernandez-Bremauntz, "Commuters' Exposure to Carbon Monoxide in the MAMC" (Ph.D. diss., Imperial College, University of London, London, 1991).
11 Gonzalez-Garcia, note 8 above.
12 Departamento del Distrito Federal, Programa de Emergencia, Proyectos, note 3 above.
13 Departamento del Distrito Federal, *La Contaminacion Atmosferica en la Zona Metropolitana de la Ciudad de Mexico y las Medidas Aplicadas para su Control* (Mexico DF, 1987); Gobierno de la Republica, *Programa Integral Contra la Contaminacion Atmosferica* (Mexico DF, 1990); and Gonzalez-Garcia, note 8 above.
14 Departamento del Distrito Federal, note 2 above.

On a more local scale, buildings and other structures can have a great effect upon pollutant dispersion. The "street canyon effect" occurs when the dispersion of low-level emissions by the prevailing wind is prevented by tall buildings on each side of a busy road.

Air Pollution Impacts

Air pollution can adversely affect human health, not only by direct inhalation but also indirectly by other exposure routes, such as drinking-water contamination, food contamination, and skin transfer. Many of the most common air pollutants directly affect the respiratory and cardiovascular systems. Increased mortality, morbidity, and impaired pulmonary function have been associated with elevated levels of SO_2 and SPM. Nitrogen dioxide and O_3 also affect the respiratory system; acute exposure can cause inflammatory and permeability responses, lung function decrements, and increases in airway reactivity. Ozone can irritate the eyes, nose, and throat and cause headaches. Carbon monoxide has a high affinity for hemoglobin and is able to displace oxygen in the blood, which can lead to cardiovascular and neurobehavioral effects. Lead inhibits hemoglobin synthesis in red blood cells in bone marrow, impairs liver and kidney function, and causes neurological damage.

The direct human health effects of air pollution vary according to both the intensity and the duration of exposure and also with the health status of the population exposed. Certain sectors of the population may be at greater risk – for example, the young and the elderly, those already suffering from respiratory and cardiopulmonary disease, hyper-responders, and people exercising.

At present, assessment of air quality for public health purposes consists essentially of examining ambient air quality against established guidelines. WHO's recommended air quality guidelines (see table 8.2) indicate the level and exposure time at which no adverse effects on human health are expected to occur. Many countries set their own national air quality standards, which are primarily designed to protect human health and may be legally enforced.

In addition to having human health impacts, a number of the pollutants considered in detail in this report have additional or indirect impacts on the environment. Oxides of sulfur and nitrogen are the principal precursors of acidic deposition. Long-range transport of SO_2, NO_x, and their corresponding acidic transformation products has been linked to soil and freshwater acidification, which adversely affects aquatic and terrestrial ecosystems. Sulfur dioxide, NO_2, and O_3 are phytotoxic, and O_3, in particular, has been implicated in crop losses and forest damage. Impaired visibility and damage to materials (such as nylon and rubber), buildings, and works of art are attributed to SO_2, acid sulfate aerosols, and O_3.

The long-range transport of air pollution from megacities can also have national and regional impacts. Oxides of nitrogen and sulfur in the "urban plume" can contribute to acid deposition at great distances from the city. Ozone concentrations are often elevated downwind of urban areas because of the time lag involved in photochemical processes and NO scavenging in polluted atmospheres.

Air Quality Monitoring

In the 1960s, recognition of the ubiquitous nature of pollutants such as SO_2, NO_x, CO, SPM, O_3, and lead in urban air and concern for their adverse impacts on human health prompted many national institutions to set up monitoring networks for the routine measurement of urban air quality. National air quality standards and other forms of legislation were also introduced to protect human health. In many of the more developed countries, early legislation and monitoring efforts focused on SO_2 and SPM. Since the late 1970s, however, and as motor vehicles became an increasingly important source of air pollutants, networks have typically expanded to incorporate the routine monitoring of traffic-related pollutants, such as CO, NO_x, and lead. During

Table 8.2 [*orig. table 1*]   A summary of WHO air quality guidelines.

Pollutant	Time-weighted average	Units[a]	Averaging time
Sulfur dioxide	500	$\mu g/m^3$	10 minutes
	350	$\mu g/m^3$	1 hour
	100–150[b]	$\mu g/m^3$	24 hours
	40–60	$\mu g/m^3$	1 year
Carbon monoxide	30	mg/m^3	1 hour
	10	mg/m^3	8 hours
Nitrogen dioxide	400	$\mu g/m^3$	1 hour
	150	$\mu g/m^3$	24 hours
Ozone	150–200	$\mu g/m^3$	1 hour
	100–120	$\mu g/m^3$	8 hours
Suspended particulate matter			
Black smoke	100–150[b]	$\mu g/m^3$	24 hours
	40–60[b]	$\mu g/m^3$	1 year
Total suspended particulates	150–230[b]	$\mu g/m^3$	24 hours
	60–90[b]	$\mu g/m^3$	1 year
Thoracic particles (PM$_{10}$)	70[b]	$\mu g/m^3$	24 hours
Lead	0.5–1	$\mu g/m^3$	1 year

[a] Micrograms per cubic meter of air ($\mu g/m^3$); milligrams per cubic meter of air (mg/m^3).
[b] Guideline values for combined exposure to sulfur dioxide and suspended particulate matter (they may not apply to situations where only one of the components is present).
Sources: World Health Organization, *Lead*, Environmental Health Criteria 3 (Geneva, 1977); idem, *Oxides of Nitrogen*, Environmental Health Criteria 4 (Geneva, 1977); idem, *Photochemical Oxidants*, Environmental Health Criteria 7 (Geneva, 1978); idem, *Sulfur Oxides and Suspended Particulate Matter*, Environmental Health Criteria 8 (Geneva, 1979); idem, *Carbon Monoxide*, Environmental Health Criteria 13 (Geneva, 1979); and World Health Organization, Regional Office for Europe, *Air Quality Guidelines for Europe*, European Series, no. 23 (Copenhagen, 1987).

the 1980s, urban air quality monitoring for the traditional air pollutants was also established in many less developed countries, especially those in Asia and South America.

In recent years, greater attention has been paid to the need to monitor photochemical oxidants, notably O_3, and their VOC precursors. Although the instrumentation is now well developed, relatively few countries routinely monitor O_3 as an indicator of photochemical pollution. In the case of VOCs, reliable monitoring instrumentation has only recently been developed; consequently, urban VOC data are scarce. Furthermore, greater attention is being paid to characterizing the nature of SPM. Because of the heterogeneous nature of the pollutant,

total SPM data have only limited value for health effect assessments. Therefore, monitoring objectives in some countries have been narrowed to determine specific size fractions of particulate matter – such as matter less than 10 micrometers in diameter (PM$_{10}$), which is in the respirable size range – and chemical characteristics like the presence of heavy metals, such as mercury, cadmium, and lead, or of organic matter, such as polyaromatic hydrocarbons.

Despite advances in the scale and scope of urban air quality monitoring in recent years, major difficulties in acquiring comprehensive and reliable air quality data for assessment purposes still exist. Monitoring capabilities vary markedly between countries, depending largely on national priorities and objectives and the availability of economic and human resources. Typically, resources are not available to support the monitoring of all air pollutants in a routine program, and a prioritization of key pollutants is made, taking into account the ease of measurement of the pollutant and its relative risk to human health. Furthermore, fixed-site monitoring stations only provide data about a city's air pollution levels at the specific time and place of sampling. A sparse set of observations at a few place-time coordinates does not necessarily give an accurate picture of the extent and severity of air pollution throughout the urban area. Even in the highly industrialized countries, there is generally a limit to the number of routine observation sites that can be established owing to the high cost of the equipment required. Thus, routine monitoring is often supplemented by spot surveys and short-term studies to characterize a particular problem. Increasingly sophisticated modeling techniques are being developed to complement monitoring data by providing estimates of ambient air pollution levels over wider areas.

The maintenance of a current emissions inventory is also a fundamental tool in air quality management programs. Without emissions inventories, it is essentially impossible to design appropriate and cost-effective control strategies.

Megacity monitoring

A recent survey by WHO showed that, of the 60 newly industrializing countries in the world, 34 had monitoring programs, while 15 had essentially no programs.[8] These include the countries where the majority of the megacities are located. Figure 8.5 summarizes the air quality monitoring capabilities in the 20 megacities. At present, the monitoring capabilities of countries fall into three general categories. Los Angeles, Mexico City, New York, São Paulo, Seoul, and Tokyo maintain comprehensive air quality monitoring networks that often provide real-time data on all major air pollu-

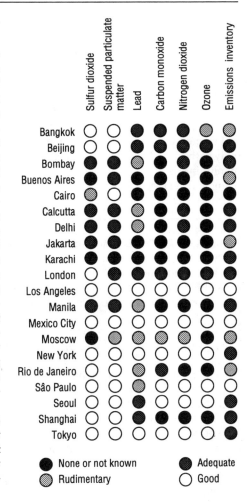

Figure 8.5 [*orig. figure 3*] Subjective assessment of the status of monitoring capabilities and availability of emissions inventories in 20 megacities.

tants. These networks incorporate adequate quality control procedures to ensure that the data are demonstrably valid. In the second category are cities with mostly marginal or just adequate air monitoring networks that measure only a few pollutants, usually at fewer sites than desirable. Included in this group are Bangkok, Beijing, Bombay, Calcutta, Delhi, London, Rio de Janeiro, and

[8] M. Schaefer, *Combating Environmental Pollution: National Capabilities for Health Protection*, WHO/PEP/91.3 (Geneva: World Health Organization, 1992).

Shanghai (see box 8.4). The third category includes the megacities with inadequate air monitoring systems that produce data of unknown quality on a few pollutants. Buenos Aires, Cairo, Jakarta, Karachi, Manila, and Moscow fall into this category. The latter two categories of cities also often lack sufficient quality assurance oversight, and so questions are raised about the reliability of the data. It is evident that air quality monitoring capabilities must be improved in many of these megacities.

Air Pollution in the Megacities

All of the major air pollutants and their sources are present in the 20 megacities, although to varying degrees. Some sources are common to all of them, while others are of importance in only certain megacities. Motor vehicle traffic is a significant source of air pollution in all of the megacities; in nearly half of the megacities, it is the single most important source. In addition to the megacities in the developed countries,

---

Box 8.4   Air Quality Monitoring in India

The database on ambient air quality status in major Indian cities is confined to the information generated by the National Environmental Engineering Research Institute (NEERI) through its National Air Quality Monitoring Network. The network started in 1978 and was discontinued in 1988 and 1989. The sites were subsequently reopened in 1990.

Ten cities take part in the network, primarily because of the location of NEERI laboratories at Ahmedabad, Bombay, Calcutta, Cochin, Delhi, Hyderabad, Jaipur, Kanpur, Madras, and Nagpur. These cities also represent a cross-section of India's different industrial, geographic, and climatic conditions. Data from the three NEERI stations in each megacity – Bombay, Calcutta, and Delhi – are also reported to the Global Environment Monitoring System (GEMS).

In each city, NEERI's monitoring stations are located at three sites that are deemed to be representative of industrial, commercial, and residential areas. In some cities, it has been necessary to relocate monitoring stations since 1978 for various reasons. Where relocation has occurred, replacement sites have been chosen that are equally representative of the designated classification. Four air quality parameters were originally chosen for measurement: suspended particulate matter (SPM) – daily, monthly, and yearly averages; sulfur dioxide (SO_2) – daily, monthly, and yearly averages; sulfation rate – monthly average; and dust fall – monthly average. Nitrogen dioxide (NO_2), ammonia, and hydrogen sulfide were also monitored where appropriate. Meteorological information was collected for the cities for use in data interpretation.

Sulfur dioxide, NO_2, and SPM samples were collected every tenth day. Sulfur dioxide samples were collected on an hourly or four-hourly batch basis for a period of 24 hours. Nitrogen dioxide was sampled using the sodium arsenite gas bubbler method. Suspended particulate matter samples were also taken on a 24-hour basis in parallel with SO_2 and NO_2 monitoring. It has been necessary for NEERI to develop its own sampling equipment specifically designed for Indian conditions. Measurements of SPM have recently focused on small respirable particulates (PM_{10}) using NEERI designed and built samplers.

Unfortunately, problems have been encountered with the data summaries. The data presented in NEERI's *Air Quality Status in Ten Cities* reports have not always matched data submitted to GEMS in other reports, and it has not been possible to resolve all of the anomalies. On the whole, it appears that the majority of these errors are typographical mistakes. However, such problems do emphasize the need for better quality assurance within the network. At present, no quality control is carried out by NEERI or the Central Pollution Control Board (CPCB), and some form of auditing system should be established.

CPCB initiated its own National Ambient Air Quality Monitoring Programme in 1985. The main function of CPCB under the 1981 Air Act is to improve the quality of air and to prevent, control, and abate air pollution throughout India. By 1990, 116 monitoring stations (out of 220 sanctioned) were being operated by the state pollution control boards and were generating data. Unfortunately, CPCB data are not directly comparable with those from the NEERI sites because of differences in sampling criteria and methodologies. CPCB has issued one data summary, *National Ambient Air Quality Statistics of India 1987–1989*, but as yet, no assessments have been made.

Very little monitoring of additional pollutants, such as carbon monoxide and ozone, is undertaken in India's cities, and there are no permanent urban monitoring stations for these pollutants. In addition, little has been done to characterize the particulate fraction or to determine atmospheric lead levels following reductions in the lead content of gasoline, although work on this has recently begun. Monitoring activities in India should be expanded to cover more pollutants and a wider geographical area. Localized air quality assessments should also be conducted to identify specific problems and to provide detailed recommendations for pollution abatement.

Bangkok, Jakarta, Manila, Mexico City, São Paulo, and Seoul also have overwhelming vehicular air pollution. As a result, emission of CO, VOCs, NO_x, and lead (where it is still used as a gasoline additive) are high. In Bangkok, Manila and Seoul, a substantial proportion of the motor vehicle fleet is diesel-powered, a characteristic that brings its own special problems, such as greater SPM, SO_2, and NO_x. In comparison, motor vehicle emissions in Beijing, Karachi, Moscow, and Shanghai are still relatively low but are increasing as motor vehicle registrations and traffic increase.

Emissions from power generation and industry are likewise a problem in almost all of the megacities. With the possible exception of Jakarta, New York, and Shanghai, basic heavy industry is generally not located in or near the megacities. The multitude of other industries – including power generation – are, however, dominant sources of air pollution, especially of SPM and sulfur and nitrogen oxides. Beijing, Bombay, Buenos Aires, Cairo, Calcutta, Seoul, and Shanghai have high industrial emissions. In Bangkok, Beijing, Manila, and Shanghai, the problem of industrial emissions is compounded by the bulk of the industrial sources being interspersed with residential areas, thus creating the potential for particularly high human exposure.

Coal and high-sulfur oil have been or are being phased out in many megacities, to be replaced by cleaner fuels containing less sulfur, such as natural gas. At present, the principal fuel in the 20 megacities is split about equally among coal, oil, and natural gas. A high level of domestic coal or biomass fuel use is a very serious health problem as there may be substantial indoor and outdoor exposures. Such use creates very high concentrations of SPM, hydrocarbons, and, in the case of coal, SO_2. Coal is the predominant fuel for both industry and energy in Delhi, Beijing, Calcutta, Shanghai, and Seoul.

The open burning of refuse is another major source of air pollution in many developing countries, but information is scarce. In several of the megacities, most notably Karachi, Mexico City, and Manila, pollution from this source is recognized.

High levels of natural wind-blown particulates also contribute to air pollution problems. Proximity to deserts or barren lands leads to high natural loadings of SPM in Beijing, Cairo, Delhi, Karachi, and Mexico City (see box 8.5).

Comparative air quality

Figure 8.6 presents a subjective overview of the relative air quality situation in the 20 megacities. To compare the cities, it was

Box 8.5   Cairo: Unbridled Dust

Cairo is the capital of Egypt and the most populous city in Africa. In 1950, Cairo had a population of 2.4 million. The United Nations estimated the 1990 population to be 9.08 million. By 2000, Cairo is expected to have a population of 11.77 million.(1)

Cairo has a desert climate characterized by very dry heat. The annual mean rainfall is only 22 millimeters. Cairo's pollution is also exacerbated by traffic congestion and low wind speeds. These factors contribute greatly to Cairo's most obvious pollution problems: suspended particulate matter (SPM).

Cairo has three Global Environment Monitoring System (GEMS) stations operated by Egypt's Ministry of Health: one in the city's central commercial area, another in a suburban industrial area, and the third in a suburban residential area. In addition, several research studies on air pollution have been carried out at the National Research Center and at Cairo University. Smoke monitoring in Cairo commenced in 1973. Total suspended particulate matter (TSP) monitoring was initiated on a regular basis at various sites in 1985. However, with the exception of the data from the three GEMS stations, only annual mean data are available; no statistics or number of observations are provided. Therefore, the data's validity is questionable. Moreover, no SPM emission inventory is available for Cairo.

Particulate pollution in Cairo stems from natural sources, such as wind-blown dusts, as well as anthropogenic sources. A large contributor to natural SPM levels is the north-easterly wind in spring, and the fresh-to-strong, hot "Khamasin" southerly wind is usually loaded with natural sand and dust. However, there are no estimates of the contribution of natural SPM to Cairo's total ambient SPM levels.

Anthropogenic sources of SPM emissions include incomplete combustion processes, industry, and traffic. Limestone quarrying and adjacent cement factories contribute huge amounts of particulates to the air around Cairo. Three large cement factories are situated at Helwãn, 24 kilometers south-east of Cairo. It is estimated that 4 percent of all the cement produced is lost as particulate emissions. In addition to cement factories, Helwãn is the site of iron and steel works, a lead and zinc smelter, coke and chemical works, fertilizer factories, and car, textile, ceramic, and brick manufacturers. Power for these industries is provided by nearby electricity-generating stations, which primarily burn heavy fuel oil and natural gas.

A second industrial complex, Shoubra El-Khayma, lies 30 kilometers northwest of Cairo. The more than 450 industries there include metallurgical works and ceramic, glass, brick, textile, and plastics manufacturing plants. A very large thermal power station provides electricity for these industries.

Smoke emissions from cars and buses in Cairo were estimated to amount to 1,200 tonnes per year in 1990.(2) This is a more than sevenfold increase since 1970. About 88 percent of these emissions come from cars. Vehicular SPM sources are believed to be relatively small compared with other SPM sources, but their impact on SPM pollution at roadside locations is probably severe. Moreover, their contribution to total SPM is likely to increase because the number of motorized trips per day is projected to rise from 3.9 million in 1980 to 12.0 million by 2000.(3) Another source of SPM in Cairo is the widespread open incineration of rubbish, over which there is little control.

The monitoring data suggest that, on the whole, annual mean smoke concentrations increased during the 1980s.(4) In 1989, annual mean smoke concentrations ranged from 40 to 190 micrograms per cubic meter of air ($\mu g/m^3$). Six out of the eight monitoring sites measured smoke concentrations above the annual guideline of 40 to 60 $\mu g/m^3$ set by the World Health Organization (WHO). At three stations, annual mean smoke levels in 1988 and 1989 even exceeded the WHO daily guideline of 125 $\mu g/m^3$.

At only two stations were smoke levels within or below the WHO annual guideline.

Annual mean TSP levels for 1987 to 1989 ranged from about 500 to 1,100 $\mu g/m^3$ – far in excess of the WHO annual guideline of 60 to 90 $\mu g/m^3$.(5) The large disparity between the black smoke and TSP data is probably caused by the large amount of suspended dust from natural sources, cement industries, and construction. Although monthly mean smoke concentrations are around 100 to 150 $\mu g/m^3$, the respective TSP levels fall between 400 and 900 $\mu g/m^3$.(6) Generally, there are slightly higher SPM levels in winter: monthly smoke levels were 10 percent higher and TSP levels were about 40 percent higher at the industrial site.

In addition to SPM, the ambient air in Cairo has very high levels of all the other major air pollutants as well. Although only limited monitoring data are available and the reliability of some of the available data is unclear, it is clear that the short-term and long-term mean pollutant concentrations regularly exceed the WHO guidelines, especially in the city center. As Cairo is faced with increasing industrial development and vehicular traffic coupled with relatively little control of emissions, the air quality will inevitably deteriorate.

1   United Nations, *Prospects of World Urbanization 1988,* Population Studies no. 112 (New York, 1989).
2   M. M. Nasralla, *Air Quality in Cairo Metropolitan Area* (Cairo; National Research Centre, Air Pollution Department, 1990).
3   A. Faiz, K. Sinha, M. Walsh, and A. Varma, *Automotive Air Pollution: Issues and Options for Developing Countries* (Washington, D.C.: World Bank, 1990).
4   Nasralla, note 2 above.
5   Ibid.
6   K. T. Hindy, S. A. Farag, and N. M. El-Taieb, "Monthly and Seasonal Trends of Total Suspended Particulate Matter and Smoke Concentration in Industrial and Residential Areas of Cairo," *Atmospheric Environment* 24B (1990): 343–53.

necessary to group them into broad categories: Those with

• serious problems, where WHO guidelines are exceeded by more than a factor of two (WHO guidelines for the most common air pollutants appear in table 8.2);

• moderate to heavy pollution, where WHO guidelines are exceeded by up to a factor of two or where WHO short-term guidelines are exceeded on a regular basis.

• low pollution, where WHO guidelines are normally met and short-term guidelines are exceeded only occasionally; and

• no data available or insufficient data for assessment.

The assessments of air pollution in figure 8.6 are based on the most recent data available. Obviously, such comparisons are very difficult owing to differences in monitoring methods, reporting procedures, and so forth. No assessment of the air quality situation has been made where data were insufficient or where the quality of the data was questionable. Not all of the data were provided by recognized monitoring networks, however, and some assessments have been based upon the results of recent studies and surveys.

Sulfur dioxide

Ambient concentrations of SO_2 have decreased dramatically in a number of the megacities. At present, however, Beijing, Mexico City, and Seoul can be considered to have serious SO_2 pollution. The ambient levels in these cities are in excess of WHO guidelines by a factor of nearly three for annual average concentrations, and peak daily concentrations exceed 700 micrograms per cubic meter of air ($\mu g/m^3$). These concentrations are well within the range at

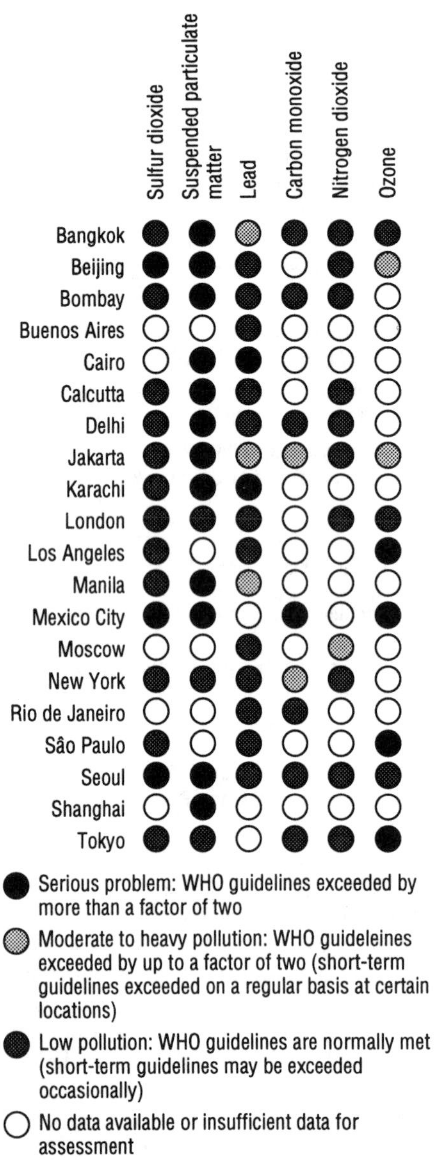

	Sulfur dioxide	Suspended particulate matter	Lead	Carbon monoxide	Nitrogen dioxide	Ozone
Bangkok	●	●	◉	●	●	●
Beijing	●	●	●	○	●	◉
Bombay	●	●	●	●	●	○
Buenos Aires	○	○	●	○	○	○
Cairo	○	●	●	○	○	○
Calcutta	●	●	●	○	●	○
Delhi	●	●	●	●	●	○
Jakarta	●	●	◉	◉	●	◉
Karachi	●	●	●	○	○	○
London	●	●	●	○	●	●
Los Angeles	●	○	●	○	○	●
Manila	●	○	◉	○	○	○
Mexico City	●	●	○	●	○	●
Moscow	○	○	●	○	◉	○
New York	●	●	●	◉	●	○
Rio de Janeiro	○	○	●	●	○	○
São Paulo	●	○	●	○	○	●
Seoul	●	●	●	●	●	●
Shanghai	○	●	○	○	○	○
Tokyo	●	●	○	●	●	●

● Serious problem: WHO guidelines exceeded by more than a factor of two

◉ Moderate to heavy pollution: WHO guidelines exceeded by up to a factor of two (short-term guidelines exceeded on a regular basis at certain locations)

● Low pollution: WHO guidelines are normally met (short-term guidelines may be exceeded occasionally)

○ No data available or insufficient data for assessment

Figure 8.6 [orig. figure 4]   Overview of air quality in 20 megacities based on a subjective assessment of monitoring data and emissions inventories.

which, in the presence of particulate matter, there are observable effects on health, such as increased mortality and morbidity of the aged and infirm. It should be noted that a large refinery in Mexico City has recently been closed, and so the city's concentrations of SO_2 should decrease markedly.

Two cities have concentrations of SO_2 near or somewhat in excess of the WHO health guidelines. In 12 of the megacities, SO_2 does not appear to be a major health threat because the concentrations are typically within WHO guidelines, although there may well be some days or even months when the levels are higher at certain locations. For example, one area in São Paulo has been slightly exceeding the guideline value, and there are periodic excursions above the guideline value during the non-monsoon seasons in Calcutta. Although London now meets the WHO annual guidelines, it still exceeds hourly and daily guideline maxima. No SO_2 data for Moscow are presented in figure 8.6 as their validity is questionable. No recent SO_2 data for Buenos Aires have been made available. Sulfur dioxide data for Cairo are scant because of problems with monitoring techniques. However, short-term studies have shown that exceedences of WHO guidelines are common because most of Cairo's power stations burn heavy fuel oil. The annual mean guidelines are exceeded by a factor of five downtown and by a factor of two in some residential areas.

Suspended particulate matter

Particulate air pollution is a very serious problem in 12 of the megacities surveyed. The ambient concentrations in these cities are persistently above the WHO guidelines both for long-term averages and for peak concentrations. The annual average levels across these cities are typically in the range of 200 to 600 $\mu g/m^3$ and peak concentrations are frequently above 1,000 $\mu g/m^3$. In cities that also have high levels of SO_2, there are notable effects on health. In some of these cities, including Beijing and Seoul, SPM decreases during the non-heating (warm) season; in others, such as Shanghai, high concentrations of SPM persist year round. In Cairo and some other cities, wind-blown dust greatly influences the level of SPM.

In the second group of cities, concentrations of SPM can be in excess of WHO guidelines by a factor of as much as two. However, there are also some monitoring stations in these cities that record concentrations within or near the WHO guidelines,

Table 8.3 [*orig. table 2*]   Lead content of gasoline.

City	Lead content of gasoline (grams per liter)[a]	Comments	City	Lead content of gasoline (grams per liter)[a]	Comments
Bangkok	0.15		Mexico City	0.54	
Beijing	0.4–0.8	80 percent unleaded gasoline	Moscow	0	No leaded fuel sales in city
Bombay	0.15		New York	0.026	95 percent unleaded gasoline
Buenos Aires	0.6–1.00				
Cairo	0.80		Rio de Janeiro	0.45	Ethanol and gasohol used
Calcutta	0.10				
Delhi	0.18		São Paulo	0.45	40 percent ethanol, 60 percent gasoline/ gasohol
Jakarta	0.6–0.73	1987 levels			
Karachi	1.5–2.00				
London	0.15	> 33 percent unleaded gasoline	Seoul	0.15	
Los Angeles	0.026	>95 percent unleaded gasoline	Shanghai	0.40	
			Tokyo	0.15	>95 percent unleaded gasoline
Manila	1.16				

[a] Lead content may be average lead content of gasoline or limit value.

Note: Many countries have recently introduced (or will soon introduce) sales of unleaded petrol.

Sources: M. P. Walsh, personal communication, Alexandria, Va., 1992; National Environmental Engineering Research Institute, *Delhi*, vol. 1 of *Air Pollution Aspects of Three Indian Megacities* (Nagpur, India, 1991); and L. Murley, ed., *Clean Air Around the World*, 2nd ed. (Brighton, UK: International Union of Air Pollution Prevention Associations, 1991).

which indicates the existence of some relatively clean areas.

Of the 20 megacities, 3 – London, New York, and Tokyo – by and large meet the WHO guidelines. All of these have undergone massive and costly air pollution reduction programs, and, where necessary, sources are equipped with effective controls.

Lead

Airborne lead is closely associated with the density of motor vehicle traffic using leaded fuel and with the concentration of lead additives in the fuel. Table 8.3 presents a summary of the latter and shows that the concentrations range from 0 to 2 grams per liter. In light of their gasoline's lead content, it is not surprising to see that airborne lead concentrations in Cairo and Karachi are high – well above the WHO annual guide-

line of 1 μg/m³. There are very few routine network data on lead in ambient air. Most information comes from special investigations into lead's effects on the mental development of infants and children.

The available information on airborne lead in Bangkok, Jakarta, Manila, and Mexico City shows that it is somewhat above the WHO guideline. In these cities, programs are under way or planned to reduce the amount of lead in gasoline.

Twelve other cities for which some data on ambient levels of lead are available have levels that are low and well within the WHO guideline. Among these, London, Los Angeles, New York, Seoul, and Tokyo are moving toward totally lead-free gasoline. In Rio de Janeiro and São Paulo, ambient lead levels are low because of reliance on alcohol-powered vehicles and reduced lead levels in gasoline (see box 8.6). The relatively low traffic density in Beijing and Moscow results in levels lower than 0.5 µg/m³. In Bombay, Calcutta, and Delhi, the majority of the spot samples were below the WHO guideline, although some samples were higher. No data are available for Shanghai and Tokyo, although the latter is thought to have very little lead pollution because of its unleaded gasoline.

Box 8.6 São Paulo: Alcoholic Automobiles

The city of São Paulo, part of the Greater São Paulo Area (GSPA), is approximately 60 kilometers from the south-east coast of Brazil. The total population of GSPA was 18.42 million in 1990 – more than 10 percent of Brazil's total population. GSPA's rapid population increase over the past decades is noteworthy: There were 5 million people in 1960, 8 million in 1970, and 12.5 million in 1980. By 2000, 23.6 million people are expected to live in GSPA.(1)

GSPA is the economically most important region in Brazil. In 1988, GSPA consumed 1.5 million tonnes of gasoline, 1.7 million tonnes of diesel oil, 1.6 million tonnes of fuel oil, 1.7 million tonnes of hydrated ethanol, 0.6 million tonnes of liquid petroleum gas, and 33.8 terawatts of electricity.(2) However, because 95 percent of electric power generation in São Paulo State is hydroelectric, the air pollution impact of electricity generation is negligible.(3)

Emissions restrictions have successfully reduced pollution from many industrial sources. Although stationary sources still contribute 69 percent of São Paulo's anthropogenic emissions of suspended particulate matter (SPM), mobile sources, mostly motor vehicles, contribute the vast majority of anthropogenic emissions of sulfur dioxide (73 percent), carbon monoxide (94 percent), oxides of nitrogen (82 percent), and non-methane hydrocarbons (70 percent).(4)

There were 4 million vehicles registered in GSPA in 1990, of which more than 3 million were registered in São Paulo. These latter included approximately 1.2 million alcohol-fueled vehicles, 2.2 million gasoline-fueled, and 300,000 with diesel engines.(5) The pervasion of alcohol-fueled vehicles is a result of Brazil's National Alcohol Programme,

Table 8.4   Emission limits for Brazilian alcohol and gasoline-fueled light-duty vehicles.

Type of emission	Date effective from	Remarks	Carbon monoxide	Oxides of nitrogen (grams per kilometer)	Hydrocarbons	Aldehydes (percent)	Idling carbon monoxide
				Emission Limits			
Exhaust	1 Jun. 1988	New vehicle configurations	24	2	2.1		3
	1 Jan. 1989	Minimum of 50% of sales	24	2	2.1		3
	1 Jan. 1990	100% of sales except light-duty trucks	24	2	2.1		3
	1 Jan. 1992	Only light-duty trucks	24	2	2.1	0.15	3
	1 Jan. 1992	100% of sales except light-duty trucks	12	1.4	1.2	0.15	2.5
	1 Jan. 1997	All light-duty vehicles	2	0.6	0.3	0.15	0.5
Evaporative	1 Jan. 1990	All light-duty vehicles	6 grams per test				
Crankcase	1 Jan. 1988	All light-duty vehicles	Emissions shall be nil under any engine operating conditions.				

Source: L. Murley, ed., *Clean Air Around the World*, 2nd edn (Brighton, U.K.: International Union of Air Pollution Prevention Associations, 1991).

which it launched in 1975 to reduce oil imports.(6) The first ethanol-fueled cars appeared in 1980. By 1990, more than 90 percent of the cars had alcohol-based engines. Sixty percent of São Paulo's light-vehicle population runs on gasoline or "gasohol" – a mixture of gasoline and anhydrous ethanol – and almost 40 percent runs on ethanol.

The combustion of gasohol and hydrous ethanol (in redesigned Otto cycle engines) is thought to have affected the ambient concentrations of air pollutants in São Paulo. For example, lead pollution fell from a level of between 1.0 and 1.1 micrograms per cubic meter of air ($\mu g/m^3$) in 1976 to between 0.1 and 0.6 $\mu g/m^3$ in 1983, before the government limited the lead content of gasoline to 0.4 milliliters of tetraethyl lead per liter.(7) And lead levels have apparently stayed low despite the rapid growth in vehicle numbers.

Alcohol combustion is also thought to increase ambient concentrations of aldehydes. Two special campaigns to determine aldehyde levels ran from July 1980 to June 1981 and from January 1985 to February 1986.(8) Aldehydes are volatile organic compounds and, as such, precursors of ozone smog. In 1989, monitoring stations around São Paulo frequently recorded one-hour ozone concentrations above 400 $\mu g/m^3$.(9) The World Health Organization's (WHO's) one-hour guideline is just 150 to 200 $\mu g/m^3$.

To reduce emissions from alcohol, gasoline, and diesel engines in motor vehicles, Brazil's National Environmental Council established in May 1986 an automotive emission control program called PROCONVE.(10) The program mandates progressively more stringent emissions limits for both heavy- and light-duty vehicles (see table 8.4). PROCONVE is aimed at reducing emissions of suspended particulate matter, carbon monoxide, oxides of nitrogen, hydrocarbons, and aldehydes as well as ambient concentrations of ozone. All of these pollutants pose serious problems for São Paulo because their ambient concentrations regularly exceed WHO guidelines.(11) Despite PROCONVE, however, an overall decrease in motor vehicle emissions will be difficult to achieve in São Paulo because of increasing motor vehicle registrations and distances traveled.

1 United Nations Environment Programme, *United Nations Environment Programme Environmental Data Report,* 2nd ed. (Oxford, England: Blackwell Publishers, 1989); United Nations, *Prospects of World Urbanization 1988,* Population Studies no. 112 (New York, 1989); and Companhia de Technologia de Saneamento Ambiental (CETESB), *Relatório de Qualidade do Ar no Estado de São Paulo – 1989* (São Pualo, Brazil, 1990).
2 Companhia Energética de São Paulo, *Estatisticas Energéticas Municipais e Regionais: Estado de São Paulo 1988* (São Pualo, Brazil, 1989).
3 Companhia Energética de São Paulo, *Balanço Energético do Estado de São Paulo 1986* (São Paulo, Brazil, 1989).
4 CETESB, note 1 above.
5 Ibid.
6 A. de Oliveira, "Reassessing the Brazilian Alcohol Programme," *Energy Policy* 19 (1991): 47–55.
7 C. M. Q. Orsini and L. C. S. Boveres, "Investigations on Trace Elements of the Atmospheric Aerosol of São Paulo, Brazil," in *Proceedings of the 5th International Clean Air Congress 1980* (Argentina, 1982), 247–55; and CETESB, *Relatório de Qualidade do Ar no Estado de São Paulo – 1988* (São Paulo, Brazil, 1989).
8 CETESB, note 1 above.
9 Ibid.
10 Ibid.; and A. Szwarc and G. M. Branco, *Automotive Emissions: The Brazilian Control Programme,* SAE Technical Paper no. 871073 (São Paulo, Brazil: CETESB, 1987).
11 CETESB, notes 1 and 7 above.

Carbon monoxide

Concentrations of CO depend particularly on the proximity of the sampling site to main traffic corridors. Comparisons between monitoring results for the 20 megacities are therefore difficult to make. Measurements made in Mexico City indicate very high concentrations of CO at different monitoring sites. Recent data on exposures of commuters show one-hour concentrations of up to 67 milligrams per cubic meter of air (mg/m^3), which is well above both the national standard and the WHO guidelines. In London, Los Angeles, and several other cities, the concentrations are not as high, but the air quality standards are being exceeded quite frequently. For instance, at some locations in Los Angeles, standards are exceeded on 15 percent of the days in the year. In Bombay, the results to date indicate that, by and large, the WHO air quality guidelines are being met, except in areas of high traffic density.

The cities with insufficient data to make an assessment are Beijing, Beunos Aires, Calcutta, Karachi, Manila, and Shanghai. Where high traffic densities occur, like in Mexico City, road-side and in-traffic concentrations could well be very high.

Nitrogen dioxide

For most of the megacities, NO_2 data are scant and often available only from special studies, not from continuous routine measurements. None of the 20 megacities for which data were available shows extremely high NO_2 concentrations. In 10 of the cities, NO_2 concentrations are generally near the guideline values, although periodic exceedences occur, particularly of the 1-hour and 24-hour maximum concentrations. For example, the data from Mexico City show that the hourly maxima range up to $700~\mu g/m^3$. In Los Angeles, hourly averages up to $500~\mu g/m^3$ are measured. The WHO hourly guideline is just $400~\mu g/m^3$.

In cities such as Moscow and São Paulo, the measured concentrations are also relatively high. There are six cities for which either no data were available or the data were grossly insufficient. For these, no eval-uation was possible. It could well be that, where there are no high traffic volumes, as in Buenos Aires and Manila, the levels of NO_2 are high at least in parts of these cities.

In cities where the use of natural gas or kerosene for domestic purposes is prevalent, the exposures of the inhabitants will be higher because of the oxides of nitrogen produced indoors in addition to the ambient levels.

Ozone

Ozone concentrations are high in Los Angeles, Mexico City, São Paulo, and Tokyo. In 1988, certain areas of Los Angeles, Mexico City, and Tokyo exceeded the national air quality standards on 50, 70 and 12 percent of the days of the year, respectively. In Mexico City, hourly O_3 concentrations reached $600~\mu g/m^3$ and at times peaked at $900~\mu g/m^3$ – four times the maximum WHO guideline. Beijing, Jakarta, and New York also show concentrations above the guideline, although less frequently. In London, Bangkok, and Seoul, measurements of O_3 so far have indicated levels normally within the guidelines but with occasional excursions above them. Of the 10 cities for which data are not available, Bombay, Cairo, Calcutta, Delhi, and Manila probably have high levels of O_3 because of their heavy traffic and sunshine.

Air Quality Trends

Figure 8.6 clearly shows that a high level of SPM is the major problem affecting the megacities as a group. It is most severe in those megacities where high concentrations of SO_2 occur simultaneously. Next in severity are levels of O_3, lead, and CO. Many of these cities must also be considered to have potentially serious NO_2 pollution; however, the data are too incomplete to reveal levels that regularly exceed the WHO guidelines. Data for O_3 are even less complete.

Nevertheless, it is clear that air pollution has escalated greatly in most of the megacities over the past decade because of rapid population growth, industrialization, and increasing energy use. Current predictions point to a substantial worsening of the situation in those cities where major air pollution control measures are not being implemented. These megacities could well see from 75- to 100-percent increases in their air pollution concentrations over the next decade. For example, in a city where the annual mean SPM concentration is currently 600 µg/m^3, this figure could escalate to more than 1,000 µg/m^3 over the next decade if current trends persist. Nitrogen dioxide concentrations are also expected to increase substantially in all of the megacities because of the forecast increase in motor vehicle traffic. Consequently, the effects on health will be magnified. For example, in cities where air pollution has thus far affected only highly exposed subgroups, the general population may now experience adverse health impacts.

In some of the megacities studied, however, the severe air pollution conditions observed today could have been much worse if certain control measures had not already been introduced. For example, in Beijing, Delhi, Seoul, and Shanghai, controls have slowed the rise in air pollution levels, which, in some cases, have stabilized before they reached the high levels that were found in London 40 years ago. Over the past 30 years, London, Los Angeles, New York, and Tokyo have reduced their air pollution dramatically (see box 8.7).

From the information collected for this report, it can be seen that history repeats itself. A number of cities have approached the same degree of pollution that was present, for example, in London in the 1950s, before "clean air" legislation was introduced. This relationship between the level of development and environmental pollution is illustrated in figure 8.7.

Box 8.7   Tokyo: Under Control

Tokyo's 12 million people account for about 10 percent of Japan's total population. The Tokyo/Yokohama agglomeration had an estimated population of 20.52 million in 1990, and its projected population for 2000 is 21.32 million.(1)

Tokyo has a large number of light industries that manufacture such goods as textiles, toiletries, electrical products, and cameras. The neighboring Yokohama-Kawasaki district has more of the heavy industry, specializing in chemicals, machinery, metallurgy, petroleum refining, automobiles, ships, and fabricated metal products. A center for iron and steel, petroleum refining, petrochemicals, electric power, and other heavy industries is also located in the northeast of Tokyo.

The bulk of Tokyo's power comes from thermal stations, but approximately 15 percent comes from hydroelectric stations. Electricity is also generated by nuclear power stations. Fossil fuel consumption in Japan remained relatively stable between 1975 and 1987.

Motor vehicle traffic has increased rapidly over the last 30 years. In 1960, the number of registered motor vehicles was about 500,000. By 1970, it was more than 2 million, and it reached 4.4 million in 1990 – an almost tenfold increase in 30 years.(2)

Monitoring of air pollutants began in the 1960s and early 1970s. The Environment Agency of Japan has set up a national air monitoring network. The network is spread over 15 regions and includes more than 1,000 stations.(3) Within the Tokyo metropolitan area, there are 35 automated ambient air monitoring stations, 20 of which are in the city center and 15 in the suburbs. At these stations, sulfur dioxide (SO_2), carbon monoxide (CO), suspended particulate matter (SPM), nitrogen dioxide (NO_2), nitric oxide (NO), and photochemical oxidants are monitored continuously. In addition, 32 car exhaust monitoring stations measure CO, NO_2, and NO. This large monitoring network has helped quantify the large improvements in air quality effected by the country's emissions control laws.

For instance, in Tokyo, annual mean SO_2 concentrations were about 160 micrograms per cubic meter of air ($\mu g/m^3$) in 1965 and 1966. With a strong initial effort to curb emissions through fuel desulfurization and the use of emission gas treatment technologies for mobile and stationary sources, those high levels were cut dramatically from 1966 to 1973. From 1974 to 1979, SO_2 annual mean levels ranged from 50 to 80 $\mu g/m^3$ in the city center and stayed at about 50 $\mu g/m^3$ in the suburbs.(4) Levels continued to decrease, and, by 1988, the annual SO_2 concentrations ranged from 18 to 26 $\mu g/m^3$, which was well below the annual mean guideline of 40 to 60 $\mu g/m^3$ set by the World Health Organization (WHO). The mean value from Tokyo's municipal monitoring stations was just 13 $\mu g/m^3$ (see figure 8.8).

Likewise, SPM pollution decreased steadily from 1975 to 1989. While concentrations from 1975 to 1977 were about 55 to 75 $\mu g/m^3$, SPM levels had fallen below 60 $\mu g/m^3$ at all monitoring sites by 1989.(5) There were no significant differences between SPM pollution levels in the city center and in the suburbs.

Although lead measurement is not part of the routine national air monitoring program, it is known that lead emissions were reduced dramatically when unleaded gasoline was introduced. More than 95 percent of the gasoline used in Tokyo is unleaded.(6) Analyses of particulate samples taken in Tokyo since 1966 have shown a downward trend for reported metals, including lead.(7)

Ambient concentrations of CO have also dropped dramatically over the past three decades.(8) Annual average CO levels were high during the 1960s. Since then, average CO concentrations have declined because of strict control legislation. While CO levels were still between 2.3 and 3.5 milligrams per cubic meter of air (mg/m^3) in the early 1970s, the average concentration in 1988 and 1989 was just 1 mg/m^3. Thus,

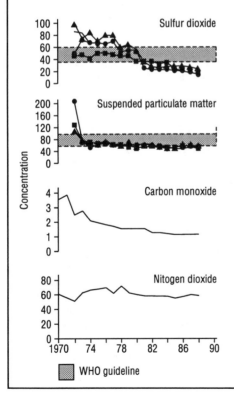

Figure 8.8   Annual mean concentrations of sulfur dioxide, suspended particulate matter, carbon monoxide, and nitrogen dioxide in Tokyo.

Note: Concentrations of sulfur dioxide, suspended particulate matter, and nitrogen dioxide are expressed in micrograms per cubic meter of air. Concentrations of carbon monoxide are in milligrams per cubic meter of air.
Sources: Environment Agency, Government of Japan, Quality of the Environment in Japan (Tokyo, 1988); T. Komeiji, K. Aoki, I. Koyama, and T. Okita, "Trends of Air Quality and Atmospheric Deposition in Tokyo," Atmospheric Environment 24A (1990): 2099–103; and Air Monitoring Division, Tokyo Metropolitan Government, Summary of Results of Regular Monitoring for Air Pollution in the FY 1989 (Tokyo, 1989).

it can be assumed that short-term CO values in Tokyo were below the WHO eight-hour guideline of 10 mg/m^3.

Although the control of other air pollutants has been successful in Tokyo, NO$_2$ concentrations have not been reduced during the last 20 years. Since monitoring began, annual mean concentrations have been around 60 μg/m^3.(9) Annual means have been around 65 μg/m^3 in the city center and 50 μg/m^3 in the suburbs. This lack of success is likely due to the rapid increase in motor vehicle traffic. However, it seems unlikely that the present WHO 24-hour guideline of 150 μg/m^3 is exceeded in Tokyo.

Data show that annual average levels of photochemical oxidants such as ozone (O$_3$) have decreased since the late 1960s.(10) Annual mean concentrations of O$_3$ were as high as 70 to 90 μg/m^3 from 1968 to 1970 but dropped to about 40 μg/m^3 in 1978. Since then, levels have been fairly stable. However, O$_3$ levels still rise above the national standards at many monitoring sites.

On days when any one-hour O$_3$ concentration exceeds 240 μg/m^3, an "oxidant warning" is given. At this level, adverse health effects may occur in a significant proportion of the exposed population. In the whole of Japan the number of "oxidant warning days" per year has varied between 59 and 171. No trend has emerged since 1977. In Tokyo, a warning was issued on 45 days in 1988. As the national one-hour warning standard of 240 μg/m^3 is higher than the WHO one-hour average guideline of 150 to 200 μg/m^3, it can be assumed that the WHO guidelines are exceeded even more frequently.

Nevertheless, for a commercial city with a population of 12 million, the air in Tokyo is very good. Tokyo is a recognized example of how air pollution in an industrial megacity can be controlled. In the 1960s, rapid population growth, industrialization, and motorization helped make environmental pollution a major social problem. Then, the municipal government began to promote various measures, such as controlling emissions from factories and developing sewage and waste disposal plants.(11) To reduce emissions of air pollutants, the metropolitan government issued regulations and standards for industries to follow. As a result, emissions and ambient levels of SO$_2$, SPM, and CO were reduced dramatically. In particular, the reduction of SO$_2$ through flue-gas treatment technologies in stationary sources has shown the potential of technological emissions control. The metropolitan government has also been implementing controls on emissions depending on the height of individual chimneys, fuel regulations, and total emission regulations.(12)

The projected urban and industrial growth will limit the effectiveness of these existing control programs, however. An alternative might be to use different energy sources. Because cars are the largest source of NO$_2$ and CO emissions, changes in fuel type might aid in reducing these levels. Additional solutions include car-pooling and increasing use of railways.

1   United Nations, *Prospects of World Urbanization 1988,* Population Studies no. 112 (New York, 1989).

2   Tokyo Metropolitan Government, *Automobile Pollution Control Plan: Towards a Better Living Environment in Tokyo* (Tokyo, 1989).

3   Government of Japan, Environment Agency, *Quality of the Environment in Japan* (Tokyo, 1988).

4   Ibid.; and T. Komeiji, K. Aoki, I. Koyama, and T. Okita, "Trends of Air quality and Atmospheric Deposition in Tokyo," *Atmospheric Environment* 24A (1990): 2099–103.

5   Government of Japan, note 3 above.

6   L. Murley, ed., *Clean Air Around the World,* 2nd ed. (Brighton, U.K.: International Union of Air Pollution Prevention Associations, 1991).

7   Komeiji, Aoki, Koyama, and Okita, note 4 above.

8 Tokyo Metropolitan Government, Air Monitoring Division, *Summary of the Results of Regular Monitoring for Air Pollution in the FY 1989* (Tokyo, 1989).
9 Ibid.
10 Ibid.
11 Tokyo Metropolitan Government, *Protecting Tokyo's Environment* (Tokyo, 1990).
12 Government of Japan, Environment Agency, *Japanese Performance of Energy Conservation and Air Pollution Control* (Tokyo, 1990).

Before rapid industrial development takes place, air pollution stems mainly from domestic sources and light industry; concentrations are generally low and increase slowly as population increases. As industrial development and energy use increase, air pollution levels rise rapidly. Then urban air pollution becomes a serious public health concern, and emission controls are introduced. Owing to the complexity of the situation, an immediate improvement in air quality cannot generally be achieved; at best, the situation is stabilized, and serious air pollution persists for some time. Several of the megacities studied are now at the point where controls must be implemented without delay. The introduction of emission controls has usually been followed by a staged reduction of air pollution as controls take effect. The earlier that integrated air quality management plans are put into effect, the lower the maximum pollution levels will be. This is especially important for those cities of developing countries that are not yet of the size and complexity of present-day megacities.

Control Strategies

Pollution control strategies rely upon local, national, and regional authorities setting air quality and emissions standards. By requiring adherence to such standards, governments can encourage industry to develop new and better technologies. In theory, this approach has been adopted by most of the megacities. However, in many cases, such strategies are not applied because some countries lack the financial and human resources to enforce compliance with standards.

Control technologies generally involve modifying the fuel or combustion technique or removing pollutants from flue gases. The type and location of the source – be it mobile or static, indoor or outdoor – and the overall cost-effectiveness of the different techniques influence the ultimate choice of control method.

Precombustion control techniques involve the use of low-pollutant fuels. Fuel cleaning is widely used to reduce the sulfur, dust, and ash content of coal. Distillate oils are often used in urban areas because of their

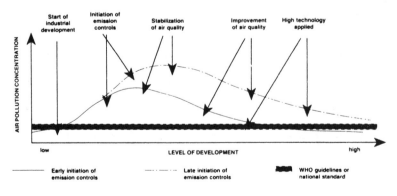

Figure 8.7 [*orig. figure 5*]   Development of air pollution problems in cities according to their developmental status.

reduced sulfur content. Many nations have reduced atmospheric emissions by using natural gas and nuclear power. In addition the use of certain gasoline additives, such as tetraethyl lead, has been reduced and even eliminated in several countries. Pre-combustion control techniques are often the simplest and most cost-effective method of reducing emissions.

Combustion modification techniques, such as low NO_x burners and fluidized bed combustion, have been employed in some countries to decrease SO_2 and NO_x emissions. However, modern combustion technology is relatively expensive and therefore beyond the reach of many developing countries.

Postcombustion control involves the removal of pollutants from flue gases and vehicle exhausts. There are a number of methods for removing SPM from flue gases, including gravity settling chambers, cyclones, spray chambers, bag filters, and electrostatic precipitators. A combination of these techniques is usually employed on larger coal-fired boilers. Flue-gas desulfurization is increasingly being adopted by industrialized nations to meet emissions standards and to achieve internationally agreed targets. Catalytic reduction is at present the best available technique for reducing NO_x emissions. The use of three-way catalytic converters to control motor vehicle pollutants has resulted in significant reductions of NO_x, CO, and hydrocarbon emissions from new vehicles. Also, the introduction of simple chimneys and vents to domestic stoves and heaters has greatly improved indoor air quality in developing regions.

The introduction of new manufacturing processes has also reduced industrial emissions. An obvious example is the use of low-temperature hydrometallurgical techniques, which reduce the SO_2 emissions associated with traditional metal-smelting methods. Many countries have adopted – mainly on economic grounds – energy conservation measures, which have effectively limited energy consumption and demand and improved generation and distribution.

Conclusions and Recommendations

From the data collected for this report, it is clear that air pollution is widespread in the 20 megacities: Each has at least 1 major pollutant that exceeds WHO health guidelines; 14 have at least 2; and 7 have 3 or more major pollutants that exceed the guidelines.

A high level of SPM is the most prevalent form of pollution. The health guidelines are exceeded in 17 megacities, and, in 12 of these, the values are double the guidelines. Sulfur dioxide and ozone pollution follow in terms of prevalence and severity. High concentrations of both SO_2 and SPM frequently occur in 5 of the megacities. This combination is particularly hazardous to health as it has been shown to increase mortality and morbidity.

High-sulfur coal and oil are being phased out in industry and power generation in many megacities. As a result, the upward trend in pollution has been reversed in 10 megacities. However, high domestic use of coal or biomass fuels is still a serious problem in 5, and it results in high human exposures to SPM, sulfur oxides, and carcinogenic polycyclic aromatic hydrocarbons.

Motor vehicle traffic – which leads to high concentrations of CO, NO_2, and O_3 – is a major source of air pollution in all of the megacities, and in half of them, it is the most important source.

The current capabilities of the municipal and national authorities to monitor air quality in the megacities and to collect information on their sources and emissions are inadequate. Only 6 of the 20 megacities have satisfactory monitoring networks and data-handling capabilities. Moreover, there is no systematic collection of information on the health risks and effects of air pollution in most of the megacities.

Despite these deficiencies, it is clear that, in many of the world's megacities, as well as in other cities, air pollution is a major health and environmental concern. This concern is increasing and should command high priority for action. National efforts to deal with this increasing problem are under way

Always based on Money

in many of these cities, but such efforts must be strengthened. Global and regional initiatives aimed at counteracting global warming and transboundary air pollution, through measures such as cleaner production and energy conservation, will also reduce the emissions of many of the air pollutants of local concern. Additional strategies will, however, be needed to deal with air pollution within the cities. The WHO Commission on Health and Environment's Panel on Energy identified the following as possible strategic elements: "urban and transportation planning, particularly in fast-growing cities, provision of safe and convenient mass transport, control of emissions from vehicles (especially automobiles) and industries, a switch to cleaner fuels, emphasis on district heating with co-generation plants, stringent controls on new industrial operations, and appropriate siting of power generating plants in relation to residential areas."[9]

Apart from these specific suggestions, several more general recommendations can be made in light of these findings on air pollution in the world's megacities. First, air quality management should be developed and implemented as a matter of urgency in those megacities where strategic planning is weak or nonexistent. Short-term, feasible approaches to reducing existing air pollution should be implemented as soon as possible. These include energy conservation, motor vehicle inspection and maintenance programs, phasing out lead in gasoline, promotion of mass transit, and establishing attractive alternatives to open burning of refuse. Over the long run, increased emphasis should be placed on preventive measures in the development of management strategies to improve air quality, such as urban and transportation planning and the introduction of "clean" technologies.

There is also an immediate need to improve the monitoring and emissions inventory capabilities of the megacities. These are prerequisites for sound air pollution management strategies aimed at protecting public health. Therefore, simple and rapid techniques for air quality monitoring and assessment should be investigated and applied where appropriate.

In the coming century, the number of megacities will continue to increase. Between 1970 and 2000, the number of cities with populations of more than 10 million will have increased sixfold, from 4 in 1970 to 24 in 2000. In 1970, there were 35 cities of more than 3 million people; today, there are 69 of them; and in 2000, the United Nations estimates, there will be 85. By 2025, this last number could have doubled.

Of the 69 cities that currently have more than 3 million people, 45 are in developing countries, and many are likely to become megacities. UNEP and WHO believe that, by studying past and present air pollution problems and air quality management strategies, the megacities of the future may be able to avoid many of the problems faced by megacities today. Thus, this report should be considered by authorities in all cities to avoid repeating mistakes of the past. In this regard, it is hoped that this assessment will contribute substantially to the formulation of national, regional, and global policies for the prevention and control of urban air pollution.

---

[9] World Health Organization, *WHO Commission on Health and Environment, Panel on Energy* (Geneva, 1992).

9

ON-ROAD VEHICLE EMISSIONS: REGULATIONS, COSTS, AND BENEFITS*

Stuart P. Beaton, Gary A. Bishop, Yi Zhang, Lowell L. Ashbaugh, Douglas R. Lawson, and Donald H. Stedman

In response to continuing pollution problems, in 1990 Congress passed the Clean Air Act Amendments, many parts of which deal with motor vehicles. Motor vehicles are the primary source of urban carbon monoxide (CO) and are an important source of the hydrocarbons (HC) and oxides of nitrogen (NO_x) that are responsible for the formation of photochemical smog and ground-level ozone (1). Cost estimates for implementing the act's mobile source provisions range up to $12 billion annually (2). Thus it is important to analyze the scientific basis for these legislative programs (1).

For any region violating air quality standards, a State Implementation Plan (SIP) must be submitted to and approved by the U.S. Environmental Protection Agency (EPA). In approving the SIP, EPA grants credit for each portion of the plan, including new vehicle emission standards, new fuels, vehicle scrappage, and inspection and maintenance (IM) programs, on the basis of predictions from a spreadsheet computer model (3). This EPA model treats all cars of a given model year as having the same adometer reading, the same annual mileage accumulation, and an equal likelihood of emission control system problems. The model has had little success in predicting urban on-road vehicle emissions (4), and the use of unverified computer models as the sole guide for public policy decisions is controversial (5). Calvert et al. (6) criticized the model and recommended in-use surveillance programs to identify high-emission vehicles and to monitor progress in emissions reduction. We present such a study and discuss its policy implications.

The California Study

During the summer of 1991, we placed an on-road remote sensor of exhaust CO and HC emissions (7) at various urban locations throughout California. We identified 66,053 different vehicles for which we collected 91,679 records with valid HC and CO

* Originally published in *Science*, 1995, vol. 268, pp. 991–3.

measurements (8). The emission distribution is highly skewed; the half of the fleet with the lowest emissions contributed less than 10% of the CO and HC, while a few high-emission vehicles dominated the mean values. In this instance, 7% of the vehicles accounted for 50% of the on-road CO emissions, and 10% of the vehicles accounted for 50% of the on-road HC emissions. These vehicles we call gross polluters. About 5% of the vehicles were gross polluters for both HC and CO (9).

It is often assumed that gross polluters are simply old vehicles. In fact, all model years of vehicles we have measured on the road include some proportion of gross polluters, as shown in figure 9.1. We found that the highest emitting 20% of the newest cars were worse polluters than the lowest emitting 40% of vehicles from any model year, even those from model years before the advent of catalytic converters (1970 and earlier). These data are typical of CO and HC results across the United States and at many other locations worldwide (10): Differences in emissions within a model year are greater than differences between the averages of the various model years. Correlation of

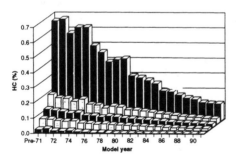

Figure 9.1 [orig. figure 1] Average HC exhaust emissions of the vehicles measured in California, reported as the HC percentage equivalent to propane and corrected for water and any excess air present in the exhaust. The emissions for each model year were sorted and divided into five groups (quintiles). The average emissions of the five quintiles for each model year are plotted from front to back, lowest to highest. Pre-71 includes all 1970 and older vehicles.

average on-road emissions and vehicle age shows that these results cannot be dismissed as random samples of normal vehicle behavior (11).

For 2 weeks during the California study, we operated two remote sensors on Rosemead Boulevard in South El Monte near Los Angeles (12). Vehicles identified by these sensors as gross polluters were immediately pulled over, and a voluntary California Smog Check – an emission control system inspection and tailpipe test – was administered. The remote sensors measured 58,063 unique vehicles, and we obtained Smog Check data on 307 of these vehicles. Of these, 126 (41%) showed deliberate tampering, and another 77 (25%) had defective or missing equipment that may not have been the result of tampering (for example, missing air pump belts). Overall, 282 (92%) failed the inspection even though all had valid registrations. Thus, less than 8% of the vehicles identified as gross polluters by remote sensing passed an immediate roadside test. When random pullover studies were carried out by the California Bureau of Automotive Repair and the California Air Resources Board, approximately 60% of the vehicles passed the roadside test, whether or not they were registered in a region of California with a scheduled IM program (13). Of the vehicles inspected in the present study, 76 were recruited for an immediate IM240 test (a loaded-mode dynamometer test) and their emissions were compared with EPA-recommended pass-fail values. All but three vehicles failed the IM240 test, and these three had already failed the roadside Smog Check (14). These data show that vehicles identified as gross polluters by the remote sensors were poorly maintained or had been tampered with (the majority apparently illegally); they were not a subset of normally maintained vehicles that were temporarily emitting more pollutants because their engines were cold or they were accelerating hard (11). If we could have inspected all 3271 gross polluters identified by the remote sensors, we estimate that only 266 of the 58,063 vehicles (0.5%) would have passed the roadside inspection.

This study indicates that the large difference between the majority of cars and the few gross polluters (the front-to-back difference in figure 9.1 is caused by poor maintenance or tampering. The smaller dependence of average emissions on vehicle age may be the combined result of deterioration, older technology, and poorer maintenance of older vehicles.

Implications for Emission Control Policy and Cost Effectiveness

On-road emissions must be controlled to reduce ambient pollutants. The skewed distributions shown in figure 9.1 imply that policies that treat all vehicles equally, or that target new vehicles, are likely to be less cost-effective than those that recognize the overriding importance of maintenance and that target poorly maintained vehicles regardless of their age. We used the California data to estimate the cost effectiveness of several proposed or hypothetical programs. Our only assumption in the cost analysis was that the data in figure 9.1

adequately represent all of California.

Tighter new car emission standards exemplify a national policy that attempts to achieve benefits by further reducing the already low emissions of new vehicles, at a cost of approximately $2 billion annually (2). It is plausible, based on reasonable assumptions regarding price increases and demand elasticity (neither of which are included in the EPA model), that such a policy could actually lead to increased on-road emissions (15).

Transportation control measures such as mandated employee car pools exemplify a policy that treats all vehicles as equal. These measures are modeled by the EPA as removing an average car from daily use. The actual improvement will be lower, because a car used for commuting is usually the best maintained car available to the driver (16). If the removed vehicle resembles the median rather than the average, then the actual CO and HC reductions are one-quarter and one-half of the EPA-predicted reductions, respectively (8).

Table 9.1 [orig. table 1] Estimated costs and benefits of various mobile source HC and CO emission reduction strategies as applied to the California fleet measured in 1991.

Action	Millions of vehicles affected	Percent reduction		Estimated cost (billions) of dollars	Percent reduction per billion dollars spent	
		HC	CO		HC	CO
Switch to reformulated fuels*	20 (100%)	17	11	1.5	11	7.3
Scrap pre-1980 vehicles	3.2 (16%)	33	42	2.2	15	19
Scrap pre-1988 vehicles	14.6 (73%)	44	67	17	2.6	3.9
Repair worst 20% of vehicles	4 (20%)	50	61	0.88	57	69
Repair worst 40% of vehicles	8 (40%)	68	83	1.76	39	47

* Reformulated fuels were estimated to cost consumers an extra $0.15 per gallon, or $75 per year for a 20-mpg car driven 10,000 miles per year. Scrappage costs per vehicle were conservatively estimated at $700 for pre-1980 cars and $1000 to $2000 for cars built from 1980 to 1988. Average repair costs were estimated at $200 per vehicle.

Alternative and reformulated fuel programs also treat all vehicles equally, regardless of their state of repair. The estimated costs and pollution reduction benefits of California's reformulated fuel program (17) are shown in table 9.1. By comparison, EPA data from a fleet of 84 vehicles testing various fuel formulations (18) indicate that if a single gross polluter (a 1984 Nissan Sentra) is repaired so that it matches the average emissions of the three other 1984 Nissan Sentras in this fleet, the resulting reduction in HC and CO emissions is more than can be obtained by fuel alterations to all 84 vehicles combined.

Scrappage programs operate on the assumption that older vehicles are more likely to be gross polluters. The industries that buy and scrap these older vehicles are typically in need of pollutant emission credits, and their participation in these programs may be supported by taxpayer dollars. The vehicle scrapped may have received little or no use, and the emission status of its replacement is unknown. In the EPA model, a newer replacement vehicle – even if it is just 1 year newer – is always considered to produce lower emissions than the older vehicle. The data in figure 9.1 show that this often may not be the case. Only if the older vehicle was a gross polluter for its model year is the replacement vehicle likely to be a low emitter. For instance, in the most recent Unocal program (19), a 1977 Toyota was scrapped that met the 1993 new car standards for exhaust emissions.

Scrapping all vehicles older than 1980 and replacing them with vehicles whose average age and emissions match those of the newer remainder of the fleet would cost $2.2 billion. We calculate that the benefits of this action would be a 33% reduction in HC emissions and a 42% reduction in CO emissions. Stated another way, this is a 15% reduction in total HC emissions and a 19% reduction in total CO emissions per billion dollars spent. If all vehicles older than 1988 were scrapped and replaced in the same way – an extreme measure (20) – the result would be a 44% reduction in HC emissions and a 67% reduction in CO emissions at a cost of $17 billion. If IM

programs controlled emissions as intended (13), scrappage programs would be less cost-effective than estimated here, because the emissions of older vehicles would not be as high.

In a targeted repair program, the worst 20% of all vehicles from each model year (the back row in figure 9.1) would be repaired to achieve the average emissions of the remaining 80% of the same model year. This action would result in a 50% reduction in HC emissions and a 61% reduction in CO emissions. Scrappage programs pay $700 to $1000 per vehicle, whereas the repairs, necessary to move a gross polluter to the lower emission categories average around $200 (21). The result is a 57% reduction in HC emissions and a 69% reduction in CO emissions per billion dollars spent. The full cost of this identification and repair program could be raised over a 4-year period by means of an annual $11 fee per vehicle (22). Because the pullover study showed that about half of the gross polluters had been illegally tampered with, the cost of the program would be cut in half if the owners of these vehicles were required to pay for their own repairs. Even if repair costs were as high as $400 per vehicle, the targeted, subsidized repair program is still estimated to be the most cost-effective option.

Conclusion

Vehicle exhaust emission measurements show that most vehicles, when properly maintained, are relatively small contributors to exhaust pollution in comparison to poorly maintained vehicles. Although poor maintenance correlates with increasing vehicle age, different states of maintenance among vehicles of a given model year far outweigh the average effect of age. Because of this factor, regulatory policies based on a computer model that targets all vehicles equally, without recognizing the overriding importance of individual maintenance, may not be cost-effective or may be ineffective.

References and Notes

1 National Academy of Sciences. *Rethinking the Ozone Problem in Urban and Regional Air Pollution* (National Academy Press, Washington, DC, 1991).

2 A. J. Krupnick and P. R. Portney, *Science* 252, 522 (1991).

3 E. L. Glover and D. J. Brzezinski, "MOBILE4 Exhaust Emission Factors and Inspection/ Maintenance Benefits for Passenger Cars" (Tech. Rep. EPA-AA-TSS-I/M-89-3, U.S. Environmental Protection Agency, Washington, DC, 1989).

4 E. M. Fujita et al., *J. Air Waste Manage. Assoc.* 42, 264 (1992): E. M. Fujita, J. C. Chow, Z. Liu, *Environ. Sci. Technol.* 28, 1633 (1994): W. R. Pierson, A. W. Gertier, R. L. Bradow, *J. Air Waste Manage. Assoc.* 40, 1495 (1990).

5 N. Oreskes, K. Shrader-Frechette, and K. Belitz [*Science* 263, 641 (1994)] argue that large computer models with multiple inputs should probably never be considered "validated." In Pennsylvania the EPA model was run with 27,000 inputs specific for each county. H. C. Scherrer and D. B. Kittelson (Society of Automotive Engineers Tech. Pap. Ser. 940302, February 1994) have examined the impact of the centralized emission program implemented in 1991 in the Minneapolis–St. Paul metropolitan area. On 23 March 1995, testifying before the Subcommittee on Oversight and Investigations, Committee on Commerce of the U.S. House of Representatives, they stated "What Minnesota has implemented is a mandated program that works in the computer models used by EPA to make public policy – not in the real world. This lack of linkage between EPA's model and real-world measurements leads to inappropriate policy decisions and wastes scarce resources. If we want to maintain public support for programs that claim to reduce air pollution, those programs must do what they claim in the real world, not just in the virtual world of the computer modeler."

6 J. G. Calvert, J. B. Heywood, R. F. Sawyer, J. H. Seinfeld, *Science* 261, 37 (1993).

7 The remote sensor takes a 0.5-s snapshot of the infrared absorption of CO_2, CO, and HC in the exhaust gases behind a moving vehicle to determine the percent of these gases in the undiluted exhaust. A video camera records each license plate number so that the make and model year can be determined through state motor vehicle records [G. A. Bishop, J. R. Starkey, A. Ihlenfeldt, W. J. Williams, D. H. Stedman, *Anal. Chem.* 61, 671A (1989);

L. L. Ashbaugh et al., in (23), pp. 885–989; D. R. Lawson, P. J. Groblicki, D. H. Stedman, G. A. Bishop, P. L. Guenther, *J. Air Waste Manage. Assoc.* 40, 1096 (1990)].

8 D. H. Stedman et al., "On-Road Remote Sensing of CO and HC Emissions in California" (Final Report, Contract A032–093, California Air Resources Board Research Division, February 1994).

9 Diesel-powered vehicles with ground-level exhaust systems were also measured by the remote sensor; they are always low emitters of CO and HC because the diesel combustion process occurs in the presence of excess air. We pulled over several gross polluters that were registered as diesel powered (and thus exempt from the IM program) but were actually gasoline powered and lacked any of the required emission controls. Heavy-duty diesel vehicles with elevated exhaust systems were not measured.

10 G. A. Bishop and D. H. Stedman, *Encyclopedia of Energy Technology and the Environment* (Wiley, New York, 1995), pp. 349–369.

11 T. C. Austin, F. J. DiGenova, T. R. Carlson, "Analysis of the Effectiveness and Cost-Effectiveness of Remote Sensing Devices" (Sierra Research Inc., Report to U.S. Environmental Protection Agency, May 1994).

12 This portion of the study was carried out with the cooperation of the California Bureau of Automotive Repair, the California Air Resources Board, and the EPA Mobile Source Emissions Research Branch.

13 D. R. Lawson, *J. Air Waste Manage. Assoc.* 43, 1567 (1993); *ibid.* 44, 121 (1994).

14 K. T. Knapp, in (23) pp. 871–884.

15 Increased vehicle prices may lead to decreased new car purchases, which may lead in turn to an older fleet. On average, older fleets produce higher emissions. Chrysler Corporation has stated that it will impose a surcharge that could exceed $2000 on all its new vehicles sold in California in 1998; the surcharge would allow Chrysler to offer a $45,000 electric minivan for around $18,000 so that Chrysler could sell enough of them to satisfy state laws ("Chrysler Electric Surcharge," *Rocky Mountain News*, 1 October 1994, p. 63A). Even electric vehicles, whose tailpipe emissions are zero, require electric power generation and have expensive batteries that require eventual replacement. Electric vehicle owners who continue to drive with weak batteries will cause traffic congestion, leading to an increase in the per mile

emissions of the gasoline-powered vehicles in the fleet. This possibility is not considered in any model of future emissions from fleets containing realistically maintained electric vehicles.

16 R. M. Rueff, *J. Air Waste Manage. Assoc.* 42, 921 (1992).

17 California Environmental Protection Agency, "Air Resources Board Staff Report, Proposed Amendments to the California Phase 2 Reformulated Gasoline Regulations, including Amendments Providing for the Use of a Predictive Model" (22 April 1994).

18 OXY-EF-2 data provided by the Mobile Source Division, U.S. Environmental Protection Agency, Ann Arbor, MI.

19 "SCRAP, a Clean Air Initiative from UNOCAL" (Unocal Co., 1991): "SCRAP2" (Report to South Coast Air Quality Management District, 1994).

20 This is a conservative estimate. We assume that all pre-1980 vehicles can be purchased for only $700 each, the next 8 million newer vehicles for $1000 each, and the newest 3.4 million vehicles for $2000 each.

21 G. A. Bishop, D. H. Stedman, J. E. Peterson, T. J. Hosick, P. L. Guenther, *J Air Waste Manage. Assoc.* 42, 2 (1993).

22 The cost of on-road monitoring is assumed to be $20 per gross polluter identified, on the basis of a cost of $0.50 per remote sensing test and a two-test identification requirement. Economies of scale may bring testing costs down to $0.16 per test when only gross polluters need to be identified [W. Harrington and V. D. McConnell, in *Cost Effective Control of Urban Smog* (Federal Reserve Bank of Chicago, November 1993), pp. 53–75].

23 J. C. Crow and D. M. Ono, eds., *PM10 Standards and Nontraditional Particulate Source Controls. AWMA/EPA International Specialty Conference, Vol. II* (Air and Waste Management Association, Pittsburgh, PA, 1992).

24 Supported by the California Air Resources Board, EPA, the American Petroleum Institute, the Coordinating Research Council, and the Colorado Office of Energy Conservation. The opinions expressed are solely those of the authors.

10(A)

CLIMATE CHANGE:
THE INTERGOVERNMENTAL
PANEL ON CLIMATE CHANGE
SCIENTIFIC ASSESSMENT –
POLICYMAKERS' SUMMARY*

J. T. Houghton, G. J. Jenkins, and J. J. Ephiraums

We are certain of the following:

- there is a natural greenhouse effect which already keeps the Earth warmer than it would otherwise be.
- emissions resulting from human activities are substantially increasing the atmospheric concentrations of the greenhouse gases: carbon dioxide, methane, chlorofluorocarbons (CFCs) and nitrous oxide. These increases will enhance the greenhouse effect, resulting on average in an additional warming of the Earth's surface. The main greenhouse gas, water vapour, will increase in response to global warming and further enhance it.

We calculate with confidence that:

- some gases are potentially more effective than others at changing climate, and their relative effectiveness can be estimated. Carbon dioxide has been responsible for over half the enhanced greenhouse effect in the past, and is likely to remain so in the future.
- atmospheric concentrations of the long-lived gases (carbon dioxide, nitrous oxide and the CFCs) adjust only slowly to changes in emissions. Continued emissions of these gases at present rates would commit us to increased concentrations for centuries ahead. The longer emissions continue to increase at present day rates, the greater reductions would have to be for concentrations to stabilise at a given level.
- the long-lived gases would require immediate reductions in emissions

* Originally published in Houghton, J. T., Jenkins, G. J. and Ephiraums, J. J. (1990) *Climate Change: The IPCC Scientific Assessment.* Cambridge: Cambridge University Press.

from human activities of over 60% to stabilise their concentrations at today's levels; methane would require a 15–20% reduction.

Based on current model results, we predict:

- under the IPCC Business-as-Usual (Scenario A) emissions of greenhouse gases, a rate of increase of global mean temperature during the next century of about 0.3°C per decade (with an uncertainty range of 0.2°C to 0.5°C per decade); this is greater than that seen over the past 10,000 years. This will result in a likely increase in global mean temperature of about 1°C above the present value by 2025 and 3°C before the end of the next century. The rise will not be steady because of the influence of other factors.
- under the other IPCC emission scenarios which assume progressively increasing levels of controls, rates of increase in global mean temperature of about 0.2°C per decade (Scenario B), just above 0.1°C per decade (Scenario C) and about 0.1°C per decade (Scenario D).
- that land surfaces warm more rapidly than the ocean, and high northern latitudes warm more than the global mean in winter.
- regional climate changes different from the global mean, although our confidence in the prediction of the detail of regional changes is low. For example, temperature increases in Southern Europe and central North America are predicted to be higher than the global mean, accompanied on average by reduced summer precipitation and soil moisture. There are less consistent predictions for the tropics and the Southern Hemisphere.
- under the IPCC Business as Usual emissions scenario, an average rate of global mean sea level rise of about 6cm per decade over the next century (with an uncertainty range of 3–10cm per decade), mainly due to thermal expansion of the oceans and the melting of some land ice. The predicted rise is

about 20cm in global mean sea level by 2030, and 65cm by the end of the next century. There will be significant regional variations.

There are many uncertainties in our predictions particularly with regard to the timing, magnitude and regional patterns of climate change, due to our incomplete understanding of:

- sources and sinks of greenhouse gases, which affect predictions of future concentrations.
- clouds, which strongly influence the magnitude of climate change.
- oceans, which influence the timing and patterns of climate change.
- polar ice sheets which affect predictions of sea level rise.

These processes are already partially understood, and we are confident that the uncertainties can be reduced by further research. However, the complexity of the system means that we cannot rule out surprises.

Our judgement is that:

- Global-mean surface air temperature has increased by 0.3°C to 0.6°C over the last 100 years, with the five global-average warmest years being in the 1980s. Over the same period global sea level has increased by 10–20cm. These increases have not been smooth with time, nor uniform over the globe.
- The size of this warming is broadly consistent with predictions of climate models, but it is also of the same magnitude as natural climate variability. Thus the observed increase could be largely due to this natural variability; alternatively this variability and other human factors could have offset a still larger human-induced greenhouse warming. The unequivocal detection of the enhanced greenhouse effect from observations is not likely for a decade or more.
- There is no firm evidence that climate has become more variable over the last few decades. However, with an

increase in the mean temperature, episodes of high temperatures will most likely become more frequent in the future, and cold episodes less frequent.

- Ecosystems affect climate, and will be affected by a changing climate and by increasing carbon dioxide concentrations. Rapid changes in climate will change the composition of ecosystems; some species will benefit while others will be unable to migrate or adapt fast enough and may become extinct. Enhanced levels of carbon dioxide may increase productivity and efficiency of water use of vegetation. The effect of warming on biological processes, although poorly understood, may increase the atmospheric concentrations of natural greenhouse gases.

To improve our predictive capability, we need:

- to *understand* better the various climate-related processes, particularly those associated with clouds, oceans and the carbon cycle.
- to *improve* the systematic observation of climate-related variables on a global basis, and further investigate changes which took place in the past.
- to *develop* improved models of the Earth's climate system.
- to *increase* support for national and international climate research activities, especially in developing countries.
- to *facilitate* international exchange of climate data.

Introduction: What is the Issue?

There is concern that human activities may be inadvertently changing the climate of the globe through the enhanced greenhouse effect, by past and continuing emissions of carbon dioxide and other gases which will cause the temperature of the Earth's surface to increase – popularly termed the "global warming". If this occurs, consequent changes may have a significant impact on society.

The purpose of the Working Group I report, as determined by the first meeting of IPCC, is to provide a scientific assessment of:

(1) the factors which may affect climate change during the next century, especially those which are due to human activity.
(2) the responses of the atmosphere–ocean–land–ice system.
(3) current capabilities of modelling global and regional climate changes and their predictability.
(4) the past climate record and presently observed climate anomalies.

On the basis of this assessment, the report presents current knowledge regarding predictions of climate change (including sea level rise and the effects on ecosystems) over the next century, the timing of changes together with an assessment of the uncertainties associated with these predictions.

This Policymakers Summary aims to bring out those elements of the main report which have the greatest relevance to policy formulation, in answering the following questions:

- What factors determine global climate?
- What are the greenhouse gases, and how and why are they increasing?
- Which gases are the most important?
- How much do we expect the climate to change?
- How much confidence do we have in our predictions?
- Will the climate of the future be very different?
- Have human activities already begun to change global climate?
- How much will sea level rise?
- What will be the effects on ecosystems?
- What should be done to reduce uncertainties, and how long will this take?

This report is intended to respond to the practical needs of the policymaker. It is neither an academic review, nor a plan for a new research programme. Uncertainties attach to almost every aspect of the issue, yet policymakers are looking for clear guidance from scientists; *hence authors have been asked to provide their best-estimates wherever possible, together with an assessment of the uncertainties.*

This report is a summary of our understanding in 1990. Although continuing research will deepen this understanding and require the report to be updated at frequent intervals, basic conclusions concerning the reality of the enhanced greenhouse effect and its potential to alter global climate are unlikely to change significantly. Nevertheless, the complexity of the system may give rise to surprises.

What Factors Determine Global Climate?

There are many factors, both natural and of human origin, that determine the climate of the earth. We look first at those which are natural, and then see how human activities might contribute.

What natural factors are important?

The driving energy for weather and climate comes from the Sun. The Earth intercepts solar radiation (including that in the short-wave, visible, part of the spectrum); about a third of it is reflected, the rest is absorbed by the different components (atmosphere, ocean, ice, land and biota) of the climate system. The energy absorbed from solar radiation is balanced (in the long term) by outgoing radiation from the Earth and atmosphere; this terrestrial radiation takes the form of long-wave invisible infrared energy, and its magnitude is determined by the temperature of the Earth-atmosphere system.

There are several natural factors which can change the balance between the energy absorbed by the Earth and that emitted by it in the form of longwave infrared radiation; these factors cause the *radiative forcing* on climate. The most obvious of these is a change in the output of energy from the Sun. There is direct evidence of such variability over the 11-year solar cycle, and longer period changes may also occur. Slow variations in the Earth's orbit affect the seasonal

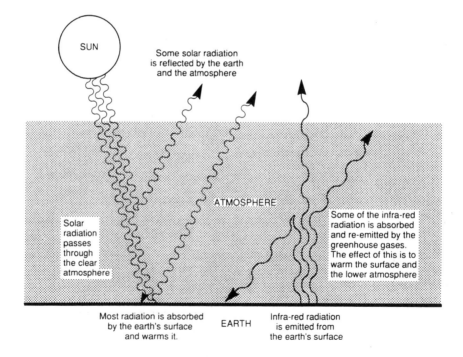

Figure 10.1 [*orig. figure 1*]   A simplified diagram illustrating the greenhouse effect.

and latitudinal distribution of solar radiation; these were probably responsible for initiating the ice ages.

One of the most important factors is the *greenhouse effect*; a simplified explanation of which is as follows. Short-wave solar radiation can pass through the clear atmosphere relatively unimpeded. But long-wave terrestrial radiation emitted by the warm surface of the Earth is partially absorbed and then re-emitted by a number of trace gases in the cooler atmosphere above. Since, on average, the outgoing long-wave radiation balances the incoming solar radiation, both the atmosphere and the surface will be warmer than they would be without the greenhouse gases (figure 10.1).

The main natural greenhouse gases are not the major constituents, nitrogen and oxygen, but water vapour (the biggest contributor), carbon dioxide, methane, nitrous oxide, and ozone in the troposphere (the lowest 10–15km of the atmosphere) and stratosphere.

Aerosols (small particles) in the atmosphere can also affect climate because they can reflect and absorb radiation. The most important natural perturbations result from explosive volcanic eruptions which affect concentrations in the lower stratosphere. Lastly, the climate has its own *natural variability* on all timescales and changes occur without any external influence.

How do we know that the natural greenhouse effect is real?

The greenhouse effect is real; it is a well understood effect, based on established scientific principles. We know that the greenhouse effect works in practice, for several reasons.

Firstly, the mean temperature of the Earth's surface is already warmer by about 33°C (assuming the same reflectivity of the earth) than it would be if the natural greenhouse gases were not present. Satellite observations of the radiation emitted from the Earth's surface and through the atmosphere demonstrate the effect of the greenhouse gases.

Secondly, we know the composition of the atmospheres of Venus, Earth and Mars are very different, and their surface temperatures are in general agreement with greenhouse theory.

Thirdly, measurements from ice cores going back 160,000 years show that the Earth's temperature closely paralleled the amount of carbon dioxide and methane in the atmosphere (see figure 10.2). Although we do not know the details of cause and effect, calculations indicate that changes in these greenhouse gases were part, but not all, of the reason for the large (5–7°C) global temperature swings between ice ages and interglacial periods.

How might human activities change global climate?

Naturally occurring greenhouse gases keep the Earth warm enough to be habitable. By increasing their concentrations, and by adding new greenhouse gases like

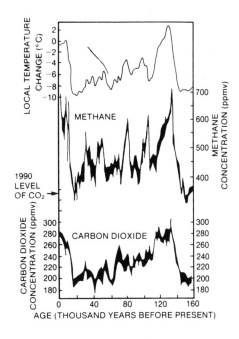

Figure 10.2 [*orig. figure 2*]   Analysis of air trapped in Antarctic ice cores shows that methane and carbon dioxide concentrations were closely correlated with the local temperature over the last 160,000 years. Present day concentrations of carbon dioxide are indicated.

chlorofluorocarbons (CFCs), humankind is capable of raising the global-average annual-mean surface-air temperature (which, for simplicity, is referred to as the "global temperature"), although we are uncertain about the rate at which this will occur. Strictly, this is an *enhanced* greenhouse effect – above that occurring due to natural greenhouse gas concentrations: the word "enhanced" is usually omitted, but it should not be forgotten. Other changes in climate are expected to result, for example changes in precipitation, and a global warming will cause sea levels to rise; these are discussed in more detail later.

There are other human activities which have the potential to affect climate. A change in the albedo (reflectivity) of the land, brought about by *desertification or deforestation* affects the amount of solar energy absorbed at the Earth's surface. Human-made *aerosols*, from sulphur emitted largely in fossil fuel combustion, can modify clouds and this may act to lower temperatures. Lastly, changes in *ozone in the stratosphere* due to CFCs may also influence climate.

What are the Greenhouse Gases and why are they Increasing?

We are certain that the concentrations of greenhouse gases in the atmosphere have changed naturally on ice-age time-scales, and have been increasing since pre-industrial times due to human activities. Table 10.1 summarizes the present and pre-industrial abundances, current rates of change and present atmospheric lifetimes of greenhouse gases influenced by human activities. Carbon dioxide, methane, and nitrous oxide all have significant natural and human sources, while the chlorofluorocarbons are only produced industrially.

Two important greenhouse gases, water vapour and ozone, are not included in this table. Water vapour has the largest greenhouse effect, but its concentration in the troposphere is determined internally within the climate system, and, on a global scale, is not affected by human sources and sinks. Water vapour will increase in response to global warming and further enhance it: this process is included in climate models. The concentration of ozone is changing both in the stratosphere and the troposphere due to human activities, but it is difficult to quantify the changes from present observations.

For a thousand years prior to the industrial revolution, abundances of the greenhouse gases were relatively constant. However, as the world's population increased, as the world became more industrialized and as agriculture developed, the abundances of the greenhouse gases increased markedly. Figure 10.3 illustrates this for carbon dioxide, methane, nitrous oxide and CFC-11.

Since the industrial revolution the combustion of fossil fuels and deforestation have led to an increase of 26% in carbon

Table 10.1 [*orig. table 1*]   Summary of key greenhouse gases affected by human activities.

	Carbon Dioxide	Methane	CFC-11	CFC-12	Nitrous Oxide
Atmospheric concentration	ppmv	ppmv	pptv	pptv	ppbv
Pre-industrial (1750–1800)	280	0.8	0	0	288
Present day (1990)	353	1.72	280	484	310
Current rate of change per year	1.8 (0.5%)	0.015 (0.9%)	9.5 (4%)	17 (4%)	0.8 (0.25%)
Atmospheric lifetime (years)	(50–200)†	10	65	130	150

ppmv = parts per million by volume;
ppbv = parts per billion (thousand million) by volume;
pptv = parts per trillion (million million) by volume.
†   The way in which CO_2 is absorbed by the oceans and biosphere is not simple and a single value cannot be given; refer to the main report for further discussion.

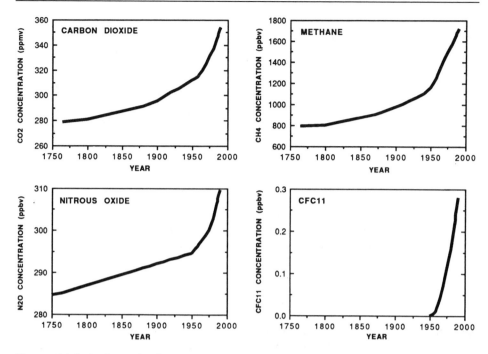

Figure 10.3 [*orig. figure 3*]   Concentrations of carbon dioxide and methane, after remaining relatively constant up to the 18th century, have risen sharply since then due to man's activities. Concentrations of nitrous oxide have increased since the mid-18th century, especially in the last few decades. CFCs were not present in the atmosphere before the 1930s.

dioxide concentration in the atmosphere. We know the magnitude of the present day fossil-fuel source, but the input from deforestation cannot be estimated accurately. In addition, although about half of the emitted carbon dioxide stays in the atmosphere, we do not know well how much of the remainder is absorbed by the oceans and how much by terrestrial biota. Emissions of chlorofluorocarbons, used as aerosol propellants, solvents, refrigerants and foam blowing agents, are also well known; they were not present in the atmosphere before their invention in the 1930s.

The sources of methane and nitrous oxide are less well known. Methane concentrations have more than doubled because of rice production, cattle rearing, biomass burning, coal mining and ventilation of natural gas; also, fossil fuel combustion may have also contributed through chemical reactions in the atmosphere which reduce the rate of removal of methane. Nitrous

oxide has increased by about 8% since pre-industrial times, presumably due to human activities; we are unable to specify the sources, but it is likely that agriculture plays a part.

The effect of ozone on climate is strongest in the upper troposphere and lower stratosphere. Model calculations indicate that ozone in the upper troposphere should have increased due to human-made emissions of nitrogen oxides, hydrocarbons and carbon monoxide. While at ground level ozone has increased in the Northern Hemisphere in response to these emissions, observations are insufficient to confirm the expected increase in the upper troposphere. The lack of adequate observations prevents us from accurately quantifying the climatic effect of changes in tropospheric ozone.

In the lower stratosphere at high southern latitudes ozone has decreased considerably due to the effects of CFCs, and there are indications of a global-scale decrease which,

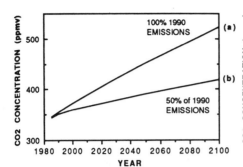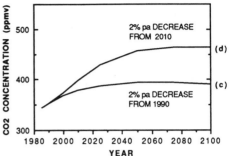

Figure 10.4 [*orig. figure 4*]   The relationship between hypothetical fossil fuel emissions of carbon dioxide and its concentration in the atmosphere is shown in the case where (a) emissions continue at 1990 levels, (b) emissions are reduced by 50% in 1990 and continue at that level, (c) emissions are reduced by 2% pa from 1990, and (d) emissions, after increasing by 2% pa until 2010, are then reduced by 2% pa thereafter.

while not understood, may also be due to CFCs. These observed decreases should act to cool the Earth's surface, thus providing a small offset to the predicted warming produced by the other greenhouse gases. Further reductions in lower stratospheric ozone are possible during the next few decades as the atmospheric abundances of CFCs continue to increase.

Concentrations, lifetimes and stabilisation of the gases

In order to calculate the atmospheric concentrations of carbon dioxide which will result from human-made emissions we use computer models which incorporate details of the emissions and which include representations of the transfer of carbon dioxide between the atmosphere, oceans and terrestrial biosphere. For the other greenhouse gases, models which incorporate the effects of chemical reactions in the atmosphere are employed.

The atmospheric lifetimes of the gases are determined by their sources and sinks in the oceans, atmosphere and biosphere. Carbon dioxide, chlorofluorocarbons and nitrous oxide are removed only slowly from the atmosphere and hence, following a change in emissions, their atmospheric concentrations take decades to centuries to adjust fully. Even if all human-made emissions of carbon dioxide were halted in the year 1990,

about half of the increase in carbon dioxide concentration caused by human activities would still be evident by the year 2100.

In contrast, some of the CFC substitutes and methane have relatively short atmospheric lifetimes so that their atmospheric concentrations respond fully to emission changes within a few decades.

To illustrate the emission-concentration relationship clearly, the effect of hypothetical changes in carbon dioxide fossil fuel emissions is shown in figure 10.4: (a) continuing global emissions at 1990 levels; (b) halving of emissions in 1990; (c) reductions in emissions of 2% per year (pa) from 1990 and (d) a 2% pa increase from 1990–2010 followed by a 2% pa decrease from 2010.

Continuation of present day emissions are committing us to increased future concentrations, and the longer emissions continue to increase, the greater would reductions have to be to stabilise at a given level. If there are critical concentration levels that should not be exceeded, then the earlier emission reductions are made the more effective they are.

The term "*atmospheric stabilisation*" is often used to describe the limiting of the concentration of the greenhouse gases at a certain level. The amount by which human-made emissions of a greenhouse gas must be reduced in order to stabilise at present day concentrations, for example, is shown in table 10.2. For most gases the reductions would have to be substantial.

How will greenhouse gas abundances change in the future?

We need to know future greenhouse gas concentrations in order to estimate future climate change. As already mentioned, these concentrations depend upon the magnitude of human-made emissions and on how changes in climate and other environmental conditions may influence the biospheric processes that control the exchange of natural greenhouse gases, including carbon

Table 10.2 [*orig. table 2*]   Stabilisation of atmospheric concentrations. Reductions in the human-made emissions of greenhouse gases required to stabilise concentrations at present day levels.

Greenhouse Gas	Reduction Required
Carbon Dioxide	>60%
Methane	15–20%
Nitrous Oxide	70–80%
CFC-11	70–75%
CFC-12	75–85%
HCFC-22	40–50%

Note that the stabilisation of each of these gases would have different effects on climate, as explained in the next section.

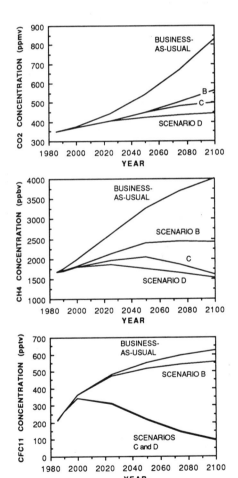

Figure 10.5 [*orig. figure 5*]   Atmospheric concentrations of carbon dioxide, methane and CFC-11 resulting from the four IPCC emissions scenarios.

dioxide and methane, between the atmosphere, oceans and terrestrial biosphere – the greenhouse gas "feedbacks".

Four scenarios of future human-made emissions were developed by Working Group III. The first of these assumes that few or no steps are taken to limit greenhouse gas emissions, and this is therefore termed Business-as-Usual (BaU). (It should be noted that an aggregation of national forecasts of emissions of carbon dioxide and methane to the year 2025 undertaken by Working Group III resulted in global emissions 10–20% higher than in the BaU scenario). The other three scenarios assume that progressively increasing levels of controls reduce the growth of emissions; these are referred to as scenarios B, C, and D. Future concentrations of some of the greenhouse gases which would arise from these emissions are shown in figure 10.5.

Greenhouse gas feedbacks

Some of the possible feedbacks which could significantly modify future greenhouse gas concentrations in a warmer world are discussed in the following paragraphs.

The net emissions of carbon dioxide from terrestrial ecosystems will be elevated if higher temperatures increase respiration at a faster rate than photosynthesis, or if plant populations, particularly large forests, cannot adjust rapidly enough to changes in climate.

A net flux of carbon dioxide to the atmosphere may be particularly evident in warmer conditions in tundra and boreal regions where there are large stores of carbon. The opposite is true if higher abundances of carbon dioxide in the atmosphere enhance the productivity of natural ecosystems, or if there is an increase in soil moisture which can be expected to stimulate plant growth in dry ecosystems and to increase the storage of carbon in tundra peat. The extent to which ecosystems can sequester increasing atmospheric carbon dioxide remains to be quantified.

If the oceans become warmer, their net uptake of carbon dioxide may decrease because of changes in (i) the chemistry of carbon dioxide in seawater, (ii) biological activity in surface waters, and (iii) the rate of exchange of carbon dioxide between the surface layers and the deep ocean. This last depends upon the rate of formation of deep water in the ocean which, in the North Atlantic for example, might decrease if the salinity decreases as a result of a change in climate.

Methane emissions from natural wetlands and rice paddies are particularly sensitive to temperature and soil moisture. Emissions are significantly larger at higher temperatures and with increased soil moisture; conversely, a decrease in soil moisture would result in smaller emissions. Higher temperatures could increase the emissions of methane at high northern latitudes from decomposable organic matter trapped

in permafrost and methane hydrates.

As illustrated earlier, ice core records show that methane and carbon dioxide concentrations changed in a similar sense to temperature between ice ages and interglacials.

Although many of these feedback processes are poorly understood, it seems likely that, overall, they will act to increase, rather than decrease, greenhouse gas concentrations in a warmer world.

Which Gases are the Most Important?

We are certain that increased greenhouse gas concentrations increase radiative forcing. We can calculate the forcing with much more confidence than the climate change that results because the former avoids the need to evaluate a number of poorly understood atmospheric responses. We then have a base from which to calculate the relative effect on climate of an increase in *concentration* of each gas in the present-day atmosphere, both in absolute terms and relative to carbon dioxide. These relative effects span a wide range; methane is about 21 times more effective, molecule-for-molecule, than carbon dioxide, and CFC-11 about 12,000 times more effective. On a kilogram-per-kilogram basis, the equivalent values are 58 for methane and about 4,000 for CFC-11, both relative to carbon dioxide.

The total radiative forcing at any time is the sum of those from the individual greenhouse gases. We show in figure 10.6 how this quantity has changed in the past (based on observations of greenhouse gases) and how it might change in the future (based on the four IPCC emissions scenarios). For simplicity, we can express total forcing in terms of the amount of carbon dioxide which would give that forcing; this is termed the *equivalent carbon dioxide concentration*. Greenhouse gases have increased since pre-industrial times (the mid-18th century) by an amount that is radiatively equivalent to about a 50% increase in carbon dioxide, although carbon dioxide itself has risen by only 26%; other gases have made up the rest.

The contributions of the various gases to

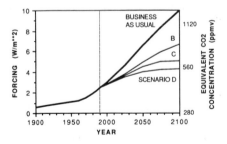

Figure 10.6 [*orig. figure 6*] Increase in radiative forcing since the mid-18th century, and predicted to result from the four IPCC emissions scenarios, also expressed as equivalent carbon dioxide concentrations.

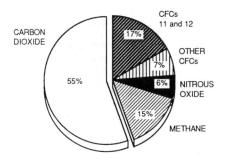

Figure 10.7 [*orig. figure 7*]   The contribution from each of the human-made greenhouse gases to the change in radiative forcing from 1980 to 1990. The contribution from ozone may also be significant, but cannot be quantified at present.

the total increase in climate forcing during the 1980s is shown as a pie diagram in figure 10.7; carbon dioxide is responsible for about half the decadal increase. (Ozone, the effects of which may be significant, is not included).

How can we evaluate the effect of different greenhouse gases?

To evaluate possible policy options, it is useful to know the relative radiative effect (and, hence, potential climate effect) of equal emissions of each of the greenhouse gases. The concept of relative Global Warming Potentials (GWP) has been developed to take into account the differing times that gases remain in the atmosphere.

This index defines the time-integrated warming effect due to an instantaneous release of unit mass (1 kg) of a given greenhouse gas in today's atmosphere, relative to that of carbon dioxide. The relative importances will change in the future as atmospheric composition changes because, although radiative forcing increases in direct proportion to the concentration of CFCs, changes in the other greenhouse gases (particularly carbon dioxide) have an effect on forcing which is much less than proportional.

The GWPs in table 10.3 are shown for three time horizons, reflecting the need to consider the cumulative effects on climate over various time scales. The longer time horizon is appropriate for the cumulative effect; the shorter timescale will indicate the response to emission changes in the short term. There are a number of practical difficulties in devising and calculating the values of the GWPs, and the values given here should be considered as preliminary. In addition to these direct effects, there are indirect effects of human-made emissions arising from chemical reactions between the various constituents. The indirect effects on stratospheric water vapour, carbon dioxide and tropospheric ozone have been included in these estimates.

Table 10.3 indicates, for example, that the effectiveness of methane in influencing climate will be greater in the first few decades after release, whereas emission of the longer-lived nitrous oxide will affect climate for a much longer time. The lifetimes of the proposed CFC replacements range from 1 to 40 years: the longer lived replace-

Table 10.3 [*orig. table 3*]   Global Warming Potentials. The warming effect of an emission of 1kg of each gas relative to that of CO_2. These figures are best estimates calculated on the basis of the present day atmospheric composition.

	TIME HORIZON		
	20 yr	100 yr	500 yr
Carbon dioxide	1	1	1
Methane (including indirect)	63	21	9
Nitrous oxide	270	290	190
CFC-11	4500	3500	1500
CFC-12	7100	7300	4500
HCFC-22	4100	1500	510

Global Warming Potentials for a range of CFCs and potential replacements are given in the full text.

ments are still potentially effective as agents of climate change. One example of this, HCFC-22 (with a 15 year lifetime), has a similar effect (when released in the same amount) as CFC-11 on a 20 year time-scale; but less over a 500 year time-scale.

Table 10.3 shows carbon dioxide to be the least effective greenhouse gas per kilogramme emitted, but its contribution to global warming, which depends on the product of the GWP and the amount emitted, is largest. In the example in table 10.4, the effect over 100 years of emissions of greenhouse gases in 1990 are shown relative to carbon dioxide. This is illustrative; to compare the effect of different emission projections we have to sum the effect of emissions made in future years.

There are other technical criteria which may help policymakers to decide, in the event of emissions reductions being deemed necessary, which gases should be considered. Does the gas contribute in a major way to current, and future, climate forcing? Does it have a long lifetime, so earlier reductions in emissions would be more effective than those made later? And are its sources and sinks well enough known to decide which could be controlled in practice? Table 10.5 illustrates these factors.

How Much do we Expect Climate to Change?

It is relatively easy to determine the direct effect of the increased radiative forcing due to increases in greenhouse gases. However, as climate begins to warm, various processes act to amplify (through positive feedbacks) or reduce (through negative feedbacks) the warming. The main feedbacks which have been identified are due to changes in water vapour, sea-ice, clouds and the oceans.

The best tools we have which take the above feedbacks into account (but do not

Table 10.4 [*orig. table 4*]   The relative cumulative climate effect of 1990 man-made emissions.

	GWP (100yr horizon)	1990 emissions (Tg)	Relative contribution over 100yr
Carbon dioxide	1	26000†	61%
Methane*	21	300	15%
Nitrous oxide	290	6	4%
CFCs	Various	0.9	11%
HCFC-22	1500	0.1	0.5%
Others*	Various		8.5%

* These values include the indirect effect of these emissions on other greenhouse gases via chemical reactions in the atmosphere. Such estimates are highly model dependent and should be considered preliminary and subject to change. The estimated effect of ozone is included under "others." The gases included under "others" are given in the full report.
† 26 000 Tg (teragrams) of carbon dioxide = 7 000 Tg (= 7 Gt) of carbon.

Table 10.5 [*orig. table 5*]   Characteristics of greenhouse gases.

Gas	Major Contributor?	Long lifetime?	Sources known?
Carbon dioxide	yes	yes	yes
Methane	yes	no	semi-quantitatively
Nitrous oxide	not at present	yes	qualitatively
CFCs	yes	yes	yes
HCFCs, etc.	not at present	mainly no	yes
Ozone	possibly	no	qualitatively

include greenhouse gas feedbacks) are three-dimensional mathematical models of the climate system (atmosphere–ocean–ice–land), known as General Circulation Models (GCMs). They synthesise our knowledge of the physical and dynamical processes in the overall system and allow for the complex interactions between the various components. However, in their current state of development, the descriptions of many of the processes involved are comparatively crude. Because of this, considerable uncertainty is attached to these predictions of climate change, which is reflected in the range of values given; further details are given in a later section.

The estimates of climate change presented here are based on:

(i)   the "best-estimate" of equilibrium climate sensitivity (i.e. the equilibrium temperature change due to a doubling of carbon dioxide in the atmosphere) obtained from model simulations, feedback analyses and observational considerations (see box 10.2: "What tools do we use?")

(ii)   a "box-diffusion-upwelling" ocean-atmosphere climate model which translates the greenhouse forcing into the evolution of the temperature response for the prescribed climate sensitivity. (This simple model has been calibrated against more complex atmosphere-ocean coupled GCMs for situations where the more complex models have been run).

How quickly will global climate change?

a. If emissions follow a Business-as-Usual pattern

Under the IPCC Business-as-Usual (Scenario A) emissions of greenhouse gases, the average rate of increase of global mean temperature during the next century is estimated to be about 0.3°C per decade (with an uncertainty range of 0.2°C to 0.5°C). This will result in a likely increase in global mean temperature of about 1°C above the present

---

Box 10.1   Estimates for Changes by 2030
            (IPCC Business-as-Usual scenario; changes from pre-industrial)
The numbers given below are based on high resolution models, scaled to be consistent with our best estimate of global mean warming of 1.8°C by 2030. For values consistent with other estimates of global temperature rise, the numbers below should be reduced by 30% for the low estimate or increased by 50% for the high estimate. Precipitation estimates are also scaled in a similar way.

Confidence in these regional estimates is low

Central North America (35°–50°N 85°–105°W)
The warming varies from 2 to 4°C in winter and 2 to 3°C in summer. Precipitation increases range from 0 to 15% in winter whereas there are decreases of 5 to 10% in summer. Soil moisture decreases in summer by 15 to 20%.

Southern Asia (5°–30°N 70°–105°E)
The warming varies from 1 to 2°C throughout the year. Precipitation changes little in winter and generally increases throughout the region by 5 to 15% in summer. Summer soil moisture increases by 5 to 10%.

Sahel (10°–20°N 20°W–40°E)
The warming ranges from 1 to 3°C. Area mean precipitation increases and area mean soil moisture decreases marginally in summer. However, throughout the region, there are areas of both increase and decrease in both parameters throughout the region.

Southern Europe (35°–50°N 10°W–45°E)
The warming is about 2°C in winter and varies from 2 to 3°C in summer. There is some indication of increased precipitation in winter, but summer precipitation decreases by 5 to 15%, and summer soil moisture by 15 to 25%.

Australia (12°–45°S 110°–115°E)
The warming ranges from 1 to 2°C in summer and is about 2°C in winter. Summer precipitation increases by around 10%, but the models do not produce consistent estimates of the changes in soil moisture. The area averages hide large variations at the sub-continental level.

Box 10.2   What Tools Do We Use to Predict Future Climate, and How Do We Use Them?

The most highly developed tool which we have to predict future climate is known as a *general circulation model or GCM*. These models are based on the laws of physics and use descriptions in simplified physical terms (called parameterisations) of the smaller-scale processes such as those due to clouds and deep mixing in the ocean. In a climate model an atmospheric component, essentially the same as a weather prediction model, is coupled to a model of the ocean, which can be equally complex.

Climate forecasts are derived in a different way from weather forecasts. A weather prediction model gives a description of the atmosphere's state up to 10 days or so ahead, starting from a detailed description of an initial state of the atmosphere at a given time. Such forecasts describe the movement and development of large weather systems, though they cannot represent very small scale phenomena; for example, individual shower clouds.

To make a climate forecast, the climate model is first run for a few (simulated) decades. The statistics of the model's output is a description of the model's simulated climate which, if the model is a good one, will bear a close resemblance to the climate of the real atmosphere and ocean. The above exercise is then repeated with increasing concentrations of the greenhouse gases in the model. The differences between the statistics of the two simulations (for example in mean temperature and interannual variability) provide an estimate of the accompanying climate change.

The long term change in *surface air temperature* following a doubling of carbon dioxide (referred to as the *climate sensitivity*) is generally used as a benchmark to compare models. The range of results from model studies is 1.9 to 5.2°C. Most results are close to 4.0°C but recent studies using a more detailed but not necessarily more accurate representation of cloud processes give results in the lower half of this range. Hence the models' results do not justify altering the previously accepted range of 1.5 to 4.5°C.

Although scientists are reluctant to give a single best estimate in this range, it is necessary for the presentation of climate predictions for a choice of best estimate to be made. Taking into account the model results, together with observational evidence over the last century which is suggestive of the climate sensitivity being in the lower half of the range, (see section: "Has man already begun to change global climate?") a value of climate sensitivity of 2.5°C has been chosen as the best estimate. Further details are given in Section 5 of the report.

In this Assessment, we have also used much simpler models, which simulate the behaviour of GCMs, to make predictions of the evolution with time of global temperature from a number of emission scenarios. These so-called box-diffusion models contain highly simplified physics but give similar results to GCMs when globally averaged.

A completely different, and potentially useful, way of predicting patterns of future climate is to search for periods in the past when the global mean temperatures were similar to those we expect in future, and then use the past spatial patterns as *analogues* of those which will arise in the future. For a good analogue, it is also necessary for the forcing factors (for example, greenhouse gases, orbital variations) and other conditions (for example, ice cover, topography, etc.) to be similar; direct comparisons with climate situations for which these conditions do not apply cannot be easily interpreted. Analogues of future greenhouse-gas-changed climates have not been found.

We cannot therefore advocate the use of palaeo-climates as predictions of regional climate change due to future increases in greenhouse gases. However, palaeo-climatological information can provide useful insights into climate processes, and can assist in the validation of climate models.

value (about 2°C above that in the pre-industrial period) by 2025 and 3°C above today's (about 4°C above pre-industrial) before the end of the next century.

The projected temperature rise out to the year 2100, with high, low and best-estimate climate responses, is shown in figure 10.8. Because of other factors which influence climate, we would not expect the rise to be a steady one.

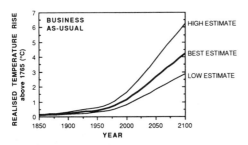

Figure 10.8 [*Orig. figure 8*] Simulation of the increase in global mean temperature from 1850–1990 due to observed increases in greenhouse gases, and predictions of the rise between 1990 and 2100 resulting from the Business-as-Usual emissions.

The temperature rises shown above are *realised* temperatures; at any time we would also be *committed* to a further temperature rise toward the equilibrium temperature (see box 10.3: "Equilibrium and Realised Climate Change"). For the BaU "best-estimate" case in the year 2030, for example, a further 0.9°C rise would be expected, about 0.2°C of which would be realised by 2050 (in addition to changes due to further greenhouse gas increases); the rest would become apparent in decades or centuries.

Even if we were able to stabilise emissions of each of the greenhouse gases at present day levels from now on, the temperature is predicted to rise by about 0.2°C per decade for the first few decades.

The global warming will also lead to increased global average precipitation and evaporation of a few percent by 2030. Areas of sea-ice and snow are expected to diminish.

b. If emissions are subject to controls
Under the other IPCC emission scenarios which assume progressively increasing

Box 10.3   Equilibrium and Realised Climate Change

When the radiative forcing on the earth-atmosphere system is changed, for example by increasing greenhouse gas concentrations, the atmosphere will try to respond (by warming) immediately. But the atmosphere is closely coupled to the oceans, so in order for the air to be warmed by the greenhouse effect, the oceans also have to be warmed; because of their thermal capacity this takes decades or centuries. This exchange of heat between atmosphere and ocean will act to slow down the temperature rise forced by the greenhouse effect.

In a hypothetical example where the concentration of greenhouse gases in the atmosphere, following a period of constancy, rises suddenly to a new level and remains there, the radiative forcing would also rise rapidly to a new level. This increased radiative forcing would cause the atmosphere and oceans to warm, and eventually come to a new, stable, temperature. A commitment to this *equilibrium* temperature rise is incurred as soon as the greenhouse gas concentration changes. But at any time before equilibrium is reached, the actual temperature will have risen by only part of the equilibrium temperature change, known as the *realised* temperature change.

Models predict that, for the present day case of an increase in radiative forcing which is approximately steady, the realised temperature rise at any time is about 50% of the committed temperature rise if the climate sensitivity (the response to a doubling of carbon dioxide) is 4.5°C and about 80% if the climate sensitivity is 1.5°C. If the forcing were then held constant, temperatures would continue to rise slowly, but it is not certain whether it would take decades or centuries for most of the remaining rise to equilibrium to occur.

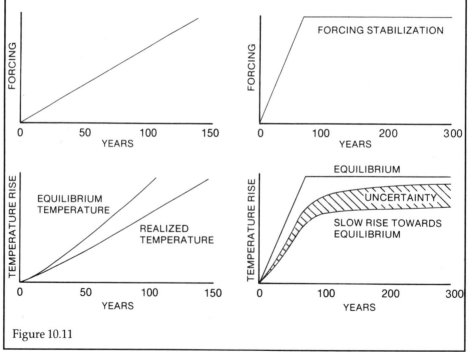

Figure 10.11

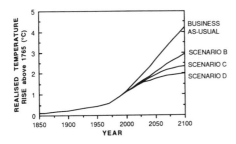

Figure 10.9 [*orig. figure 9*] Simulations of the increase in global mean temperature from 1850–1990 due to observed increases in greenhouse gases, and predictions of the rise between 1990 and 2100 resulting from the IPCC Scenarios B, C and D emissions, with the Business-as-Usual case for comparison.

levels of controls, average rates of increase in global mean temperature over the next century are estimated to be about 0.2°C per decade (Scenario B), just above 0.1°C per decade (Scenario C) and about 0.1°C per decade (Scenario D). The results are illustrated in figure 10.9, with the Business-as-Usual case shown for comparison. Only the best-estimate of the temperature rise is shown in each case.

The indicated range of uncertainty in

global temperature rise given above reflects a subjective assessment of uncertainties in the calculation of climate response, but does not include those due to the transformation of emissions to concentrations, nor the effects of greenhouse gas feedbacks.

What will be the patterns of climate change by 2030?

Knowledge of the global mean warming and change in precipitation is of limited use in determining the impacts of climate change, for instance on agriculture. For this we need to know changes regionally and seasonally.

Models predict that surface air will warm faster over land than over oceans, and a minimum of warming will occur around Antarctica and in the northern North Atlantic region.

There are some continental-scale changes which are consistently predicted by the highest resolution models and for which we understand the physical reasons. The warming is predicted to be 50–100% greater than the global mean in high northern latitudes in winter, and substantially smaller than the global mean in regions of sea-ice in summer. Precipitation is predicted to increase on average in middle and

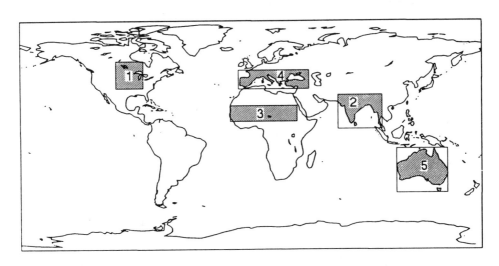

Figure 10.10 [*orig. figure 10*]   Map showing the locations and extents of the five areas selected by IPCC.

high latitude continents in winter (by some 5–10% over 35–55°N).

Five regions, each a few million square kilometres in area and representative of different climatological regimes, were selected by IPCC for particular study (see figure 10.10). Box 10.1 gives the changes in temperature, precipitation and soil moisture, which are predicted to occur by 2030 on the Business-as-Usual scenario, as an average over each of the five regions. There may be considerable variations within the regions. In general, confidence in these regional estimates is low, especially for the changes in precipitation and soil moisture, but they are examples of our best estimates. We cannot yet give reliable regional predictions at the smaller scales demanded for impacts assessments.

How will climate extremes and extreme events change?

Changes in the variability of weather and the frequency of extremes will generally have more impact than changes in the mean climate at a particular location. With the possible exception of an increase in the number of intense showers, there is no clear evidence that weather variability will change in the future. In the case of temperatures, assuming no change in variability, but with a modest increase in the mean, the number of days with temperatures above a given value at the high end of the distribution will increase substantially. On the same assumptions, there will be a decrease in days with temperatures at the low end of the distribution. So the number of very hot days or frosty nights can be substantially changed without any change in the variability of the weather. The number of days with a minimum threshold amount of soil moisture (for viability of a certain crop, for example) would be even more sensitive to changes in average precipitation and evaporation.

If the large-scale weather regimes, for instance depression tracks or anticyclones, shift their position, this would effect the variability and extremes of weather at a particular location, and could have a major effect. However, we do not know if, or in what way, this will happen.

Will storms increase in a warmer world?

Storms can have a major impact on society. Will their frequency, intensity or location increase in a warmer world?

Tropical storms, such as typhoons and hurricanes, only develop at present over seas that are warmer than about 26°C. Although the area of sea having temperatures over this critical value will increase as the globe warms, the critical temperature itself may increase in a warmer world. Although the theoretical maximum intensity is expected to increase with temperature, climate models give no consistent indication whether tropical storms will increase or decrease in frequency or intensity as climate changes; neither is there any evidence that this has occurred over the past few decades.

Mid-latitude storms, such as those which track across the North Atlantic and North Pacific, are driven by the equator-to-pole temperature contrast. As this contrast will probably be weakened in a warmer world (at least in the Northern Hemisphere), it might be argued that mid-latitude storms will also weaken or change their tracks, and there is some indication of a general reduction in day-to-day variability in the mid-latitude storm tracks in winter in model simulations, though the pattern of changes vary from model to model. Present models do not resolve smaller-scale disturbances, so it will not be possible to assess changes in storminess until results from higher resolution models become available in the next few years.

Climate change in the longer term

The foregoing calculations have focussed on the period up to the year 2100; it is clearly more difficult to make calculations for years beyond 2100. However, while the timing of a predicted increase in global temperatures has substantial uncertainties, the prediction that an increase will eventually occur is more certain. Furthermore, some model calculations that have been extended beyond 100 years suggest that, with continued increases in greenhouse climate forcing, there could be significant changes in the ocean circulation, including a decrease in North Atlantic deep water formation.

Other factors which could influence future climate

Variations in the output of *solar energy* may also affect climate. On a decadal time-scale solar variability and changes in greenhouse gas concentration could give changes of similar magnitudes. However, the variation in solar intensity changes sign so that over longer time-scales the increases in greenhouse gases are likely to be more important. *Aerosols* as a result of volcanic eruptions can lead to a cooling at the surface which may oppose the greenhouse warming for a few years following an eruption. Again, over longer periods the greenhouse warming is likely to dominate.

Human activity is leading to an increase in aerosols in the lower atmosphere, mainly from sulphur emissions. These have two effects, both of which are difficult to quantify but which may be significant particularly at the regional level. The first is the direct effect of the aerosols on the radiation scattered and absorbed by the atmosphere. The second is an indirect effect whereby the aerosols affect the microphysics of clouds leading to an increased cloud reflectivity. Both these effects might lead to a significant regional cooling; a decrease in emissions of sulphur might be expected to increase global temperatures.

Because of long-period couplings between different components of the climate system, for example between ocean and atmosphere, the Earth's climate would still vary without being perturbed by any external influences. This *natural variability* could act to add to, or subtract from, any human-made warming; on a century time-scale this would be less than changes expected from greenhouse gas increases.

How Much Confidence do we have in Our Predictions?

Uncertainties in the above climate predictions arise from our imperfect knowledge of:

- future rates of human-made emissions;
- how these will change the atmospheric concentrations of greenhouse gases;
- the response of climate to these changed concentrations.

Firstly, it is obvious that the extent to which climate will change depends on the rate at which greenhouse gases (and other gases which affect their concentrations) are emitted. This in turn will be determined by various complex economic and sociological factors.

Secondly, because we do not fully understand the sources and sinks of the greenhouse gases, there are uncertainties in our calculations of future concentrations arising from a given emissions scenario. We have used a number of models to calculate concentrations and chosen a best estimate for each gas. In the case of carbon dioxide, for example, the concentration increase between 1990 and 2070 due to the Business-as-Usual emissions scenario spanned almost a factor of two between the highest and lowest model result (corresponding to a range in radiative forcing change of about 50%).

Furthermore, because natural sources and sinks of greenhouse gases are sensitive to a change in climate, they may substantially modify future concentrations (see earlier section: "Greenhouse gas feedbacks"). It appears that, as climate warms, these feedbacks will lead to an overall increase, rather than decrease, in natural greenhouse gas abundances. For this reason, climate change is likely to be greater than the estimates we have given.

Thirdly, climate models are only as good as our understanding of the processes which they describe, and this is far from perfect. The ranges in the climate predictions given above reflect the uncertainties due to model imperfections; the largest of these is cloud feedback (those factors affecting the cloud amount and distribution and the interaction of clouds with solar and terrestrial radiation), which leads to a factor of two uncertainty in the size of the warming. Others arise from the transfer of energy between the atmosphere and ocean, the atmosphere and land surfaces, and between the upper and deep layers of the ocean. The treatment of sea-ice and convection in the models is also crude. Nevertheless, for reasons given in box 10.4, we have substantial confidence that models can predict at least the broad-scale features of climate change.

Furthermore, we must recognise that our imperfect understanding of climate processes (and corresponding ability to model them) could make us vulnerable to surprises; just as the human-made ozone hole over Antarctica was entirely unpredicted. In particular, the ocean circulation, changes in which are thought to have led to periods of comparatively rapid climate change at the end of the last ice age, is not well observed, understood or modelled.

Will the Climate of the Future be very Different?

When considering future climate change, it is clearly essential to look at the record of climate variation in the past. From it we can learn about the range of natural climate variability, to see how it compares with what we expect in the future, and also look for evidence of recent climate change due to man's activities.

Climate varies naturally on all time-scales from hundreds of millions of years down to the year-to-year. Prominent in the Earth's history have been the 100,000 year glacial–interglacial cycles when climate was mostly cooler than at present. Global surface temperatures have typically varied by 5–7°C through these cycles, with large changes in

---

Box 10.4   Confidence in Predictions from Climate Models

What confidence can we have that climate change due to increasing greenhouse gases will look anything like the model predictions? Weather forecasts can be compared with the actual weather the next day and their skill assessed; we cannot do that with climate predictions. However, there are several indicators that give us some confidence in the predictions from climate models.

When the latest atmospheric models are run with the present atmospheric concentrations of greenhouse gases and observed boundary conditions their simulation of present climate is generally realistic on large scales, capturing the major features such as the wet tropical convergence zones and mid-latitude depression belts, as well as the contrasts between summer and winter circulations. The models also simulate the observed variability; for example, the large day-to-day pressure variations in the middle latitude depression belts and the maxima in interannual variability responsible for the very different character of one winter from another both being represented. However, on regional scales (2,000km or less), there are significant errors in all models.

Overall confidence is increased by atmospheric models' generally satisfactory portrayal of aspects of variability of the atmosphere, for instance those associated with variations in sea surface temperature. There has been some success in simulating the general circulation of the ocean, including the patterns (though not always the intensities) of the principal currents, and the distributions of tracers added to the ocean.

Atmospheric models have been coupled with simple models of the ocean to predict the equilibrium response to greenhouse gases, under the assumption that the model errors are the same in a changed climate. The ability of such models to simulate important aspects of the climate of the last ice age generates confidence in their usefulness. Atmospheric models have also been coupled with multi-layer ocean models (to give coupled ocean-atmosphere GCMs) which predict the gradual response to increasing greenhouse gases. Although the models so far are of relatively coarse resolution, the large-scale structures of the ocean and the atmosphere can be simulated with some skill. However, the coupling of ocean and atmosphere models reveals a strong sensitivity to small-scale errors which leads to a drift away from the observed climate. As yet, these errors must be removed by adjustments to the exchange of heat between ocean and atmosphere. There are similarities between results from the coupled models using simple representations of the ocean and those using more sophisticated descriptions, and our understanding of such differences as do occur gives us some confidence in the results.

ice volume and sea level, and temperature changes as great as 10–15°C in some middle and high latitude regions of the Northern Hemisphere. Since the end of the last ice age, about 10,000 years ago, global surface temperatures have probably fluctuated by little more than 1°C. Some fluctuations have lasted several centuries, including the Little Ice Age which ended in the nineteenth century and which appears to have been global in extent.

The changes predicted to occur by about the middle of the next century due to increases in greenhouse gas concentrations from the Business-as-Usual emissions will make global mean temperatures higher than they have been in the last 150,000 years.

The *rate of change* of global temperatures predicted for Business-as-Usual emissions will be greater than those which have occurred naturally on Earth over the last 10,000 years, and the rise in sea level will be about three to six times faster than that seen over the last 100 years or so.

Has Man already begun to Change the Global Climate?

The instrumental record of *surface temperature* is fragmentary until the mid-nineteenth century, after which it slowly improves. Because of different methods of measurement, historical records have to be harmonised with modern observations, introducing some uncertainty. Despite these problems we believe that a real warming of the globe of 0.3°C–0.6°C has taken place over the last century; any bias due to urbanisation is likely to be less than 0.05°C.

Moreover since 1900 similar temperature increases are seen in three independent data sets: one collected over land and two over the oceans. Figure 10.12 shows current estimates of smoothed global-mean surface temperature over land and ocean since 1860. Confidence in the record has been increased by their similarity to recent satellite measurements of mid-tropospheric temperatures.

Although the overall temperature rise

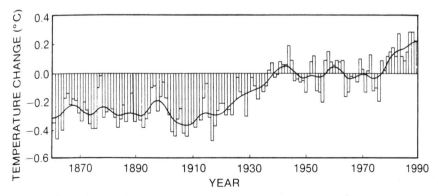

Figure 10.12 [*orig. figure 11*]   Global-mean combined land-air and sea-surface temperatures. 1861–1989, relative to the average for 1951–80.

has been broadly similar in both hemispheres, it has not been steady, and differences in their rates of warming have sometimes persisted for decades. Much of the warming since 1900 has been concentrated in two periods, the first between about 1910 and 1940 and the other since 1975; the five warmest years on record have all been in the 1980s. The Northern Hemisphere cooled between the 1940s and the early 1970s when Southern Hemisphere temperatures stayed nearly constant. The pattern of global warming since 1975 has been uneven with some regions, mainly in the northern hemisphere, continuing to cool until recently. This regional diversity indicates that future regional temperature changes are likely to differ considerably from a global average.

The conclusion that global temperature has been rising is strongly supported by the retreat of most *mountain glaciers* of the world since the end of the nineteenth century and the fact that global *sea level* has risen over the same period by an average of 1 to 2mm per year. Estimates of thermal expansion of the oceans, and of increased melting of mountain glaciers and the ice margin in West Greenland over the last century, show that the major part of the sea level rise appears to be related to the observed global warming. This apparent connection between observed sea level rise and global warming provides grounds for believing that future warming will lead to an acceleration in sea level rise.

The size of the warming over the last century is broadly consistent with the predictions of climate models, but is also of the same magnitude as natural climate variability. If the sole cause of the observed warming were the human-made greenhouse effect, then the implied climate sensitivity would be near the lower end of the range inferred from the models. The observed increase could be largely due to natural variabilty; alternatively this variability and other man-made factors could have offset a still larger man-made greenhouse warming. The unequivocal detection of the enhanced greenhouse effect from observations is not likely for a decade or

more, when the commitment to future climate change will then be considerably larger than it is today.

Global-mean temperature alone is an inadequate indicator of greenhouse-gas-induced climatic change. Identifying the causes of any global-mean temperature change requires examination of other aspects of the changing climate, particularly its spatial and temporal characteristics – the man-made climate change "signal". Patterns of climate change from models such as the Northern Hemisphere warming faster than the Southern Hemisphere, and surface air warming faster over land then over oceans, are not apparent in observations to date. However, we do not yet know what the detailed "signal" looks like because we have limited confidence in our predictions of climate change patterns. Furthermore, any changes to date could be masked by natural variability and other (possibly man-made) factors, and we do not have a clear picture of these.

How much will Sea Level Rise?

Simple models were used to calculate the rise in sea level to the year 2100; the results are illustrated in figures 10.13, 10.14 and 10.15. The calculations necessarily ignore any long-term changes, unrelated to greenhouse forcing, that may be occurring but cannot be detected from the present data on land ice and the ocean. The sea level rise expected from 1990–2100 under the IPCC

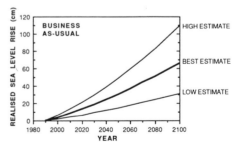

Figure 10.13 [*orig. figure 12*]   Sea level rise predicted to result from Business-as-Usual emissions, showing the best-estimate and range.

Business-as-Usual emissions scenario is shown in figure 10.13. An average rate of global mean sea level rise of about 6cm per decade over the next century (with an uncertainty range of 3–10cm per decade). The predicted rise is about 20cm in global mean sea level by 2030, and 65cm by the end of the next century. There will be significant regional variations.

The best estimate in each case is made up mainly of positive contributions from thermal expansion of the oceans and the melting of glaciers. Although, over the next 100 years, the effect of the Antarctic and Greenland ice sheets is expected to be small, they make a major contribution to the uncertainty in predictions.

Even if greenhouse forcing increased no further, there would still be a commitment to a continuing sea level rise for many decades and even centuries, due to delays in climate, ocean and ice mass responses. As an illustration, if the increases in greenhouse gas concentrations were to suddenly stop in 2030, sea level would go on rising from 2030 to 2100, by as much again as from 1990–2030, as shown in figure 10.14.

Predicted sea level rises due to the other three emissions scenarios are shown in figure 10.15, with the Business-as-Usual case for comparison; only best-estimate calculations are shown.

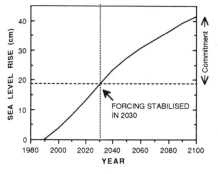

Figure 10.14 [*orig. figure 13*] Commitment to sea level rise in the year 2030. The curve shows the sea level rise due to Business-as-Usual emissions to 2030, with the additional rise that would occur in the remainder of the century even if climate forcing was stabilised in 2030.

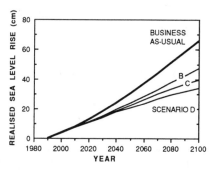

Figure 10.15 [*orig. figure 14*]   Model estimates of sea-level rise from 1990–2100 due to all four emissions scenarios.

The West Antarctic Ice Sheet is of special concern. A large portion of it, containing an amount of ice equivalent to about 5m of global sea level, is grounded far below sea level. There have been suggestions that a sudden outflow of ice might result from global warming and raise sea level quickly and substantially. Recent studies have shown that individual ice streams are changing rapidly on a decade-to-century time-scale; however this is not necessarily related to climate change. Within the next century, it is not likely that there will be a major outflow of ice from West Antarctica due directly to global warming.

Any rise in sea level is not expected to be uniform over the globe. Thermal expansion, changes in ocean circulation, and surface air pressure will vary from region to region as the world warms, but in an as yet unknown way. Such regional details await further development of more realistic coupled ocean-atmosphere models. In addition, vertical land movements can be as large or even larger than changes in global mean sea level; these movements have to be taken into account when predicting local change in sea level relative to land.

The most severe effects of sea level rise are likely to result from extreme events (for example, storm surges) the incidence of which may be affected by climatic change.

What will be the Effect of Climate Change on Ecosystems?

Ecosystem processes such as photosynthesis and respiration are dependent on climatic factors and carbon dioxide concentration in the short term. In the longer term, climate and carbon dioxide are among the factors which control ecosystem structure, i.e., species composition, either directly by increasing mortality in poorly adapted species, or indirectly by mediating the competition between species. Ecosystems will respond to local changes in temperature (including its rate of change), precipitation, soil moisture and extreme events. Current models are unable to make reliable estimates of changes in these parameters on the required local scales.

Photosynthesis captures atmospheric carbon dioxide, water and solar energy and stores them in organic compounds which are then used for subsequent plant growth, the growth of animals or the growth of microbes in the soil. All of these organisms release carbon dioxide via respiration into the atmosphere. Most land plants have a system of photosynthesis which will respond positively to increased atmospheric carbon dioxide ("the carbon dioxide fertilization effect") but the response varies with species. The effect may decrease with time when restricted by other ecological limitations, for example, nutrient availability. It should be emphasized that the carbon content of the terrestrial biosphere will increase only if the forest ecosystems in a state of maturity will be able to store more carbon in a warmer climate and at higher concentrations of carbon dioxide. We do not yet know if this is the case.

The response to increased carbon dioxide results in greater efficiencies of water, light and nitrogen use. These increased efficiencies may be particularly important during drought and in arid/semi-arid and infertile areas.

Because species respond differently to climatic change, some will increase in abundance and/or range while others will decrease. Ecosystems will therefore change in structure and composition. Some species

may be displaced to higher latitudes and altitudes, and may be more prone to local, and possibly even global, extinction; other species may thrive.

As stated above, ecosystem structure and species distribution are particularly sensitive to the rate of change of climate. We can deduce something about how quickly global temperature has changed in the past from palaeo-climatological records. As an example, at the end of the last glaciation, within about a century, temperature increased by up to 5°C in the North Atlantic region, mainly in Western Europe. Although during the increase from the glacial to the current interglacial temperature simple tundra ecosystems responded positively, a similar rapid temperature increase applied to more developed ecosystems could result in their instability.

What should be done to Reduce Uncertainties and How Long will this Take?

Although we can say that some climate change is unavoidable, much uncertainty exists in the prediction of global climate properties such as the temperature and rainfall. Even greater uncertainty exists in predictions of regional climate change, and the subsequent consequences for sea level and ecosystems. The key areas of scientific uncertainty are:

- *clouds*: primarily cloud formation, dissipation, and radiative properties, which influence the response of the atmosphere to greenhouse forcing;

- *oceans*: the exchange of energy between the ocean and the atmosphere, between the upper layers of the ocean and the deep ocean, and transport within the ocean, all of which control the rate of global climate change and the patterns of regional change;

- *greenhouse gases*: quantification of the uptake and release of the greenhouse gases, their chemical reactions in the atmosphere, and how these may be influenced by climate change;

- *polar ice sheets*: which affect predictions of sea level rise.

Studies of land surface hydrology, and of impact on ecosystems, are also important.

To reduce the current scientific uncertainties in each of these areas will require internationally coordinated research, the goal of which is to improve our capability to observe, model and understand the global climate system. Such a program of research will reduce the scientific uncertainties and assist in the formulation of sound national and international response strategies.

Systematic long-term *observations* of the system are of vital importance for understanding the natural variability of the Earth's climate system, detecting whether man's activities are changing it, parameterising key processes for models, and verifying model simulations. Increased accuracy and coverage in many observations are required. Associated with expanded observations is the need to develop appropriate comprehensive global information bases for the rapid and efficient dissemination and utilization of data. The main observational requirements are:

(i) the maintenance and improvement of observations (such as those from satellites) provided by the World Weather Watch Programme of WMO.

Box 10.5   Deforestation and Reforestation

Man has been deforesting the Earth for millennia. Until the early part of the century, this was mainly in temperate regions, more recently it has been concentrated in the tropics. Deforestation has several potential impacts on climate: through the carbon and nitrogen cycles (where it can lead to changes in atmospheric carbon dioxide concentrations), through the change in reflectivity of terrain when forests are cleared, through its effect on the hydrological cycle (precipitation, evaporation and runoff) and surface roughness and thus atmospheric circulation which can produce remote effects on climate.

It is estimated that each year about 2 Gt of carbon (GtC) is released to the atmosphere due to tropical deforestation. The rate of forest clearing is difficult to estimate; probably until the mid-20th century, temperate deforestation and the loss of organic matter from soils was a more important contributor to atmospheric carbon dioxide than was the burning of fossil fuels. Since then, fossil fuels have become dominant; one estimate is that around 1980, 1.6 GtC was being released annually from the clearing of tropical forests, compared with about 5 GtC from the burning of fossil fuels. If all the tropical forests were removed, the input is variously estimated at from 150 to 240 GtC; this would increase atmospheric carbon dioxide by 35 to 60 ppmv.

To analyse the effect of reforestation we assume that 10 million hectares of forests are planted each year for a period of 40 years, i.e., 4 million km^2 would then have been planted by 2030, at which time 1 GtC would be absorbed annually until these forests reach maturity. This would happen in 40–100 years for most forests. The above scenario implies an accumulated uptake of about 20 GtC by the year 2030 and up to 80 GtC after 100 years. This accumulation of carbon in forests is equivalent to some 5–10% of the emission due to fossil fuel burning in the Business-as-Usual scenario.

Deforestation can also alter climate directly by increasing reflectivity and decreasing evapotranspiration. Experiments with climate models predict that replacing all the forests of the Amazon Basin by grassland would reduce the rainfall over the basin by about 20%, and increase mean temperature by several degrees.

(ii)   the maintenance and enhancement of a programme of monitoring, both from satellite-based and surface-based instruments, of key climate elements for which accurate observations on a continuous basis are required, such as the distribution of important atmospheric constituents, clouds, the Earth's radiation budget, precipitation, winds, sea surface temperatures and terrestrial ecosystem extent, type and productivity.

(iii)   the establishment of a global ocean observing system to measure changes in such variables as ocean surface topography, circulation, transport of heat and chemicals, and sea-ice extent and thickness.

(iv)   the development of major new systems to obtain data on the oceans, atmosphere and terrestrial ecosystems using both satellite-based instruments and instruments based on the surface, on automated instrumented vehicles in the ocean, on floating and deep sea buoys, and on aircraft and balloons.

(v)   the use of palaeo-climatological and historical instrumental records to document natural variability and changes in the climate system, and subsequent environmental response.

The *modelling* of climate change requires the development of global models which couple together atmosphere, land, ocean and ice models and which incorporate more realistic formulations of the relevant processes and the interactions between the different components. Processes in the biosphere (both on land and in the ocean) also need to be included. Higher spatial resolution than is currently generally used is required if regional patterns are to be predicted. These models will require the largest computers which are planned to be available during the next decades.

Understanding of the climate system will be developed from analyses of observations and of the results from model simulations. In addition, detailed studies of particular processes will be required through targetted observational campaigns. Examples of such field campaigns include combined observational and small-scale modelling studies for different regions, of the formation, dissipation, radiative, dynamical and microphysical properties of clouds, and ground-based (ocean and land) and aircraft measurements of the fluxes of greenhouse gases from specific ecosystems. In particular, emphasis must be placed on field experiments that will assist in the development and improvement of sub grid-scale parametrizations for models.

The required program of research will require unprecedented international co-operation, with the World Climate Research Programme (WCRP) of the World Meteorological Organization and International Council of Scientific Unions (ICSU), and the International Geosphere-Biosphere Programme (IGBP) of ICSU both playing vital roles. These are large and complex endeavours that will require the involvement of all nations, particularly the developing countries. Implementation of existing and planned projects will require increased financial and human resources; the latter requirement has immediate implications at all levels of education, and the international community of scientists needs to be widened to include more members from developing countries.

The WCRP and IGBP have a number of ongoing or planned research programmes, that address each of the three key areas of scientific uncertainty. Examples include:

- *clouds*:
  International Satellite Cloud Climatology Project (ISCCP);
  Global Energy and Water Cycle Experiment (GEWEX).
- *oceans*:
  World Ocean Circulation Experiment (WOCE);
  Tropical Oceans and Global Atmosphere (TOGA).
- *trace gases*:
  Joint Global Ocean Flux Study (JGOFS);
  International Global Atmospheric Chemistry (IGAC);
  Past Global Changes (PAGES).

As research advances, increased understanding and improved observations will lead to progressively more reliable climate predictions. However considering the complex nature of the problem and the scale of the scientific programmes to be undertaken we know that rapid results cannot be expected. Indeed further scientific advances may expose unforeseen problems and areas of ignorance.

Time-scales for narrowing the uncertainties will be dictated by progress over the next 10–15 years in two main areas:

- Use of the fastest possible computers, to take into account coupling of the atmosphere and the oceans in models, and to provide sufficient resolution for regional predictions.
- Development of improved representation of small-scale processes within climate models, as a result of the analysis of data from observational programmes to be conducted on a continuing basis well into the next century.

10(B)

CLIMATE CHANGE: RADIATIVE FORCING OF CLIMATE – EXECUTIVE SUMMARY*

J. T. Houghton, L. G. Meira Filho, J. Bruce, Lee Hoesung, B. A. Callander, E. Haites, N. Harris, and K. Maskell

Introduction

In its first Scientific Assessment of Climate Change in 1990 the IPCC concluded that the increase of greenhouse gas concentrations due to human activities would result in a warming of the Earth's surface. "Radiative forcing" is the name given to the effect which these gases have in altering the energy balance of the Earth–atmosphere system and, using this concept, the 1990 report introduced a tool for policymakers, the Global Warming Potential, which allowed the relative warming effect of different gases to be compared. Other factors, natural and human, also cause radiative forcing. The 1990 report not only examined these factors but also reviewed a wide range of information on how climate has behaved in the past and how it might change in the future as a result of human influence.

The 1992 IPCC Supplementary report reviewed the key conclusions of the 1990 report and affirmed the basic understanding of climate change contained in the 1990 report. It did, however, provide more detail on two sources of *negative* radiative forcing – depletion of ozone in the stratosphere, and aerosols derived from anthropogenic emissions.

The scope of the present report covers only those factors which cause radiative forcing of climate change, and includes updated values of Global Warming Potentials. The full range of topics related to climate, including the *response* of climate to radiative forcing, will be covered in the

* Originally published in Houghton, J. T., Meira Filho, L. G., Bruce, J., Hoesung Lee, Callander, B. A., Haites, E., Harris, N. and Maskell, K. (1995) *Climate Change 1994: Radiative Forcing of Climate, Executive Summary.* Cambridge: Cambridge University Press, pp. 11–14.

second IPCC Scientific Assessment, scheduled for publication in 1995.

Major New Results since IPCC 1992

These new findings add to the detail of our knowledge but do not substantially change the essential results concerning radiative forcing of climate which appeared in the 1990 or the 1992 IPCC scientific assessments.

- *Revised values of Global Warming Potentials (GWPs)* – compared to GWPs listed in the 1992 IPCC report most values are larger by typically 10 to 30%. The uncertainties in the new GWPs are typically ±35%.
- *Revised methane GWP* – includes both direct and indirect effects. While the product of the revised GWP for methane and the current estimated annual anthropogenic emissions is significantly less than that of carbon dioxide over a 100-year time horizon, it is comparable over a 20-year time horizon.
- *Stablisation of atmospheric carbon dioxide concentrations* – a range of carbon cycle models indicates that stabilisation of atmospheric carbon dioxide concentration at all considered levels between one and two times today's concentrations (that is to say, between 350 and 750 ppmv[1]) could be attained only with global anthropogenic emissions that eventually drop to substantially below 1990 levels.
- *Improved estimation of forcing by aerosols* – model calculations indicate that the negative radiative forcing from sulphate aerosols and aerosols from biomass burning, when globally-averaged, may be a significant fraction of the positive radiative forcing caused by anthropogenic greenhouse gases since

the pre-industrial era. However, the estimates of the aerosol radiative forcing are highly uncertain; moreover, the forcing is highly regional and cannot be regarded as a simple offset to greenhouse gas forcing.
- *Recent low growth rate of carbon dioxide concentration is not unusual[2]* – between 1991 and 1993 the rate of increase in the atmospheric concentrations of carbon dioxide slowed substantially compared to the average rates of increase over the previous decade. However, the modern observational record for carbon dioxide since the 1950s contains other periods of similarly low growth rates. In the latter half of 1993, the carbon dioxide growth rates increased.
- *Sharp reduction in methane growth rate* – the rate of increase of the atmospheric abundance of methane has declined over the last decade, slowing dramatically in 1991 to 1992, though with an apparent increase in the growth rate in late 1993.
- *Climatic impact of Mt. Pinatubo* – the eruption of Mt. Pinatubo in June 1991 produced a large, transient increase of stratospheric aerosols which resulted in a surface cooling over about 2 years estimated from observations to be about 0.4°C, consistent with model simulations which predicted a global mean cooling of 0.4 to 0.6°C.
- *Global carbon budget* – New estimates of terrestrial carbon uptake during the 1980s have better quantified the known sinks, particularly forest regrowth in the Northern Hemisphere.

Sources of Radiative Forcing and their Magnitude

Anthropogenic and natural factors cause radiative forcing of various magnitudes and

[1] 1 ppmv = 1 part per million by volume.
[2] In the sense that anomalies in the growth rate of atmospheric carbon dioxide are not unusual. The anomaly of the early 1990s does have some unusual features in terms of its magnitude, duration and its coincidence with the decrease in the growth rate of methane, but it remains too early to identify either the causes of the 1990s' anomaly or its significance to the long-term growth of carbon dioxide.

of different signs. The concept of radiative forcing enables us to compare the potential effects of different factors, though care must be taken where these factors have large seasonal or regional variation.

First, we consider the gases carbon dioxide, methane, nitrous oxide and the halocarbons which have increased through human activities and which are well-mixed throughout the atmosphere.

- The increase in carbon dioxide (CO_2) since the pre-industrial era (from about 280 to 356 ppmv) makes the largest individual contribution to greenhouse gas radiative forcing: 1.56 Wm^{-2}, consistent with previous IPCC reports.
- The increase of methane (CH_4) since pre-industrial times (from 0.7 to 1.7 ppmv) contributes about 0.5 Wm^{-2} to radiative forcing.
- The increase in nitrous oxide (N_2O) since pre-industrial times (from about 275 to about 310 ppbv[3]) contributes about 0.1 Wm^{-2} to radiative forcing.
- The observed concentrations of halocarbons, including CFCs 11, 12, 113, 114, 115, methyl-chloroform and carbon tetrachloride, have resulted in a direct radiative forcing of about 0.3 Wm^{-2}.
- The atmospheric concentrations of a number of HCFCs and HFCs, which are being used as substitutes for halocarbons controlled under the Montreal Protocol, have increased substantially. Their combined contribution to radiative forcing is, however, still less than 0.05 Wm^{-2} because of their low atmospheric concentrations.

Second, we consider changes in concentrations of ozone and aerosols which are believed to contribute significantly to radiative forcing. Patterns of historical change in these constituents are strongly regional in character, leading to two important consequences: (i) estimates of their globally-averaged radiative forcing are less certain than for the well-mixed gases (because the patterns of change are not well-quantified)

and (ii) any negative forcing due to aerosols cannot be regarded as a simple offset to the effect of greenhouse gases (because the regional patterns of the forcings are different). Nevertheless, we report such estimates in order to provide a broad indication of their relative magnitude.

- Limited observations and model simulations suggest that tropospheric ozone in the Northern Hemisphere has increased since pre-industrial times resulting in a global average radiative forcing of 0.2 to 0.6 Wm^{-2}.
- Halocarbon-induced depletion of ozone in the stratosphere has resulted in a negative global average radiative forcing of about −0.1 Wm^{-2}. This has occurred mainly since the late 1970s over which period it has been of similar magnitude, but opposite sign, to the forcing caused by the halocarbons. Prior to the onset of significant ozone depletion the radiative forcing due to the halocarbons was between +0.1 and 0.2 Wm^{-2}.
- Anthropogenic particles in the atmosphere, derived from emissions of sulphur dioxide and from biomass burning, exert a net negative radiative forcing. The *direct* forcings, globally averaged, are probably in the ranges −0.25 to −0.9 Wm^{-2} for sulphate aerosols and −0.05 to −0.6 Wm^{-2} for aerosols from biomass burning. The *indirect* effect of aerosols, due to their effect on cloud properties, may cause a further negative forcing of a magnitude similar to the direct effect. The forcing shows large regional variations, with the largest values in industrialised regions in the Northern Hemisphere.

Third, we consider natural factors which can also exert positive or negative radiative forcings.
- Since about 1850 a change in the Sun's output may have resulted in a positive radiative forcing estimated at between 0.1 and 0.5 Wm^{-2}.
- Some volcanic eruptions, such as that

[3] 1 ppbv = 1 part per billion (thousand million) by volume.

of Mt. Pinatubo in June 1991, result in a short-lived (a few years) increase in aerosols in the stratosphere, causing a large (about -4 Wm^{-2} in the case of Mt. Pinatubo) but short-lived negative radiative forcing of climate. The effect of the Mt. Pinatubo eruption has been detected in the observed temperature record.

Trends in Greenhouse Gas and Aerosol Concentrations

- Over the decade 1980 to 89 the atmospheric abundance of CO_2 increased at an average rate of about 1.5 ppmv (0.4% or 3.2 billion tonnes of carbon) per year as a result of human activities, equivalent to approximately 50% of anthropogenic emissions over the same period.
- The rate of increase of the atmospheric abundance of methane has declined over the last decade, slowing dramatically in 1991 to 1992, though with an apparent increase in the growth rate in late 1993. The average trend over 1980 to 1990 is about 13 ppbv (0.8% or 37 million tonnes of methane) per year.
- The atmospheric abundance of nitrous oxide increased at an average annual rate (1980 to 1990) of about 0.75 ppbv (0.25% or 3.7 million tonnes of nitrogen) per year. The observations indicate that the growth rate varied during this period.
- The rates of increase of atmospheric concentrations of several major ozone-depleting halocarbons have fallen, demonstrating the impact of the Montreal Protocol and its amendments and adjustments. The total amount of organic chlorine in the troposphere increased by only 1.6% in 1992, about half of the rate of increase (2.9%) in 1989.
- The monitoring network for tropospheric ozone is sparse, making detection of global trends difficult. Since the 1960s concentrations of tropospheric ozone have almost certainly increased over large parts of

the Northern Hemisphere but trends during the 1980s were small and of variable sign.
- Anthropogenic aerosol and precursor emissions have increased over the past 150 years, but while local trends (positive and negative) in concentrations are evident, no clear picture emerges of a contemporary *global* trend in atmospheric concentrations of anthropogenic aerosols in the size range important for radiative forcing.

The Stabilisation of Greenhouse Gas Concentrations

Several carbon cycle models have been used to study the implications for future atmospheric concentrations of carbon dioxide, of a range of global anthropogenic *emission* scenarios. The same models have been used to study the broad implications, in terms of *emissions*, of stabilising carbon dioxide *concentrations* in the range 350 ppmv (near current levels) to 750 ppmv. Differences in projected concentrations and emissions between models are typically ±15%; additional uncertainties arise from the various assumptions and simplifications used. The following results emerge:

- If carbon dioxide emissions were maintained at today's levels, they would lead to a nearly constant rate of increase in atmospheric concentrations for at least two centuries, reaching about 500 ppmv (approaching twice the pre-industrial concentration) by the end of the 21st century.
- A stable level of carbon dioxide concentration at values up to 750 ppmv can be maintained only with anthropogenic emissions that eventually drop below 1990 levels.
- To a first approximation the eventual stabilised concentration is governed more by the accumulated CO_2 emissions from now until the time of stabilisation, and less by the exact path taken to reach stabilisation. This means that, for example, for a given stabilisation scenario, higher emissions in

early decades imply lower emissions later on. For the range of arbitrary stabilisation cases studied, accumulated emissions to the end of the 21st century were between 300 and 430 GtC[4] for stabilisation at 350 ppmv, between 880 and 1060 GtC for stabilisation at 550 ppmv, and between 1220 and 1420 GtC for stabilisation at 750 ppmv. For comparison the corresponding accumulated emissions for IPCC IS92[5] emission scenarios are 770 to 2190 GtC.

If methane emissions were maintained at today's levels, atmospheric concentrations would effectively stabilise within 50 years at about 1900 ppbv, 11% higher than at present. Conversely, a reduction in annual methane emissions to levels about 35 million tonnes (roughly 10% of anthropogenic emissions) below current levels would stabilise concentrations at today's levels. (This calculation assumes that natural sources and atmospheric losses of methane are not affected by changing climate and atmospheric composition over the next century.)

If emissions of nitrous oxide were maintained at today's levels, atmospheric concentrations would effectively stabilise after several centuries at about 400 ppbv, 30% higher than at present and 50% above pre-industrial levels. Conversely a reduction of more than 50% of anthropogenic sources would stabilise concentrations at today's level of about 310 ppbv.

In contrast to the long-lived greenhouse gases, aerosols and tropospheric ozone are rapidly removed from the atmosphere and stabilisation of precursors would lead quickly to stable atmospheric concentrations.

The predictions of changes in atmospheric chlorine loading indicate that the depletion of stratospheric ozone should peak within the next decade and then slowly recover during the first half of the next century.

Global Warming Potential

Revised GWPs have been calculated. Furthermore, GWPs have been calculated for a number of new species, in particular hydrochlorofluorocarbons (HCFCs), hydrofluorocarbons (HFCs) and perfluorocarbons (PFCs).

The GWP concept is difficult to apply to short-lived species (for example, oxides of nitrogen, non-methane hydrocarbons and aerosols). New tools need to be developed to characterise their radiative forcing.

---

[4] 1 GtC = 1 billion tonnes of carbon.
[5] In 1992 IPCC produced six scenarios, termed IS92a–f, for future emissions of greenhouse gases and their precursors.

11

FUTURE SEA LEVEL RISE: ENVIRONMENTAL AND SOCIO-POLITICAL CONSIDERATIONS*

Richard A. Warrick and Atiq A. Rahman

1 Introduction: Sea Level Variations and Impact Issues

Discussions of future sea level rise often give the impression that sea level is, and has been, stable. Nothing could be further from the truth. The level of the sea surface varies across a broad range of time and space scales for many reasons. These include very long-term changes in ocean basin volume (from, for example, sea floor spreading or sedimentation) and medium-term changes in ocean mass from variations in groundwater, surface water or land-based ice. There are also shorter-term dynamic changes due to oceanographic (e.g. ocean current) or meteorological (e.g. atmospheric pressure) factors on the local or regional scale. The schema in figure 11.1 is suggestive of a number of such factors.

As a result of global warming, mean sea level (MSL) may change for two reasons: expansion of the oceans due to higher sea temperatures, and changes in land-based ice. The latter would include changes in the mass of the large Greenland and Antarctic ice sheets, and of smaller ice caps and mountain glaciers.

Various projections of future sea level change are summarized in table 11.1. In general, all agree that as a result of global warming, MSL should rise. The most recent projections depict a rise of less than one metre before the end of the next century – a change to which many communities might easily adapt. For instance, the best estimate given by the report of the Intergovernmental Panel on Climate Change (IPCC) is an average rate of rise of 6 millimetres per year (mm/yr) over the next century, if no actions are taken to stem the emissions of greenhouse gases (Warrick and Oerlemans, 1990).

None the less, the large uncertainties around most projections suggest that the possibility of a much larger rise cannot be ruled out. Whether slow or fast, however,

* Originally published in I. M. Mintzer (ed.) (1992) *Confronting Climate Change: Risks, Implications and Responses*. Cambridge: Cambridge University Press, pp. 97–112.

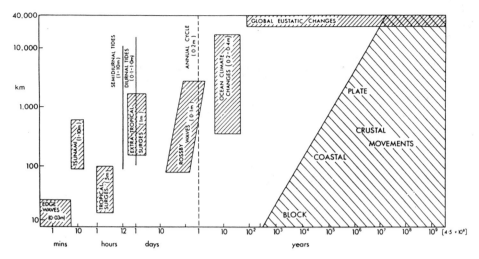

Figure 11.1 [*orig. figure 1*]   The level of the sea surface varies across a broad range of time and space scales. This figure shows some of the physical and geological processes responsible for sea-level changes, with indications of the length (vertical) and time-scales (horizontal) for each. Global processes have a length scale of 40,000 km.
Source: Pugh, 1992.

the crucial factor in many situations will not simply be the rate of global sea level rise, but the circumstances at the waters' edge, the underlying rate of vertical land movement, the response of natural systems, and the way in which each community chooses to meet the challenge.

At coastal locations, relative sea level (RSL; sea level relative to land) is affected by vertical movements of the land as well as changes in the actual sea surface elevation. Large-scale crustal movements can be regionally important. In parts of Scandinavia and Hudson's Bay, for example, relative sea level is falling one metre per century as a result of "isostatic rebound", i.e. crustal uplift due to the disappearance of the large continental ice sheets following the last glaciation (Peltier, 1986). In some areas like the East Coast of North America and Southeast Britain, sea level is still rising from land subsidence due to collapse of the "forebulge" at the margins of the last glacial advance (Emery and Aubrey, 1991). In other areas, land accretion or degradation processes interact with changes in the sea surface and with crustal uplift/subsidence to affect a dynamic sea level balance.

For instance, in the Mississippi Delta, RSL change is determined by the difference between deltaic subsidence and global MSL rise on the one hand, and land accretion from sedimentation on the other (Day et al., 1992). In short, the meaning of global MSL rise for specific locations must be interpreted in relation to the mix of other factors which can affect relative sea level in complex ways (for recent reviews see NRC, 1990; Emery and Aubrey, 1991; Warrick et al., 1992).

The risks of sea level rise affect many countries, with potentially important international complications. With their emissions of CO_2 and other greenhouse gases, the industrialised nations of the world cause the bulk of the global warming problem. Warming, in turn, affects ice sheets at the extreme high latitudes, glaciers in remote mountain areas, and ocean temperatures worldwide, raising global sea level. The threat of environmental and socio-economic impacts is worrisome to many small island states and low-lying river deltas in developing countries, many of which have so far contributed negligibly to greenhouse gas emissions. Thus, the political issue of attributing responsibility – who is to blame?

Table 11.1 [*orig. table 1*]   Estimates of future global sea level changes (in centimetres). *Source*: Warrick, 1992; as modified from Warrick and Oerlemans, 1990 and Raper et al., 1991.

	Contributing factors			Total rise[a]			
	Thermal expansion	Alpine	Greenland	Antarctica	Best estimate	Range[l]	To (year)
Gornitz (1982)	20		20 (combined)		40		2050
Revelle (1983)	30	12	13		71[b]		2080
Hoffman et al. (1983)	28 to 115	28 to 230 (combined)				56 to 345	2100
						26 to 39	2025
PRB (1985)	c	10 to 30	10 to 30	−10 to 100		10 to 160	2100
Hoffman et al. (1986)	28 to 83	12 to 37	6 to 27	12 to 220		1 to 367	2100
						0 to 21	2025
Robin (1986)[d]	30 to 60[d]	20 ±12[d]	to +10[d]	to −10[d]	80	25 to 165	2080
Thomas (1986)	28 to 83	14 to 35	9 to 45	13 to 80	100	60 to 230	2100
Villach (1987) Jaeger (1988)[d]					30	−2 to 51	2025
Raper et al. (1991)	4 to 18	2 to 19	1 to 4	−2 to 3	21[d]	5 to 44[a]	2030
Oerlemans (1989)					20	0 to 40	2025
Van der Veen (1988)[h]	8 to 16	10 to 25	0 to 10	−5 to 0		28 to 66	2085
Warrick and Oerlemans (IPCC, 1990)[i]	28 to 66	8 to 20	3 to 23	−7 to 0	66	31 to 110	2100
Wigley & Raper (1992)[k]					46[k]	3 to 124[l]	2100

[a] from the 1980s.
[b] total includes additional 17cm for trend extrapolation.
[c] not considered.
[d] for global warming of 3.5°C.
[f] extreme ranges, not always directly comparable.
[g] internally consistent synthesis of components.
[h] for a global warming of 2.4°C.
[i] estimated from global sea level and temperature change from 1800–1980 and global warming of 3.5 ± 2.0°C for 1980–2080.
[j] for IPCC "Business as usual" forcing scenario only.
[k] for IPCC Policy Scenario B.
[l] for all IPCC forcing scenarios.

– arises and lies at the heart of difficult decisions regarding who should bear the burden of greenhouse gas emission reductions, and whether compensation to the "victims" should be forthcoming.

Few countries serve to illustrate these issues better than Bangladesh. Bangladesh is one of the poorest of the world's Least Developed Countries (LDCs), with an average per capita GNP of US$170 and a low per capita use of fossil fuels. Bangladesh rates poorly on indicators of development: less than one-third of the population is literate, and the population of approximately 11.5 million people (in 1990) is growing at an annual rate of about 2.2%. The densely populated, low-lying country of about 144,000 km^2 consists largely of the delta of the three major rivers of the world, namely the Ganges, the Brahmaputra and the Meghna, and is the victim of recurrent, climate-related natural disasters: the devastating floods of 1987 and 1988 and the unprecedented cyclone of April 1991 are examples. For these reasons, Bangladesh may be one of the countries least responsible for the causes of climate change and most vulnerable to the effects, particularly sea level rise (Rahman and Huq, 1989).

In this chapter, the case of Bangladesh illustrates the issues we wish to examine. The chapter is divided into two parts. In the first, we outline some considerations for local environmental and socio-economic impact assessments, in light of the dynamic responses of natural and human systems. In the second part, we consider the global geopolitical dimensions, particularly with regard to the issue of attribution of responsibility for the enhanced greenhouse effect and future sea level rise.

2 Environmental and Socio-economic Impacts: Some Considerations[1]

Too often, in the interests of expediency, projections of global mean sea level are applied statically for purposes of local impact assessment. By this, we mean that a projected rise – say, one or two metres – is simply added to mean sea level and the shape of the "new" coastline laid out. The adverse environmental effects of salt intrusion (into fresh groundwater, surface waters, wetlands or soils), erosion (of beaches and cliffs) or innundation (permanent or temporary storm surge flooding) are estimated and evaluated separately, if at all.

For example, a number of such studies have been carried out for Bangladesh. Milliman et al. (1989) and Broadus (1992), for instance, postulated an inundation of up to one-third of the country due to 1 m and 3 m rises in relative sea level by estimating the area below the 1 m and 3 m contour lines. The population, land, agricultural production, etc., within the affected area were then calculated as a basis for estimating impacts. A recent study by the Bangladesh Centre of Advanced Studies in association with the Centre for Global Change, University of Maryland (Ali and Huq, 1991) followed a similar approach for a 1 m rise, albeit in a more detailed manner. Such studies provide a first rough approximation of potential impacts. However, it is extremely unlikely that the effect will be so straightforward. The real world works in vastly different, more complex and dynamic ways.

2.1 Dynamic influence of local natural systems

The actual physical impacts on Bangladesh are difficult to predict, because the coastal system will respond dynamically as sea level rises. Huge quantities of sediment (1.5–2.5 billion tonnes per year) are carried by the rivers – whose combined flood level can exceed 140,000 m^{3}s^{-1} – into Bangladesh from the whole of the Himalayan drainage system, including the countries of Nepal, China and India. About two-thirds of this sediment goes into the Bay of Bengal and, over the long term, causes land subsidence within the Delta region. In some coastal

[1] In this section we choose not to reiterate the range of possible impacts of sea level rise. These have been summarised elsewhere (e.g. Titus, 1992; Teggart et al., 1990) and exemplified in numerous case studies (e.g. Frascette, 1991; Titus et al., 1990).

areas, the sediment rates are sufficient to more than compensate for deltaic subsidence and the land accretes; elsewhere, insufficient sediment results in erosion.

The result is that, naturally, the coastal configuration undergoes large changes, as shown in figure 11.2. Against this large natural variation, it is unclear whether a global sea level rise of, say, 6 mm/year would, in fact, even be discernible. Furthermore, Brammer (1989; 1992) believes that, due to changes in river gradients, sedimentation and drainage, the primary impact of global MSL on Bangladesh will be an increase in flooding in the depressed basins *upstream* rather than only at the coast or within the tidal limits. Thus, the fate of Bangladesh is unlikely to follow simple predictions of an inexorably retreating coastline. Predicting the effects of global sea level rise on Bangladesh depends critically on an understanding of accretion and

erosion rates and how they interact with sea level variations.

A similar set of relationships exists in the marshlands of the East Coast of the United States. In many locations, sedimentation rates have kept up with sea level rise, aided to a large extent by the marshland vegetation (Stevenson et al., 1986). If the rate of relative sea level rise becomes too high, however, salt stress and other factors would adversely affect the health of plants, thereby reducing sedimentation rates and exacerbating the changes. In this case, the picture that emerges is one of a dynamic, non-linear process, a central element of which is biological and thus dependent on the health of local life forms.

Biological health is also important for many low-lying, coral-fringed island-states, like the Republic of Maldives. The Maldives would not necessarily simply be "drowned" by rising sea level – provided that the rise is

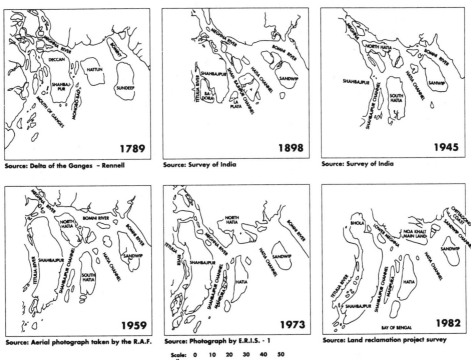

Figure 11.2 [*orig. figure 2*]   The historical development of the lower Meghna estuary, on the Bay of Bengal, shown over two centuries. Against this large natural variation in coastal configuration at the regional level, it is unclear whether a global sea level rise of six mm per year would be discernible.

not too fast. Healthy coral can grow and keep up, maintaining the coral platform necessary for the dissipation of wave energy and the accretion of sediments, in effect compensating for the change in sea level. For the Maldives, the coral growth rate is not known precisely, but may range between 3 mm/yr and 10 mm/yr (Woodroffe, 1989; Edwards, 1989). If the threshold speed is exceeded, or if ocean temperatures become too warm, the coral would die, with the possibility of accelerating erosion and land loss. Again, it appears that dynamic, non-linear processes are at work.

In these cases and others, accurately projecting the rate and effects of RSL rise requires understanding of the interrelated roles of a number of complex environmental feedbacks.

2.2 Dynamic influence of local human activity

A related point to emphasize is that, in many cases, human interference in the dynamic processes can change local sea level by amounts comparable to those projected as a result of future global warming. The Mississippi River Delta is once again a good example. As described by Day et al. (1992), the Delta was roughly in a state of dynamic balance (or slowly growing) for thousands of years before the present century. Since then, levee construction along the banks of the Mississippi River, the construction of smaller scale dykes, the damming of distributaries, and the development of numerous canals and diversions have effectively starved the wetlands of needed freshwater and sediments. Now, RSL is rising at a rate of 1 metre per century and up to 100 km² of wetlands are lost each year (Gagliano et al., 1981).

In the Republic of Maldives, pollution and mining of coral for construction purposes are threatening to disrupt the islands' natural capacity to respond to rising sea level (Pernetta and Sestini, 1989). The coral fringing the capital island of Male, where most Maldivians currently reside, is effectively dead, leaving the island vulnerable to rising sea level. Mining of coral elsewhere in the Maldives continues in response to high demand for scarce building material, largely for breakwaters for docks to support a growing tourism industry and for construction of Western-style houses in preference to traditional styles.

In Bangladesh, it has long been purported, but not proved, that deforestation in the headwaters of the Ganges–Brahmaputra–Meghna river system has affected runoff, sediment flow and deposition rates, with consequent changes in coastlines and inland flooding. The possible effects of this deforestation, however, may be minor compared to effects that could accrue from a number of interventions being considered under the Flood Action Plan (FPCO, 1991). The comprehensive plan includes major structures and extensive embankments. Fears have been expressed that such interventions in the major river systems could have unexpected adverse effects on sedimentation and hydrological processes that could exacerbate the effects of future sea level rise (Huq and Rahman, 1990).

In these cases, the effects of local human interference – as a result of settlement in the area, and not in response to any threat of global warming or sea level rise – have to be understood. Every realm of human activity takes place within a cultural-historical context, with its own shared perception of both the natural resources and the alternative technologies available. Understanding of the "human ecology" of coastal environments is a fundamental element in predicting changes in, and seeking adequate solutions to, local sea level rise.

2.3 Manipulating dynamic processes to favourable advantage

The final point is that the complex dynamics of this human-environmental system can be manipulated to favourable advantage. The threat of future global MSL rise can provoke actions and policies that nurture the ability of natural systems to respond (by reducing RSL rise), or the ability of human systems to respond (by reducing the hazards posed by RSL rise).

Often, this can be accomplished indirectly by altering the management contexts in which decisions are made or by broadening the range of perceived choices in resource use. For example, eliminating the dikes and canals that are obstacles to freshwater and sediment flow could help re-establish land accretion processes in selected areas of the Mississippi Delta region. This could be promoted through incentives for less intensive land uses (like wetland and wildlife reserves) that preclude the need for such obstacles (Day et al., 1992). For the Maldives, it has been suggested that the best sea defence may be achieved by putting a stop to coral mining and encouraging the development of alternative building materials, thus preserving the health of coral to respond to slowly rising seas (Commonwealth Group of Experts, 1989).

The potential options for manipulating the natural systems in Bangladesh are less clear. It has been suggested that *controlled* flooding, rather than strict reliance on flood prevention through embankments and other engineering structures, might be one way of directing sediment and ensuring land accretion at critical coastal areas in the face of sea level rise (Mahtab, 1989). However, it must be admitted that, because of the complex dynamic processes involved, the uncertainties regarding the consequences of this and other more conventional interventions in the hydrological system are currently immense.

So, too, are the uncertainties regarding the environmental and socio-political effects of sea level rise on the coastal region of Bangladesh. Certainly, the spectre, often put forth, of millions of migrating homeless fleeing a retreating coastline and spilling over international borders – "eco-migration" – is not a foregone conclusion, maybe not even a realistic one. It is becoming clear that climate and sea level change must be viewed as one set of components amongst many interrelated factors – poverty, indebtedness, high population growth and meagre resources – that define vulnerability. Whether the future brings an inflammation of festering international tensions and regional conflicts may depend as much on the socio-economic conditions of the population-at-risk as on the rate of sea level rise. The more desperate the con-

ditions, the larger the impacts and subsequent ramifications are likely to be.

For this reason, the most efficacious strategy for countries like Bangladesh – for reducing possible adverse impacts in the face of large uncertainty – lies in modifying the *human* systems, e.g., by accelerating social and economic development. It is unlikely that the main threat to Bangladesh stems directly from sea level creeping slowly higher. Rather, the cause for alarm arises principally from the prospect of more frequent, severe extreme events causing storm surge inundation as a result of higher sea level and/or increased severe tropical storm activity. As during the cyclone of April 1991, many thousands lost their lives due to a lack of opportunities for alternative livelihoods in less hazardous locations, inadequate communication and transportation systems, inadequate or too few shelters, and a reluctance to leave their few worldly goods unattended, even at the risk of death. Development creates *options* for vulnerable inhabitants who currently occupy lands susceptible to the long- and short-term environmental changes, such as sea level rise and storm surges. In short, development reduces poverty – which reduces vulnerability.

2.4 *The need for integrated assessments*

Effectively changing the pattern of human activities requires, foremost, a thorough interdisciplinary knowledge of the interplay of human activities and nature, especially as they pertain to the specific coastal situations at hand. Unfortunately, in many parts of the world which are vulnerable to sea level rise, such knowledge is meagre, and the uncertainties about future RSL rise and its effects are large. The next step forward is to move away from the simplistic, static approaches to impact assessments – rife in the literature – and to develop methods of integrated assessment that take account of the dynamic influences of both human and natural systems on relative sea level and its effects on the coastal environment.

3 Socio-Political Issues in International Response

The prospect of rising sea level is certain to motivate certain countries, like Bangladesh, to endorse (at least in principle) proposals for reducing greenhouse gas emissions and slowing climatic change. But although Bangladesh contains over 2% of the world's population and is one of the world's most densely populated countries, the country has one of the lowest fossil fuel consumption rates per capita. Most of the people consume non-commercial energy such as fuelwood, which is mostly used for cooking. Lighting in rural areas is confined to only 2 hours per night and is derived mostly from kerosene. Rural electrification exists but is available to less then 10% of the households. Consequently, Bangladesh contributes a miniscule 0.06% of the world's annual emissions of carbon into the atmosphere. Surely, the burden of CO_2 emissions reduction should fall elsewhere.

In the global context of environment and development the case of Bangladesh thus highlights delicate international issues. Who is responsible for the enhanced greenhouse effects and the prospective rise in sea level? Who, therefore, should accept responsibility for the costs of reducing greenhouse gas emissions? By extended implication, who should pay for damages accruing from any sea level rise? Such issues are clouded by the uncertainties regarding the rates of rise and the effects of emission policies in slowing the changes. To clarify the issues, we must first address the question of accountability with a clear understanding of the scientific and political realities.

3.1 Accounting for the enhanced greenhouse effect: procedural difficulties

Lately, the question of the attribution of responsibility for the enhanced greenhouse effect has been hotly debated. The debate has been sparked by the somewhat contentious publication by the World Resources Institute (WRI, 1990) of country-by-country estimates of greenhouse gas emissions and their combined radiative forcing effects, and the equally contentious

critique by the Centre for Science and Environment in New Delhi (CSE: Agarwal and Narain, 1991; also see McCully, 1991). We shall not review the relevant literature since this has been done elsewhere (Hammond et al., 1991; *Environment*, 1991; *Nature* 349, 1991). Rather, let us first elucidate some of the difficulties involved in resolving the question of responsibility. Then, we shall offer a range of estimates that we feel is reasonable.

3.1.1 Inadequacy of scientific knowledge

First, in some critical areas the current scientific knowledge concerning greenhouse gas sources, sinks and fluxes is meagre. For example, most carbon cycle models fail to "balance," predicting current atmospheric concentrations that are too high compared to observations – the "missing sink" problem. Estimates of current carbon emissions from changing land use (including deforestation) differ by as much as a factor of ten. With respect to methane, there is a slim base of knowledge upon which to estimate local emissions. Together, such problems create large uncertainties in attempts to disaggregate GHG forcing changes on a country-by-country basis.

3.1.2 The problem of acceptable accounting procedures

There is also a lack of general agreement regarding the accounting procedures for allocating greenhouse gas emissions, concentrations and consequent changes in radiative forcing on a country-by-country basis. Moreover, the choice of procedures can be strongly influenced by socio-political judgements.

For example, should each country's annual greenhouse gas emissions be calculated net or gross – that is, with or without taking into account the capacity of the ocean, atmosphere and land to "absorb" emissions (the "sinks")? If on a net basis, how should the absorptive capacity be allocated in proportion to the country's emission rate (e.g. by WRI, 1990) or weighted on a per capita basis (e.g. by Agarwal and Narain, 1991)? A per capita allocation scheme reduces the alleged Third World responsibility, particularly for relatively high greenhouse gas-emitting, highly populated countries (e.g. China and India). (See further Grubb et al., 1992.)

Furthermore, there is the issue of entitlement to any *growth* in the absorptive capacity. For instance, if the CO_2 "fertilization" effect increases the annual biospheric uptake of atmospheric CO_2, who is entitled to claim it – those countries with the highest emission rates, those with the positively affected vegetation, or every person equally?

There is also the question of whether accounting should be based on current emissions, or cumulative (e.g. from pre-industrial times). Some would argue (Smith, 1991) that the cumulative approach is technically more correct, because of the long residence times of most greenhouse gases in the atmosphere, and a fairer way to estimate responsibility for the enhanced greenhouse effect. But one conceptual difficulty is that actual greenhouse gas emissions from earlier periods have been largely "washed out" of the atmosphere: a molecule of CO_2 emitted in 1800 is unlikely to be in the atmosphere today. Another difficulty with the cumulative approach is that most emission data for earlier periods are scanty; for example, Third World fossil fuel carbon emissions can be estimated fairly accurately back only to 1950. Data on carbon emissions from land use change (deforestation) are poor for any period, even today. Notwithstanding these difficulties, the cumulative approach would attribute less responsibility for the enhanced greenhouse effect to the Third World.

3.1.3 Ethical considerations

Finally, there are some purely moral/ethical aspects that should be considered. For example, to paraphrase Agarwal and Narain

(1991), does it serve the cause of global justice to assign equal weight to units of "climate warming potential" from greenhouse gases, regardless of origin? That would make a unit of methane from a Third World farm equal to a unit of carbon dioxide from a gas-guzzler in the industrialized West. Another such issue relates to production versus consumption: To whom should emissions be allocated, the country of origin or the country that eventually consumes the product for which the greenhouse gas was emitted? This issue becomes increasingly important as more energy-intensive "metal-bashing" activities (such as manufacture of car components) move to developing nations, with the productive output designated for export to the West.

One thing is clear: there are no objective answers to these questions. But there may be cooperative political agreements that can address these issues.

3.2 How much of the enhanced greenhouse effect can be blamed on the Third World?

Bearing in mind the above issues, we now suggest a method for allocating the shares of responsibility. We do so by apportioning the radiative forcing changes (from IPCC; Shine et al., 1990) from the increased concentrations of the major greenhouse gases according to the fraction of total emissions contributed by countries of the developing world.[2] Our data come largely from the World Resources Institute (1990) and *Trends '90* (Boden et al., 1990).

Two approaches were followed. In the first, calculations were based on the relative emissions and radiative forcing changes which occurred over the decade of the 1980s – a time of high growth in energy use rates for the developing nations. But contemporary trends do not tell the whole story: less than one-quarter of the full radiative forcing

[2] We chose not to make allocations on a per capita basis, since we are interested in investigating the world-regional differences. Nor did we consider allocating the absorptive capacity (sinks) according to region or population, because of the large uncertainties in greenhouse gas sinks and, importantly, because of the lack of procedural guidance for doing so, as discussed above. Thus we avoid the "minefield" and keep the approach simple. We consider the "developing world" to be all countries with the exception of Western and Eastern Europe, the former Soviet Union, the United States, Canada, Japan, Australia and New Zealand.

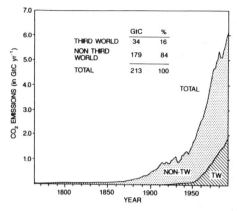

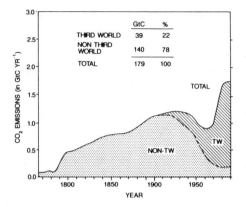

Figure 11.3 [*orig. figure 3*]   Comparison of fossil fuel-derived carbon dioxide emissions from Third World ("TW") and other ("non-TW") countries since 1765. On a cumulative basis, the non-TW countries have accounted for more than four-fifths of total emissions.

Figure 11.4 [*orig. figure 4*]   Comparison of carbon dioxide emissions from land use changes (deforestation) since 1765, as derived from an inverse carbon cycle model. Emissions are divided between Third World ("TW") and other ("non-TW") countries. On a cumulative basis, the total carbon emission from deforestation nearly matches that from fossil fuel burning. These results suggest that the non-TW countries have been responsible for about three-quarters of the cumulative deforestation-related carbon emissions.

change (since pre-industrial time) occurred during the 1980s. Accordingly, we also considered the cumulative changes in greenhouse gas emissions and radiative forcing since 1765, at the dawn of the industrial era.

As mentioned above, for the cumulative approach the lack of historical emission data is a problem. For fossil fuel carbon emissions from the developing world prior to 1950, we linearly interpolated to zero in 1900 (figure 11.3). To infer the past deforestation contribution, we used a balanced *inverse* carbon cycle model (Wigley, 1992), that is, a model that calculates the CO_2 emission history required to produce a given history of atmospheric CO_2 concentrations. Assuming a current biospheric carbon emission rate of 1.6 gigatonnes of carbon (GtC) per year (Bolin et al., 1986; Houghton et al., 1990), the model results suggest that, on a cumulative basis, the total carbon emission from deforestation nearly matches that from fossil fuel burning (figure 11.4). Interestingly, developing versus developed world partitioning (albeit highly judgemental) suggests that the

industrialized nations have been responsible for at least three-quarters of cumulative deforestation carbon emissions, presumably from large-scale deforestation in temperate latitudes during the last century.[3] For methane and CFC emissions, we adopt the WRI estimates (for 1987) and assume that the ratio between developed and developing nations applies to earlier periods.

Tables 11.2 and 11.3 show the relative contributions of the two groups of nations for, respectively, the current and cumulative approaches. These estimates suggest that the developing world is responsible for 27–35% of the enhanced greenhouse effect.[4]

---

[3] We assume that net deforestation in the developing world regions was zero prior to 1900.

[4] In comparison, the World Resources Institute (WRI, 1990) obtains 44% (assuming a high Third World contribution from deforestation) and Agarwal and Narain (1991) obtain about 20% (with a lower deforestation estimate and absorptive capacity allocated on a per capita basis).

Table 11.2 [*orig. table 2*]   Estimated contributions to radiative forcing change (in WM^{-2}) during the 1980s.

Greenhouse gas	Third	Non-third	Total
CO_2	0.123 41%	0.177 59%	0.30
(fossil fuel)	(0.063) (21%)	(0.169) (56%)	
(deforestation)	(0.060) (20%)	(0.008) (03%)	
CH_4	0.046 57%	0.034 43%	0.08
CFCs	0.013 10%	0.117 90%	0.13
N_2O	0.006 20%	0.024 80%	0.03
Total forcing change	0.188 35%	0.352 65%	0.54 100%

Source: Shine et al., 1990; Boden et al., 1990; WRI, 1990.

Table 11.3 [*orig. table 3*]   Estimated contributions to radiative forcing change (in WM^{-2}) from 1765 to 1990.

Greenhouse gas	Third	Non-third	Total
CO_2	0.28 19%	1.22 81%	1.50
(fossil fuel)	(0.13) (9%)	(0.68) (45%)	
(deforestation)	(0.15) (10%)	(0.54) (36%)	
CH_4	0.32 57%	0.24 43%	0.56
CFCs	0.03 10%	0.26 90%	0.29
N_2O	0.02 20%	0.08 80%	0.10
Total forcing change	0.65 27%	1.8 73%	2.45 100%

Source: Shine et al., 1990; Boden et al., 1990; WRI, 1990.

3.3   Future sea level rise "commitment": who's responsible?

These estimates allow us to make some judgements about responsibility for possible sea level rise. We start by asking, how much will *future* sea level rise due to changes in greenhouse gas concentrations that have *already* taken place? We shall call this the mean sea level rise "commitment" (a variation on the term introduced by Mintzer, 1987).[5] The commitment results from lags in the climate system, including the response of the oceans and land-ice; one could think of it as the MSL rise that is "in the pipeline" and essentially unavoidable.[6]

In order to estimate the magnitude of the commitment, we used the same model (with minor modifications), developed by Wigley and Raper (1987; 1992; also Wigley et al., 1991), that provided the time-dependent climate and sea level changes for the IPCC scenarios (Bretherton et al., 1990; Warwick and Oerlemans, 1990).[7] From greenhouse gas concentration changes, the model generates estimates of global-mean temperature

[5] For forecasting purposes, we define the term here as the difference between model-predicted global MSL rise by 1990 (our cut-off date for current observational data) and a specified future date, with the model being forced by observed increases in greenhouse gas concentrations to 1990, with no additional increases thereafter.

[6] This interpretation is given some credence by the fact that global CO_2 emissions would have to be reduced immediately by 60–80% just to achieve a stabilisation of CO_2 concentration (Watson et al., 1990).

[7] The core model is a box-upwelling diffusion energy-balance model is forced by greenhouse gas concentrations changes (from 1765 to 2100) with model parameters that can be "tuned" to reflect uncertainties in the climate sensitivity to greenhouse forcing, the penetration of heat from the surface to deeper layers of the ocean, upwelling rates and other processes. The model assumptions were the same as the "best estimate" values used by IPCC.

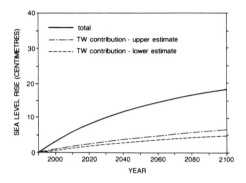

Figure 11.5 [*orig. figure 5*]   The present commitment to future sea level rise. Even if all new additions to greenhouse gas concentrations were shut off tomorrow, the sea level would continue to rise. Because of lags in the climate system, a certain "commitment" to future sea level rise is unavoidable – a consequence of concentration changes that have already taken place. This chart shows the unavoidable sea level rise to which we were "committed" by 1990, and the proportion of that rise attributes to Third World ("TW") countries. Greenhouse gas emissions by the developing world have committed us to a mean sea level rise of only 2.7–3.5 cm by 2030, and 5.0–6.5 cm by 2100.

change and oceanic thermal expansion. The temperature changes drive a global alpine glacier ice-melt model and determine Greenland and Antarctic ice sheet contributions to global MSL.

The resulting MSL rise commitment is shown in figure 11.5. To isolate the "committed" rise from any rise due to further increases in greenhouse gas concentrations, we "shut off" additional changes in radiative forcing in the model in 1990. Even so, mean sea level continues to rise sharply during the first few decades, at a (declining) rate of about 3 mm/yr, which is approximately 1.5–3.0 times faster than the average rate of MSL rise observed over the past 100 years.

During these decades, heat continues to penetrate to, and be transported within, the deeper layers of the modelled ocean. This retards surface warming and causes further thermal expansion of the oceans. Land ice

changes continue to take place, both from the model-predicted warming attained by 1990 and from additional warming thereafter (still not including any new increases in greenhouse gas concentrations, however). The 1990–2030 sea level rise is about 10 cm from already-existing commitments – more than half that expected under the "business-as-usual" reference scenario (see Section 3.4 below). Near the end of the 21st Century, the curve has flattened considerably, returning to a pre-1990 rate of MSL rise (about 1.5 mm/yr). The 1990–2100 sea level rise commitment is 18.5 cm.

These figures, if accurate, are the bottom limit; they assume that emissions are reduced sufficiently to stabilize atmospheric concentrations of greenhouse gases at today's levels – an unlikely prospect. But we make the distinction anyway, because the "committed" rise is one way of analysing past responsibility. As shown in figure 11.5, greenhouse gas emissions by the developing world have committed us to a MSL rise of only 2.7–3.5 cm by 2030 and 5.0–6.5 cm by 2100. This rate is well below the overall rate of global sea level rise experienced during the last 100 years. Clearly, the developing world, which contains about three-quarters of the world's population and which may be most vulnerable, has contributed to future sea level rise in only a minor way so far.

3.4 Future "un-committed" sea level rise: how much?

We now consider future sea level rise to which we are *not* committed – that is, the portion of a projected future rise which could conceivably be prevented. Let us define this "un-committed" sea level rise as the difference, at any future date, between the committed rise (as above) and any projection resulting from a baseline emissions scenario that gives positive changes in radiative forcing.

For the baseline scenario, we have elected to use a modified version of the IPCC Business-as-Usual scenario (BaU) (Swart, 1991). In comparison to the previous BaU case, this NEWBaU scenario, as we shall call it, has slightly higher CO_2 emissions than its predecessor (but this is partly offset by our

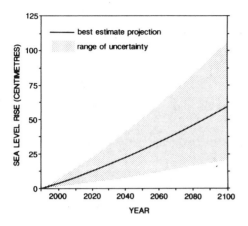

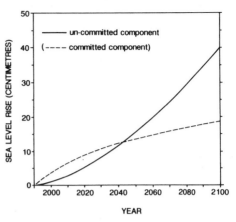

Figure 11.6 [*orig. figure 6*]   Estimates of sea level rise between 1990 and 2100, for a modified version of the IPCC Business-as-Usual emissions scenario. This produces a best estimate (solid line) of a 58 cm rise by the year 2100, with a range of uncertainty from 21 cm to 105 cm.

Figure 11.7 [*orig. figure 7*]   Committed and un-committed (preventable) components of the sea level rise estimates shown in figure 11.6. By the year 2020, more than half of the projected rise is still un-preventable. It is not until the year 2040 that the un-committed component overtakes and surpasses the committed component; the reductions in sea level rise due to cutting emissions today will accrue only after many years into the future.

use of a carbon cycle model with biospheric feedbacks that gives lower CO_2 concentration changes) and includes larger CFC production cuts as per the London amendment to the Montreal Protocol. As shown in figure 11.6, this results in a 1990–2100 mean sea level rise "best estimate" of 58 cm, with a range of uncertainty from 21 cm to 105 cm. The range of uncertainty is large because the uncertainties in the land-ice response are compounded by the uncertainties in the global-mean temperature change.

Figure 11.7 shows both the "un-committed" and "committed" components of the NEWBaU best-estimate projection. It is evident that, even with the strongest political will, there is little hope of immediately slowing the rate of MSL rise; most of this rise has already been determined by past changes in greenhouse gas forcing. There is comparatively little uncommitted sea level rise during the first part of the 21st Century; for example, by the year 2020 more than half of the projected rise is still un-preventable. It is not until the year 2040 that the uncommitted component overtakes and surpasses the committed sea level rise. Decision makers thus face a largely unprece-

dented situation in the field of environmental issues: they must consider actions with potentially large short-term costs, but with benefits which are uncertain and which would accrue only after the passage of many decades into the future.

3.5 Unilaterally setting one's house in order: to what avail?

We have avoided making any direct attribution of responsibility for future un-committed MSL rise. We do not think it entirely proper to be pointing fingers at groups of countries which are guilty of greenhouse gas emissions only in hypothetical scenarios. Nevertheless, we would like to examine and evaluate some prescriptions that can be found in the literature that do imply such responsibility.

3.5.1 CO_2 emission reduction: the OECD nations acting alone

First, we take the assertion by Agarwal and Narain (1991) that it is time for the West to stop "preaching environmental constraints

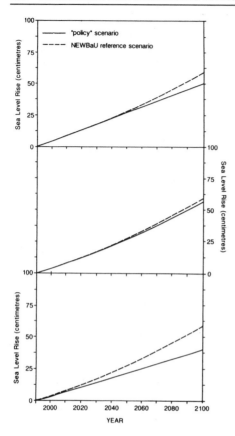

Figure 11.8 [*orig. figure 8*]   The effects of three hypothetical emission "policies" in reducing the rate of sea level rise: reducing OECD carbon emissions to zero by 2083 (top panel); stopping deforestation by 2010 and re-afforestion thereafter (middle panel); and halving the growth of global emissions of carbon dioxide, methane and nitrous oxide (bottom panel). All projections are compared to the modified IPCC Business-as-Usual scenario (dashed line). Neither strong OECD action nor strong deforestation controls would, alone, make much difference from the Business-as-Usual case. The lower panel suggests that concerted global cooperation would be required to make a significant reduction in the rate of sea level rise.

and conditionalities to the developing countries . . . [and to] first set their own house in order" (p. 23). Aside from the moral and ethical imperatives of this statement (with which we agree), there is the practical issue: what difference would it make to projections of global MSL rise?

For the purpose of a sensitivity test, we apply a scenario in which the OECD countries reduce all fossil fuel-related CO_2 emissions to zero by the year 2083. For other greenhouse gases and all other countries including developing nations, the assumptions remain the same as NEWBaU. The results are shown in figure 11.8 (top panel). As compared to the NEWBaU case, such a Draconian policy would make little difference to the projected MSL rise – about 6 cm prevented by 2100 – if the rest of the world carried on its merry way.

Of course, in the NEWBaU reference case, CO_2 emissions were assumed to grow much more slowly in the OECD than in the rest of

the world, particularly the developing countries. None the less, in general, it might be concluded that a policy of the West "setting its own house in order" cannot, alone, do the trick.

3.5.2 Stopping deforestation in the developing world

Second, let us examine an assertion frequently made in the West: that the developing world must cease deforestation in order to slow global warming and sea level rise. Currently, action by developing nations depends largely on their ability to halt deforestation. In the NEWBaU reference scenario, deforestation carries on until tropical forests are depleted by 2100. What difference would it make if, instead, deforestation ceased altogether by the year 2010, and the biosphere, through reafforestation, became a sink for 1 GtC per year by 2025 and for the remainder of the 21st Century?

The results are shown in Figure 11.8 (middle panel). In terms of future MSL rise, deforestation – or not – makes little difference, about 1 cm by 2100. In the context of future climate change, the deforestation issue is largely a "red herring" – notwithstanding the fact that "every little bit counts," and that there are indisputably sound environmental, social and economic reasons for putting forest resource management on a sustainable basis. It would be a tragedy to lose most of the world's rain-forests, but it would not elevate the oceans much.

4 The need for International Cooperation

In the lexicon of the global change issue, "one-worldism" has come under some heavy criticism lately by some who see the concept as a means of unduly shifting blame and responsibility for the enhanced green-house effect to undeserving countries, like Bangladesh. There may be more than a grain of truth in the assertion. But, on the other hand, from a practical standpoint "unilater-alism" would appear to be an ineffectual basis for slowing sea level rise. As we showed above, relying solely on a region or sub-set of nations, like the industrialized West, to solve the problem is tantamount to postponing action. Can developing coun-tries like Bangladesh afford the delay?

Let us speculate on some possible geopolitical consequences. The combination of seal-level rise and changing climate, super-imposed on increasing population density and poverty in Bangladesh, could increase the political tensions that already exist in the region. While global climate change in general might render life difficult, sea level rise could render life in some areas im-possible. This could apply not just to Bangladesh, but to island-states like the Maldives or to the low-lying coasts of Southeast Asia where mangrove swamps that have supported coastal communities for centuries could cease to be productive. Aside from increased cries for financial aid, there could be growing migration pressure. Even where people remain, sea level rise could change the productivity of coastal environments that have traditionally supported food production, recreation, transportation and other important activities.

Of course, it would not be right to assume that global sea level rise would automati-cally lead to local sea level rise at the same rate, as discussed earlier. Nor would it auto-matically lead to increased political conflict

or major migration. But it certainly has the potential for exacerbating existing tensions that have, in the past, contributed to such regional unrest and disruption. These polit-ical stresses, distracting any opportunity for mutual aid, could make dealing with the underlying problem of sea level rise more difficult. Thus, while the uncertainties regarding future sea level rise are very large, so, too, are the potential geopolitical conse-quences of doing nothing to slow it.

Because of the large "commitment", the opportunities for slowing the sea level rise predicted by current models are somewhat limited, particularly during the next several decades. In the longer term, a larger portion of the rise is "preventable," but will require a major deviation from the "business-as-usual" course. For instance, in order to keep the model predicted sea level rise to about 40 cm by 2100 (a reduction of only about half of the "un-committed" sea level rise compo-nent) it is necessary to cut by half the projected growth of global CO_2, CH_4 and N_2O emissions, as compared to the NEWBaU scenario (figure 11.8, bottom panel). Clearly, such cuts are not possible without suppressing the high rate of growth of fossil fuels in the developing world exhibited over the last several decades (around 4.5% per annum). The crux of the problem is how to do this without also suppressing social, economic and techno-logical development.

In short, international cooperation and participation in reducing emissions of greenhouse gases will be required in order to alter substantially a "business-as-usual" path of sea level rise in the future. Although most of the responsibility for the future sea level rise *commitment* falls directly into the lap of the most industrialised countries of the world, concerted action by those coun-tries alone will do little to slow the overall rate of future rise. This means that, if the most dangerous outcomes are to be avoided, all countries, including developing coun-tries, must cooperate in an international effort to reduce the rate of growth in green-house gas emissions. This fact must be grasped clearly by those in the Third World. At the same time, the industrialized nations must take greater responsibility for fostering "trajectories" of international development that can achieve dual objectives: improving the economic conditions of those in the South and ensuring future energy paths that are compatible with international goals of greenhouse gas emission reductions for the world as a whole.

Acknowledgements
Research support provided by the EPOCH Programme of the Commission of the European Communities is gratefully acknowledged. The authors wish to thank T. M. I. Wigley and S. C. B. Raper for use of their models. P. M. Kelly provided helpful comments on early drafts of this chapter.

References
Agarwal, A. and S. Narain, 1991: *Global Warming in an Unequal World: A Case of Environmental Colonialism.* Centre for Science and Environment, New Delhi.

Ali, S. I. and S. Huq, 1991: International Sea Level Rise: A National Assessment of Effects and Possible Responses for Bangladesh, prepared for Centre for Climate Change. University of Maryland, Maryland, USA.

Boden, T. A., P. Kanciruk and M. P. Farrell (eds), 1990: *Trends '90: A Compendium of Data on Global Change.* Carbon Dioxide Information Analysis Center, Oak Ridge National Laboratory, Oak Ridge, Tennessee.

Bolin, B., B. R. Doos, J. Jäger and R. A. Warrick (eds), 1986: *The Greenhouse Effect, Climatic Change and Ecosystems.* John Wiley and Sons, Chichester.

Brammer, H., 1989: Monitoring the evidence of the greenhouse effects and its impact on Bangladesh. In *Proceedings of the Conference on the Greenhouse Effect and Coastal Area of Bangladesh*, ed. H. Moudud. A. Rahman et al. Dhaka, Bangladesh.

Brammer, H., 1992: The complexities of detailed impact assessment for the Ganges–Brahmaputra–Meghna delta of Bangladesh. In *Climate and Sea Level Change: Observations, Projections and Implications*, ed. R. A. Warrick, E. M. Barrow and T. M. L. Wigley. Cambridge University Press, Cambridge.

Bretherton, F. P., K. Bryan and J. D. Woods, 1990: Time dependent greenhouse-gas-induced climate change. In *Climate Change: The IPCC Scientific Assessment*, ed. J. T. Houghton. G. J. Jenkins and J. J. Ephraums. Cambridge University Press, Cambridge.

Broadus, J. M. 1992: Possible impacts of and adjustments to sea level rise: the cases of Bangladesh and Egypt. In *Climate and Sea Level Change: Observations, Projections and Implications*, ed. R. A. Warrick, E. M. Barrow and T. M. L. Wigley. Cambridge University Press, Cambridge.

Commonwealth Group of Experts, 1989: *Climate Change: Meeting the Challenge.* Commonwealth Secretariat, London.

Day, J. W., W. H. Conner, R. Costanza, G. P. Kemp and I. A. Mendelssohn, 1992: Impacts of sea level rise on coastal ecosystems. In *Climate and Sea Level Change: Observations, Projections and Implications*, ed. R. A. Warrick, E. M. Barrow and T. M. L. Wigley. Cambridge University Press, Cambridge.

Edwards, A. J., 1989: The implications of sea-level rise for the Republic of Maldives. *Report to the Commonwealth Expert Group on Climate Change and Sea Level Rise.* Centre for Tropical Coastal Management Studies, University of Newcastle-upon-Tyne.

Emery, K. O. and D. G. Aubrey, 1991: *Sea Levels, Land Levels and Tide-Gauges.* Springer Verlag. New York.

FPCO (Flood Plain Coordinating Organization), 1989: *Flood Action Plan.*

Frascetto, R. (ed.), 1991: *Impact of Sea Level Rise on Cities and Regions.* Marsilio Editori, Venice.

Gagliano, S. M., K. J. Meyer and K. M. Wicker, 1981: Land loss in the Mississippi River deltaic plain. *Trans Gulf Coast Assoc. Geol. Soc.*, 31, 295–300.

Hammond, A. L., E. Rodenburg and W. R. Moonmaw, 1991: Calculating national accountability for climate change. *Environment*, 33(1), 11.

Houghton, J. T., G. J. Jenkins and J. I. Ephraums (eds.), 1990: *Climate Change: The IPCC Scientific Assessment.* Cambridge University Press, Cambridge.

Huq, S. and A. A. Rahman, 1990: Global warming and Bangladesh: implications and response. In *Responding to the Threat of Global Warming: Options for the Pacific*, eds., D. G. Street and T. A. Siddiqi, Argonne National Laboratory, Argonne, USA. 1990.

Mahtab, F. U., 1989: *Effect of Climate Change and Sea Level Rise on Bangladesh.* Report for the Commonwealth Secretariat.

McCully, P., 1991: Discord in the greenhouse: how WRI is attempting to shift the blame for global warming. *The Ecologist*, 21(4), 157–165.

Milliman, J. D., J. M. Broadus and F. Gable, 1989: Environmental and economic impacts of rising sea level and subsiding deltas: the Nile and

Bengal examples. *Ambio*, 18(6), 340–345.

Mintzer. I., 1987: *A Matter of Degrees: Potential for Controlling the Greenhouse Effect.* World Resources Institute Research Report No. 5. World Resources Institute, Washington, D.C.

NRC (National Research Council), 1990: *Sea-Level Change.* National Academy Press, Washington, D.C.

Peltier, W. R., 1986: Deglaciation-induced vertical motion of the North American continent and transient lower mantle rheology. *J. Geophys. Res.*, 91, 9099–9123.

Pernetta, J. C. and G. Sestini, 1989: The Maldives and the impact of expected climatic changes, UNEP Regional Seas Reports and Studies No. 104. UNEP, Nairobi.

Pugh, D., 1992: Improving sea level data. In *Climate and Sea Level Change: Observations, Projections and Implications*, ed. R. A. Warrick, E. M. Barrow and T. M. L. Wigley. Cambridge University Press, Cambridge.

Rahman, A. A. and S. Huq, 1989: Greenhouse effect and Bangladesh: a conceptual framework. In *Proceedings of the Conference on the Greenhouse Effect and Coastal Area of Bangladesh*, eds. Moudud, H. et al. Dhaka.

Shine, K. P., R. G. Derwent, D. J. Wuebbles and J. J. Morcrette, 1990: Radiative forcing of climate. In *Climate Change: The IPCC Scientific Assessment*, ed. J. T. Houghton, G. J. Jenkins and J. J. Ephraums. Cambridge University Press, Cambridge.

Smith, K. R., 1991: Allocating responsibility for global warming: the Natural Debt Index. *Ambio*, 20.

Stevenson, J. C., L. G. Ward and M. S. Kearney, 1986: Vertical accretion in marshes with varying rates of sea level rise. In *Estuarine Variability*, ed. D. A. Wolfe. Academic Press, New York.

Swart, R., 1991: Personal communication.

Titus, J. G., 1992: Regional impacts and responses. In *Climate and Sea Level Change: Observations, Projections and Implications*, ed. R. A. Warrick, E. M. Barrow and T. M. L. Wigley. Cambridge University Press, Cambridge.

Titus, J. G., R. Wedge, N. Psuty and J. Fancher

(eds.), 1990: *Changing Climate and the Coast*, Vol. 1 and 2. Report to the Intergovernmental Planet on Climate Change from the Miami Conference on Adaptive Responses to Sea Level Rise and Other Impacts of Global Climate Change. US Environmental Protection Agency, Washington, D.C.

Warrick, R. A., and J. Oerlemans, 1990: Sea level rise. In *Climate Change: The IPCC Scientific Assessment*, ed. J. T. Houghton, G. J. Jenkins and J. J. Ephraums. Cambridge University Press, Cambridge.

Warrick, R. A., E. M. Barrow and T. M. L. Wigley (eds.), 1992: *Climate and Sea Level Change: Observations, Projections and Implications.* Cambridge University Press, Cambridge.

Watson, R. T., H. Rodhe, H. Oeschger and U. Siegenthaler, 1990: Greenhouse gases and aerosols. In *Climate Change: The IPCC Scientific Assessment*, ed. J. T. Houghton, G. J. Jenkins and J. J. Ephraums. Cambridge University Press, Cambridge.

Wigley, T. M. L., 1992: A simple inverse carbon cycle model. *Global Biogeochemical Cycles* (in press).

Wigley, T. M. L. and S. C. B. Raper, 1987: Thermal expansion of sea water associated with global warming. *Nature*, 330, 127–131.

Wigley, T. M. L. and S. C. B. Raper, 1992: Future changes in global mean temperature and sea level. In *Climate and Sea Level Change: Observations, Projections and Implications*, ed. R. A. Warrick, E. M. Barrow and T. M. L. Wigley. Cambridge University Press, Cambridge.

Wigley, T. M. L., T. Holt and S. C. B. Raper, 1991: *STUGE (an Interactive Greenhouse Model): User's Manual.* Climatic Research Unit, University of East Anglia, Norwich.

Woodroffe, C., 1989: Maldives and sea-level rise: an environmental perspective. Department of Geography, University of Wollongong (unpublished).

WRI (World Resources Institute), 1990: *World Resources 1990–1991 – A Guide to the Global Environment.* Oxford University Press, Oxford.

12

RESPONDING TO THE EL NIÑO SOUTHERN OSCILLATION*

Maryam Golnaraghi and Rajiv Kaul

The El Niño Southern Oscillation (ENSO) – a natural physical phenomenon caused by the interactions of ocean and atmosphere in the tropical Pacific – is associated with extensive ecological and economic disasters, such as the destruction of fisheries in the Pacific Ocean, the failure of Indian monsoons, and droughts and torrential floods in various regions of the world. Until the first half of this century, the term El Niño was used to refer to the occurrence of a local warm current off the Peruvian coast. This definition changed in the 1960s when new scientific ideas related the local current to large-scale interactions between the tropical Pacific and the atmosphere. El Niño is a phase of the Southern Oscillation when the trade winds are weak and the sea surface temperatures in the equatorial Pacific increase by 1° to 4°C.[1] The impacts of ENSO – which occurs every three to seven years and is the most prominent signal in year-to-year natural climate variability – are felt worldwide, owing to disruptions of atmospheric general circulation and associated climate patterns around the world.

Societies whose wealth is derived pri-marily from agricultural and hydrological resources are particularly vulnerable to dramatic interannual climate disruptions. Cumulative losses incurred by nations from damage to infrastructure and decrease in industrial and agricultural production during the 1982–1983 ENSO – the most destructive event of this century – amounted to an estimated $13 billion. The event has been blamed for droughts in countries from India to Australia, floods from Ecuador to New Zealand, and fires in West Africa and Brazil. On the Colombian and Ecuadorian coasts, heavy rain caused rivers to overflow, resulting in erosion, flooding, and mudslides. This led to the isolation of production and population centers and therefore to losses in the production of consumer goods, and the storage of supplies.[2]

A historical look at the parallel evolution in the scientific knowledge and social perception of the ENSO phenomenon reveals how effective social responses to interannual climate change depend on scientific progress; how societies vulnerable to ENSO apply present scientific knowledge

* Originally published in *Environment*, 1995, vol. 37, no. 1, pp. 16–20, 38–44.
[1] The term La Niña is applied when sea surface temperatures in the central and eastern tropical Pacific are unusually low and when the trade winds increase.
[2] S. R. Jordan, "Impact of ENSO Events in the South Eastern Pacific Region with Special Reference to the Interaction of Fishing and Climate Variability." in M. H. Glantz, R. W. Katz, and N. Nicholls, *Teleconnections Linking Worldwide Climate Anomalies* (Cambridge, U.K.: Cambridge University Press, 1991), chapter 13.

to domestic needs; and how other factors such as unsustainable use of natural resources increase nations' vulnerability to ENSO-related climate impacts. At present, scientists are capable of forecasting the onset of ENSO up to one year in advance through sea surface temperature signals. Despite high scientific uncertainty associated with current forecasts, this information holds important societal value. Peru and Brazil, among other nations, provide examples of how policymakers have used forecast information to manage socioeconomic responses with reasonable success.

Discovering ENSO

In 1891, Luis Carranza, president of the Lima Geographical Society, brought to the attention of a group of distinguished geographers the extension of a warm current flowing between the ports of Paita and Pacasmayo. This current was named El Niño, the Christ child, by Paita sailors because it appeared just after Christmas.[3]

The earliest recorded ENSO event was four and a half centuries ago in 1525 to early 1526, as reported primarily by Spanish explorers and military personnel in northwestern South America. Subsequent, more conclusive scientific studies have confirmed these observations, and geomorphological, paleontological, and archeological studies indicate ENSO-related natural disasters for the past 40,000 years.[4]

Scientific History

The 20th century has seen three distinct periods of scientific progress in understanding ENSO. From 1900 to the 1960s, scientific understanding was developed through the efforts of individuals, whereas from the 1960s until 1982, individuals moved toward scientific collaboration

supported by national programs. Since 1982, scientists have been cooperating internationally. The scientific responses to ENSO during this century have moved from a "reaction" toward a "prediction" of events.

India's catastrophic droughts and famines of 1877 and 1899 led to efforts to understand the reason for the failure of the monsoons that had brought about these disasters. In 1904, Sir Gilbert Walker, Director-General of the Observatories in India, investigated how to predict the annual variabilities of the monsoons. Although his efforts did not result in the ability to predict monsoon failure, he discovered a correlation in the patterns of atmospheric pressure at sea level in the tropics, the ocean temperature, and rainfall fluctuations across the Pacific Ocean, which he named the Southern Oscillation. His statistical studies showed that the primary characteristic of the Southern Oscillation is a seesaw in atmospheric pressure at sea level between the southeastern subtropical Pacific and the Indian Ocean. Walker pioneered the applications of statistical methods for understanding the concept of teleconnections, the connections between climate conditions around the globe.[5] In a series of papers published between 1923 and 1937, he confirmed that monsoons are part of a global phenomenon. But because no theoretical evidence existed for linking worldwide climate anomalies, Walker's work was greeted with much skepticism. Scientific interest in this phenomenon faded away until 1957–1958, when "exceptionally" anomalous oceanographic and atmospheric conditions were revealed in data routinely collected in the tropical Pacific by merchant ships. Weak trade winds, the westward extension of the El Niño current beyond the International Date Line, and severe rainfall in the normally arid regions of the central tropical Pacific were noted. These observations raised scientific attention and were

[3] S. G. Philander, *El Niño, La Niña, and the Southern Oscillation* (San Diego, Calif.: Academic Press, 1990).
[4] W. H. Quinn, R. S. Bradly, and P. Jones, eds., "The Historical Record of El Niño Events," in *Climate Science Since AD 1500* (London: Chapman and Hall, 1992).
[5] B. G. Brown and R. W. Katz, "Use of Statistical Methods in the Search for Teleconnections: Past, Present and Future," in M. H. Glantz, R. W. Katz, and N. Nicholls, eds., *Teleconnections Linking Worldwide Climate Anomalies* (Cambridge, U.K.: Cambridge University Press, 1991), chapter 12.

instrumental in revealing the link between El Niño and the Southern Oscillation.[6]

The Walker Circulation

Jacob Bjerknes, of the University of California at Los Angeles, suggested that the unusual conditions observed for the ocean and atmosphere in 1957–1958 were not unique and occurred interannually. He showed that the fluctuations in sea surface temperature and rainfall are associated with large-scale variations in equatorial trade winds, which in turn affect the Southern Oscillation. Bjerknes proposed a physical mechanism that linked El Niño with the Southern Oscillation. During normal conditions, dry air sinks over the cold water of the eastern tropical Pacific and flows westward along the equator as part of the trade winds. The air is moistened as it moves toward the warm waters of the western tropical Pacific. There it rises in towering rain clouds before its return flow into the upper troposphere. Bjerknes named this process the Walker Circulation. He noted that the sea surface temperature gradients – the cold waters along the Peruvian coast and the warm waters in the western tropical Pacific – are necessary for the atmospheric gradients that drive this circulation. Interactions between the ocean and the atmosphere set up a positive feedback mechanism, where an initial change in the ocean could alter the meteorological conditions, which in turn result in further oceanic changes that reinforce the initial conditions.

The Kelvin Wave Hypothesis

The studies of Klaus Wyrtki in the 1970s, based on analyses of tide gauge data, indicate that the development of the exceptionally warm current off the coast of Peru is a consequence of changes in the circulation of the tropical Pacific Ocean in response to changes in the winds that drive the circulation.[7] Wyrtki, of the University of Hawaii, proposed the Kelvin Wave Hypothesis, an interpretation of the fundamental ocean dynamics associated with the onset of the warm episodes, which could be tested through field experiments and simulations with numerical models. It provides the explanation for the coupling of the decreasing trade winds and equatorial wave activities in the tropical upper layers in the ocean forced by wind and the evolution of the sea surface temperature anomalies during the onset of the warm episode.[8]

Scientists began to compile the historical records of surface winds and sea surface temperature collected from merchant ships and started to conduct statistical analyses and numerical experiments using highly simplified models.

This research documented the basinwide extent of the ENSO phenomenon – the times of year that it begins, peaks, and ends; its correlation with the Indian monsoons; and how the different stages of El Niño correspond with swings of the Southern Oscillation.

Existing national programs began to enhance the scientific collaboration for understanding ENSO and its impact on climate anomalies. Severe socioeconomic consequences of the 1972–1973 and 1976–1977 ENSO events raised public attention and led to the establishment of two new major programs in the United States to study year-to-year climate variability as a common goal – the National Climate Program and the World Climate Research Program.[9] By the end of the 1970s, ENSO had emerged as a major research topic in the new National Climate Program.

In addition to national programs in the

[6] E. M. Rasmussen and J. M. Wallace, "Meteorological Aspects of ENSO," *Science* 222 (1983): 1195–1202.
[7] K. Wyrtki, "Fluctuations of the Dynamic Topography in the Pacific Ocean," *Journal of Physical Oceanography* 5 (1975a): 450–59; and K. Wyrtki, "El Niño – The Dynamic Response of the Pacific Ocean to Atmospheric Forcing," *Journal of Physical Oceanography* 5 (1975b): 572–84.
[8] A. J. Busalacchi and J. J. O'Brien, "Interannual Variability of the Equatorial Pacific in the 1960s," *Journal of Geophysical Research* 86 (1981): 10901–07.
[9] National Research Council, *TOGA: A Review of Progress and Future Opportunities* (Washington, D.C.: National Academy Press, 1990).

United States, the countries most vulnerable to the event addressed the phenomenon by initiating multinational regional cooperations. In 1974, national governments in the southeast Pacific joined research efforts with the Regional Program for the Study of the El Niño Phenomenon (ERFEN). ERFEN, which provides in situ observational results to the scientific community, has coordinated the activities of Peru, Chile, Colombia, and Ecuador to study the oceanographic, meteorological, and biological aspects of ENSO.

International scientific cooperation

Although scientific collaborations to study ENSO and its impact on the global climate had begun in the 1970s, it was only in 1982 that an international scientific effort was initiated. A scientific steering group was established by the World Meteorological Organization/International Council of Scientific Unions Joint Steering Committee and the Scientific Committee on Oceanic Research/Intergovernmental Commission Committee on Climate Changes and the Oceans to evaluate the design of an international research program for studying the processes in the tropical ocean and its interactions with global climate.

The initial stages of this scientific planning coincided with the strongest warm episode of the century, the 1982–1983 ENSO, which brought the most devastating climate conditions to various regions in the world. This episode caught the scientific community by surprise because it had not exhibited a typical "onset phase." These unusual conditions led scientists to conclude that only through a program continuously monitoring the ocean, the atmosphere, and their interaction would they be assured of observing the coupled atmospheric-oceanic processes during the critical onset phase of these episodes.

In 1985, after the world had suffered billions of dollars in damages due to the 1982–1983 ENSO, the World Climate Research Program launched a 10-year international research program called Tropical Ocean and Global Atmosphere (TOGA).[10] TOGA was successful in attaining many of its goals and in fostering cooperation in the international scientific research community to work on problems of global importance. In addition, the institutional framework provided by TOGA has enhanced the domestic scientific capacity of less developed nations. International collaborations have led to various informal channels for provision and exchange of data and forecast information. This has given scientists from less developed nations the capacity to advise their decisionmakers.

Scientists have proposed CLIVAR (Climate Variability) – a 15-year follow-up program to TOGA – to begin in January 1995. Its objectives are to determine the limits of predictability, to extend the predictive skill of seasonal and interannual climate variation from the tropics to other regions, and to develop an operational climate forecasting system capable of delivering forecast information internationally.

Socioeconomic History

Climate anomalies induced by ENSO have been responsible for severe worldwide socioeconomic damage. This realization has come in three stages: Until the 1972–1973 event, El Niño was perceived as affecting only local communities and industries along the eastern Pacific coast near Peru; then, from 1972–1973 to 1982–1983, ENSO-related climate anomalies were recognized as the cause of natural disasters worldwide; and from 1982–1983 onward, nations began to realize the need for national programs that use scientific information in policy planning. Throughout this time, although the impacts of the event have remained consistent, political and economic developments in social structure have changed the way the phenomenon is managed. Nations understand that to mitigate the social costs of ENSO-induced climate anomalies, policymaking should be anticipatory rather than reactive.

---

[10] See National Research Council, note 9 above; and National Research Council, *U.S. Participation in the TOGA Program: A Research Strategy* (Washington, D.C.: National Academy Press, 1986).

El Niño: local impacts and responses

The Peruvian shore is barren desert land that borders a cold productive ocean home to many forms of fish and other marine life. Locals had long noticed that every few years during the early calendar months, an exceptionally warm current accompanied by very heavy rains penetrated southward and moderated the normally cold sea surface temperatures. During such times, the desert land was converted into abundant green pastures, and crops such as cotton, bananas, and coconut could be grown in regions that had no vegetation during normal years. To local communities, these years were known as *años de abundancia*, or years of abundance. At the same time, birds and marine life that thrived under the dry, cold conditions disappeared, and new species appeared (see box 12.1).

Early this century, the primary climate-related socioeconomic impact of El Niño was believed to be on the production of guano in Peru. For more than 1,000 years, guano – composed chiefly of the droppings of sea birds – has been mined on the rocky coast for use as fertilizer in traditional Peruvian practice. During the late 19th century, however, mining was so intense

---

Box 12.1   How El Niño Affects Biodiversity

The waters off the western coast of South America are among the most productive in the world. The equatorward winds along the coastline drive the surface waters off shore and result in the upwelling of the cold waters, which are rich in nutrients, including nitrates, silicates, and phosphates. The continuous injection of the nutrients into the surface layers where light conditions are ideal results in high rates of primary production. Herbivores graze on the abundant phytoplankton, causing the organic matter to move along the marine food chain.

During El Niño, waters stirred up are from shallower, nutrient-poor depths. The rate of primary production is highly reduced, thereby disrupting the food chain as well as contributing to the failure of reproduction in some species. Many bird and marine species usually present along the western coast of South America temporarily disappear and other species, such as yellow and black water snakes, appear in the region. For example, the Peruvian anchovy, which prefers relatively cool waters of 16° to 18°C, is concentrated in small pockets of cold water in the region, making it more accessible to predators, especially humans. As a result, the number of anchovies is reduced significantly. During moderate El Niño years some survive, but during a severe phase they disappear altogether. This was the case in 1983, when the anchovy catch off Peru was less than 1 percent of that 10 years before.

Unlike the anchovy, other species such as shrimp and scallop thrive in the conditions induced by El Niño. During the El Niño of 1982–1983, shrimp and scallop populations along the coast of Peru rapidly grew. While the shrimp were probably carried southward by the current, the scallops already existed locally and became abundant in the favorable conditions.[1]

These ecological changes are not confined to the western coast of South America. During the 1982–1983 El Niño, severe damage resulting in the reproductive failure of marine species, birds, and other species were observed and investigated along the California coast, in the central equatorial Pacific, as well as on the South American continent.[2]

1   R. T. Barber and F. P. Chavez, "Biological consequences of El Niño," *Science* 222 (1986): 1203–10.
2   S. G. Philander, *El Niño, La Niña, and the Southern Oscillation* (San Diego, Calif.: Academic Press, 1990).

that 150-foot mountains of guano – which had collected over 25 centuries – were removed in only decades.[11] Between 1848 and 1875, 20 million tons were exported to Europe and the United States. By the early 1900s, the exploitation of ancient guano beds was found to be a major source of wealth and regional revenues for the Peruvians living along the coast.

In 1909, the Peruvian government, recognizing that the resource was being depleted, established the Guano Administration to manage the annual amounts removed and to protect the birds and their eggs from predation.[12] As a result, extraction of guano decreased. Miners of the resource could export only the quantity renewed annually.

Regulating the guano industry made it increasingly vulnerable to the impacts of El Niño. During an El Niño event, which caused the biological productivity of the Peruvian waters to be greatly reduced, the anchoveta (anchovies) that provided 95 percent of the birds' diet, would sharply decrease. Large numbers of birds died and the annual production of guano, which determined the quantity exported, would drop. Prior to the 1909 regulations, the effects of El Niño on the lucrative industry had been less important. In the El Niño years of 1925, 1939, and 1957, there were significant reductions in revenues from guano exports.[13]

But by the 1950s, the local communities – aided by the Guano Administration – succeeded in managing the mining of this resource so that it would remain renewable. The 255,000 tons of guano the birds produced in 1953 was sold at roughly $50 per ton, amounting to a gross annual value of $11.2 million. Even though the commodity regained value to those who mined, processed, and exported it, in the 1950s and 1960s, Peruvian economic interests shifted from the export of guano to that of pelagic fish, such as anchoveta. One domestic factor contributing to this shift was the 1948 presidential election of Manuel Odria, who actively promoted a policy of development through the export of Peruvian resources. Exports of resources would bring the nation foreign exchange, which would provide the means to purchase technology for the development of national industries. It was the new president's policies that provided the impetus for export-oriented exploitation of anchoveta.[14]

Furthermore, international demands for anchoveta grew after the collapse of the California sardine industry in the 1950s. The annual catches of more than 600,000 tons that the Californian fishing industry had seen in the 1940s dropped to less than 150,000 tons by 1953. This decline provided a huge market for the Peruvian anchoveta, a protein-rich, high-value ingredient of fish meal, which is used in animal feed.[15] By the 1970s, anchoveta fishing and processing were important sectors in Peru's national economy, making up one-third of Peruvian exports.

In 1969–1970, the total fish meal industry was worth $350 million, employing 20,000 full-time fishermen and 3,000 factory workers in 125 fish meal plants. Peru became the world's number one fishing nation (by weight), producing 42 percent of the world's total catch.[16] In addition to the economic

[11] G. Paulik, "Anchovies, Birds and Fishermen in the Peru Current," in M. H. Glantz and J. D. Thompson, eds., *Resource Management and Environmental Uncertainty: Lessons from Coastal Upwelling Fisheries* (New York: John Wiley, 1981).

[12] R. Murphy, "Guano and the Anchoveta Fishery," in M. H. Glantz and J. D. Thompson, note 11 above.

[13] Ibid.

[14] M. H. Glantz, "Floods, Fires and Famine: Is El Niño to Blame?" *Oceanus* 27, no. 2 (1984): 14–19.

[15] J. Vondruska, "Postwar Production, Consumption and Prices of Fish Meal," in Glantz and Thompson, note 11 above.

[16] N. Nicholls, "The El Niño/Southern Oscillation Phenomenon," in M. H. Glantz, R. W. Katz, and M. Krenz, eds., *Impact: Climate Crisis, the UNEP NCAR Report on Economic and Societal Impacts Associated with the 1982–83 Worldwide Climate Anomalies* (Nairobi, Kenya: United Nations Environment Programme and National Center for Atmospheric Research, 1985), 2–10.

benefits, the "booming" fish meal industry had particular social significance. Historically, Peru had been a rigid society with a landed aristocracy, where opportunities to increase wealth were restricted to members of the wealthy. The development of the fish meal industry provided an explosive chance for the underprivileged groups to become wealthy as well.

Lucrative exports combined with inefficient management by the national government led to overfishing. Reminiscent of the experience of the guano industry some decades earlier, overfishing of anchoveta was depleting the resource, and its related industries became more vulnerable to El Niño-induced fluctuations. In two decades, anchoveta catches had grown by 27,400 percent. After the 1972–1973 event, the anchoveta fisheries collapsed.

El Niño years were no longer perceived as the years of abundance. Throughout the century, economic activities and social structures shifted from traditional practices of limited agriculture and fishing to the over-mining of guano and later to the exploitation of anchoveta. National development policies and international market pressures contributed to these shifts. The climate-related impacts of El Niño, which had traditionally benefited local communities by converting the barren coastal land into abundant green pastures, began to adversely affect the national socioeconomic well-being.

1972–1973 ENSO: global Impacts and responses

The 1972–1973 event changed the common perception of the phenomenon: El Niño was increasingly seen as a phenomenon that induced climate anomalies around the world. During the 1972–1973 episode, droughts occurred in eastern Australia, northeastern Brazil, East Africa, and the Soviet Union, while floods and severe storms happened in southeastern Brazil, northwestern Peru, on the Pacific coast of

Australia, and in the Philippines. The co-incidence of these worldwide climate disturbances precipitated a reaction in the international policy community. Policy-makers blamed the global scale of destruction on El Niño. They ignored the implications of a lacking national infrastructure and the effect of exploitative social and economic policies that had increased the vulnerability of nations to the climate-related impacts.

After the 1972–1973 event, the Peruvian anchoveta fisheries failed to reach their previously high production levels. By 1973, Peru's production of fish meal was diminished to 420,000 tons – the level of production in 1960. To prevent overfishing, the socialist Peruvian government nationalized the fishing industry in 1973.[17] Nonetheless, the high levels of anchoveta production attained before the 1972–1973 event have never been recovered. The Peruvian economy had become dependent on anchoveta fishing, which led to losses in associated industries like shipbuilding, gear manufacturing, net producing, and fish processing.[18]

Although ENSO-related droughts and floods affected aggregate world agricultural production, uncoordinated policy responses enhanced the social costs of reduced food supplies. Shortages in production were exacerbated by unanticipated shifts in the pattern of food supplies. For example, losses from the decreased production of anchoveta affected the world market of protein for animal feed that was dependent on Peruvian exports. As international supplies of fish meal were reduced, soybeans became a lucrative substitute. The increased demand for soybeans made farmers substitute them for corn and wheat, which further contributed to the international food shortage.[19] In 1973, a severe drought in the Soviet Union led to decreases in the production of grain. The United States exported an extraordinarily large amount of grain to the Soviet Union that year, generating a

[17] See Glantz, note 14 above.
[18] See Glantz, note 14 above.
[19] See Glantz, note 14 above.

shortage in the food exports available to other parts of the world. The combination of ENSO-induced climate anomalies and the uncoordinated reaction of policymakers contributed to the international food crisis. For the first time since World War II, per-capita food production declined.

Several factors combined to generate the public perception that climate anomalies induced by ENSO had global dimensions:

- Scientific evidence linked the local El Niño current to global climate anomalies.
- International interest arose due to the severe socioeconomic repercussions of the collapse of the Peruvian anchoveta industry on the regional economy and global fish meal markets.
- Droughts and floods occurring in various parts of the world as well as the international food shortage had global socioeconomic effects.

The attention of policymakers was directed to the socioeconomic impacts of climate variation and the need for further information to predict and manage responses to interannual climate change. However, although the 1972–1973 event changed previous perceptions of El Niño, national responses to manage ENSO-related socioeconomic impacts did not emerge until the 1980s. While the scientific community progressed throughout the 1970s toward understanding and forecasting certain aspects of ENSO, it took another major event in 1982–1983 – probably the most destructive of this century – to break the inertia of policymakers.

1982–1983 ENSO: Global Turning Point

The 1982–1983 event marked an important turning point in the management of social responses to ENSO. Throughout the 1980s, countries began to establish national programs and to use scientific under-

standing of the phenomenon as a guide to policy planning. A look at the experiences of Brazil and Peru – nations that suffered major losses – reveals how some less developed nations are socially and economically affected by climate variability. Because of their relatively successful social responses to the ENSO events of 1986–1987 and 1991–1992, respectively, these two nations provide valuable case studies for demonstrating the social benefits of incorporating forecast information into ENSO-related policymaking.

Peru: Managing the 1986–1987 ENSO

Peru was the first nation in South America to use forecast information in its domestic planning. The estimated loss of $2 billion in damages to agriculture, industry, transportation, and other economic sectors during the 1982–1983 event led the government to set up a network incorporating scientific expertise into the policy process. The scientific community's prediction that the events of 1982–1983 would not be repeated in the following year was presented to the heads of agrarian organizations, banking officials, and the Minister of Agriculture. The 1983–1984 agrarian plan incorporated this information and yields were high. In Peru, production in each economic sector is based on an interrelated set of national plans. Agricultural production goals are set by the Ministry of Agriculture and the national committees of non-governmental agrarian organizations, and determine water distribution in irrigated areas, fertilizer prices, interest rates on loans from the agrarian bank, and prices of agricultural products.[20]

The monitoring of ENSO is carried out by a special national committee, the National Studies of the El Niño Phenomenon (ENFEN), in consultation with modelers in the United States and other parts of the world.

Since 1984, ENFEN has presented a fore-

---

[20] P. Lagos and J. Buizer, "El Niño and Peru: A Nation's Response to Interannual Climate Variability." in S. K. Majumdar, G. S. Forbes, E. W. Miller, and R. F. Schmaltz, eds., *Natural and Technological Disasters, Causes, Effects and Preventive Measures* (Pennsylvania Academy of Science, 1992), 221–38.

178 PREDICTING AND MANAGING ATMOSPHERIC CHANGE

cast every November to the farming community, the head of the nongovernmental farming organization, and government representatives, who decide on a production strategy based on expectations of rainfall. The forecast presents one of the following possibilities for the coming year: normal weather conditions, slightly wetter and warmer than normal; ENSO condition; or a cold event – conditions cooler and drier than normal. This information helps policymakers and farmers determine the combination of crops to be sown.

A forecast issued by ENFEN in the fall of 1986 used information from U.S. models and ocean and atmospheric data from the Pacific to predict a mild event. The information was delivered to government agencies through the channels established in 1984 and was spread throughout the nation by the popular media. Consequently, aggregate yields of rice and cotton were higher than during the previous ENSO year when no forecast information was used, revealing its benefit to agricultural practices.[21]

Despite successful management in 1986–1987, no institutional channels exist to directly feed climate information to the Ministry of the Presidency, which is the nation's most influential office in national planning.[22] ENFEN made an official forecast of the 1991–1992 ENSO by early October 1991. Even though national newspapers and television and radio stations allotted much attention to the potential impacts of the forthcoming climate anomalies and this information was used by the Ministry of Agriculture, bureaucracy prevented the information from reaching higher levels of government.

Brazil: Ceara's Management of the 1991–1992 ENSO

Brazil was severely affected by the climate-related impacts of the 1982–1983 ENSO. Approximately 88 percent of the Brazilian northeast – an area of 1.54 million square kilometers, comprising 8.2 percent of the nation's territory – was affected by drought.[23] The 1983 share of agriculture in gross national product fell from a normal 18.2 percent in 1978 to 7.6 percent.[24] Fourteen million people – primarily farmers with limited resources and small holdings – were affected, most losing their entire production and thereby worsening the socioeconomic disparities between rich and poor. However, the economic repercussions were not limited to these small landholders. In 1983, the national government provided assistance for 2.8 million people, which, along with price increases of food products by 300 percent, led to the rise in national indebtedness, both in the public and private sectors. On average, during the drought, 35 percent of a family's expenditure was required for debt repayments.

It became clear during this episode that the Brazilian northeast – in particular, the states of Ceara, Rio Grande do Norte, Parahyba, and Sergi-Pi – is prone to interannual droughts. The poor majority of this primarily rural region depends on subsistence farming, growing beans, cotton, and maize, and exists alongside profit-oriented sugar cane plantations, cattle ranches, and nut and food processing plants. The economic disparities in the region have led to the popular belief that efforts to manage water resources and agriculture occur for the benefit of wealthy land owners.[25] Such social conditions add complexities to the

[21] Ibid.

[22] J. J. O'Brien, *Workshop Report on the Economic Impact of ENSO Forecasts on the American, Australian, and Asian Continents* (Tallahassee: Florida State University, 1993).

[23] J. G. Gasques and A. Magalhaes, "Climate Anomalies and Their Impacts in Brazil During the 1982–83 ENSO Event," in M. H. Glantz, R. W. Katz, and M. Krenz, note 16 above, pages 30–36.

[24] Brazil-U.S. Collaboration in Science and Technology, *Workshop on ENSO and Seasonal to Interannual Climate Variability: Socio-Economic Impacts, Forecasting and Application to the Decision-Making Process* (Fortaleza and Florianopolis, Brazil: Inter-American Institute, 1993).

[25] See O'Brien, note 22 above.

role of the state government in implementing regional policies.

Ceara's Foundation for Meteorological and Hydrological Resources (FUNCEME), a government agency in the northeast state of Ceara, has taken the lead in Brazil in providing forecast information. Founded in 1988, the agency receives information from scientists in the United States, France, and Germany and provides "management support for decision making related to droughts based on scientific information in the areas of meteorology, water resources, and environmental data."[26]

In 1991, Ceara's governor Ciro Gomes convened a meeting of his secretariat to plan anticipatory policies after receiving an ENSO-related drought forecast. To alert local people of the forthcoming climate anomalies, the governor and members of FUNCEME organized a grassroots campaign: They traveled to the state's interior to explain that FUNCEME was a reliable source of information and that farmers would benefit from their recommendations. Farmers were informed that despite ENSO, crops could be planted but they had to follow a "firm planting time" and plant "appropriate seeds."[27] The state agencies guided the farmers in planting cotton, beans, and corn using seeds of short growing cycles, less productive but more resistant to water stress.[28] The agencies also provided financial support to agricultural production by providing low-interest loans to farmers through the region's banks.

This campaign also addressed issues of water resources conservation. Based on the forecast, the governor decided to invest $13 million in the construction of a new dam on the Pacajus river earlier than originally planned. In December 1991, the government began to control the amount of water consumed in Fortaleza. Under normal levels of consumption, water reserves would have been used up by December 1992.

Controlling the usage of water made them last until April 1993.

The combined approach of managing food production and water consumption and convincing poorer farmers that the government worked not only for the rich reaped noticeable benefits. In July 1992, rainfall was 23 percent below normal, yet agricultural output was 530,000 tons – close to the region's mean output of 650,000 tons. In 1987, during a similar ENSO and without preemptive agricultural planning, the rainfall was 30 percent below average and output amounted to 100,000 tons.[29] Other states in the Brazilian northeast where no water management had occurred, such as Rio Grande do Norte, reported drought in January 1992. FUNCEME had informed these states of the impending decreases in rainfall but the information failed to generate policy responses.

Lessons From Past Responses

The success of Peru and Brazil in adopting policies to anticipate destructive impacts of ENSO can be instructive for other countries that are vulnerable to ENSO-related impacts.

Effective management programs are established when scientists and policymakers collaborate. For example, the organization of the grassroots campaign by Governor Gomes of Ceara – which sought to convince farmers that FUNCEME was a reliable source of information and to mitigate traditional social beliefs that government policies favored the wealthy – played an instrumental part in agricultural planning in this region.

Such collaborations are made possible through the creation of national institutions that provide channels for integrating scientific information into policy planning. After their respective establishment in 1984 and 1988, ENFEN and FUNCEME have

[26] A. Moura, *International Research Institute for Climate Prediction: A Proposal* (Washington, D.C.: National Oceanic and Atmospheric Administration Office of Global Programs, 1992).
[27] Ibid.
[28] See O'Brien, note 22 above.
[29] See Moura, note 26 above.

provided climate forecast information on a yearly basis.

Communicating policy recommendations to various end-users is crucial. In Peru, the media played an important role in 1986 by informing the public of the potential impacts of the forthcoming ENSO event. In Brazil, Governor Gomes's personal efforts in 1991 were the means to disseminate information to his constituency. While Peru responded on the national level, the Brazilian success was confined to the state of Ceara.

Finally, the end-users need to alter their behaviour based on policy recommendations. In Peru, prior to the 1986–1987 ENSO, recommendations by the Ministry of Agriculture guided farmers to plant rice and cotton in ratios that would result in high yields. As farmers were responsive to recommendations, the total yields were significantly higher than previous years when no policy planning had occurred.

Room for improvement

Despite reasonable success in anticipating ENSO-related impacts in Peru and Brazil, there is significant room for improvement in managing responses to future events.

The application of forecast information to regional needs is still immature. Although our examples of Peru and Brazil demonstrate successful incidents of policy planning based on forecast information, high levels of uncertainty exist.

Also, governmental planning has been restricted primarily to the agricultural sectors. Agriculture is a large part of the national economies in Peru and Brazil, but other sectors are also vulnerable to severe climate conditions. For example, during the 1982–1983 ENSO in Peru, although the losses in the agriculture business were the highest, they accounted for only 32 percent of the total costs to the nation. Losses owing to reduced consumer goods and oil production were 24 percent and damage to infrastructure was 15.2 percent of the total costs. Comparative assessments of the impact of climate information on various economic sectors would enable policymakers to further mitigate social costs.

Successful implementation of the new policies requires efficient networks to communicate information to potential end-users. In the state of Ceara, prior to the 1991–1992 ENSO, had it not been for the efforts of the governor to disseminate information to his constituency, the state would have incurred great losses. However, such individual initiatives are inefficient and unreliable as a permanent communication system.

Forging Ahead

The evolution of science and socioeconomic responses to ENSO impacts in the 20th century reveals that policy responses have lagged behind advances in scientific knowledge. Yet, though necessary, progress in science is not sufficient for success in managing social responses. The cases of Peru and Brazil demonstrate that other factors such as unsustainable use of natural resources and the globalization of economies make societies more vulnerable to interannual climate variabilities,[30] adding complexity to the management of social responses.

Effective policymaking in Peru and Brazil for managing ENSO-related impacts has been the result of factors on international to local scales: progress in the science of climate forecasting; the development of programs to direct research, analyze data, and communicate relevant information to decisionmakers; initiatives from policymakers to incorporate this information in their planning process; channels to deliver information to end-users; and change in the behavior of end-users based on policy recommendations. Although this framework is not exhaustive, it is the first step toward identifying the various scales and elements in the management of social responses to ENSO.

---

[30] In recent years, other nations – including Australia, China, Ecuador, India and Ethiopia – have begun to establish national programs to provide information to policymakers. For further information see O'Brien, note 22 above.

Acknowledgements

This work was supported by the Harvard Center for Science and International Affairs and the Consortium for International Earth Science Information Network with NASA funds under Grant NAGW–2901 and the Ford Fellowships Program at Harvard College. We acknowledge, with many thanks, the essential support of Tom Parris, Eileen Shea, Claudia Nierenberg, and James Buizer at the NOAA Office of Global Programs. Special thanks to Michael McElroy and William Clark, and the Harvard University Committee on the Environment for enabling us to present parts of this work at the July 1994 Pacific Basin Oceanography Society conference. Thanks also to James McCarthy, Mark Cane, Edward Sarachik, Michael Glantz, Allan Robinson, and Robert Keohane for helpful comments.

Part III

Reducing Land Degradation

INTRODUCTION AND GUIDE TO FURTHER READING

13 Pimental, D., Harvey, C., Resosudarmo, P., Sinclair, K., Kurz, D., McNair, M., Crist, S., Shpritz, L., Fitton, L., Saffouri, R. and Blair, R. 1995: Environmental and economic costs of soil erosion and conservation benefits. *Science*, 267, 1117–23.

This is an excellent paper which describes the environmental and economic costs of soil erosion. It is useful because it also summarizes the main erosion processes and the various ways that soil can be protected. It compares these environmental costs with the costs of soil conservation strategies, arguing that conservation is extremely cost beneficial.

14 Skole, D. and Tucker, C. 1993: Tropical deforestation and habitat fragmentation in the Amazon: satellite data from 1978 to 1988. *Science*, 260, 1905–10.

With specific reference to a portion of the Brazilian Amazon Basin, this paper shows how satellite data may be used to examine the extent and processes of deforestation. It shows how the estimated rates of deforestation are lower than previously thought, although the effects on biological diversity are considerably greater than suggested by previous calculations. The paper also provides important tables showing the characteristics of many of the world's major forests.

15 Hulme, M. and Kelly, M. 1993: Exploring the links between desertification and climate change. *Environment*, 35, 6, 4–11, 39–46.

In this paper, Hulme and Kelly discuss the various sets of processes, both as a result of climate change and human-induced land degradation, which may result in desertification. It is a particularly useful paper as it discusses the nature of climate change and human activity in the Sahara and Sahel regions.

16 Pearce, F. 1994. Rush for rock in the Highlands. *New Scientist*, 8 July, 11–12.

The possible construction of superquarries in Scotland is presently a source of much controversy. In this article, Pearce discusses the various environmental and economic factors that must be considered if superquarries are to be developed and managed in an environmentally sound manner.

17 Daily, G. C. 1995: Restoring value to the world's degraded lands. *Science*, 269, 251–354.

Using an index to describe the rate of land degradation, i.e., the potential direct instrumented value (PDIV), this paper examines the rates at which the Earth's surface is being degraded. This paper suggests strategies and the time frame required to help restore the value to the world's degraded lands.

The human species has been modifying the land throughout its existence. Prehistoric evidence shows that in Palaeolithic times the early hunter-gatherers used fire and burnt large areas of forest. Early agronomists cleared vegetation to create farmland, modified drainage and altered the soil. As civilizations developed, cities evolved and expanded, and new lands were reclaimed from wetlands and waterways. The rapid growth of population during the twentieth century has increased the pressure for more available land for agriculture, resource exploitation such as quarrying, mining and forestry, and for city expansion to cope with the increasing populations and urbanization. Highways as transportation systems have also been developed. All these processes have resulted in increased land degradation. As such, land degradation can be defined as essentially *"the substantial decrease in either or both of an area's biological productivity or usefulness due to human interference"* (Johnson and Lewis 1995, p. 2).

In recent years there has been growing concern regarding the protection of the land surface. This has initiated the development of management strategies to reduce the risks from, and the increase of, environment degradation. Environmental risk management has emerged as a result of this concern. It involves technical assessments, and the perception of the probability of particular events occurring and the resulting seriousness of the consequences of both natural and human activity. The most common method of assessment involves an environmental impact assessment (EIA). EIAs were first introduced in 1969 in the US National Environment Policy Act as a means of helping to evaluate the acceptability of projects under consideration. An EIA involves an amalgamation of studies involving predetermined approaches and integrating a variety of disciplines. This is mainly undertaken by scientists and/or engineers and economists, but increasingly sociologists are becoming involved in the process. EIAs are now undertaken in most developed countries, but legislation imposing their use and the vigour with which they are being undertaken varies

considerably. In addition, environmental audits are being used to evaluate at regular intervals the environmental performance of companies and to check that companies comply with environmental standards. In the United Kingdom in 1992, strict standards were set for environmental audits under the British Standards Institution (BS7750), ready for compliance with the European Eco-Management and Audit Scheme. These regulations often fall short of being able to protect much of the Earth's surface, although they are beginning to be considered as an essential component of proposed projects.

Part III examines the main issues associated with land-use changes as an aid to helping environmental managers appreciate some of the complex issues involved with trying to protect the value and quality of the land surface. Topics covered include: soil erosion; destruction of vegetation; desertification; quarrying; and restoring degraded land.

The destruction of the soil is one of the most serious threats to the value of land at present. Soil is degraded by many different processes, including chemical changes, such as salinization, laterization and acidification, and by compaction and leaching. These lead to changes in the physical and chemical properties of the soil which may reduce the ability of the soil to sustain the original vegetation and alter the hydrological characteristics of the ground. These changes ultimately lead to accelerated erosion and associated problems, such as landsliding and sedimentation. One of the first well-documented effects of human-influenced soil erosion was experienced in the dust-bowl region of North America during the 1920s and 1930s, a consequence of bad land management and several years of drought. The US government quickly recognized the problems and supported the programmes to protect and improve soil quality, contributing much to the development of soil science. Much of the knowledge of soil protection has stemmed from these studies. The chapter by Pimentel et al. (1995) examines both the environmental and economic costs of soil erosion. It summarizes the main processes and ways of protecting the soil, and is an essential read for environmental managers who need to consider the effects on soil of land-use changes.

The destruction of forest and other vegetation types is also another major problem, leading not only to the extermination of many precious animal and plant species, but resulting in land degradation by disrupting nutrient cycles and leading to lower soil quality and ultimately soil erosion. Throughout history, human activity has resulted in the destruction of vegetation. Bush and Fleney (1987) showed, for example, that humans were degrading forests in England as long ago as 8,900 years BP. The major phase of deforestation in Europe, for example, began in the eleventh century AD, and in the United States from the beginning of the seventh century AD. Today, the major areas of vegetation destruction are in the tropical and subtropical regions, where rainforests and savannas are being destroyed at alarming rates. This is a result of the conversion of land into ranches to raise cattle and the need for timber (Burman, 1991). The chapter by Skole and Tucker (1993) examines the extent of this problem for tropical deforestation and habitat fragmentation in the Amazon. It shows how satellite data may be used to help quantify the problem.

The next major issue involves deforestation and degradation of vegetation, which leads in turn to the degradation of soil, reducing the fertility of the land, which may ultimately lead to 'desertification'. Desertification is a complex issue, with varying opinions on the definition and nature as well as the extent of the problem and its causal mechanisms. Binns (1990) discusses the problems of defining desertification and its confusion with the term land degradation. Whilst Nelson (1988, p. 2) defines desertification as *a process of sustained land (soil and vegetation) degradation in arid, semi-arid and dry sub-humid areas, caused at least partially by man,* and reducing *productive potential to an extent which can neither be readily reversed by removing the cause nor easily reclaimed without substantial investment.* Binns (1990, p. 111) suggests that desertifi-cation should be regarded *as an extreme form of land degradation occurring where vegetation cover falls below 35 per cent on a long-term basis.* The exact cause of desertification for different regions of the world is greatly debated, with views ranging from purely climatic causes to solely a consequence of land degradation resulting from poor land use. In most cases the causes of desertification are more likely to be a combination of the two sets of factors. It is difficult to assess, however, the relative importance of each set of processes. An understanding of desertification, therefore, requires a knowledge of: the nature of the problem, biologically, pedologically and geographically; the social, economic and political environment that may have induced the problem; and the nature of climate change, natural and/or human induced. In their article, Hulme and Kelly examine both the human and natural factors which may be responsible for desertification in the Sahara and Sahel regions. This chapter is particularly useful because it highlights the uncertainties associated with the analysis of desertification and because it may be used as a model to help examine similar problems in other regions of the world.

The quality of the Earth's surface is also threatened by quarrying and mining, and their associated activities. Large areas of the Earth's surface have been degraded to obtain fossil fuels, aggregates and minerals. The consequences are wide ranging. They not only reduce the aesthetics of the regions concerned, but mining activities also create a large variety of different types of pollution. One of the most common types of pollution occurs as a result of the dust that is produced from extracting or crushing rock. This is particularly associated with open-cast mines or quarries, where building stone and aggregates are extracted. The dust creates atmospheric pollution, contaminates vegetation and water supplies, and leads to a variety of environmental health problems. Lead and arsenic are other examples of pollutants which are associated with mining activity, and they are particularly found in tin, silver, copper, zinc and lead mining. These heavy metals are discharged directly into streams or are leached from spoil waste into groundwater supplies. The problem has been particularly common in regions such as Cornwall and North Wales in the United Kingdom where tin, copper, zinc and lead have been mined over long periods. Mercury poisoning is another particularly hazardous problem. It particularly affects areas where noble metals have been or are being mined, because mercury is a by-product of the amalgamation process which is involved in the extraction of such metals. This is a major problem in parts of the Amazon basin where silver was mined in the past and gold is currently being extracted (Lacerda and Salomons, 1991; Nriagu, 1993). Waste sediment may also be discharged into streams. This causes changes of the turbidity and sometimes leads to sedimentation, which drastically alters the niches of particular organisms, and hence has profound effects on the aquatic ecosystem. The chapter by Pearce (1994) examines some of these environmental concerns, illustrated by considering issues surrounding the proposal to develop superquarries in Scotland. This article also considers the political and economic concerns related to such developments.

It is difficult to assess the geographical extent of land degradation, but the chapter by Daily (1995) attempts to quantify the present state of knowledge. In addition Daily provides suggestions for strategies which may be helpful in aiding the restoration of the value of degraded lands.

The chapters presented in Part III have highlighted the wide range of concerns associated with the degradation of the Earth's surface. Such issues should be examined and understood by environmental managers who are striving to develop sustainable environments, and who are playing an increasing role in the evaluation of projects that effect the environment.

References

Binns, T. 1990: Is desertification a myth? *Geography*, 75, 2, 106–13.
Burman, A. 1991: Saving Brazil's savannas. *New Scientist*, 2 March, 20–4.
Bush, M. B. and Fleney, J. R. 1987: The age of

British chalk grasslands. *Nature*, 329, 434–6.
Lacerda, L. D. and Salomons, W. 1991: *Mercury in the Amazon*. Dutch Ministry of Housing, Physical Planning and Environmental Report, Institute of Soil Fertility, Harem.
Nelson, R. 1988: Dryland management: the 'desertification' problem. *Environment Department Working Paper No. 8*, Washington: World Bank.
Nriagu, J. O. 1993: Legacy of mercury pollution. *Nature*, 363, 589–90.

Other references cited in text are listed in the selected readings below.

Suggested further reading

Adriano, D. C. (ed.) 1992: *Biogeochemistry of Trace Metals*. London: Lewis Publishers.

This book is an edited volume that presents recent data on the effects of trace metals, particularly their effects on soil quality, and the threat of transfer of these contaminants to consumers. It is an essential text for hydrogeologists and geochemists, soil scientists, ecologists, agronomists and toxicologists.

Blaikie, P., Cannon, T., Davis, I., Wisner, B. 1994: *At Risk*. London: Routledge.

In this book, Blaikie et al. assess and define the concept of vulnerability. The book examines the characteristics of famines and droughts, biological hazards, floods, coastal storms, earthquakes, volcanoes and landslides. It draws together practical and policy conclusions with a view to reduce disaster and promoting a safer environment. This is an essential text for all students and practitioners concerned with hazard mitigation and/or environmental impact assessment.

Cooke, R. U. and Doornkamp, J. C. 1990: *Geomorphology in Environmental Management*. Oxford: Oxford University Press, 2nd edition.

This is a very useful textbook that describes the use of geomorphology in environmental management and risk assessment. It covers topics that include: slope instability; catchment studies; erosion and weathering problems; neotectonics; aeolian environments; and glacial systems.

Eden, M. J. and Parry, J. T. (eds) 1996: *Land Degradation in the Tropics*. London: Cassell Academic.

This edited volume is a wide-ranging, coherent and scholarly account of land degradation in the tropics. It emphasizes the integration of information and theory from both the environmental and social sciences. The book illustrates the application of scholarly analysis in actual policy formulation, planning and management. A wide variety of case studies includes: degradation of tropical forests; degradation in the drier tropics; degradation in tropical wetlands; and urban and industrial degradation in the tropics. This is a useful text for planners, managers and policy-makers, as well as students of environmental science who are concerned with land degradation in the tropics.

Ellis, S. and Mellor, A. 1995: *Soils and Environment*. London: Routledge.

   Soils and Environment explores the ways in which soils are both influenced by, and themselves influence, the environment. The book also describes the analysis of soil properties, soil processes and classification. It also discusses soil–human interactions and examines the relation to land systems, environmental problems and management, soil surveys and land evaluation. This is a useful text for students of all levels who have interests in soil management.

Goudie, A. 1993: *The Human Impact: On the Natural Environment*. Oxford: Blackwell Scientific, 4th edition.

   This comprehensive and well-illustrated undergraduate text addresses the ways that human activity has changed, and is changing, the Earth's surface. Topics include: desertification, deforestation, plant and animal invasions, marine pollution, climatic change and environmental uncertainty.

Grainger. A. 1990: *The Threatening Desert: Controlling Desertification*. London: Earthscan.

   The distribution and processes of desertification are outlined. The book is particularly useful because it examines the successes and failures which have accompanied the various attempts to combat desertification as set out by the Plan of Action resulting from 1977 Nairobi United Nations Conference on Desertification. Grainger's arguments for a new International Plan of Action to help control the increasing threat of desertification is thought-provoking and useful for environmental managers.

Hewitt, K. 1996: *Regions of Risk*. Harlow: Longman.

   This book examines the various aspects of hazards, human vulnerability and disaster. It includes a variety of case studies on natural and technological hazards, as well as an examination of social violence. It emphasizes the cultural and social aspects of hazard assessment and response, as well as providing an examination of the cross-cultural differences and international scope of risk and disaster preparedness and response. This is a particularly useful text for students who are studying environmental risk assessment.

Ives, J. D. and Messerli, B. 1989: *The Himalayan Dilemma*. London: Routledge.

   This book examines the interaction between human activities and the natural environment in the Himalayas. The authors consider the problems of reconciling development and conservation in this sensitive environment, and provide some excellent references to particular case studies.

Johnson, D. L. and Lewis, L. A. 1995: *Land Degradation: Creation and Destruction*. Oxford: Blackwell Publishers.

   This clear and accessible text describes the history and current state of land degradation through an analysis of the linkages between the natural environment and human activity. It provides a wide range of examples from a variety of environmental, economic and historical settings.

Michener, W. K., Brunt, J. W. and Stafford, S. G. (eds) 1994: *Environmental Information Management and Analysis: Ecosystem to Global Scales*. London: Taylor and Francis.

   This is a useful collection of 31 papers which examines the state-of-the-art technologies for the management and analysis of environmental data. It includes papers on: successful design, implementation and effective operation of environmental information systems, and strategies on the management for massive data storage; data access; data sharing; assessing and assuring data quality; and quality control of data.

Morgan, R. P. C. 1986: *Soil Erosion and Conservation*. London: Longman.

   This book is an excellent introductory text on the magnitude, frequency, rates and mechanics of wind and water erosion. It assesses erosion hazard, and describes methods of measurement, modelling and monitoring, and strategies for erosion control and conservation practices.

Parnwell, M. and Bryant, R. (eds.) 1996: *Environmental Change in South-East Asia.* London: Routledge.

This edited book examines the interaction of people, politics and ecology. It examines the nature of the environmental degradation in relation to rapid economic growth in South-East Asia and the dilemmas facing policy-makers who seek to promote sustainable development.

Thomas, D. S. G. and Middleton, N. J. 1994: *Desertification: Exploding the Myth.* Chichester: Wiley.

A useful text that explores the origin of the 'desertification myth'. It examines how this myth has spawned multimillion dollar initiatives and is now regarded as a leading environmental issue. The book examines the political and institutional factors that created the myth, sustaining it and protecting it against scientific criticism. It is particularly useful for students and practitioners with interests and concerns in drylands.

Williams, M. A. J. and Balling, R. C. 1995: *Interactions of Desertification and Climate.* London: Edward Arnold.

The UNEP and the WMO commissioned this book to examine the current knowledge of the interactions between desertification and climate in drylands. The book provides a series of useful recommendations for future dryland management, and is, therefore, useful for all workers with interests in desertification and climate change.

13

ENVIRONMENTAL AND ECONOMIC COSTS OF SOIL EROSION AND CONSERVATION BENEFITS*

David Pimentel, C. Harvey, P. Resosudarmo, K. Sinclair, D. Kurz, M. McNair, S. Crist, L. Shpritz, L. Fitton, R. Saffouri, and R. Blair

Soil erosion is a major environmental and agricultural problem worldwide. Although erosion has occurred throughout the history of agriculture, it has intensified in recent years (1). Each year, 75 billion metric tons of soil are removed from the land by wind and water erosion, with most coming from agricultural land (2). The loss of soil degrades arable land and eventually renders it unproductive. Worldwide, about 12×10^6 ha of arable land are destroyed and abandoned annually because of nonsustainable farming practices (1), and only about 1.5×10^9 ha of land are being cultivated (3, 4). Per capita shortages of arable land exist in Africa, Asia, and Europe because of lost eroded land and the expansion of the world population to nearly 6 billion (1, 5).

To adequately feed people a diverse diet, about 0.5 ha of arable land per capita is needed (6), yet only 0.27 ha per capita is available. In 40 years, only 0.14 ha per capita will be available both because of loss of land and rapid population growth (5). In many regions, limited land is a major cause of food shortages and undernutrition (4, 7). Over 1 billion humans (about 20% of the population) now are malnourished because of food shortages and inadequate distribution (8, 9). With the world population increasing at a quarter of a million per day and continued land degradation by erosion, food shortages and malnutrition have the potential to intensify (10, 11).

The use of large amounts of fertilizers, pesticides, and irrigation help offset deleterious effects of erosion but have the potential to create pollution and health

* Originally published in *Science*, 1995, vol. 267, pp. 1117–23.

problems, destroy natural habitats, and contribute to high energy consumption and unsustainable agricultural systems. Erosion also is a major cause of deforestation: As agricultural land is degraded and abandoned, more forests are cut and converted for needed agricultural production (12).

In this article, we (i) examine the ways in which erosion reduces soil fertility and crop productivity, (ii) assess the environmental and economic costs of soil erosion, and (iii) compare various agricultural techniques and practices that reduce erosion and help conserve water and soil resources.

Erosion on Croplands and Pastures

Worldwide erosion rates. Of the world's agricultural land, about one-third is devoted to crops and the remaining two-thirds is devoted to pastures for livestock grazing (4, 13). About 80% of the world's agricultural land suffers moderate to severe erosion, and 10% suffers slight to moderate erosion (9). Croplands are the most susceptible to erosion because their soil is repeatedly tilled and left without a protective cover of vegetation. However, soil erosion rates may exceed 100 tons ha^{-1} year^{-1} in severely overgrazed pastures (14). More than half of the world's pasturelands are overgrazed and subject to erosive degradation (15).

Soil erosion rates are highest in Asia, Africa, and South America, averaging 30 to 40 tons ha^{-1} year^{-1}, and lowest in the United States and Europe, averaging about 17 tons ha^{-1} year^{-1} (16). The relatively low rates in the United States and Europe, however, greatly exceed the average rate of soil formation of about 1 ton ha^{-1} year^{-1} (the rate of conversion of parent material into soil in the A, E, and B horizons) (17). Erosion rates in undisturbed forests range from only 0.004 to 0.005 ton ha^{-1} year^{-1} (18, 19).

Erosion rates in the United States. In the last 200 years of U.S. farming, an estimated 10^8 ha (~30%) of farmland has been abandoned because of erosion, salinization, and waterlogging (13, 18, 20). Wind erosion appears to be worsening, while water erosion appears to be declining (13, 21, 22).

Croplands in the United States lose soil at an average rate of 17 tons ha^{-1} year^{-1} from combined water and wind erosion, and pastures lose 6 tons ha^{-1} year^{-1} (13). About 90% of U.S. cropland is losing soil above the sustainable rate (23, 24). About 54% of U.S. pastureland (including federal lands) is overgrazed and subject to high rates of erosion (25, 26).

The extent of U.S. soil erosion is well documented. One-half of the fertile topsoil of Iowa has been lost during the last 150 years of farming (27, 28), and loss of topsoil continues at a rate of about 30 tons ha^{-1} year^{-1} (13). Similarly, about 40% of the rich Palouse soils of the northwest United States has been lost in the past century.

During the past 50 years, the average farm size has more than doubled from 90 to 190 ha (29, 30). To create larger farms and fields, farmers have removed the grass strips, shelterbelts, and hedgerows that once protected soil from erosion (23, 24, 31). Crop specialization has also led to the use of heavier machines that damage the entire soil ecosystem (32, 33).

Erosion Processes

Erosion results from energy transmitted from rainfall and wind. Raindrops hit exposed soil with an explosive effect, launching soil particles into the air. In most areas, raindrop splash and sheet erosion are the dominant forms of erosion (34, 35). Erosion is intensified on sloping land, where more than half of the soil contained in the splashes is carried downhill.

Airborne soil particulates can be transported thousands of miles. For instance, soil particles from eroded African lands are blown as far as Brazil and Florida (36), and Chinese soil has been detected in Hawaii (37).

Factors Influencing Erosion

Erosion increases dramatically on steep cropland. Yet, steep slopes are now routinely being converted from forests for agricultural use because of the increasing needs of the human population and land degradation (1). Once under conventional cultivation, these steep slopes suffer high erosion rates: In Nigeria, cassava fields on steep (~12%) slopes lost 221 tons ha^{-1} year^{-1}, compared with an annual soil loss of 3 tons ha^{-1} year^{-1} on flat (<1%) land (38). The Philippines, where over 58% of the land has slopes greater than 11%, and Jamaica, where 52% of the land has slopes greater than 20%, exhibit soil losses as high as 400 tons ha^{-1} year^{-1} (1).

Living and dead plant biomass left on fields reduce soil erosion and water runoff by intercepting and dissipating raindrop and wind energy. In Missouri, for example, barren land lost soil at a rate 123 times that of land that was covered with sod (which lost <0.1 ton ha^{-1} year^{-1}) (39). Similarly in Oklahoma, areas without rye grass or wheat cover lost 2.5 to 4.8 times as much water as land with cover (40).

Loss of vegetative cover is particularly widespread in many third-world countries. About 60% of crop residues in China and 90% in Bangladesh are removed and burned

for fuel each year (41). In areas where fuel is scarce, even the roots of grasses and shrubs are collected (42).

Both the texture and the structure of soil influence its susceptibility to erosion. Soils with medium to fine texture, low organic matter content, and weak structural development have low infiltration rates and experience increased water runoff (35).

Erosion and Productivity

Because of erosion-associated loss of productivity and population growth, the per capita food supply has been reduced over the past 10 years and continues to fall (43). The Food and Agriculture Organization reports that the per capita production of grains, which make up 80% of the world food supply, has been declining since 1984 (44).

Crop yields on severely eroded soil are lower than those on protected soils because erosion reduces soil fertility and water

Table 13.1 [orig. table 1] Water runoff rates compared for conservaton versus conventional plantings of corn.

Treatment	Water runoff (cm depth)	Conserved water (cm)	Increased yield* (tons ha^{-1})
Corn stover mulch vs. no stover residue (110)	0.06 1.30	1.24	0.34
Rye cover mulch vs. residue burned (111)	3.9 17.4	13.5	3.4
Manure mulch vs. no manure (112)	9.0 13.1	4.1	1.1
Corn-oats-hay-hay vs. conventional continuous (113)	0.58 3.08	2.50	0.6
No-till in sod vs. conventional (114)	3.7 10.7	7.0	1.8
Level terraced vs. contour planted (115)	0.94 8.14	7.2	1.8
Dense planting vs. bare soil (116)	2.49 3.32	0.97	0.2
Reduced till vs. conventional (117)	2.1 3.6	1.5	0.4

* Increased yield based on the results of Troeh et al. (50).

availability. Corn yields on some severely eroded soils have been reduced by 12 to 21% in Kentucky, 0 to 24% in Illinois and Indiana, 25 to 65% in the southern Piedmont (Georgia), and 21% in Michigan (45–47). In several areas of the Philippines, erosion has caused declines in corn productivity as severe as 80% over the last 15 years (48).

Erosion by water and wind adversely affects soil quality and productivity by reducing infiltration rates, water-holding capacity, nutrients, organic matter, soil biota, and soil depth (33, 49, 50). Each of these factors influences soil productivity individually but also interacts with the other factors, making assessment of the impacts of soil erosion on productivity difficult.

All crops require enormous quantities of water for their growth and the production of fruit (51–53). For example, during a single growing season, a hectare of corn (yield, 7000 kg ha^{-1}) transpires about 4×10^6 litres of water (54), and an additional 2×10^6 litres ha^{-1} concurrently evaporate from the soil (55, 56).

When erosion occurs, the amount of water runoff increases, so that less water enters the soil matrix and becomes available for the crop (table 13.1). Moderately eroded soils absorb from 10 to 300 mm less water per hectare per year than uneroded soils, or between 7 to 44% of total rainfall (57–60). This degree of water loss reduces crop productivity; even a runoff rate of 20 to 30%

Table 13.2 [*orig. table 2*]   Estimated annual economic and energetic costs (per hectare) of soil and water loss from conventional corn assuming a water and wind erosion rate of 17 tons ha^{-1} year^{-1} over the long term (20 years).

Factors	Annual quantities lost	Cost of replacement (dollars)	Energetic costs (10^3 kcal)	Yield loss after 20 years of erosion (%)
Water runoff	75 mm*	30†	700‡	7*
Nitrogen	50 kg§		500‖	
Phosphorus	2 kg§	100§	3‖	8¶
Potassium	410 kg§		260‖	
Soil depth	1.4 mm*	16#	–	7**
Organic matter	2 tons*	–	–	4††
Water holding capacity	0.1mm*	–	–	2‡‡
Soil biota	–	–	–	1§§
Total on-site		146	1460	20‖‖
Total off-site		50¶¶	100	
Grand total		196#	1560	

* Table 13.3.   † The cost of replacing this much water by ground-water irrigation based on 1992 dollars (118). The value is reduced by 40% because it is assumed that water erosion accounts for 60% of U.S. erosion (119). However, if rainfall were abundant, then this replacement cost would not be necessary.   ‡ Energy required to pump ground water from a depth of 30 m (120).   § Total nutrients loss, based on the results of Troeh et al. (50).   ‖ Energy required to replace the fertilizers lost (121). ¶ Based on the total loss of 340 tons ha^{-1} of soil over 20 years and the mineralization and availability of the nutrients in this soil.   # Estimated.   ** Based on reduced productivity of about 6% per loss of 2.5 cm of soil (122).   †† Organic matter content of the soil was assumed to decline from 4 to 3% over this period, resulting in a 4% decline in productivity.   ‡‡ After the loss of 17 tons ha^{-1} year^{-1} of soil, the water holding capacity was assumed to decline 1.9 mm and productivity declined 2%; with severe erosion over time, plant-available water may decline 50 to 75% (17, 123).   §§ Reductions in soil biota were assumed to reduce infiltration of water and reduce organic matter recycling. ‖‖ Percentages do not add up because the impacts of the various factors are interdependent and some overlap exists (for example, organic matter is interrelated with water resources, nutrients, soil biota, and soil depth). This loss would occur if lost nutrients and water were not replaced.   ¶¶ Table 13.4.

of total rainfall can result in significant water shortages for crops (61). In the tropics, Lal (31) reported that erosion may reduce infiltration by up to 93%.

In addition to creating water deficiencies, soil erosion causes shortages of basic plant nutrients, such as nitrogen, phosphorus, potassium, and calcium, which are essential for crop production. A ton of fertile agricultural topsoil typically contains 1 to 6 kg of nitrogen, 1 to 3 kg of phosphorus, and 2 to 30 kg of potassium, whereas a severely eroded soil may have nitrogen levels of only 0.1 to 0.5 kg per ton (50, 62). Wind and water erosion selectively remove the fine organic particles, leaving behind large particles and stones. Eroded soil typically contains about three times more nutrients than the soil left behind (63–65).

When nutrient reserves are depleted by erosion, plant growth is stunted and crop yields decline (table 13.2). Soils that suffer severe erosion may produce 15 to 30% lower corn yields than uneroded soils (46, 52), and with fertilization, the yield reductions range from 13 to 19% (45–47). Under the current average soil erosion rates (17 tons ha^{-1} year^{-1}), the loss of nitrogen, phosphorus, and potassium can be expected to cause a long-term drop in crop yields. If soil erosion is 1 ton ha^{-1} year^{-1} or less, if crop residues are left on the land, and if nutrients are added to offset any of the nutrients removed with the crop, then soil quality and productivity will remain high and sustainable.

Organic matter, a necessary component of soil, facilitates the formation of soil aggregates, increases soil porosity, and thereby improves soil structure, water infiltration, and ultimately overall productivity (66, 67). In addition, organic matter increases water infiltration, facilitates cation exchange, enhances root growth, and stimulates the proliferation of important soil biota (34). About 95% of the nitrogen and 25 to 50% of the phosphorus is contained in organic matter (34).

Fertile topsoils typically contain about 100 tons of organic matter (or 4% of total soil weight) per hectare (68, 69). Because most of the organic matter is near the soil surface in the form of decaying leaves and stems,

erosion of topsoil results in a rapid decrease in levels of soil organic matter. Several studies have demonstrated that the soil removed by either wind or water erosion is 1.3 to 5 times richer in organic matter than the soil left behind (34, 70). The loss of 17 tons of soil per hectare by rainfall removes nearly 2 tons of organic matter (69).

Once the organic matter layer is depleted, soil productivity and crop yields decline because of the degraded soil structure and depletion of nutrients. For example, the reduction of soil organic matter from 4.3 to 1.7% lowered the yield potential for corn by 25% in Michigan (71).

Although soil biota are often ignored in assessments of the impact of erosion, they are a critical component of the soil and constitute a large portion of the soil biomass. One square meter of soil may support populations of about 200,000 arthropods and enchytraeids and billions of microbes (72, 73). A hectare of good quality soil contains an average of 1000 kg of earthworms, 1000 kg of arthropods, 150 kg of protozoa, 150 kg of algae, 1700 kg of bacteria, and 2700 kg of fungi (74). Soil biota recycle the basic nutrients required by plants (74, 75). Also, the tunneling and burrowing activities of earthworms and other soil biota enhance productivity by increasing water infiltration rates.

The erosion typical of conventional agriculture may decrease the diversity and abundance of soil organisms (76, 77), whereas practices that maintain the soil organic matter content at optimum levels favor the proliferation of soil biota (78). Thus, the simple practice of straw-mulching may increase biota threefold (79), and the application of organic matter or manure may increase earthworm and microorganism biomass as much as fivefold (80).

Soils form slowly: It takes between 200 and 1000 years to form 2.5 cm (1 inch) of topsoil under cropland conditions, and even longer under pasture and forest conditions (24, 33, 61, 81). In the United States, where 2.5 cm of soil are lost every 16.5 years, soil has been lost at about 17 times the rate at which it has formed (17). Estimates are that the average U.S. topsoil depth was about

23 cm in 1776. Today, after about 200 years of farming, the average depth has declined to about two-thirds of the original soil depth (~15 cm) (82).

Model of Erosion Effects on Crop Productivity

To assess how and to what extent erosion decreases crop productivity, it is necessary to consider the multiple factors that influence erosion rates, as well as the soil components that affect productivity. We have developed empirical models that incorporate the numerous factors affecting both erosion rates and soil productivity. The slope of the land, soil composition, and extent of vegetative cover influence the rate of erosion, and the soil depth, presence of soil biota, organic matter, water-holding capacity, and nutrient levels influence the soil's productive capacity. These factors form a complex and interdependent system. Changes in one factor subsequently affect all or many others. The models demonstrate how soil erosion causes the loss of soil nutrients, depth, biota, organic matter, and water resources and how these losses translate into reduced crop productivity. The models are based on the following set of assumptions: ~700 mm of rainfall, soil depth of 15 cm, slope of 5%, loamy soil, 4% organic matter, and soil erosion rate of 17 tons ha^{-1} $year^{-1}$. The models provide a perspective on the interdependence of the various factors associated with the ecological effects of erosion.

On the basis of empirical evidence, it appears that when soil erosion by water and wind occurs at a rate of 17 tons ha^{-1} $year^{-1}$, an average of 75 mm of water, 2 tons of organic matter, and 15 kg of available nitrogen are lost from each hectare each year (table 13.3). In addition, soil depth is reduced by 1.4 mm, the water-holding capacity is decreased by less than 0.1 mm, and soil biota populations are diminished. When combined, these losses translate into an 8% reduction in crop productivity over the short term (1 year). The loss of water and nutrients account for nearly 90% of the loss in productivity (table 13.3). This model assumes that the nutrients and water are not replaced.

Table 13.3 [*orig. table 3*]   Initial effects of factors contributing to reduced corn yield by means of soil erosion of 17 tons ha^{-1} $year^{-1}$ (10 tons ha^{-1} $year^{-1}$ by water and 7 tons ha^{-1} $year^{-1}$ by wind).

Factors	Quantities Lost	Yield Loss (%)
Water runoff	75 mm*	7*
Nitrogen†	15 kg	
Phosphorus†	0.6 kg	2.4¶
Potassium†	123 kg	
Soil depth	1.4 mm‡	0.3‡
Organic matter	2 tons‖	0.2‖
Water holding capacity	0.1 mm§	0.1§
Soil biota	–	0.1#
Total		8**

* Based on a water erosion rate of about 10 tons ha^{-1} $year^{-1}$ on 5% sloping land under conventional tillage, water loss would be nearly 100 mm (57–60, 124). A conservative loss of 75 mm was assumed, and based on this water loss, the estimated yield reduction was 7% (125–127). † Total nutrients lost are based on the results of Troeh et al. (50) but reduced as a result of the nutrients that would not be immediately available because of a shortage of time for mineralization (17).   ‡ Based on a bulk density of 1.25 g cm^{-1} and reduced yield of 6% per 2.5 cm of soil (122).   § Water-holding capacity of the soil was calculated to be reduced by 0.1 mm on the basis of the loss of 17 tons ha^{-1} $year^{-1}$ (17). ‖ Based on a 4% organic matter content of the soil and an enrichment factor of 3; the yield loss is minimal initially but is significant in the long term. The loss of N, P, and K nutrients was estimated to reduce yield by 2.4% (128). # Reductions in soil biota were assumed to reduce infiltration of water and reduce organic matter recycling but have a minimal impact on yield for a single year.   ** This estimated loss occurs after the loss of 17 tons ha^{-1} $year^{-1}$. Percentages do not add up because the impacts of the various factors are interdependent and overlap exists (for example, organic matter is interrelated with water resources, nutrients, soil biota, and soil depth).

Evaluated over the long term (20 years), empirical evidence again confirms that water and nutrient loss continue to have the greatest effect on crop productivity, accounting for 50 to 75% of the reduced productivity (65) (table 13.2). A reduction in soil depth of 2.8 cm results in a reduction in productivity of about 7%. Soil depth is particularly critical because it takes hundreds of years to replace a single centimeter of lost topsoil. The other factors, including soil biota, water-holding capacity, and soil depth, become significant in the long term. Again, this model assumes that the lost nutrients and water are not replaced; if they were replaced, then the 20% loss estimate would be reduced by one-fourth to one-third (45–47). On a yearly basis, the effects of soil erosion often can be temporarily offset by the extensive use of fertilizers, irrigation, plant breeding, and other inputs. However, the long-term cumulative loss of soil organic matter, biota, soil depth, and water-holding capacity in some cases cannot be replaced by those interventions.

Erosion Costs

Energy costs. About 6% of the total amount of energy spent in the United States is used in agriculture. Assuming an average erosion rate of 17 tons ha^{-1} year^{-1} for combined wind and water erosion, we estimate that the on-site and off-site impacts of soil erosion and associated rapid water runoff require an additional expenditure of 1.6×10^6 kcal of fossil energy per hectare per year (table 13.2). This suggests that about 10% of all the energy used in U.S. agriculture today is spent just to offset the losses of nutrients, water, and crop productivity caused by erosion. Although developed countries are currently using fossil energy-based fertilizers, pesticides, and irrigation to mask the damage of soil erosion and to maintain high crop productivity, heavy dependence on fossil fuels is a risk because fossil energy supplies are finite. Developing nations that use intensive agricultural technologies also rely intensively on the use of fossil energy-based fertilizers, pesticides, and irrigation to provide high yields (43).

On-site costs. The use of inappropriate agricultural practices and subsequent soil and water loss are responsible for significant economic and environmental on-site costs.

The major on-site costs of erosion by both water and wind are those expended to replace the lost nutrients and water (tables 13.1 and 13.2). When erosion by water and wind occurs at a rate of 17 tons ha^{-1} year^{-1}, about 75 mm of water and 462 kg of nutrients are lost per hectare (table 13.2). In the United States, if water had to be replaced, it would cost about $30 ha^{-1} year^{-1} to replace by pumping ground water for irrigation and would require the expenditure of about 70 litres of diesel fuel per hectare (assuming that water were available). An additional $100 ha^{-1} would be required for fertilizers to replace the lost nutrients (table 13.2). If the on-site and off-site costs are summed, erosion costs the United States a total of about $196 ha^{-1} (table 13.2). In other parts of the world, where irrigation is not possible or fertilizers are too costly, the price of erosion is paid in reduced food production.

In the United States, an estimated 4×10^9 tons of soil and 130×10^9 tons of water are lost from the 160×10^6 ha of cropland each year. This translates into an on-site economic loss of more than $27 billion each year, of which $20 billion is for replacement of nutrients (50) and $7 billion for lost water and soil depth (table 13.2). The most significant component of this cost is the loss of soil nutrients.

The costs of erosion are also high in other regions of the world. In Java, for example, on-farm losses of productivity related to erosion are estimated to cost the economy $315 million per year (83). The 6.6×10^9 tons of Indian soil (14) lost each year contains 5.4×10^6 tons of fertilizer worth $245 million (84). Furthermore, up to half of the amount of fertilizers applied each year in areas of India characterized by heavy rainfall during the southwest monsoon is lost as a result of ammonia volatilization and leaching (85). In Costa Rica, yearly erosion from farm and pasture land removes nutrients worth 17% of the crop value and 14% of the value of livestock products (86).

In addition to substantial economic losses of nutrients and water, erosion causes significant ecological damage. The removal of soil may affect plant composition and deplete soil biodiversity. Some studies of the effects of erosion focus only on changes in soil depth. In such studies, the importance of biodiversity, organic matter, and the other complex of interdependent variables is overlooked. As a result, Lal (87) reports that such studies significantly underestimate the impact of soil erosion. Studies on reduced soil depth report crop yield reductions of only 0.13 to 0.39% per centimeter of soil lost (88, 89).

Off-site costs. Erosion not only damages the immediate agricultural area where it occurs but also negatively affects the surrounding environment. Off-site problems include roadway, sewer, and basement siltation, drainage disruption, undermining of foundations and pavements, gullying of roads, earth dam failures, eutrophication of waterways, siltation of harbors and channels, loss of reservoir storage, loss of wildlife habitat and disruption of stream ecology, flooding, damage to public health, plus increased water treatment costs (90).

The most serious of off-site damages are caused by soil particles entering the water systems (91). Of the billions of tons of soil lost from U.S. cropland each year, about 60% is deposited in streams and rivers (13). These sediments harm aquatic plants and other organisms by contaminating the water with soil particles along with fertilizer and pesticide chemicals, which adversely alter habitat quality (92).

Siltation is a major problem in reservoirs because it reduces water storage and electricity production and shortens the lifetime and increases the maintenance costs of dams. About 880×10^6 tons of agricultural soils are deposited into American reservoirs and aquatic systems each year, reducing their flood-control benefits, clogging waterways, and increasing operating costs of water treatment facilities (4). To maintain navigable waterways, the United States annually spends over $520 million to dredge soil sediments from waterways (93).

Heavy sedimentation frequently leads to river and lake flooding (2). For example, some of the flooding that occurred in the midwestern United States during the summer of 1993 was caused by increased sediment deposition in the Mississippi,

Table 13.4 [*orig. table* 4]   Damages by wind and water erosion and the cost of erosion prevention each year.

Type of damage	Cost (millions of dollars)
*Wind erosion**	
Exterior paint	18.5
Landscaping	2,894.0
Automobiles	134.6
Interior, laundry	986.0
Health	5,371.0†
Recreation	223.2
Road maintenance	1.2
Cost to business	3.5
Cost to irrigation and conservation districts	0.1
Total wind erosion costs	9,632.5
Water erosion‡	
In-stream damage	
Biological impacts	No estimate
Recreational	2,440.0
Water-storage facilities	841.8
Navigation	683.2
Other in-stream uses	1,098.0
Subtotal in-stream	5,063.0
Off-stream effects	
Flood damages	939.4
Water-conveyance facilities	244.0
Water-treatment facilities	122.0
Other off-stream uses	976.0
Subtotal off-stream	2,318.0
Total water erosion costs	7,381.0
Total costs of wind and water erosion damage	17,013.5§
Cost of erosion prevention‖	8,400
Total costs (on and off-site)¶	44,399.0
Benefit/cost ratio	5.24

* (95–97, 129).   † Health estimates are partly based on Lave and Seskin (130).   ‡ (93, 96, 97, 129).   § Agriculture accounts for about two-thirds of the off-site effects.   ‖ See text.   ¶ The total on-site costs are calculated to be $27 billion (see table 13.3 and text).

Missouri, and other rivers located in the central United States. The combined damage of the 1993 flood to crops and homes was assessed by the government to be $20 billion (94).

Wind erosion produces significant off-site damage and costs. It is estimated that household property damage from the sandblasting of automobiles, buildings, and landscapes by blown soil particles and maintenance costs total over $4 billion per year in the United States (95–97). In addition, the removal of accumulated soil from public and private buildings, roads, and railways similarly results in costs of over $4 billion per year (95, 96).

An example of the magnitude of wind erosion is found in New Mexico, where about two-thirds of the land is used for agriculture, including grazing. The total off-site erosion costs in this state, including health and property damage, are estimated to be $465 million annually (95). If we assume similar erosion costs in the western United States, the total off-site costs from wind erosion alone could be as great as $9.6 billion each year in the United States (table 13.4).

Combined on-site and off-site effects. The cost of all off-site environmental impacts of U.S. soil erosion, most of which is from agriculture, is estimated to be about $17 billion per year (1992 dollars) (table 13.4). An additional yearly loss of $27 billion is attributed to reduced soil productivity. If off-site and on-site costs are combined, the total cost of erosion from agriculture in the United States is about $44 billion per year (table 13.4), or about $100 per hectare of cropland and pasture. This erosion cost increases production costs by about 25% each year.

Of the 75×10^9 tons of soil eroded worldwide each year (2), about two-thirds come from agricultural land. If we assume a cost of $3 per ton of soil for nutrients (50), $2 per ton for water loss (table 13.2), and $3 per ton for off-site impacts (table 13.4), this massive soil loss costs the world about $400 billion per year, or more than $70 per person per year.

Erosion Control Technologies

Reliable and proven soil conservation technologies include ridge-planting, no-till cultivation, crop rotations, strip cropping, grass strips, mulches, living mulches, agroforestry, terracing, contour planting, cover

Table 13.5 [*orig. table 5*]   Annual soil loss (tons per hectare) by crop and technology in the United States.

Technology	State	Soil loss (tons ha^{-1})
Corn		
Conventional, continuous (131)	MO	47
Conventional, plow-disk (132)	IN	47
Conventional, plow-disk (132)	OH	27
Conventional, continuous (133)	PA	20
Conservation, rotation (133)	PA	7
Conservation, contour (57)	IL	6
Conservation, no-till (134)	MS	0.3
Soybeans		
Conventional (135)	MS	36
Conservation, rotation (135)	MS	9
Conservation, no-till (67)	GA	0.02
Cotton		
Conventional (136)	MS	91
Conservation, no-till (136)	MS	1.3
Wheat		
Conventional (137)	WA	22
Conservation, mulch (138)	MS	1.7
Natural vegetation		
Undisturbed grass (18)	KS	0.07
Undisturbed forest (139)	NH	0.02

crops, and windbreaks (98). Although the specific processes vary, all conservation methods reduce erosion rates by maintaining a protective vegetative cover over the soil, which is often accompanied by a reduction in the frequency of plowing. Ridge-planting, for example, reduces the need for frequent tillage and also leaves vegetative cover on the soil surface year round, and crop rotations ensure that some part of the land is continually covered with vegetation. Each conservation method may be used separately or in combination with other erosion-control techniques. To determine the most advantageous combination of appropriate conservation technologies, the soil type, specific crop or pasture, slope, and climate (rainfall and wind intensity), as well as the socioeconomics of the people living in a particular site must be considered.

The implementation of appropriate soil and water conservation practices has the potential to reduce erosion rates from 2 to 1000-fold and water loss from 1.3 to 21.7-fold (tables 13.1 and 13.5). Conservation technologies also significantly reduce nutrient loss. For example, when corn residue cover was increased by 10, 30, and 50%, the amount of nitrogen lost in surface runoff was reduced by 68, 90 and 99%, respectively (99).

By substantially decreasing soil and nutrient loss, conservation technologies preserve the soil's fertility and enable the land to sustain higher crop yields. In many instances, the use of conservation technologies may actually increase yields (100). Contour planting, for example, has increased cotton yields by 25% (Texas), corn yields by 12.5% (Missouri), soybeans by 13% (Illinois), and wheat by 17% (Illinois) (101–103). On U.S. land with a 7% slope, yields from cotton grown in rotation increased by 30%, and erosion was reduced by nearly one-half (104). In areas where winds are strong, the establishment of tree and shrub shelterbelts helps reduce wind energy by as much as 87% and thereby decreases erosion by as much as 50% (50).

Conclusion

We estimate that it would take an investment of $6.4 billion per year ($40 per hectare for conservation) to reduce U.S. erosion rates from about 17 tons ha^{-1} year^{-1} to a sustainable rate of about 1 ton ha^{-1} year^{-1} on most cropland. To reduce erosion on pastureland, the United States would have to spend an additional $2.0 billion per year ($5 per hectare for conservation) (30, 105–107) (table

13.4). The total investment for U.S. erosion control would be about $8.4 billion per year. Given that erosion causes about $44 billion in damages each year, it would seem that a $8.4 billion investment is a small price to pay: For every $1 invested, $5.24 would be saved (table 13.4). This small investment would reduce U.S. agricultural soil loss by about 4×10^9 tons and help protect our current and future food supply.

Currently, the United States spends $1.7 billion per year in the Conservation Reserve Program to remove highly erodible land from production, and this saves about 584×10^6 tons of soil each year (108). Therefore, in this system $2.91 is invested to save 1 ton of soil, whereas in our proposed conservation system, we assume a cost of $2.10 per ton of soil saved.

When economic costs of soil loss and degradation and off-site effects are conservatively estimated into the cost/benefit analyses of agriculture, it makes sound economic sense to invest in programs that are effective in the control of widespread erosion. Human survival and prosperity depend on adequate supplies of food, land, water, energy, and biodiversity. Infertile, poor-quality land will not sustain food production at the levels required by the growing world population. We should heed President Roosevelt's (109) warning that "A nation that destroys its soils, destroys itself."

Acknowledgements

We thank the following people for reading an earlier draft of this article, for their many helpful suggestions, and in some cases for providing additional information: I. P. Abrol, D. Wen, R. H. Dowdy, H. E. Dregne, W. Edwards, M. T. El-Ashry, H. A. Elwell, R. F. Follett, D. W. Fyrear, G. E. Hallsworth, H. Hurni, T. N. Khoshoo, R. Lal, G. W. Langdale, W. B. Magrath, K. C. McGregor, G. F. McIsaac, F. Mumm von Mallinckrodt, T. L. Napier, K. R. Olson, W. Parham, D. Southgate, B. A. Stewart, M. Stocking, M. S. Swaminathan, D. L. Tanaka, G. B. Thapa, F. Troeh, P. W. Unger, A. Young, R. Bryant, M. Giampietro, T. Scott, and N. Uphoff. Special thanks go to D. Dalthorp for his many constructive comments and his time and enthusiasm dedicated to this study.

References and Notes

1 R. Lal and B. A. Stewart, *Soil Degradation* (Springer-Verlag, New York, 1990).
2 N. Myers, *Gaia: An Atlas of Planet Management* (Anchor and Doubleday, Garden City, NY, 1993).
3 P. Buringh, in *Food and Natural Resources*, D. Pimentel and C. W. Hall, Eds. (Academic Press, San Diego, 1989), pp. 69–83.
4 World Resources Institute, *World Resources 1992–1993* (Oxford Univ. Press, New York, 1992).
5 D. Pimentel, R. Harman, M. Pacenza, J. Pecarsky, M. Pimentel, *Popul. Environ.* 15, 347 (1994).
6 R. Lal, in (3), pp. 85–140.
7 D. Pimentel, Ed., *World Soil Erosion and Conservation* (Cambridge Univ. Press, Cambridge, 1993).
8 *World Can Cut Hunger Rate in Half*, World Bank press release, Washington, DC, 29 November 1993.
9 J. G. Speth, *Towards an Effective and Operational International Convention on Desertification* (International Negotiating Committee, International Convention on Desertification, United Nations, New York, 1994).
10 M. Giampietro and D. Pimentel, *The NPG Forum* (Negative Population Growth, Teaneck, NJ, October 1993).
11 A. Gore, *The Earth Times* 1, 31 (15 June 1994).
12 N. Myers, *Deforestation Rates in Tropical Forests and their Climatic Implications* (Friends of the Earth Report, London, 1989).
13 U.S. Department of Agriculture, *The Second RCA Appraisal: Soil, Water, and Related Resources on Nonfederal Land in the United States: Analysis of Conditions and Trends* (U.S. Department of Agriculture, Washington, DC, 1989).
14 R. Lal, in (7), pp. 7–26.
15 Worldwatch Institute, *State of the World 1988* (Worldwatch Institute, Washington, DC, 1988).
16 C. J. Barrow, *Land Degradation* (Cambridge Univ. Press, Cambridge, 1991).
17 F. R. Troeh and L. M. Thompson, *Soils and Soil Fertility* (Oxford Univ. Press, New York, ed. 5, 1993).
18 H. H. Bennett, *Soil Conservation* (McGraw-Hill, New York, 1939).
19 E. Roose, in *Conservation Farming on Steep Lands* (Soil and Water Conservation Society, Ankeny, IA, 1988), pp. 130–131.
20 D. Pimentel et al., *Science* 194, 149 (1976).
21 U.S. Department of Agriculture, *Agriculture and the Environment* (U.S. Department of Agriculture, Economic Research Service, Washington, DC, 1971).
22 L. K. Lee, *J. Soil Water Conserv.* 45, 622 (1990).
23 N. W. Hudson, in *Soil Erosion and Conservation in the Tropics* (Spec. Publ. 43, American Society of Agronomists, Madison, WI, 1982), pp. 121–133.
24 R. Lal, in *Quantification of the Effect of Erosion on Soil Productivity in an International Context*, F. R. Rijsberman and M. G. Wolman, Eds. (Delft Hydraulics Laboratory, Delft, Netherlands, 1984), pp. 70–94.
25 L. Hood and J. K. Morgan, *Sierra Club Bull.* 57, 4 (1972).
26 E. K. Byington, *Grazing Land Management and Water Quality* (American Society of Agronomists and Crop Scientists Society of America, Harpers Ferry, WV, 1986).
27 J. Risser, *Smithsonian* 11, 120 (March, 1981).
28 G. A. Klee, *Conservation of Natural Resources* (Prentice-Hall, Englewood Cliffs, NJ, 1991).
29 U.S. Department of Agriculture, *Agricultural Statistics 1967* (Government Printing Office, Washington, DC, 1967).
30 U.S. Department of Agriculture, *Agricultural Statistics 1992* (Government Printing Office, Washington, DC, 1992).
31 R. Lal, *Soil Erosion Problems on an Alisol in Western Nigeria and their Control* (International Institute of Tropical Agriculture, Ibadan, Nigeria, 1976).
32 F. H. Buttel, *Cornell Rural Sociology Bulletin No. 128* (Cornell University, Ithaca, NY, 1982).
33 Office of Technology Assessment, *Impacts of Technology on U.S. Cropland and Rangeland Productivity* (U.S. Congress Office of Technology Assessment, Washington, DC, 1982).
34 F. E. Allison, *Soil Organic Matter and Its Role in Crop Production* (Elsevier, New York, 1973).
35 G. R. Foster, R. A. Young, M. J. M. Ronkens, C. A. Onstad, in (52), pp. 137–162.
36 M. Simons, *New York Times*, 29 October 1992, p. A1.
37 J. R. Parrington, W. H. Zoller, N. K. Aras, *Science* 220, 195 (1983).
38 P. O. Aina, R. Lal, G. S. Taylor, Eds., *Soil Erosion: Prediction and Control* (Soil Conservation Society of America, Ankeny, IA, 1977).
39 U.S. Forest Service, *The Major Range Problems and Their Solution* (U.S. Forest Service, Washington, DC, 1936).
40 A. N. Sharpley and S. J. Smith, in *Cover Crops for Clean Water*, W. L. Hargrove, Ed. (Soil and Water Conservation Society, Ankeny, IA, 1991), p. 41.

41 Wen Dazhong, in (7), pp. 63–86.

42 L. McLaughlin, thesis, Cornell University (1991).

43 H. W. Kendall and D. Pimentel, *Ambio* 23, 198 (1994).

44 Statistical data, *FAO Q. Bull. Stat.* 5, 1 (1992).

45 W. W. Frye, S. A. Ebelhar, L. W. Murdock, R. L. Bevins, *Soil Sci. Soc. Am. J.* 46, 1051 (1982).

46 K. R. Olson and E. Nizeyimana, *J. Prod. Agric.* 1, 13 (1988); D. L. Schertz, W. C. Moldenhauert, S. J. Livingston, F. A. Weesies, E. A. Hintz, *J. Soil Water Conserv.* 44, 604 (1989), G. W. Landale, J. E. Box, R. A. Leonard, A. P. Barnett, W. G. Fleming, *ibid.* 34, 226 (1979).

47 D. L. Mokma and M. A. Sietz, *J. Soil Conserv.* 47, 325 (1992).

48 H. E. Dregne, *ibid.*, p. 8.

49 S. A. El-Swaify, W. C. Moldenhauer, A. Lo, *Soil Erosion and Conservation* (Soil Conservation Society of America, Ankeny, IA, 1985).

50 F. R. Troeh, J. A. Hobbs, R. L. Donahue, *Soil and Water Conservation* (Prentice-Hall, Englewood Cliffs, NJ, 1991).

51 National Soil Erosion–Soil Production Research Planning Committee, *J. Soil Water Conserv.* 36, 82 (1981).

52 R. F. Follett and B. A. Stewart, Eds., *Soil Erosion and Crop Productivity* (American Society of Agronomy and Crop Science Society of America, Madison, WI, 1985).

53 M. Falkenmark, in (3), pp. 164–191.

54 L. Leyton, in *Plant Research and Agroforestry*, P. A. Huxley, Ed. (International Council for Research in Agroforestry, Nairobi, Kenya, 1983), pp. 379–400.

55 R. P. Waldren, in *Crop-Water Relations*, I. D. Teare and M. M. Peet, Eds. (Wiley, New York, 1983), pp. 187–212.

56 R. H. Donahue, R. H. Follett, R. N. Tulloch, *Our Soils and Their Management* (Interstate, Danville, IL, 1990).

57 C. A. Van Doren, R. S. Stauffer, E. H. Kidder, *Soil Sci. Soc. Am. Proc.* 15, 413 (1950).

58 R. C. Wendt and R. E. Burwell, *J. Soil Water Conserv.* 40, 450 (1985).

59 R. C. Wendt, E. E. Alberts, A. T. Helmers, *Soil Sci. Soc. Am. J.* 50, 730 (1986).

60 C. E. Murphee and K. C. McGregor, *Trans. ASAE* 34, 407 (1991).

61 H. A. Elwell, *Zimbabwe Sci. News* 19, 27 (1985).

62 M. Alexander, *Introduction to Soil Microbiology* (Wiley, New York, 1977).

63 P. N. Bhatt, *Soil Conserv. Dig.* 5, 37 (1977).

64 R. Lal, in *Nitrogen Cycling in West African Ecosystems*, T. Rosswall, Ed. (Reklan and

Katalogtryck, Uppsala, Sweden, 1980), pp. 31–38.

65 A. Young, *Agroforestry for Soil Conservation* (CAB International, Wallingford, UK, 1989).

66 K. Chaney and R. S. Swift, *J. Soil Sci.* 35, 223 (1984).

67 G. W. Langdale, L. T. West, R. R. Bruce, W. P. Miller, A. W. Thomas, *Soil Technol.* 5, 81 (1992).

68 R. F. Follett, S. C. Gupta, P. G. Hunt, in *Soil Fertility and Organic Matter as Critical Components of Production Systems* (Soil Science Society of America and American Society of Agronomy, Madison, WI, 1987).

69 A. Young, *Outlook Agric.* 19, 155 (1990).

70 H. L. Barrows and V. J. Kilmer, *Adv. Agron.* 15, 303 (1963).

71 R. E. Lucas, J. B. Holtman, L. J. Connor, in *Agriculture and Energy*, W. Lockeretz, Ed. (Academic Press, New York, 1977), pp. 333–351.

72 M. Wood, *Soil Biology* (Blackie, Chapman and Hall, New York, 1989).

73 E. Lee and R. C. Foster, *Aust. J. Soil Res.* 29, 745 (1991).

74 D. Pimentel et al., *BioScience* 30, 750 (1980).

75 J. A. Van Rhee, *Plant Soil* 22, 43 (1965).

76 O. Atlavinyte, *Pedobiologia* 4, 245 (1964).

77 ——, *ibid.*, 5, 178 (1965).

78 W. S. Reid, in (52), pp. 235–250.

79 S. P. Teotia, F. L. Duky, T. M. McCalla, *Effect of Stubble Mulch on Number and Activity of Earthworms* (Nebraska Experiment Station Research Bulletin 165, Lincoln, NE, 1950).

80 G. A. E. Ricou, in *Grassland Ecosystems of the World: Analysis of Grasslands and Their Uses*, R. T. Coupland, Ed. (Cambridge Univ. Press, Cambridge, 1979), pp. 147–153.

81 N. W. Hudson, *Soil Conservation* (Cornell Univ. Press, Ithaca, NY, ed. 2, 1981).

82 P. W. Chapman, F. W. Fitch, C. L. Veatch, *Conserving Soil Resources: A Guide to Better Living* (Smith, Atlanta, GA, 1950).

83 W. Magrath and P. Arens, "The Costs of Soil Erosion on Java: A Natural Resource Accounting Approach," *Environment Department Working Paper No. 18* (World Bank, Washington, DC, 1989).

84 H. P. Chaudhary and S. K. Das, *J. Indian Soc. Soil Sci.* 38, 126 (1990).

85 M. S. Swaminathan, personal communication.

86 R. Repetto, *Sci. Am.* 266, 94 (June 1992).

87 R. Lal, in *Soil Erosion Research Methods* (Soil and Water Conservation Society, Ankeny, IA, 1988), pp. 187–200.

88 P. N. Crosson and T. T. Stout, *Productivity*

Effects of Cropland Erosion in the United States (Resources for the Future, Washington DC, 1983).

89 F. J. Pierce, R. H. Dowdy, W. A. P. Graham, *J. Soil Water Conserv.* 39, 131 (1984).

90 D. M. Gray and A. T. Leiser, *Biotechnical Slope Protection and Erosion Control* (Kreiger, Malabar, FL, 1969).

91 U.S. Department of Agriculture, *Fact Book of Agriculture* (U.S. Department of Agriculture, Office of Public Affairs, Washington, DC, 1990).

92 E. H. Clark, in *Soil Loss: Processes, Policies and Prospects*, J. M. Harlin and G. M. Bernardi, Eds. (Westview, New York, 1987), pp. 59–89.

93 E. H. Clark, *J. Soil Water Conserv.* 40, 19 (1985).

94 W. Allen, *St Louis Post Dispatch*, 27 January 1994, p. 01B.

95 P. C. Huszar and S. L. Piper, in *Off-Site Costs of Soil Erosion: The Proceedings of a Symposium* (Conservation Foundation, Washington, DC, 1985), pp. 143–166.

96 Soil Conservation Service, *Wind Erosion Report, Nov. 1992–May 1993* (Soil Conservation Service, U.S. Department of Agriculture, Washington, DC, 1993).

97 S. L. Piper, personal communication.

98 M. R. Carter, Ed., *Conservation Tillage in Temperate Agroecosystems* (Lewis, Boca Raton, FL, 1994).

99 R. G. Palis, G. Okwach, C. W. Rose, P. G. Saffigna, *Aust. J. Soil Res.* 28, 623 (1990).

100 P. Faeth, *Agricultural Policy and Sustainability: Case Studies from India, Chile, the Philippines and the United States* (World Resources Institute, Washington, DC, 1993).

101 E. Burnett and C. E. Fisher, *Soil Sci. Soc. Am. Proc.* 18, 216 (1954).

102 D. D. Smith, *J. Am. Soc. Agron.* 38, 810 (1946).

103 E. L. Sauer and H. C. M. Case, *Soil Conservation Pays Off: Results of Ten Years of Conservation Farming in Illinois* (Univ. of Illinois Agriculture Experiment Station Bulletin, Urbana, IL, 1954).

104 B. H. Hendrickson, A. P. Barnett, J. R. Carreler, W. E. Adams, *USDA Tech, Bull. No. 1281* (U.S. Department of Agriculture, Washington, DC, 1963).

105 R. H. Hart, M. J. Samuel, P. S. Test, M. A. Smith, *J. Range Manage.* 41, 282 (1988).

106 C. A. Taylor Jr., N. E. Garza Jr., T. O. Brooks, *Rangelands* 15, 53 (1993).

107 J. M. Fowler, A. Torell, G. Gallacher, *J. Range Manage.* 47, 155 (1994).

108 U.S. Department of Agriculture, *Agricultural Statistics 1991* (Government Printing Office, Washington, DC, 1991).

109 F. D. Roosevelt, *Letter from President to Governors* (White House, Washington, DC, 26 February 1937).

110 J. W. Ketcheson and J. J. Onderdonk, *Agron. J.* 65, 69 (1973).

111 S. D. Klausner, P. J. Zwerman, D. F. Ellis, *J. Environ. Qual.* 3, 42 (1974).

112 G. W. Musgrave and O. R. Neal, *Am. Geophys. Union Trans.* 18, 349 (1937).

113 J. Ketcheson, *J. Soil Water Conserv.* 32, 57 (1977).

114 R. G. Spomer, R. F. Piest, H. G. Heinemann, *Trans. ASAE* 19, 108 (1976).

115 G. C. Schuman, R. G. Spomer, R. F. Piest, *Soil Sci. Soc. Am. Proc.* 37, 424 (1973).

116 A. Mohammad and F. A. Gumbs, *J. Agric. Eng. Res.* 27, 481 (1982).

117 G. F. McIsaac and J. K. Mitchell, *Trans. ASAE* 35, 465 (1992).

118 W. Hinz, *Ariz. Farmer-Stockman* 64, 16 (1985).

119 National Academy of Sciences, *Soil and Water Quality: An Agenda for Agriculture* (National Academy of Sciences, Washington, DC, 1993).

120 J. C. Batty and J. Keller, in (121), pp. 35–44.

121 D. Pimentel, Ed., *Handbook of Energy Utilization in Agriculture* (CRC Press, Boca Raton, FL, 1980).

122 L. Lyles, *J. Soil Water Conserv.* 30, 279 (1975).

123 D. L. Schertz, W. C. Moldenhauer, S. J. Livingston, G. A. Weesies, E. A. Hintz, *ibid.*, 44, 604 (1989).

124 L. A. Kramer, *Trans. ASAE* 29, 774 (1986).

125 R. J. Hanks, in *Limitations to Efficient Water Use in Crop Production*, H. M. Taylor, W. R. Jordan, T. R. Sinclair, Eds. (American Society of Agronomy, Crop Science Society of America, and Soil Science Society of America, Madison, WI, 1983), pp. 393–411.

126 J. Shalhavet, A. Mantell, H. Bielorai, D. Shimshi, *Irrigation of Field and Orchard Crops Under Semi-Arid Conditions* (Pergamon, Elmsford, NY, 1979).

127 R. A. Feddes, in *Crop Water Requirements*, International Conference, Versailles, France, 11 to 14 September 1984 [United Nations Educational, Scientific, and Cultural Organization (UNESCO) under the auspices of the Food and Agricultural Organization and the World Meteorological Organization, 1984], pp. 221–234.

128 C. V. Kidd and D. Pimentel, Eds. *Integrated Resource Management: Agroforestry for Development* (Academic Press, San Diego, 1992).

129 U.S. Bureau of the Census, *Statistical Abstract of the United States 1990* (Government Printing Office, Washington, DC, 1992).

130 L. Lave and E. Seskin, *Air Pollution and Human Health* (Johns Hopkins Press, Baltimore, MD, 1977).

131 F. D. Whitaker and H. G. Heinemann, *J. Soil Water Conserv.* 28, 174 (1973).

132 W. C. Moldenhauer and M. Amemiya, *Iowa Farm Sci.* 21, 3 (1967).

133 P. Faeth, R. Repetto, K. Kroll, Q. Dai, G. Helmers, *Paying the Farm Bill: U.S. Agricultural Policy and the Transition to Sustainable Agriculture* (World Resources Institute, Washington, DC, 1991).

134 K. C. McGregor and C. K. Mutchler, *Trans. ASAE* 35, 1841 (1992).

135 ——, R. F. Cullum, *ibid.*, p. 1521.

136 C. K. Mutchler, L. L. McDowell, J. D. Greer, *ibid.* 28, 160 (1985).

137 L. C. Johnson and R. I. Papendick, *Northwest Sci.* 42, 53 (1968).

138 K. C. McGregor, C. K. Mutchler, M. J. M. Romkens, *Trans. ASAE* 33, 1551 (1990).

139 F. H. Bormann, G. E. Likens, T. C. Siccama, R. S. Pierce, J. S. Eaton, *Ecol. Monogr.* 44, 255 (1974).

14

Tropical Deforestation and Habitat Fragmentation in the Amazon: Satellite Data from 1978 to 1988*

David Skole and Compton Tucker

Deforestation has been occurring in temperate and tropical regions throughout history (1). In recent years, much attention has focused on tropical forests, where as much as 50% of the original extent may have been lost to deforestation in the last two decades, primarily as a result of agricultural expansion (2). Global estimates of tropical deforestation range from 69,000 km^2 year^{-1} in 1980 (3) to 100,000 to 165,000 km^2 year^{-1} in the late 1980s; 50 to 70% of the more recent estimates have been attributed to deforestation in the Brazilian Amazon, the largest continuous region of tropical forest in the world (2, 4, 5).

The area and rate of deforestation in Amazonia are not well known, nor are there quantitative measurements of the effect of deforestation on habitat degradation. We used 1:500,000 scale photographic imagery from Landsat Thematic Mapper data and a geographic information system (GIS) to create a computerized map of deforestation and evaluate its influence on forest fragmentation and habitat degradation. Areas of

deforestation were digitized into the GIS and the forest fragments and edge effects that result from the spatial pattern of forest conversion were determined.

Background

Tropical deforestation is a major component of the carbon cycle and has profound implications for biological diversity. Deforestation increases atmospheric CO_2 and other trace gases, possibly affecting climate (6, 7). Conversion of forests to cropland and pasture results in a net flux of carbon to the atmosphere because the concentration of carbon in forests is higher than that in the agricultural areas that replace them. The paucity of data on tropical deforestation limits our understanding of the carbon cycle and possible climate change (8). Furthermore, while occupying less than 7% of the terrestrial surface, tropical forests are the home to half or more of all plant and animal species (9). The primary adverse effect of tropical deforestation is massive

* Originally published in *Science*, 1993, vol. 260, pp. 1905–10.

Table 14.1 [*orig. table 1*] Tropical forest area (3) and reported tropical deforestation rates by country. The deforestation rates from the 1970s are from the Food and Agriculture Organization (FAO) (3). The 1980s data are from Meyers (2) and the World Resources Insitute (WRI) (5).

Country	Total forest area (km²)	Percent of world total	Deforestation rate, 1970s (km2)	Percent of world total	Deforestation rates, late 1980s			
					Myers (km2)	Percent of world total	WRI (km2)	Percent of world total
Brazil	3,562,800	30.7	13,600	19.7	50,000	36.1	80,000	48.4
Indonesia	1,135,750	9.8	5,500	8.0	12,000	8.7	9,000	5.4
Zaire	1,056,500	9.1	1,700	2.5	4,000	2.9	1,820	1.1
Peru	693,100	6.0	2,450	3.6	3,500	2.5	2,700	1.6
Columbia	464,000	4.0	8,000	11.6	6,500	4.7	8,200	5.0
India	460,440	4.0	1,320	1.9	4,000	2.9	15,000	9.1
Bolivia	440,100	3.8	650	1.0	1,500	1.1	870	0.5
Papua, New Guinea	337,100	2.9	210	0.3	3,500	2.5	220	0.1
Venezuela	318,700	2.7	1,250	1.8	1,500	1.1	1,250	0.8
Burma	311,930	2.7	920	1.3	8,000	5.8	6,770	4.1
Others*	2,829,930	24.4	33,300	48.3	44,100	31.8	39,610	23.9
Total	11,610,350	100.0	68,900	100.0	138,600	100.0	165,440	100.0

* Sixty-three other countries.

extinction of species including, for the first time, large numbers of vascular plant species (10).

Deforestation affects biological diversity in three ways: destruction of habitat, isolation of fragments of formerly contiguous habitat, and edge effects within a boundary zone between forest and deforested areas. This boundary zone extends some distance into the remaining forest. In this zone there are greater exposure to winds; dramatic micrometeorological differences over short distances; easier access for livestock, other nonforest animals, and hunters; and a range of other biological and physical effects. The result is a net loss of plant and animal species in the edge areas (11).

There is a wide range in current estimates of the area and rate of deforestation in Amazonia. Scientists at the Instituto Nacional de Pesquisas Espaciais (12–15) estimated a total deforested area of 280,000 km^2 as of 1988 and an average annual rate of 21,000 km^2 year^{-1} from 1978 to 1988. Other studies (2, 4, 5) have reported rates that range from 50,000 to 80,000 km^2 year^{-1} (table 14.1). Additional deforestation estimates have been made for geographically limited study areas in the southern Amazon Basin of Brazil with Landsat and meteorological satellite data (16–20).

The Amazon Basin of Brazil has been defined by law to include the states of Acre, Amapá, Amazonas, Pará, Rondônia, and Roraima plus part of Mato Grosso, Maranhão, and Tocantins and is referred to as the Legal Amazon (21). It covers an area of ~5,000,000 km^2, of which ~4,090,000 km^2 is forested, ~850,000 km^2 is cerrado or tropical savanna, and ~90,000 km^2 is water (table 14.2). Confusion has arisen among researchers regarding the stratification of the Brazilian Amazon into forest, cerrado, and water strata. A Food and Agriculture Organization (FAO)–United Nations Environmental Program (UNEP) study (3) found 3,562,800 km^2 of forest, whereas Fearnside and co-workers claim there is 4,195,660 km^2 of forest, 793,279 km^2 of cerrado (17), and 4,906,784 km^2 total (13). Meanwhile, an IBGE study (22) found 20,972 km^2 of water, 3,793,664 km^2 of forest, and 1,149,943 km^2 of cerrado for a total of 4,964,920 km^2. These differences prevent comparison of different deforestation studies.

The use of satellite data and the GIS make it possible to explicitly stratify Amazonia on the basis of cover types (22), thereby providing a means of comparison with other studies. This approach is also necessary for spatial analysis of habitat fragmentation and

Table 14.2 [orig. table 2] Predeforestation water, forest, and cerrado land cover for the Brazilian Amazon by state as used in this study. The values determined in this study were based on the IBGE vegetation map and by interpretation of satellite data (22). Areas obscured by clouds were excluded from deforestation and affected forest habitat analyses (97% of the cloud-affected data were over tropical forest).

State	Water area (km^2)	Forest area (km^2)	Cloud total (km^2)	Cerrado area (km^2)	Total area (km^2)
Acre	393	152,394	0	0	152,787
Amapá	1,188	137,444	53,566	978	139,610
Amazonas	29,842	1,531,122	94,058	14,379	1,575,343
Maranhão	1,344	145,766	13,444	114,675	261,785
Mato Grosso	4,212	527,570	8,630	368,658	900,440
Pará	49,522	1,183,571	56,807	28,637	1,261,730
Rondônia	1,462	212,214	474	24,604	238,280
Roraima	1,817	172,425	15,232	51,464	225,706
Tocantins	2,914	30,325	0	244,005	277,244
Total	92,694	4,092,831	(242,211)	847,400	5,032,925

edge effects of deforestation. Finally, GIS provides a data management tool with which we could manage large amounts of spatial data and precisely merge and geocode information from the more than 200 satellite images used in this study.

Remote Sensing

The large area of the Brazilian Amazon necessitates a straightforward and accurate method of measurement. Landsat Thematic Mapper photo products are inexpensive and of sufficient spatial and spectral resolution for the determination of deforestation. Analysis with visual interpretation techniques produces quantitative results similar to digital processing of full-resolution, multispectral data from the Thematic Mapper and SPOT (23).

We acquired 210 black and white photographic images of the entire Brazilian Amazon. They were obtained with channel five of the Landsat Thematic Mapper (1.55 to 1.75 μm) at 1:500,000 scale and were primarily from 1988 (24). We digitized the deforested areas with visual deforestation interpretation and standard vector GIS technique. The digitized scenes were projected into equal-area geographic coordinates (latitude, longitude), edge matched, and merged in the computer to form a single, seamless data-set for the entire Brazilian Amazon.

Spatial analysis of the geometry of deforestation is critical to the estimation of forest fragmentation and the edge effect. If 100 km^2 of tropical deforestation occurs as a 10 km by 10 km square and we assume that the edge effect is 1 km, the total area affected is ~143 km^2. In contrast, if the 100 km^2 of deforestation is distributed as ten strips, each 10 km by 1 km, the affected area is ~350 km^2.

We extracted forest fragments <100 km^2 that were isolated by deforestation and computed edge effects for a zone of 1 km along the boundaries. All areas of closed canopy tropical forest deforested by 1988 were delineated, including areas of secondary growth on abandoned fields and pastures where visible. Areas of long-term forest degradation along river margins in

central Amazonia were also included, as were scattered small clearings associated with rubber tappers, mining operations, airfields, and other small disturbances. All visible roads, power line right of ways, pipelines, and similar human-made features were also digitized into the GIS and treated as deforestation. We used 50 digital Landsat Multispectral Scanner (MSS) scenes from 1986 and 15 digital Thematic Mapper images from 1988 for detailed examination of Acre, Amazonas, Mato Grosso, Pará, and Rondônia.

To determine the extent of deforestation in 1978, we used the GIS to digitize maps of scale 1:500,000 from single-channel Landsat MSS data, produced jointly by the Instituto Brasiliero de Desenvolvimento Florestal (IBDF) and the Instituto de Pesquisas Espaciais (INPE) in the early 1980s (12, 23). These maps did not differentiate between forest and cerrado clearing. We compiled forest, cerrado, and water data by combining a vegetation map with analysis of Landsat images and meteorological satellite data (25). Our deforestation and affected habitat analyses for 1978 and 1988 were restricted to closed-canopy forest of the Brazilian Amazon.

Deforestation and Forest Fragmentation

Distribution of deforestation and affected habitat in the Brazilian Amazon for 1978 and 1988 was concentrated in a crescent along the southern and eastern fringe of the Amazon [a spatial pattern similar to the distribution of fires observed from thermal anomalies in data from Landsat's Advanced Very-High Resolution Radiometer (AVHRR) (20)] and along major transportation corridors in the interior of the Amazon. Deforestation increased between 1978 and 1988 (78,000 to 230,000 km^2), while the total affected habitat increased (208,000 to 588,000 km^2) (table 14.3). The total area deforested increased by a factor of two to three or more in every state except Amapa; but it is likely that the deforested area in Amapa is higher than our assessment because excessive cloud cover in this region

Table 14.3 [*orig. table 3*]   Tropical deforestation, forest isolated or cut off by deforestation, and the area of forest adversely affected by a 1-km edge effect from adjacent areas of deforestation in the Brazilian Amazon. Areas that were obscured by clouds were omitted from this analysis. Parentheses following the edge effect entries contain the ratio between a 500-m buffer and a 1,000-m buffer.

State	Deforested (km^2)	Isolated (km^2)	Edge effect (km^2)	Total (km^2)
		1978		
Acre	2,612	18	4,511	7,141
Amapá	182	0	368	550
Amazonas	2,300	36	6,498	8,834
Maranhão	9,426	705	13,120	23,251
Mato Grosso	21,134	776	25,418	47,328
Pará	30,449	2,248	49,791	82,488
Rondônia	6,281	991	17,744	25,016
Roraima	196	4	812	1,012
Tocantins	5,688	337	6,584	12,609
Total	78,268	5,115	124,846	208,229
		1988		
Acre	6,369	405	23,686 (0.517)	30,460
Amapá	210	1	689 (0.537)	900
Amazonas	11,813	474	36,392 (0.582)	48,679
Maranhão	31,952	2,123	28,147 (0.626)	62,222
Mato Grosso	47,568	2,542	71,128 (0.580)	121,238
Pará	95,075	6,837	116,669 (0.633)	218,581
Rondônia	23,998	2,408	52,345 (0.657)	78,751
Roraima	1,908	1	5,236 (0.521)	7,145
Tocantins	11,431	1,437	6,760 (0.659)	19,628
Total	230,324	16,228	341,052 (0.610)	587,604

prevented complete analysis (table 14.2). We found that 6% of closed-canopy forest had been cleared as of 1988 and ~15% of the forested Amazon was affected by deforestation-caused habitat destruction, habitat isolation, and edge effects (table 14.3).

Our analysis of the spatial pattern of deforestation found a strong tendency toward spatial concentration; areas of undisturbed tropical forest tended to be sizable (table 14.4). This is more pronounced than table 14.4 indicates because many of the large areas of undisturbed tropical forest are contiguous among states.

For the entire Brazilian Amazon, our deforestation estimate is close to, but lower than, the estimates of Fearnside et al. (13) and the INPE (15) of ~280,000 km^2 as of 1988. The difference is a result of three factors: (i) different stratification of forest, cerrado, and water; (ii) slightly different estimates of secondary growth, which is spectrally similar to intact forest in channel five; and (iii) positional accuracy, interpretation, and boundary generalization. We estimate that ~30,000 km^2 of the difference is from a different evaluation of the forest-cerrado boundaries in Mato Grosso and Tocantins. By comparison, our analysis suggests that deforestation estimates based on coarse-resolution meteorological satellite data in the southern Amazon of Brazil have overestimated deforestation by ~50% (18, 23).

The average deforestation rate in the closed-canopy forests from 1978 to 1988 (~15,000 km^2 year^{-1}) (table 14.3) is higher than the rate from 1975 to 1978 (3) but considerably lower than recent estimates (2, 4, 5, 20). Our estimates can be used in assessments of net flux of carbon from land

Table 14.4 [*orig. table 4*]   Spatial characteristics of isolated and remaining tropical forest within the Legal Amazon as determined by analysis of 1988 Landsat Thematic Mapper imagery. Isolated forest refers to areas of forest <100 km² surrounded by deforestation. Remaining forest refers to tropical forest that has not been deforested and includes both isolated and larger areas of forest. Many of the largest remaining areas of tropical forest are contiguous among states. Areas affected by clouds were omitted from this analysis.

State	Isolated forest		Undisturbed remaining forest		Range of areas (km²)	
	Area (km²)	Polygons (no.)	Area (km²)	Polygons (no.)	Minimum	Maximum
Acre	405	603	146,025	605	<1	139,215
Amapá	1	2	83,676	3	<1	83,675
Amazonas	474	464	1,425,253	465	<1	1,424,779
Maranhão	2,123	1,035	100,554	1,042	<1	70,057
Mato Grosso	2,542	2,016	478,619	2,027	<1	471,792
Pará	6,837	4,030	1,032,194	4,032	<1	1,021,263
Rondônia	2,408	1,587	187,743	1,588	<1	185,335
Roraima	1	2	155,326	6	<1	152,414
Tocantins	1,437	493	18,894	508	<1	6,982
Total	16,228	10,232	3,628,284	10,276		

clearing and biomass burning in the Brazilian Amazon. Current estimates of these fluxes have largely been based on model calculations with deforestation values much higher than we report. In addition, many deforested areas are in stages of regrowth following abandonment (26). If regrowth is widespread, estimates of the net flux of carbon should be further reduced because carbon accumulates in regrowing biomass.

The preponderance of affected habitat results from proximity to areas of deforestation (~341,000 km² for a 1-km edge effect) and not from isolation of forest (~15,000 km²) or deforestation per se (~230,000 km²). While the rate of deforestation averaged ~15,000 km² year⁻¹ in the Brazilian Amazonia from 1978 to 1988, the rate of habitat fragmentation and degradation was ~38,000 km² year⁻¹. Implications for biological diversity are not encouraging and provide added impetus for the minimization of tropical deforestation.

References and Notes

1   R. P. Tucker and J. F. Richards, *Global Deforestation and the Nineteenth Century World Economy* (Duke Univ. Press, Durham, NC, 1983); J. F. Richards, *Environment* 26, 6 (1984); M. Williams, *Prog. Hum. Geogr.* 13, 176 (1989); in *The Earth as Transformed by Human Action*, B. L. Turner II et al., Eds. (Cambridge Univ. Press, Cambridge, 1990), pp. 179–201.

2   N. Myers, *Clim. Change* 19, 3 (1991).

3   "Los Recursos Forestales de la America Tropical," 32/6. 1301–78–04, *Tech. Rep. No. 1* (Food and Agriculture Organization of the United Nations, Rome, 1981); "Forest Resources of Tropical Africa," 32/6. 1301–78–04. *Tech. Rep. No. 2* (Food and Agriculture Organization of the United Nations, Rome, 1981); "Forest Resources of Tropical Asia," 32/6. 1301–78–04. *Tech. Rep. No. 3* (Food and Agriculture Organization of the United Nations, Rome, 1981).

4   "The Forest Resources of the Tropical Zone by Main Ecological Regions," *Report to the United Nations Conference on Environment and Development by the Forest Resource Assessment 1990 Project* (Food and Agriculture Organization of the United Nations, Rome, 1992). The FAO Forest Assessment 1990 Project has produced several reports, and estimates from them have varied considerably. The recent release of another report [P. Aldhous, *Science* 259, 1390 (1993)] provides slightly different estimates than those reported in 1992.

5 *World Resources 1990–91: A Report by the World Resources Institute in Collaboration with the United Nations Environment Program and The United Nations Development Program* (Oxford Univ. Press, New York, 1990).

6 J. H. C. Gash and W. J. Shuttleworth, *Clim. Change* 19, 123 (1991); R. A. Houghton et al., *Nature* 316, 617 (1985); R. A. Houghton and D. L. Skole, in *The Earth as Transformed by Human Action*, B. L. Turner II et al., Eds. (Cambridge Univ. Press, Cambridge, 1990), pp. 393–408; ——, D. S. Lefkowitz, *J. For. Ecol. Manage.* 38 173 (1991); M. Keller, D. J. Jacob, S. C. Wofsy, R. C. Harris, *Clim. Change* 19, 139 (1991); E. Salati, in *The Geophysiology of Amazonia: Vegetation and Climate Interaction*, R. E. Dickinson, Ed. (Wiley Interscience, New York, 1987), pp. 273–296, —— and C. A. Nobre, *Clim. Change* 19, 177 (1991); E. Salati and P. B. Vose, *Ambio* 12, 67 (1983); *Science* 225, 129 (1984); J. Shukla, C. Nobre, P. Sellers, *ibid.* 247, 1322 (1990).

7 R. A. Houghton, *Clim. Change* 19, 99 (1991).

8 *Climate Change 1992: the Supplementary Report to the IPCC Scientific Assessment*, J. T. Houghton, B. A. Callander, S. K. Varney, Eds. (Intergovernmental Panel on Climate Change, Cambridge Univ. Press, Cambridge, 1992).

9 E. O. Wilson, in *Biodiversity*, E. O. Wilson and F. M. Peters, Eds. (National Academy Press, Washington, DC, 1988), pp. 3–20; A. H. Gentry, *Proc. Natl. Acad. Sci. U.S.A.* 85, 156 (1988). The immense biological diversity of tropical forests is difficult to comprehend. For example, ten selected 1-ha plots in Borneo contained 700 species of trees and 1 ha of tropical Peru contained 300 tree species. By comparison, 700 tree species occur in all of North America. "Species" is used to mean organisms that can breed freely with each other and conversely, cannot breed freely with other species.

10 G. T. Prance and T. S. Elias, Eds., *Extinction is Forever* (New York Botanical Garden, New York, 1982); R. Lewin, *Science* 234, 14 (1986); T. L. Erwin, *Coleopt. Bull.* 36, 74 (1982); *Bull. Entomol. Soc. Am.* 30, 14 (1983); P. R. Ehrlich and E. O. Wilson, *Science* 253, 758 (1991); E. O. Wilson and F. M. Peters, Eds., *Biodiversity* (National Academy Press, Washington, DC, 1988); W. V. Reid and K. R. Miller, *Keeping Options Alive: The Scientific Basis for Conserving Biodiversity* (World Resources Institute, Washington, DC, 1989); E. O. Wilson, *Issues Sci. Technol.* 2, 20 (fall 1985); T. L. Erwin, in *Biodiversity*,

E. O. Wilson and F. M. Peters, Eds. (National Academy Press, Washington, DC, 1988), pp. 123–129; R. M. May, *Science* 241, 1441 (1988); A. H. Knoll, in *Extinctions*, M. H. Nitecki, Ed. (Univ. of Chicago Press, Chicago, 1984), pp. 21–68.

11 J. F. Franklin and R. T. T. Forman, *Landscape Ecol.* 1, 5 (1987); L. D. Harris, *The Fragmented Forest: Island Biogeographic Theory and the Preservation of Biotic Diversity* (Univ. of Chicago Press, Chicago, 1984); L. D. Harris, *Conserv. Biol.* 2, 330 (1988); D. H. Janzen, *Oikos* 41, 402 (1983); T. E. Lovejoy et al., in *Extinctions*, M. H. Nitecki, Ed. (Univ. of Chicago Press, Chicago, 1984), pp. 295–325; T. E. Lovejoy et al., in *Conservation Biology: The Science of Scarcity and Diversity*, M. E. Soule, Ed. (Sinauer, Sunderland, MA, 1988), pp. 257–285; D. S. Wilcove, C. H. McLellan, A. P. Dobson, *ibid.*, pp. 237–256; D. A. Saunders, R. J. Hobbs, C. R. Margules, *Conserv. Biol.* 5, 18 (1991); J. Terborgh, *Science* 193, 1029 (1976); B. A. Wilcox, in *Conservation Biology an Evolutionary-Ecological Perspective*, M. E. Soule and B. A. Wilcox, Eds. (Sinauer, Sunderland, MA, 1980), pp. 95–117; K. H. Redford, *BioScience* 42, 412 (1992).

12 A. T. Tardin et al., *Rep. 411–NTE/142* (Instituto Nacional de Pesquisas Espaciais, São Jose dos Campos, Brazil, 1979); A. T. Tardin et al., "Subprojeto Desmatamento," *IBDF/CNPq-INPE* (Instituto de Pesquisas Espaciais, São Jose dos Campos, Brazil, 1980).

13 P. M. Fearnside, A. T. Tardin, L. G. M. Meira, *Deforestation Rate in Brazilian Amazonia* (National Secretariat of Science and Technology, Brasilia, Brazil, 1990).

14 A. T. Tardin and R. P. da Cunha, *Report INPE–5015–RPE/609* (Instituto de Pesquisas Espaciais, São José dos Campos, Brazil, 1990).

15 *Deforestation in Brazilian Amazonia* (Instituto Nacional de Pesquisas Espaciais, São José dos Campos, Brazil, 1992).

16 D. J. Mahar, *Government Policies and Deforestation in Brazil's Amazon Region* (World Bank, Washington, DC, 1989).

17 P. M. Fearnside, *Environ. Conserv.* 17, 213 (1990).

18 A. M. Cross, J. J. Settle, N. A. Drake, R. T. M. Paivinen, *Int. J. Remote Sensing* 12, 1119 (1991); J. P. Malingreau and C. J. Tucker, *Ambio* 17, 49 (1988).

19 R. F. Nelson and B. N. Holben, *Int. J. Remote Sensing* 7, 429 (1986); R. F. Nelson, N. Horning, T. A. Stone, *ibid.* 8, 1767 (1987); C. J. Tucker, B. N. Holben, T. E. Goff, *Remote Sensing Environ.* 15, 255 (1984); G. M.

Woodwell, R. A. Houghton, T. A. Stone, R. F. Nelson, W. Kovalick, *J. Geophys. Res.* 92, 2157 (1987).

20 A. W. Setzer and M. C. Pereira, *Ambio* 20, 19 (1991); A. W. Setzer, *Relatorio INPE-4534-RPE/565* (Instituto Nacional de Pesquisas Espaciais, São José dos Campos, Brazil, 1988).

21 *Anuario Estatistico do Brasil 1991* (Fundacão Instituto Brasileiro de Geografia e Estatística, Rio de Janeiro, Brazil, 1991), vol. 51, pp. 1–1024.

22 *Mapa de Vegetacão do Brasil* (Fundacão Instituto Brasileiro de Geografia e Estatística, Rio de Janeiro, Brazil, 1988).

23 D. L. Skole, thesis, University of New Hampshire (1992).

24 Images: 7 from 1989, 175 from 1988, 8 from 1987, and 20 from 1986. All data from the Brazilian Landsat receiving station. The exact boundary between intact forest and deforested land was digitized in the Universal Transverse Mercator projection and then edited and error-checked with use of clear velum plots of the line-work overlaid on each photographic image. Each Landsat scene contained coordinate control points in decimal degree units, such that each scene could be geographically registered within precise tolerances and mosaicked together. For digitization, vertices were placed approximately every 50 m of ground position. Tests of positional accuracy in digitizing followed those of R. Dunn, R. Harrison, and J. C. White [*Int. J. Geograph. Inf. Syst.* 4, 385 (1990)] and indicated encoding; hence, area-estimation errors were less than 3% (23). The variance associated with interpretation and de-lineation of boundaries between intact forest and deforested areas was less than 10% overall. Further accuracy assessment was made in test sites established in Rondonia, where fragmentation was very high. An explicit spatial comparison between our estimate of deforestation and the same derived from high-resolution (20-m resolution) SPOT satellite imagery was highly correlated ($r^2 = 0.98$; $y = 1.11x - 57.358$). Additional ground checking and verification was done in eastern Para state (north of Manaus) and along the Rio Negro, both in Amazonas.

25 Fundamental to our analysis was a specified representation for water, cerrado or savanna, and forest for the Brazilian Amazon. We used a vegetation map (23) that was augmented by Landsat Thematic Mapper and meteorological satellite imagery for more accurate depiction of cerrado and water. This GIS representation is available upon request.

26 R. B. Buschbacher, *BioScience* 36, 22 (1986); C. Uhl, R. B. Buschbacher, E. A. S. Serrao, *J. Ecol.* 76, 663 (1988); R. B. Buschbacher, C. Uhl, E. A. S. Serrao, *ibid.*, p. 682.

27 This work was supported by National Aeronautics and Space Administration's mission to planet Earth and the Eos Data Information System's Landsat Pathfinder Program. We acknowledge S. Tilford and W. Huntress for initiating this research, W. Chomentowski for assistance in developing the satellite and GIS database, and A. Nobre for his assistance in interpreting the satellite data. G. Batista, M. Heinicke, and T. Grant assisted with the GIS representation of forest, water, and cerrado.

15

EXPLORING THE LINKS BETWEEN DESERTIFICATION AND CLIMATE CHANGE*

Mike Hulme and Mick Kelly

More than 100 countries are suffering the consequences of desertification, or land degradation in dryland areas.[1] Loss of productivity and other social, economic, and environmental impacts are directly affecting the perhaps 900 million inhabitants of these nations. There is also concern that the environmental impact of dryland degradation may be felt further afield. Some have suggested that this impact might even be felt worldwide.

The first international effort to address desertification occurred in 1977, when the United Nations Conference on Desertification (UNCOD) recognized that desertification was a major environmental problem with high human, social, and economic costs. The conference adopted the Plan of Action to Combat Desertification (PACD), a 20-year, worldwide program to arrest further dryland degradation. Sixteen years later, after several reviews, PACD has achieved little success.[2] A second phase in

the international response to desertification began at the United Nations Conference on Environment and Development (UNCED) in June 1992. It was agreed then that a Convention to Combat Desertification should be ready for signing and ratification by June 1994 and that an Intergovernmental Negotiating Committee on Desertification (INC-D) should be established to guide this process. It has also been decided that projects to mitigate land degradation in drylands will qualify for allocations from the Global Environment Facility (GEF),[3] but only insofar as the projects pertain to the GEF goals of protecting the global environment by reducing greenhouse-gas emissions, preserving biodiversity, and protecting international waters.

A significant obstacle to the work of INC-D is that desertification is a difficult word to define. In 1991, the UN Environment Programme (UNEP) defined desertification as "land degradation in arid,

---

* Originally published in *Environment*, 1993, vol. 35, no. 6, pp. 4–11, 39–46.
[1] UN Environment Programme, *World Atlas of Desertification* (Sevenoaks, U.K.: Edward Arnold, 1992).
[2] R. S. Odingo, "Implementation of the Plan of Action to Combat Desertification (PACD) 1978–1991," *Desertification Control Bulletin* 21 (1992):6–14.
[3] The Global Environment Facility was established in 1990 to help developing nations respond to global environmental change insofar as this response will reduce the global impact of the problems.

semi-arid and dry sub-humid areas resulting mainly from adverse human impact."[4] Just one year later, UNCED adopted the definition of "land degradation in arid, semi-arid and dry sub-humid areas resulting from various factors including climatic variations and human activities."[5] The different emphasis placed on climate variation in these two definitions is indicative of the disagreement that exists concerning the relative importance of the various causes of dryland degradation. This disagreement may appear to be an impractical, academic issue for the large numbers of people dependent on drylands for their livelihood, but misconceptions and arguments must be resolved for any adequate response to desertification to be made.

The first step in discussing such issues must be to define the most important terms and thereby avoid confusion. The terms *climate change* and *climate variation* are used here to indicate climate variability and trends arising from both natural and anthropogenic causes. The term *global-mean warming* indicates climate change resulting from greenhouse-gas emissions. *Desertification* is taken here to mean land degradation in dry-land regions, or the permanent decline in the potential of the land to support biological activity and, hence, human welfare. Desertification should not be confused with drought or desiccation. *Drought* refers to a period of two years or more with below-average rainfall, and *desiccation* is aridification resulting from a dry period lasting a decade or more.[6]

Climate change undoubtedly alters the frequency and severity of drought and can cause desiccation in various regions of the world. It does not necessarily follow, however, that drought and desiccation will, by themselves, induce, or even contribute

to, desertification. Whether or not desertification occurs depends upon the nature of resource management in these dryland regions. Identifying the contribution of climate variation to desertification is not a simple matter, and the difficulties are compounded by the possibility that desertification itself may generate climate change.

Does Climate Change Cause Desertification?

The definitions of desertification adopted by UNEP in 1991 and by UNCED in 1992 both implicitly link climate change and the assessment of the extent of desertification. Because arid, semi-arid, and dry subhumid areas are climatically defined,[7] any change in climate that results in an expansion or contraction of these areas is likely to change the formal, measured extent of the problem. For example, when an arid area becomes extremely dry or hyperarid, because of climate change, the area defined as being prone to desertification decreases because hyperarid areas are not included in the accepted definition. Conversely, when a humid area converts to subhumid, the defined area within which desertification is considered possible increases.

That climates do change over the decades has now been established beyond dispute. In the African Sahel, for example, annual rainfall during the most recent three decades has been between 20 and 40 percent less than it was from 1931 to 1960. Table 15.1 illustrates this change in a different way. Within contiguous Africa, there has been a net shift of land area toward aridity, especially toward hyperaridity, and a consequent net loss of semi-arid and dry subhumid land. Overall, areas prone to desertification have decreased from 52.4 percent of mainland

[4] UN Environment Programme, *Status of Desertification and Implementation of the UN Plan of Action to Combat Desertification*, UNEP/GCSS.III/3 (Nairobi: UN Environment Programme, 1991).
[5] United Nations, "Managing Fragile Ecosystems: Combating Desertification and Drought," chapt. 12 of *Agenda 21* (New York: United Nations, 1992), pt. 2.
[6] A. Warren and M. M. Khogali, *Assessment of Desertification and Drought in the Sudano-Sahelian Region: 1985 to 1991* (New York: UN Development Programme and UN Sudano-Sahelian Office, 1992).
[7] In the 1992 UNEP desertification assessment, a simple moisture index (the ratio of precipitation to potential evapotranspiration) was used to define these boundaries.

Table 15.1 [*orig. table 1*]   Change in areas of moisture zones in contiguous Africa.

Moisture zone	Mean land area from 1931 to 1960		Mean land area from 1961 to 1990		Net change between two periods	
	(millions of hectares)	(percentage of total)[a]	(millions of hectares)	(percentage of total)	(millions of of hectares)	(percent)
Hyperarid	450.8	15.1	501.5	16.8	+50.7	+1.7
Arid	676.9	22.7	680.0	22.8	+3.1	+0.1
Semi-arid	620.9	20.8	606.9	20.3	−14.0	−0.5
Dry subhumid	264.4	8.9	250.0	8.4	−14.4	−0.5
Humid	972.4	32.6	947.0	31.7	−25.4	−0.9

[a] Percentages total to 100.1 because of rounding.
Source: M. Hulme, R. Marsh, and P.D. Jones, "Global Changes in a Humidity Index Between 1931–60 and 1961–90," *Climate Research* 2 (1992): 1–22.

Africa to 51.5 percent between these two 30-year periods – a reduction of 25.3 million hectares. The amount of hyperarid land, however, has increased by more than 50 million hectares.

Determining the precise contribution of climate change to the problem of desertification is not an easy matter. When resource management failure has occurred, there is no doubt that climate variation can aggravate the problem. But separating out the interrelated impacts of climatic and human factors is extremely difficult. Some progress has, however, been made. To cite one example, Compton Tucker and his colleagues at the U.S. National Aeronautics and Space Administration's Space Flight Center in Maryland have used a satellite index of active vegetative cover to determine the extent of the Sahara Desert between 1980 and 1989.[8] Their analysis shows that very substantial interannual variations exist in the extent and quality of surface vegetation in dryland regions. Because much of the vegetation response detected by the satellite is caused by changes in rainfall, much (but not all) of the variability in the extent of the Sahara is due to interannual rainfall variations (see figure 15.1). Thus, the index can be used to discriminate between the degrada-tion of vegetation cover caused by rainfall and that which is due to other factors, most notably failures in resource management. Figure 15.1 relates the estimates of change in the extent of the Sahara to independently derived annual rainfall data for the years from 1980 to 1989. Linear regression analysis indicates that a considerable amount of the year-to-year variation in areal extent – 83 percent – can be explained by the rainfall data. This is a statistically significant relationship. The relationship does, however, leave some residual variability in the extent of the Sahara unexplained, as shown by the lowest curve in figure 15.1. This residual component of the variability has tended to increase over the past decade, and statistical analysis suggests that it amounts to an average annual increase in the extent of the Sahara of about 41,000 square kilometers per year. This is equivalent to an average annual areal increase not directly related to annual rainfall variations of almost 0.5 percent from 1980 onwards, which would amount to almost a 5-percent increase in total area over the decade. This trend could be the result of the cumulative impact of a series of dry years on vegetation recovery. For example, a particular rainfall amount in 1989 may generate less vegetation than the same

[8] The index, known as the Normalized Difference Vegetation Index (NDVI), is derived from the visible and near infrared sensors on polar-orbiting satellites and is an indicator of the photosynthetic vigor of surface biomass. See C. J. Tucker, H. E. Dregne, and W. W. Newcomb, "Expansion and Contraction of the Sahara Desert from 1980 to 1990," *Science* 253 (1991):229–301.

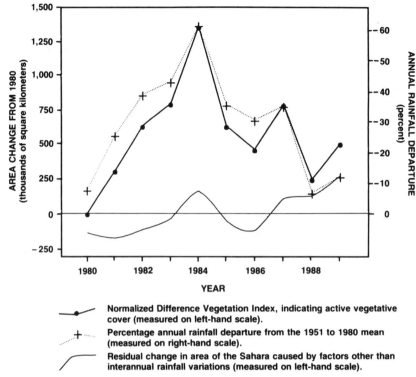

Figure 15.1 [*orig. figure 1*]   Change in the area of the Sahara from 1980 to 1989.
Sources: For the area of the Sahara, see C. J. Tucker, H. E. Dregne, and W. W. Newcomb, "Expansion and Contraction of the Sahara Desert from 1980 to 1990," Science 253 (1991):229–301. The other data are from the Climatic Research Unit in Norwich, England.

amount in 1980 because of the drought years preceding 1989. Alternatively, the increase in extent may well be due to a deterioration of vegetative cover caused by human activity. Clearly, the relative contributions of human activity and climate change to desertification will vary from region to region and from time to time. Separating out the relative roles of these factors to identify the most appropriate response in any particular situation – and to accord each factor due weight in the desertification convention – is a pressing challenge.

Can Desertification Change Climate?

Separating cause and effect is rendered more difficult by the fact that desertification may, in turn, affect both local climate and climates further afield. Recently, Bob Balling, Jr., of Arizona State University has suggested that surface air temperature has increased significantly in desertified regions owing to changes in land cover and that this effect has substantially affected global-mean temperature.[9] Desertification is likely to lead to reductions in surface soil moisture, which result in more energy available to heat the air (sensible energy) because less goes to evaporate water (latent energy). While it is conceivable that the warming of desertified areas may have been great enough to produce a measurable increase in global-mean temperature (see box 15.1), the influence would have been small compared to the potential impact of an enhanced greenhouse effect. The question of whether desertification has had or will have a

[9] R. C. Balling, Jr., "Impact of Desertification on Regional and Global Warming," *Bulletin of the American Meteorological Society* 72 (1991):232–34.

detectable effect on global climate is, nevertheless, a critical one. If a clear influence could be established, it might be argued that dryland degradation should be classed as a global environmental problem in its own right. At present, though, the evidence of a substantial effect must be considered extremely weak.

Box 15.1   Has Desertification "Contaminated" the Global-mean Temperature Record?

Bob Balling, Jr., of Arizona State University recently suggested that, during the 20th century, desertified areas have warmed by about 0.5°C relative to nondesertified areas.(1) When averaged globally, he argues, this warming significantly "contaminates" the global-mean temperature record and so complicates the search for a warming signal, or component, caused by enhancement of the greenhouse effect.(2) However, Balling's identification of areas that are severely desertified and nondesertified is derived from a map of desertification prepared for the UN Conference on Desertification back in 1977. That map has been superseded by more accurate data collected for the recent assessment by the UN Environment Programme (UNEP).(3) Moreover, Balling's calculated desertification warming signal of 0.5°C per 100 years is based on only a small subset of these "desertified" areas. He extrapolates from this subset to the global scale by suggesting that more than 30 percent of all land (including 90 percent of all drylands) is prone to this desertification warming signal.

UNEP's 1992 global assessment of desertified areas estimated, however, that only 20 percent of dryland regions were seriously degraded. This figure suggests that only 6 percent of all land may exhibit a desertification warming signal, rather than the 30 percent assumed by Balling. The potential bias is, therefore, comparable in magnitude to other biases that may affect the global temperature record. Finally, it should be noted that the differential warming between desertified and neighboring nondesertified areas found by Balling at a resolution of 5° latitude/longitude may simply indicate that these areas differ in their sensitivity to climate variability and may not be evidence of warming caused by desertification per se. Thus, although Balling may be correct in principle, he has greatly overstated his case, and his analysis provides no convincing evidence that warming caused by desertification has substantially affected the global-mean temperature record.

1   R. C. Balling, Jr., "Impact of Desertification on Regional and Global Warming," *Bulletin of the American Meteorological Society* 72 (1991):232–34. The ideas expressed in this article have received a moderate amount of attention in certain circles skeptical of the influence of greenhouse-gas emissions on global climate and have been introduced into discussions of the links between climate change and desertification. See, for example, UN Sudano-Sahelian Office and UN Development Programme, *GEF and Desertification: UNSO/UNDP Workshop, Nairobi, 28–30 October 1992* (New York: UNSO/UNDP, 1992).
2   T. M. L. Wigley, G. I. Pearman, and P. M. Kelly, "Indices and Indicators of Climate Change: Issues of Detection, Validation and Climate Sensitivity," in I. M. Mintzer, ed., *Confronting Climate Change: Risks, Implications and Responses* (Cambridge, England: Cambridge University Press, 1992), 85–96. For a discussion of other potential sources of bias in the global-mean temperature record, see C. K. Folland, T. Karl, and K. Ya. Vinnikov, "Observed Climate Variations and Change," in J. T. Houghton, G. J. Jenkins, and J. J. Ephraums, eds., *Climate Change: The IPCC Scientific Assessment* (Cambridge, England: Cambridge University Press, 1990), 195–238; and C. K. Folland et al., "Observed Climate Variability and Change," in J. T. Houghton, B. A. Callander, and S. K. Varney, eds., *Climate Change 1992: The Supplementary Report to the IPCC Scientific Assessment* (Cambridge, England: Cambridge University Press, 1992), 135–70.
3   UN Environment Programme, *World Atlas of Desertification* (Sevenoaks, U.K.: Edward Arnold, 1992).

There is a better-established, if less direct, link, however, between dryland degradation and global-mean warming through the influence of desertification on the sources and sinks of greenhouse gases. Progressive desertification of drylands in the tropics and elsewhere is likely to reduce a potential carbon sink by reducing the carbon sequestered or stored in these ecosystems. As vegetation dies and soil is disturbed, emissions of carbon dioxide will increase. Desertification may also affect emissions of other greenhouse gases. For example, nitrous oxide emissions might increase because of greater fertilizer use. Methane production may increase in poorly fed cattle. On the other hand, because dry soils are methane sinks, desertification might reduce the gas's atmospheric concentration. There remains a large measure of uncertainty about the relative magnitudes of the various sources and sinks for carbon dioxide and the other greenhouse gases,[10] and such uncertainty adds to the difficulty of quantifying the precise effect of desertification on global-mean warming.

The problem of the "missing" carbon sink illustrates this difficulty: It is impossible to balance the global carbon budget on the basis of current understanding of the major carbon sources and sinks; a certain amount of carbon released into the atmosphere cannot be accounted for. Several possibilities might account for the discrepancy, including a substantial carbon dioxide and/or nitrogen fertilization effect on plant growth; a larger uptake of carbon by the oceans than has previously been thought

likely; greater carbon sequestering as a result of recent reforestation programs in northern midlatitudes; and a larger carbon-storing capacity of annual grasses in tropical and subtropical regions. Desertification is clearly relevant to this last possibility as it would alter the effectiveness of a sink that may prove more important than is now estimated.

Although the importance of the net carbon flux associated with desertification is impossible to quantify at this time, a rough order of magnitude can be estimated. Data from the 1980s indicate that atmospheric carbon dioxide accounts for about 55 percent of all greenhouse-gas forcing.[11] Of the global carbon emissions, the net biospheric contribution (that is, the non-industrial component, largely resulting from land-use changes) is variously estimated at between 10 percent and 30 percent (or between 5 percent and 15 percent of total greenhouse forcing). The bulk of these biospheric emissions results from tropical deforestation; land conversion in dryland areas is a minor contributor.[12] When considering the net contribution of dryland regions to greenhouse-gas sources and sinks, a distinction should be made between the contribution arising from nondegrading changes in land use (almost certainly the primary contribution) and that from desertification (the secondary contribution). (An example of a sustainable land-use change is the conversion of a dryland from shrubs to grassland with no subsequent degradation in soil quality.) Therefore, desertification's contribution of carbon to greenhouse forcing and,

[10] R. T. Watson, H. Rodhe, H. Oeschger, and U. Siegenthaler, "Greenhouse Gases and Aerosols," In J. T. Houghton, G. J. Jenkins, and J. J. Ephraums, eds., *Climate Change: The IPCC Scientific Assessment* (Cambridge, England: Cambridge University Press, 1990), 12–40; and R. T. Watson, L. G. Meira Filho, E. Sanhueza, and A. Janetos, "Greenhouse Gases: Sources and Sinks," in J. T. Houghton, B. A. Callander, and S. K. Varney, eds., *Climate Change 1992: The Supplementary Report to the IPCC Scientific Assessment* (Cambridge, England: Cambridge University Press, 1992), 25–46.

[11] K. P. Shine, R. G. Derwent, D. J. Wuebbles, and J. -J. Morcrette, "Radiative Forcing of Climate," in Houghton, Jenkins, and Ephraums, eds., note 10 above, pages 41–68. The results of more recent research suggest that the carbon dioxide contribution to the total anthropogenic climate forcing may be higher when the effects of emissions of sulfur compounds and ozone depletion are considered. See Watson, Meira Filho, Sanhueza, and Janetos, note 10 above; and T. M. L. Wigley and S. C. B. Raper, "Implications of Revised IPCC Emissions Scenarios," *Nature* 357 (1992):293–300.

[12] A. F. Bouwman, "Land Use Related Sources of Greenhouse Gases," *Land Use Policy*, April 1990, 154–64.

hence, to global-mean warming is almost certainly less than a few percent of the global total. It is not possible at this time to estimate net emissions of other greenhouse gases resulting from dryland degradation.

Despite the relative unimportance of desertification as a direct contributor to global warming, there is a clear need to improve understanding of the desertification-related sources and sinks of greenhouse gases. Greater knowledge is important for both scientific and political reasons. The scientific reason is that improved projections of future atmospheric carbon dioxide concentrations and estimates of future rates of global-mean warming depend on better quantification of where and how fast atmospheric carbon is sequestered. The same is true for all the major greenhouse gases. The political reason is that national inventories of greenhouse-gas sources and sinks will be an important element in future negotiations connected with the UN Framework Convention on Climate Change.

By restoring a terrestrial carbon sink and reducing direct emissions of carbon, a reversal of desertification could measurably contribute to reducing global greenhouse forcing but would not by itself lead to a significant reduction in future global-mean warming. The contribution of dry-land degradation is too small to have a substantial impact. Nevertheless, because most dryland countries have relatively low industrial carbon emissions, desertification could be a major element of such nations' individual net carbon budgets. In such cases, arresting desertification should be considered a priority action. Only a modest stimulation or protection of carbon sinks in such countries would offset a significant proportion of their other emissions of carbon.

Desiccation in the Sahel

No region has been at the center of the debate over the causal links between climate change, desiccation, and desertification more than the African Sahel. Over the past 25 years, the Sahel has undergone severe desiccation and increasing deterioration of the soil quality and vegetative cover (see box 15.2). More than in any other dryland region in the world, it is the simultaneous occurrence of these phenomena in the Sahel that raises questions about the links between climate change and desertification and, in particular, about the cause of the sustained decline in rainfall. Arguments over these questions date back to the mid 1970s, when the meteorology behind the Sahelian crisis of 1972 and 1973 was first discussed.[13] Current ideas about the causes of the desiccation have crystalized around two central themes: internal biogeophysical feedback mechanisms within Africa associated with land-cover changes, such as desertification; and global circulation changes associated with particular patterns of heat distribution in the oceans (ocean temperature departures from the historical mean). With regard to the ultimate cause of the oceanic changes, two possibilities present themselves: They may be a manifestation of quasi-periodic natural fluctuations in ocean circulation, the result of natural climate variability; and/or they may be a response of the ocean system to anthropogenic greenhouse-gas and sulfate aerosol forcing of the climate system. There is some evidence for each of these three causal agents, and each of them has very different implications in terms of both the appropriate remedial actions and the political repercussions for the INC-D negotiations.

Land-cover changes in Africa

The idea that modification of land-cover characteristics in dryland regions might affect regional rainfall was first proposed by

[13] M. H. Glantz, "The Value of a Long-Range Weather Forecast for the Sahel," *Bulletin of the American Meteorological Society* 58 (1977):150–58; P. J. Lamb, "Large-Scale Tropical Atlantic Surface Circulation Patterns Associated with Sub-Saharan Weather Anomalies," *Tellus* 30 (1978):240–51; and S. E. Nicholson, "The Nature of Rainfall Fluctuations in Subtropical West Africa," *Monthly Weather Review* 108 (1980):473–87.

Box 15.2   The Desiccation of the Sahel

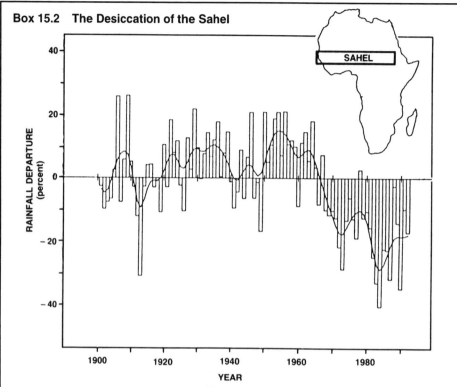

Figure 15.2   Annual rainfall departure index for the Sahel from 1901 to 1992, expressed as a percent departure from the mean annual rainfall between 1951 and 1980.

Note: Up to 60 rainfall stations contributed data to this region series. The smooth curve results from applying a filter that emphasizes variations on a time-scale longer than 10 years.

Within recent years, the climate of the Sahel has exhibited a continuing trend toward desiccation. For about 25 years, rainfall has been substantially lower than it was during the first seven decades of the century (see figure 15.2). This desiccation represents the most substantial and sustained change in rainfall for any region in the world ever recorded by meteorological instruments. Individual years, such as 1984 and 1990, have seen rainfall totals drop to less than 50 percent of those received during the 1930s, 1940s, and 1950s. Contrasting two successive 30-year periods (from 1931 to 1960 and from 1961 to 1990), the rainfall decline over this region has been between 20 and 40 percent. Although this magnitude of desiccation is unprecedented in the instrumental record, it is harder to assess how unusual it is for the longer-term history of the Sahel. Using a combination of lake levels, landscape descriptions, and historical accounts, Sharon Nicholson at Florida State University has shown that recurrent droughts enduring from one to two decades have been a feature of the Sahelian climate over the last few centuries.(1) Quantifying the severity and precise duration of such droughts, however, is impossible. Examining the historical levels of Lake Chad, one of the great inland lakes of Africa, which lies toward the south of the Sahel, may provide some clues. A comparison of the current decline in the level of Lake Chad to its historical variations suggests that the present desiccation in the Sahel is at least as severe as anything experienced during the last millennium.

1   S. E. Nicholson, "Climatic Variations in the Sahel and Other African Regions During the Past Five Centuries," *Journal of Arid Environments* 1 (1978):3–24.

Joseph Otterman, an environmental scientist at Tel Aviv University, in 1974 and arose from his empirical work in the Negev Desert.[14] His initial contention was that bared, high-reflecting soils would increase surface albedo (reflectivity), reduce convective processes, and thus decrease rainfall. Around the same time, Jule Charney, a meteorologist at the Massachusetts Institute of Technology, was developing his bio-geophysical hypothesis that land-cover changes, primarily around the Sahara, could enhance aridity.[15] Charney's proposed mechanism involved a desertification-induced change in the vertical energy flux in the atmosphere over dryland regions. Charney's mechanism was subsequently criticized because of his omission of the role of soil moisture and the absence of any discussion of latent/sensible heat ratios.[16]

Charney's hypothesis received considerable attention because it, or some variant, would provide an apparent explanation for self-reinforcing drought (that is, desiccation) in dryland regions. According to this hypothesis, an initial change in land-cover characteristics occurs in association with desertification. The initial change may involve a change or removal of vegetation and/or a deterioration in soil quality and moisture-holding capacity. The land-cover change is then amplified as land surface-atmosphere interaction suppresses rainfall, either by reducing surface moisture or by increasing atmospheric subsidence. Lower rainfall, in turn, increases moisture stress on vegetation, lowers soil moisture levels, and further reduces rainfall amounts, thereby closing the feedback loop. The significance of this hypothesis for the present discussion is that, if such land-cover changes can account for the rainfall decline in the Sahel or even for a significant proportion of that decline, then it is the complex matrix of processes leading to desertification in recent decades that is responsible for the Sahel's desiccation.

A substantial amount of effort has been directed over the last 15 years to refining Charney's basic hypothesis and to examining the sensitivity of regional rainfall to large-scale changes in land cover through climate modelling experiments. These experiments have been performed for various regions, including the Amazon, the Sahara, and Tropical Africa.[17] Such experiments have also addressed a wide range of physical mechanisms for desertification-induced desiccation by modelling interactions among surface albedo, soil moisture and evaporation, and changes in surface roughness and vegetation.[18] These experiments clearly show that large-scale conversion of land-cover characteristics can generate climate change on local and regional scales.

There appears, however, to be a fundamental difficulty in attributing the recent desiccation in the Sahel to land-cover changes on the basis of these model

[14] J. Otterman, "Baring High-Albedo Soils by Overgrazing: A Hypothesised Desertification Mechanism," *Science* 186 (1974):531–33.

[15] J. G. Charney, "Dynamics of Deserts and Drought in the Sahel," *Quarterly Journal of the Royal Meteorological Society* 101 (1975):193–202.

[16] See, for example, S. B. Idso, "A Note on Some Recently Proposed Mechanisms of Genesis of Deserts," *Quarterly Journal of the Royal Meteorological Society* 103 (1977):369–70.

[17] R. E. Dickinson and A. Henderson-Sellers, "Modelling Tropical Deforestation: A Study of GCM Land-Surface Parameterisations," *Quarterly Journal of the Royal Meteorological Society* 114 (1988):439–62; W. M. Cunnington and P. R. Rowntree, "Simulations of the Saharan Atmosphere: Dependence on Moisture and Albedo," *Quarterly Journal of the Royal Meteorological Society* 112 (1986):971–99; M. F. Mylne and P. R. Rowntree, "Modelling the Effects of Albedo Change Associated with Tropical Deforestation," *Climate Change* 21 (1992):317–43; and Y. C. Sud and M. J. Fenessey, "A Study of the Influence of Surface Albedo on July Circulation in Semi-Arid Regions Using the GLAS GCM," *Journal of Climatology* 2 (1982):105–25.

[18] Y. C. Sud and M. J. Fenessey, "Influence of Evaporation in Semi-Arid Regions on the July Circulation: A Numerical Study," *Journal of Climatology* 4 (1984):393–98; and J. Lean and D. A. Warrilow, "Simulation of the Regional Climatic Impact of Amazon Deforestation," *Nature* 342 (1989):411–13.

experiments. Observational evidence of the marked, large-scale, sustained changes in surface albedo in dryland regions that are introduced into most model experiments remains weak.[19] (Surface albedo is the measure of land-cover characteristics used in these experiments.) The albedo increases caused by desertification that have been observed are on the order of 25 to 50 percent,[20] and yet a doubling of albedo is used in many model experiments. Moreover, the observed changes have been localized in extent and often short-term, rather than widespread and sustained as assumed in the modeling studies. All of the modeling experiments that have displayed substantial regional rainfall reduction as a response to land-cover changes have been "sensitivity" rather than "simulation" experiments; rather than imposing observed perturbations to surface vegetation, soil moisture, and so on, they have imposed arbitrarily determined changes, which in all cases have been much larger than those that have actually been observed. Although these experiments are important to understanding how the various physical systems are linked, it is dangerous to draw the conclusion from their results that observed land-cover changes have accounted for observed rainfall changes in the recent past.

A surer way to proceed is to conduct simulation experiments, which impose known perturbations on the model (for example, the observed soil moisture conditions in a given year) and then to examine whether the model reproduces the observed rainfall anomaly of that year. The most impressive set of such simulation experiments has been completed at the British Meteorological Office. The investigators simultaneously perturbed both ocean temperatures and initial soil moisture con-

ditions in a manner consistent with observations.[21] They concluded that ocean temperature forcing appears to dominate the effects of the land surface moisture feedback. Although this work confirms that land surface feedback can play a part in generating self-sustaining drought, the role of this mechanism is secondary to that of variability within the wider climate system.

In light of current empirical and modeling evidence, then, it appears that desertification is not, in itself, a primary cause of the recent desiccation in the Sahel. The degradation of both soil and vegetative cover in dryland regions could well have contributed to the rainfall decline, but this contribution cannot have accounted for anything more than a small fraction of the observed trend. If there is severe and sustained degradation of a substantial dryland area over the next few decades, however, the significance of this internal feedback mechanism may well increase. Over the next 50 years, though, it is more likely that land-cover changes in the humid and subhumid regions of the tropics will lead to substantial changes in regional climate than will those occurring in dryland areas.

Natural changes in ocean circulation

The British Meteorological Office's experiments confirmed the importance of a set of natural mechanisms of climate change that appear to be responsible for the Sahelian desiccation. These mechanisms involve links between Sahelian drought and sea surface temperature (SST) anomalies in the neighboring Atlantic Ocean and other oceans. Research has shown that there is a significant correlation on the interannual time-scale between higher-than-normal SSTs south of West Africa and reduced Sahelian rainfall.[22] In the early 1980s, it was

[19] M. Hulme, "Is Environmental Degradation Causing Drought in the Sahel?" *Geography* 74 (1989):38–46.
[20] S. I. Rasool, "On Dynamics of Deserts and Climate," in J. T. Houghton, ed., *The Global Climate* (Cambridge, England: Cambridge University Press, 1984), 107–20.
[21] D. P. Rowell et al., "Causes and Predictability of Sahel Rainfall Variability," *Geophysical Research Letters* 19 (1992):905–08.
[22] Lamb, note 13 above; and J. M. Lough "Atlantic Sea Surface Temperatures and Weather in Africa" (Ph.D. diss., University of East Anglia, 1980).

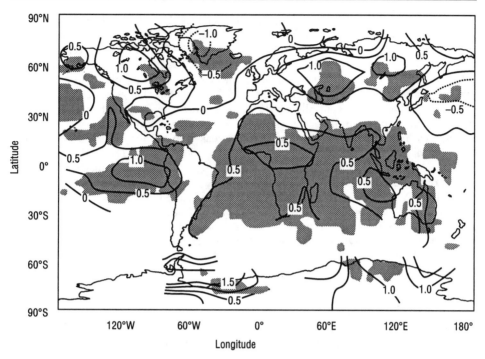

Figure 15.3 [*orig. figure 2*]   Global field of annual surface air temperature departures associated with drought in the Sahel (shaded areas indicate where the areas are statistically significant).

argued that a change in atmospheric circulation was affecting both the SST pattern and rainfall over the Sahel. In other words, the SST pattern was just an indicator of the processes affecting Sahelian rainfall rather than the primary cause. It is now thought, however, that the SST pattern may well be the direct cause of the shift in the atmospheric circulation that subsequently affects Sahelian rainfall.

During the mid 1980s, Chris Folland and his colleagues at the Meteorological Office confirmed this statistical correlation between local variations in ocean temperatures and African rainfall on a time-scale of years to decades and found evidence of a broader relationship between Sahelian rainfall and worldwide ocean temperatures.[23] They demonstrated that differences in SST anomalies between the Northern and Southern Hemispheres, most marked in the Atlantic sector, were related to Sahelian rainfall. Higher temperatures south of the equator and lower temperatures north of the equator (see figure 15.3) were associated with lower rainfall over much of northern tropical Africa. Modeled simulations of the effects of these observed SST anomaly patterns confirmed this association. The success of these model experiments in simulating the observed Sahelian rainfall anomalies suggested that the SST anomaly pattern was the direct cause of the rainfall anomalies. This work has since been extended, confirming the original empirical and model results, and the relationship has provided the basis of an experimental seasonal forecasting scheme.[24]

[23] C. K. Folland, D. E. Parker, and F. E. Kates, "Worldwide Marine Temperature Fluctuations, 1856–1981," *Nature* 310 (1984):670–73; and C. K. Folland, T. N. Palmer, and D. E. Parker, "Sahel Rainfall and Worldwide Sea Temperatures, 1901–1985," *Nature* 320 (1986):602–7.
[24] C. K. Folland, J. A. Owen, M. N. Ward, and A. W. Colman, "Prediction of Seasonal Rainfall in the Sahel Region Using Empirical and Dynamical Methods," *Journal of Forecasting* 10 (1991):21–56.

The physical basis of this relationship appears to lie in a disturbance to the meridional, Hadley circulation of the atmosphere over the Atlantic/Africa sector that is induced by the pattern of contrasting hemispheric ocean temperature anomalies. The Hadley circulation exerts a controlling influence on African rainfall patterns. It determines, in part, the position of the Intertropical Convergence Zone – specifically, the extent of its annual north-south migration, which, in turn, affects the strength of the southwesterly airflow originating in the tropical Atlantic that brings the Sahel much of its rain (see box 15.3).[25] Thus, the link between lower rainfall in the Sahel and a particular pattern of SST anomalies in the oceans has been well established. The initial cause of this SST pattern, however, has yet to be determined. The spatial scale of the phenomenon may provide some evidence of the cause. Although temperatures in the Atlantic Ocean are probably the dominant influence on Sahelian rainfall, the recently observed pattern of SST anomalies in the Atlantic sector is part of a much larger trend in surface air temperatures. Large-scale warming has affected both hemispheres since the late 19th century and has resulted in a net global-mean warming of about 0.5°C (see the top graph in figure 15.4). There has been a clear difference, however, in the warming rates of the two hemispheres during recent decades, with the Northern Hemisphere warming more slowly than the Southern Hemisphere (see the second graph in figure 15.4).

This relationship between conditions in the Atlantic/Africa sector and worldwide climatic trends is not confined to temperature; a link also exists between the desiccation of the Sahel in recent decades and global rainfall fluctuations. Although the rainfall deficit in the Sahel is the most striking rainfall change of recent decades, rainfall has been lower in many parts of the northern tropics and sub-tropics but has increased at higher latitudes (see figure 15.5).[26] The Sahel's desiccation could be considered a regional manifestation of the global shift in the climate system that has occurred since the 1950s.

The outstanding question in this line of reasoning concerns the initial cause of the temperature change that gives rise to the rainfall disturbance. The observed ocean temperature pattern may well be a manifestation of natural climate variability. For example, it could be the result of a reduction in the northward transport of heat in the Atlantic Ocean.[27] More recently, Alayne Street-Perrott and Alan Perrott at the University of Oxford have hypothesized that this reduction in heat transport may be the result of a freshening (reduction of salinity) of the surface waters of the northern North Atlantic.[28] By stabilizing the water column, the freshening reduces deep convection in the North Atlantic and the compensatory surface inflow of water and, hence, heat from the south. Although natural climatic variability is undoubtedly a possible cause of the observed abnormal pattern in ocean temperature, there is equally convincing, albeit equally circumstantial, evidence to suggest that the abnormal pattern may be linked to global-mean warming.

Is there a greenhouse-related mechanism

---

[25] For a more detailed discussion of the proposed mechanism, see C. K. Folland and J. A. Owen, "GCM Simulation and Prediction of Sahel Rainfall Using Global and Regional Sea Surface Temperatures," in *Modelling the Sensitivity and Variations of the Ocean-Atmosphere System*, WMO/TD no. 254 (Geneva: World Meteorological Organization, 1988), 107–15.

[26] R. S. Bradley et al., "Precipitation Fluctuations over Northern Hemisphere Land Areas Since the Mid-19th Century," *Science* 237 (1987):171–75.

[27] R. E. Newell and J. Hsiung, "Factors Controlling Free Air and Ocean Temperature of the Last 30 Years and Extrapolation to the Past," in W. H. Berger and L. D. Labeyrie, eds., *Abrupt Climatic Change: Evidence and Implications* (Dordrecht, the Netherlands: Reidel, 1987), 67–87.

[28] F. A. Street-Perrott and R. A. Perrott, "Abrupt Climate Fluctuations in the Tropics: The Influence of Atlantic Ocean Circulation," *Nature* 343 (1990):607–12.

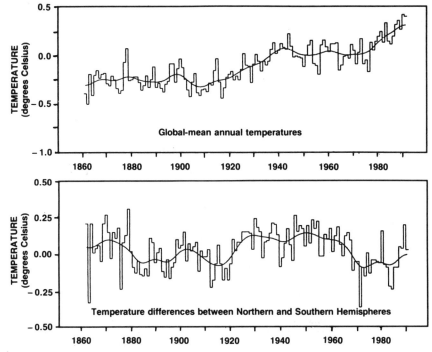

Figure 15.4 [*orig. figure 3*]   Temperature fluctuations from 1861 to 1991.
Note: The zero points on the vertical scales represent the mean global temperature from 1951 to 1980. In the lower graph, positive values indicate that the Northern Hemisphere was warmer than the Southern Hemisphere, and negative values indicate the reverse.
Source: C. K. Folland et al., "Observed Climate Variability and Change," in J. T. Houghton, B. A. Callendar, and S. K. Varney, eds., *Climate Change 1992: The Supplementary Report to the IPCC Scientific Assessment* (Cambridge, England: Cambridge University Press, 1992), 135–70.

Box 15.3   Why Is The Sahel Dry?

The Sahel of Africa possesses a monsoonal climate – that is, the climate of the region exhibits a very strong seasonality with a nearly 180° reversal of the prevailing surface wind direction between the wet and dry halves of the year. The winter, or dry, monsoon lasts from October through April and is characterized by northerly or northeasterly surface winds circulating clockwise around the Saharan anticyclone. These winds are extremely dry and lead to no rainfall during these months. The summer, or wet, monsoon commences sometime between April and June and arrives progressively from the south. The moisture in these rain-bearing southerly or southwesterly surface winds originates mostly over the Atlantic Ocean and, to a lesser extent, the Indian Ocean. The sea surface temperatures of these oceans, therefore, exert important control over the rainfall in the Sahel by altering both the moisture characteristics and the vigor of the wet monsoon flow into northern tropical Africa. The wet monsoon varies in duration from five or six months in the southern Sahel (about 10°N) to only a month or two in the far north (about 16°N). This variation in duration creates a very tight gradient in annual rainfall – from less than 100 millimeters north of 16°N to more than 800 millimeters south of about 10°N. Because of this tight gradient, relatively small inter-annual variations either in the northward penetration or moisture load of the wet monsoon or in the strength of the atmospheric disturbances that lead to the rain outfall result in relatively large variations in the total volume of rainfall received by a locality. Consequently, the rainfall of the Sahel is highly variable from year to year.

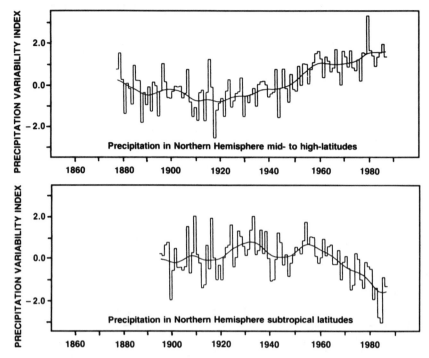

Figure 15.5 [*orig. figure 4*]   Precipitation fluctuations in the Northern Hemisphere.
Note: The zero point on the vertical scale of the upper graph represents the mean precipitation from 1877 to 1986. The zero point on the lower graph represents the mean precipitation from 1895 to 1986. The scale is an index of large-scale variations in precipitation, which takes into account the marked changes in mean rainfall that occur from place to place and from month to month.
Source: R. S. Bradley et al., "Precipitation Fluctuations over Northern Hemisphere Land Areas Since the Mid-19th Century," *Science* 237 (1987): 171–75.

that could account for the interhemispheric temperature contrast associated with Sahelian desiccation? Recent research has indicated that there are two alternative – or, more likely, complementary – mechanisms that could induce a temperature difference between the hemispheres.

First, the emission of sulfur compounds as a result of human activity (specifically, fossil fuel combustion) increases the amount of sulfate aerosols in the atmosphere. These aerosols reflect solar radiation, both directly and by altering cloud albedo. Any increase in their concentration in the atmosphere is, therefore, likely to have a cooling effect, offsetting greenhouse warming. Estimates

of the scale of this effect vary, but it is considered possible that sulfur dioxide emissions may have reduced the level of warming that rising greenhouse-gas concentrations might have effected by a significant amount.[29] As most sulfur emissions come from the Northern Hemisphere and the sulfate aerosols have a short residence time in the atmosphere, this cooling effect would be largely confined to the Northern Hemisphere. As a result, greenhouse-gas-induced warming would be offset in the Northern Hemisphere but relatively unaffected in the Southern Hemisphere, thereby inducing a differential warming rate and a temperature contrast between the two hemi-

[29] I. Isaksen, V. Ramaswamy, H. Rodhe, and T. M. L. Wigley, "Radiative Forcing of Climate," in Houghton, Callander, and Varney, eds., note 10 above, pages 47–68; and Wigley and Raper, note 11 above.

spheres. Estimates of the scale of this effect are not inconsistent with the observed differential between the warming trends in the two hemispheres.[30]

A second greenhouse-related factor exists, particularly in the Atlantic Ocean, that may have caused different rates of warming in the hemispheres. Recent time-dependent ("transient") experiments of enhanced greenhouse-gas forcing using general circulation models (GCMs) that incorporate a dynamic ocean model have suggested that the rate of warming may be retarded in the northern North Atlantic sector and over the Southern Ocean around Antarctica.[31] In these areas, the sinking of dense, saline water masses results in a localized increase in the effective heat capacity and, therefore, in the thermal inertia of the ocean. The increase in thermal inertia slows the warming of the overlying air as heat is drawn down into the ocean. This process induces a meridional gradient in temperature in the Atlantic Ocean north of 60°S, which is not unlike the temperature pattern associated with lower rainfall in the Sahel (see figure 15.3). This mechanism may amplify, or even be triggered by, changes in the temperature field and atmospheric circulation that are induced by the effects of sulfate aerosols.

Thus, a physically plausible argument can be advanced linking the recent desiccation in the Sahel to global-mean warming. However, just as it is impossible to ascribe with any certainty the observed global-mean warming to enhancement of the greenhouse effect,[32] neither can the interhemispheric temperature contrast be attributed with

confidence to greenhouse-gas-plus-sulfate forcing or to any other greenhouse-related mechanism. For now, the evidence must be considered circumstantial.

The future of the Sahel

GCM experiments have recently been conducted to determine whether significant rainfall changes over northern tropical Africa may result from future greenhouse-gas forcing. (Unfortunately, none of these experiments incorporates the sulfate aerosol effect, and only one allows the ocean circulation to respond realistically to greenhouse-gas forcing.) The experiments indicate that global-mean warming should lead to an overall increase in global-mean precipitation because evaporation over the warmer oceans increases the moisture content of the atmosphere.[33] The distribution of rain and snowfall, however, will also be determined by changes in the atmospheric circulation and by other climatic factors.

Figure 15.6 shows a composite estimate of the percentage change in mean annual rainfall that may accompany each 1°C rise in global-mean temperature induced by greenhouse-gas forcing. The composite is based on a set of seven GCM experiments.[34] An increase in rainfall is apparent in most areas, but annual rainfall decreases over the Mediterranean, North Africa, and a large part of the Sahel, especially the Western Sahel. The effect is most marked over the southwestern margins of the Sahara, in Mauritania and northern Mali and Niger. In the latter two areas, annual rainfall decreases by more than 6 percent for every 1°C of global-mean warming. If global-mean

---

[30] See, for example, Wigley and Raper, note 11 above.

[31] W. L. Gates, P. R. Rowntree, and Q-C. Zeng, "Validation of Climate Models," in Houghton, Jenkins, and Ephraums, eds., note 10 above, pages 93–130; and W. L. Gates et al., "Climate Modelling, Climate Prediction and Model Validation," in Houghton, Callander, and Varney, eds., note 10 above, pages 97–134.

[32] T. M. L. Wigley and T. P. Barnett, "Detection of the Greenhouse Effect," in Houghton, Jenkins, and Ephraums, eds., note 10 above, pages 243–55.

[33] J. F. B. Mitchell, S. Manabe, V. Meleshko, and T. Tokioka, "Equilibrium Climate Change and Its Implications for the Future," in Houghton, Jenkins, and Ephraums, eds., note 10 above, pages 137–64.

[34] Climatic Research Unit, *A Scientific Description of the ESCAPE Model* (Norwich, England: Climatic Research Unit, 1992); and P. M. Kelly and M. Hulme, *Climate Scenarios for the SADCC Region* (Norwich, England: Climatic Research Unit/London and International Institute for Environment and Development, 1992).

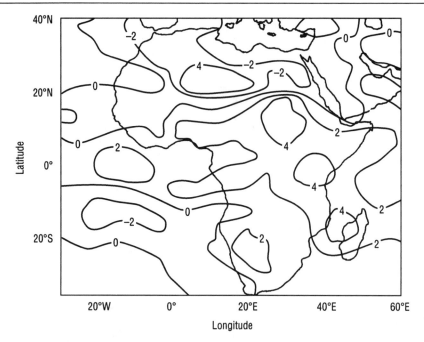

Figure 15.6 [*orig. figure 5*] Percentage change in mean annual rainfall over Africa per degree Celsius of global-mean warming resulting from increased greenhouse-gas forcing.

Note: The percentage change in mean annual rainfall is standardized by global-mean warming. The increase in greenhouse-gas forcing is as predicted by an ensemble of seven global climate modeling experiments.

Source: P. M. Kelly and M. Hulme, *Climate Scenarios for the SADCC Region* (Norwich, England: Climatic Research Unit/London and International Institute for Environment and Development, 1992).

warming follows the 1992 projections of the Intergovernmental Panel on Climate Change,[35] rainfall in these areas decreases by between 6 and 30 percent by 2100. This is, however, a slow rate of decline, as long as it occurs progressively, in comparison with the 20 to 40 percent decline in rainfall experienced by the Sahel during recent decades.

The current performance of climate models in estimating regional rainfall patterns is considered quite weak. Model-to-model differences in predicted rainfall changes over northern Africa are large, and individual models predict a complex spatial pattern of change. Nevertheless, the model composite does show a well-defined reduction in rainfall over much of the Sahel. Compounded by increased temperatures (which can be predicted with greater confidence), lower rainfall would inevitably

cause substantial reductions in soil moisture availability.

Negotiations for the Convention to Combat Desertification will be complicated by the technical and scientific uncertainties underlying many aspects of the desertification issue. It is to be hoped that the negotiators will be assisted by the kind of technical support that was provided by the Intergovernmental Panel on Climate Change during the negotiations for the Framework Convention on Climate Change. Although varying degrees of uncertainty surround the links between climate change, desiccation in dryland regions, and desertification, this brief assessment has shown that there are intrinsic links that should not be ignored (see figure 15.7). Even the rather speculative link between global-mean warming and desiccation in the Sahel warrants serious consideration on a pre-

[35] Houghton, Callander, and Varney, eds., note 10 above; and Wigley and Raper, note 11 above.

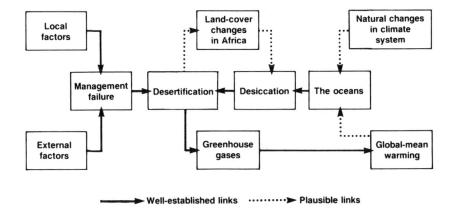

Figure 15.7 [*orig. figure 6*]   The matrix of cause and effect surrounding desertification and the role of climate change.
Note: Desertification is the result of resource management failure, the product of both local factors, such as population pressure and inequity, and external factors, such as the state of the global economy, commodity prices, and the burden of debt. Desertification is aggravated by climate change – desiccation – which may be the result of natural mechanisms with the climate system, such as ocean-atmosphere feedback; by desertification itself through, for example, surface-atmosphere interaction; or possibly by global-mean warming. Finally, desertification contributes to global-mean warming through its effect on the sources and sinks of greenhouse gases, such as carbon dioxide. Many uncertainties affect assessment of the relative role of these various factors.

cautionary basis because of the serious implications for those living in this dryland area.

The area prone to desertification is, by definition, determined by climatic conditions and, hence, by climate change. It is more difficult to determine the precise balance between the human and climatic factors that lead to desertification. There is, however, no doubt that, against a background of resource management failure in dryland regions, climate change will aggravate the problem. It is also clear that, by modifying surface characteristics, desertification can induce significant changes in local temperature. But it is far less likely that the global-mean temperature has been affected to any significant extent by dryland degradation.

Progressive desertification of the dryland tropics and other areas is likely to reduce a potential carbon sink by reducing the carbon stored in these ecosystems. Moreover, as vegetation dies and soil is disturbed, desertification increases emissions of carbon dioxide. Desertification may also increase or decrease emissions of other green-

house gases. Although desertification contributes only a small percentage of all greenhouse-gas emissions, understanding of desertification-related sources and sinks of greenhouse gases should still be improved and desertification rates reduced to enable dryland countries to offset growth in their emissions of other greenhouse gases.

The hypothesized link between the recent desiccation of the Sahel and an interhemispheric contrast in ocean temperature is supported by empirical studies, model simulations, and theoretical argument. The initial cause of the interhemispheric temperature contrast has yet to be determined. It may be the result of natural climate variability or a manifestation of global-mean warming resulting from greenhouse-gas and other anthropogenic emissions. Both possibilities are physically plausible, and neither possibility is contradicted by the available data.

Finally, on the basis of current evidence, it appears likely that the role of desertification in causing (via climate change and land-cover changes) the recent desiccation in the Sahel is very much secondary to that of forcing by ocean temperature anomalies. The rainfall decline estimated to have resulted from past degradation of both soil and vegetative cover in Africa is only sufficient to account for a small fraction of the observed desiccation in the Sahel.

In the future, however, the relative importance of these various factors may change. For example, a sustained degradation of a substantial dryland area may well increase the significance of desertification as a causative agent. On the other hand, climate model experiments suggest that rainfall over the Sahel may decrease even more as global-mean warming develops. In this case, the role of greenhouse-gas emissions in reducing rainfall in the Sahel would become more prominent. Whatever the complexity of these mechanisms, their investigation cannot be considered to be solely of academic interest; as the inhabitants of the Sahel well know, the trend of reduced rainfall continues.

Acknowledgements

The authors wish to thank Sarah Granich for her contribution to this article and Tim O'Riordan for his helpful comments. The article is based on a technical report prepared for the Overseas Development Administration in London, but it does not necessarily reflect the administration's views.

16

RUSH FOR ROCK IN THE HIGHLANDS*

Fred Pearce

When Kurt Larson sails home from work each night, he passes swans, seals and dolphins in the waters of Loch Linnhe in western Scotland. That, he claims, is a measure of how clean his workplace is.

Not bad, since he runs Britain's largest quarry, the first "superquarry". The 250-hectare estate at Glensanda near Fort William, formerly a home just to deer, now produces 5 million tonnes a year of granite, blasted in a series of giant steps from the hills above the loch. "Eventually we will produce 15 million tonnes a year, and we have enough rock to keep us going for a hundred years or more," says Larson.

Glensanda opened in the late 1980s and looks like being the first of a series of mountain-moving operations on Scotland's coast. Any week now, the Scottish Office will rule on a controversial proposal, already approved by the Western Isles Council, for a superquarry on the Hebridean island of Harris that will gouge the heart out of Roineabhal mountain, leaving behind a new loch. And elsewhere, a committee of the Highland Regional Council has singled out a site near Durness on the wild northern coast of Scotland. Consultants Ove Arup

will submit a feasibility study to the council in April.

Scotland's lochs are surrounded by massive reserves of hard rocks that are suitable for crushing into aggregate to feed the incessant demands of road builders and cement makers throughout the British Isles and on the continent. But proponents of superquarries argue that developers must act now because large coastal sites are also being developed in Spain, Ireland and Norway, and the market will only ever sustain a small number of European superquarries.

Hard rocks are increasingly being traded internationally, much like other more valuable minerals. The trigger for this trade is the increasing shortage of land in densely populated areas of Europe, such as southeastern England, and parts of northern Germany and the Netherlands, which is squeezing out local sand and gravel quarrying for local construction needs. This trend has combined with the growing technical and economic feasibility of replacing sand and gravel with crushed rock to create a fast growing market for imported rock delivered by sea.

* Originally published in *New Scientist*, 1994, pp. 11–12.

Crushing Demand

A wide variety of rocks fit the bill for crushing into aggregates, including most igneous rocks such as granite, metamorphic rocks such as quartzites and gneisses, plus compacted limestones and sandstones.

In Britain, demand for aggregates has trebled in 30 years and the government believes it will rise by a further two-thirds by 2010. Early in 1993, the Department of the Environment concluded that England could no longer meet its own demand for aggregates, either from sand and gravel or the limestone hills of the Peak District and Mendips. Coastal superquarries in Scotland or abroad would have to make up the difference, said the department.

That decision followed a report to the government by Ove Arup in 1991 which concluded that "between 15 and 20 superquarry development opportunities exist in Norway, Scotland and Spain". Within weeks of the report's publication, Redland, one of Britain's largest aggregate firms, applied to the Western Isles Council for permission to start quarrying at Roineabhal mountain, near Rodel.

The Rodel development looks likely to cost around £50 million. And any company investing that sort of money in a coastal superquarry will have its eyes on markets far beyond southeast England. Most of Glensanda's output, for example, never touches British soil once it leaves the shore of Loch Linnhe. "Our first customers were in Texas," says Larson. "There is no hard rock to be found along the US eastern coast, so we supplied rocks to Houston for making road bases, asphalt and railroad ballast," he says. "Then came the Channel Tunnel. Now two-thirds of our output goes to northern Europe, mostly Germany."

Germ of an Idea

The idea of Scottish superquarries was conceived by Colin Gribble, a geologist at the University of Glasgow, and Ian Wilson, a Scottish entrepreneur specialising in minerals. It arose from the government scheme, dreamt up in the late 1960s, to build

London's third airport at Maplin Sands on the Essex coast. Wilson proposed building the runways and massive sea walls out of rock from a quarry found by Gribble at Peterhead on Scotland's east coast.

The Maplin project sank without trace. But Wilson thought Scottish hard rock could supply other needs down south. In 1976, he persuaded Ralph Verney, then an adviser to the environment secretary, to recommend to the government that a large scale study be carried out into the creation of "mammoth coastal quarries" in Scotland.

At the time, most mining companies opposed the idea, saying it was too expensive. The government got cold feet, but Wilson persuaded the Scottish Office to commission him and Gribble to write a report on the prospects for coastal quarries, which was published in 1980.

While Gribble has never had an interest in the commercial side of the quarries, Wilson has an eye for the main chance. Before writing the report, he travelled the Scottish coast, buying the mineral rights to likely reserves of hard rocks beside deep-water berths. "Wherever I found a good site, I negotiated long leases for the minerals," says Wilson.

His final report named 16 potential sites, mostly on the Scottish western coast and islands, for many of which he owned the mineral rights – a fact he declared in the report. Today, those sites, and two others, whose rights Wilson acquired later, form the shortlist for future superquarries.

One of his first purchases was for the rocks beneath land near Rodel. Later, Wilson sold his idea to Redland for a quarry on the site to produce some 600 million tonnes of rock. For each tonne removed, Wilson will receive a royalty.

In the 1980 report, Wilson and Gribble named Rodel as the best site in Scotland for a superquarry. Its reserve of anorthosite, a coarse-grained igneous rock, was "unique" in meeting the highest criteria for most European uses for hard rocks, they said. Anorthosite is extremely hard, strong and resistant to weathering. Equally important, Rodel can accommodate a harbour 24 metres deep alongside the quarry. The economics of superquarries are crucially dependent on keeping down transport costs through direct loading onto large ships.

The sheer size of superquarries and the cost of developing them means that few will ever be developed. "I don't think that more than five will be used in Scotland," says Gribble. "There is Glensanda and probably Harris, one looks likely in the inner Hebrides, another is probable in the Shetlands, where there are good deep berths such as Sullom Voe, and another is possible on the coastal mainland."

At present, the Highland Regional Council wants to develop at Durness and beside Loch Eriboll, two connected sites on the north coast for which Wilson owns the mineral rights. Durness contains the most northerly outcrop of limestone in Britain, and the Loch Eriboll site contains gneiss, feldspar and quartz. If this site is not given the go-ahead, Wilson could also offer the council Kentallen, close to Glensanda, or Loch Glencoul, south of Durness.

The superquarry at Durness and Eriboll emerged last summer from a short list considered by a council working party. The decision to focus on one site was controversial within that group, and others also see it as curious. One of the council's main criteria was local support. The group's minutes record that there had been "a consultation with the Community Council who were in favour of the proposal". But the leader of a local opposition group, Iris Wallace, refutes this. "The Community Council has never voted on the proposal and we know very little about it," she says.

Gribble, doesn't believe the Durness/Eriboll superquarry will ever happen. "I spent some time looking at the site, and I'd be very surprised if the Arup study came out in favour," he says. "Geologically it is highly complex, which would create problems, and there are several sites of special scientific interest (SSSIs) close by."

The limestone area contains a large SSSI and faces onto lochs that have been put forward for international protection under the Ramsar Convention. The quarrying would also take place above both the Smoo Caves, a tourist attraction and another SSSI

containing rare limestone plants, and would be near the town's water source, which could be destroyed.

Green Objections

Environmentalists are horrified at the prospect of superquarries at such pristine sites. They argue that ballast water taken from seas elsewhere in the world and discharge off Scotland could introduce foreign organisms which could destroy native species. And they also believe superquarries would be unnecessary if the construction industry recycled more demolition rubble (Forum, 23 October 1993). The environmental argument also has a nationalist tinge. Greens view the development of superquarries as another case of the English exploiting Scotland's natural resources. "Oil, gas, plantation softwood, surplus electricity are all sent south. Now it is to be Scottish mountains," says Kevin Dunion, director of Friends of the Earth Scotland.

But Wilson argues that superquarries will do far less damage to the environment than working equivalent amounts of sand, gravels or limestone elsewhere. A typical Scottish superquarry could contain a cubic kilometre of mountain that could yield 2.6 billion tonnes of aggregate. "Six superquarries on the Scottish coast are preferable to 45 large inland quarries in the Mendips, Leicestershire or the Peak District, or even 450 sand and gravel pits in southeast England," Wilson says.

Moreover, he argues that superquarries could be the catalyst needed for economic regeneration in depopulated areas of Scotland. This raises the question of whether superquarries should be cut off from, or integrated with, local communities.

Larson's employer, Foster Yeoman, which operates the Glensanda quarry, developed it as a remote "island site". It is miles from the nearest house and can only be reached by water. Larson says "the fact that we have no neighbours is a great advantage". But Wilson says that "this gave maximum benefit to the company, but not to the community". He believes superquarries should be near communities, providing both jobs and the prospect that rock ships could offer a cheap "piggyback" for exports to Europe of products from fish farms and other enterprises.

Scotland, says Wilson, "has a world-class natural resource in its coastal rocks". But unless it exploits the resource soon, it could lose out in the growing international trade. Beyond Europe, superquarries are being developed in Mexico and Canada, he says, "some of which have stated they have designs on the British market".

17

RESTORING VALUE TO THE WORLD'S DEGRADED LANDS*

Gretchen C. Daily

Rehabilitation of the world's degraded lands is important for several reasons. First, increasing crop yields is crucial to meeting the needs of the growing human population (1) for food, feed, biomass energy, fiber, and timber (in the absence of a massive increase in the equity of global resource distribution) (2). Second, anthropogenic changes in land productivity have deleterious impacts on major biogeochemical cycles that regulate greenhouse gas fluxes and determine Earth's total energy balance (3). Third, biodiversity preservation depends, in part, on increasing yields on human-dominated land to alleviate pressure to convert remaining natural habitat (4). And fourth, land is frequently a limiting factor of economic output, and its degradation threatens to undermine economic development in poor nations (5, 6) and social stability globally (7).

Here I estimate the rate at which potential direct instrumental value (PDIV) could be restored to degraded lands from a biophysical (as opposed to socioeconomic) perspective. PDIV is the capacity of land to supply humanity with direct benefits only, such as agricultural, forestry, industrial, and medicinal products. It does not incorporate indirect values [for example, ecosystem services (8)], option values, or nonuse values

(9) and is thus a conservative measure of value. PDIV is not the same as potential net primary production (NPP), and may even vary inversely with it; for example, average NPP in agricultural systems is typically lower (and DIV higher) than in the natural systems they replace (10). Because PDIV depends on complex and variable factors such as human knowledge and preferences, it is impossible to quantify precisely.

Below I make rough approximations of changes in PDIV on the basis of global surveys of human-induced land degradation. Case histories of recovery from natural or human-induced disturbances are reviewed in order to derive estimates of the time required to restore PDIV to presently degraded lands. Finally, potentially illuminating projections are offered of future changes in PDIV.

Global Extent and Severity of Land Degradation

Land degradation refers to a reduction in the capacity of land to supply benefits to humanity. It results from an intricate nexus of social, economic, cultural, political, and biophysical forces operating across a broad spectrum of time and spatial scales (11).

* Originally published in *Science*, 1995, vol. 269, pp. 350–4.

Here I consider only the biophysical agents of degradation that trace directly to human land use since 1945, although other proximate biophysical agents, such as air pollution (12), stratospheric ozone depletion (13), and climate changes (14), are also important.

The geographic distribution of degraded land is poorly documented; even less well documented is the severity of degradation, which is typically judged qualitatively (15). The onset of degradation is often masked by intensification of land use that compensates, in the short run, for declines in the natural underpinnings of productivity; however, intensification usually exacerbates degradation, as do natural positive feedbacks (such as the concentration of soil resources by shrubs) (16). Global assessments have been undertaken of degradation of soils (in all biomes), drylands, and tropical forest lands.

Soil degradation. The extent of soil degradation induced by human activity since 1945 was evaluated as ~2 billion ha, or 17% of Earth's vegetated land, in a recent study sponsored by the United Nations Environment Program (UNEP) (17). Of this, ~750 million ha (38%) are classified as lightly degraded (defined as exhibiting a small decline in agricultural productivity and retaining full potential for recovery); ~910 million ha (46%) are moderately degraded (exhibiting a great reduction in agricultural productivity; amenable to restoration only through considerable financial and technical investment); ~300 million ha (15%) are severely degraded (offering no agricultural utility under local management systems; reclaimable only with major international assistance); and ~9 million ha (0.5%) are extremely degraded (incapable of supporting agriculture and unreclaimable).

The percent of area affected seems regionally to be independent of ecological zone or economic status; for example, it is 20%, 22%, and 23% in Asia, Africa, and Europe, respectively. The direct causes of these forms of degradation (and estimates of the relative importance of each) are overgrazing (35%), deforestation (30%), other agricultural activities (28%), overexploitation for fuel wood (7%), and bioindustrial activities (1%).

Global rates of change in soil degradation are unknown. the UNEP study constitutes the first standardized global assessment and is the baseline for planned future monitoring on a decadal basis.

Drylands degradation. UNEP has also carried out a series of generally accepted global assessments of desertification (15). In the most recent assessment, desertification refers to land degradation in arid, semiarid, and dry subhumid areas (hereafter called drylands) resulting mainly from adverse human impact (18). Desertification is distinct from natural oscillations of vegetation productivity that occur at desert fringes (19); hyperarid deserts are not considered to be at risk of desertification and are excluded from assessments thereof (18).

The total desertified drylands area amounts to ~3.6 billion ha, or 70% of global drylands area (excluding hyperarid regions). Roughly 2.6 billion ha thereof exhibit no soil degradation, but have reduced crop yields, livestock forage, and woody biomass for fuel and building material (20). Of rangelands, which make up 88% of the drylands area, ~1.223 billion ha (27%) are degraded slightly or not at all; ~1.267 billion ha (28%) are moderately degraded; ~1.984 billion ha (44%) are severely degraded; and ~72 million ha (1.6%) are very severely degraded. Degradation classes are roughly comparable with those defined in the soil survey, and the principal direct causes of degradation are the same (18, 21). That rate of abandonment of drylands due to degradation is probably ~9 to 11 million ha year^{-1} (22). Rates of degradation seem to be accelerating, particularly in developing nations (23).

Tropical moist forest degradation. Land degradation in tropical moist forest afflicts ~427 million ha (24). The present global annual rate of tropical forest clearing (defined as depletion of forest cover to less than 10% in all types of tropical forest) is ~15.4 million ha year^{-1} (25) and is projected to accelerate (26, 27). In addition, an area of roughly equal size is disrupted, but not cleared outright, through selective logging and shifting cultivation (26). The extent to which clearing and disruption precipitate land degradation is unknown. Rates of abandonment of recently cleared areas, especially in hilly or mountainous regions, of as high as 75 to 100% are indicative of one extreme (26, 28). In general, probably only ~50% of the tropical forest land cleared each year expands the area yielding agricultural benefits, whereas the other half replaces abandoned lands (29).

Total degraded area. As a crude but conservative estimate of the total degraded area, I use the sum of (i) areas affected by soil degradation, (ii) drylands with vegetation degradation but no soil degradation, and (iii) degraded tropical moist forest lands, that is, ~5.0 billion ha (30). This amounts to ~43% of Earth's vegetated surface.

Time Required to Restore PDIV

There have been few attempts to rehabilitate degraded land on a large scale. Possible rates of recovery can be inferred from studies of succession on land that has experienced volcanic eruption, shifting cultivation, continuous agricultural production followed by abandonment, or reclamation. The time required to restore PDIV varies tremendously with ecosystem type, history and spatial pattern of land use, the degree of alteration of climatic factors, and the types of benefits ultimately desired – those derived from crop cultivation as compared to those derived from extractive exploitation of natural vegetation, for example.

Volcanic eruption. Following volcanic eruption, the regeneration of lost top- and subsoil may be the limiting process with respect to time and difficulty of rehabilitation. At one extreme, the rate of topsoil formation is especially rapid on volcanic ash; a mere 100 years after the 1883 eruption of Krakatau, for example, soil 25 cm deep had formed on a daughter island, Rakata (31). More typical soil formation rates are ~1 cm per 100 to 400 years, however. At such rates it takes ~3000 to 12,000 years to develop sufficient soil to form productive land (32).

Rates of colonization and succession are comparatively swift in the absence of natural impediments (33). Rakata serves as a

model for the recovery of a presumably sterilized site 40 km from species source pools (34). Many generalist groups with high dispersal capabilities became reestablished during the first 50 years after eruption. However, important taxa with lower dispersal capabilities, more specialized resource requirements, or higher trophic positions remain poorly represented even today (35). Similarly, 23 vascular plant species were present on Surtsey two decades after its birth (of 450 on the Iceland mainland 35 km away), but only a few had become widely established (36).

Shifting cultivation. Shifting (swidden) cultivation generally involves slashing and burning of forest patches to create temporary fields that are harvested in a rotation between brief periods of cultivation and longer periods of fallowing. Cultivation typically lasts 1 to 3 years, during which a combination of declining soil fertility, competition from weeds, and pest or pathogen outbreak conspires to diminish yields sharply (37). The plot is then left fallow. Long-term studies of recovery of productive potential in swidden systems are few (38), but fallow periods required to make a system sustainable are ~20 years (ranging between 5 and 40 years) in the humid tropics and may be considerably longer elsewhere (39).

Abandoned cropland and pasture. Rates and paths of natural succession vary widely on abandoned land formerly under continuous agricultural production. The chief commonality is the nonlinear relation between the intensity and duration of land use and the time required for recovery after abandonment. Factors influencing succession on old-fields (land abandoned after some combination of cropping and pasturing) are extremely complex, but the severity of erosion, initial floristic composition, and character of the ex situ seed source are paramount (40). In some areas, initial reestablishment of climax species was observed after 40 years of abandonment; in contrast, highly eroded fields experienced little succession during that period (41).

The conversion to pasture of up to ~43 million ha of Amazon rain forest over the past three decades (42) caused rapid declines in productivity and land abandonment after only 4 to 8 years of use (43). Extrapolation of rates of biomass accumulation and succession over 8 years since abandonment suggests that sites with a history of light use (20% of now-abandoned pasture) could reach forest stature in 100 years, those of moderate use (~70%) in 200 years, and those of heavy use (less than 10%) in 500 years or more (44). These estimates assume no further human impact. In many situations worldwide, recovery of productivity on abandoned land is prevented by burning (45) or episodic human exploitation of regrowth as it occurs.

Even without continued human disruption, however, regrowth of forest may not occur at all (as in the case of fire-climax grasslands) (46, 47). For example, an agricultural area of ~3.5 million hectares in eastern Amazonia that was abandoned in the early part of this century had little vegetation aside from scrub and brush 50 years later (48). In India, trees have failed to establish in abandoned, desertified areas adjacent to sacred forest groves despite ample seed sources (49).

Reclamation. Experience in reclamation of degraded areas, although limited, indicates unequivocally that human intervention may be effective (even essential) in ensuring a path and rate of succession that would achieve substantive improvements at time scales relevant to society (50). The potential for accelerating recovery is difficult to assess, as most degraded areas with known

histories have not yet recovered. Moreover, recovery is nonlinear (with respect to time), and intervention can only accelerate some phases of the process.

Where land is suited to direct human use and has not been stripped of topsoil, substantial recovery may be achieved in as few as 3 to 5 years with intensive management (51) but more typically may take 20 years (52). However, recovery of self-sustaining, mature ecosystems in areas unsuited for intensive agriculture may take 100 years or more.

Projections of Future Land Productivity

Despite great uncertainties, I venture crude estimates of the present global loss of PDIV and possible future changes therein. The light, moderate, severe, and extreme degradation classes are assumed to correspond to a residual PDIV of 90%, 75%, 50%, and 0%, respectively (table 17.1, column 1). These values are conservative in that severely (as well as extremely) degraded land is generally abandoned (17).

It is further assumed that the distribution among classes of the ~5 billion hectares of degraded land is proportional to that of degraded land in the global soil survey (summarized above), for which the data appear most reliable (table 17.1, column 3) (53). On the basis of the foregoing evaluation of natural and human-accelerated recovery rates, rough rehabilitation times are proposed for each class of land (table 17.1,

Table 17.1 [*orig. table 1*]   Estimated severity of global land degradation under three different scenarios (A, B, and C) 25 years into the future (2020). Scenario A, degradation arrested immediately; scenario B, conservative rates of growth of degradation; scenario C, accelerated rates of growth of degradation. The percent total degraded land is given in parentheses.

Severity of degradation (% PDIV)	Time required to restore PDIV (years)	Degraded land (10^6 ha)			
		1995	A	B	C
Light (90)	3–10	1900 (38)	1150 (59)	3130 (40)	4360 (41)
Moderate (75)	10–20	2300 (46)	0 (0)	3530 (45)	4760 (45)
Severe (50)	50–100	750 (15)	750 (38)	1042 (13)	1335 (13)
Extreme (0)	>200	50 (1)	50 (3)	69 (1)	88 (1)

Table 17.2 [orig. table 2] Estimated rates of change in degradation classes (10^6 year^{-1}) used in scenario B (tables 17.1 and 17.3).

Severity of degradation	Drylands	Tropical moist forest	Total
Light	35.0*	14.2†	49.2
Moderate	35.0*	14.2†	49.2
Severe	9.4‡	2.3§	11.7
Extreme	0.6‡	0.15‖	0.75

* These rates are derived from the mean rate of land degradation from 1945 to 1990, assuming that all currently degraded land was so rendered during that period (5.0×10^9 ha per 45 years = 111×10^6 ha year^{-1}). From this conservative estimate (given that rates of degradation have accelerated) is subtracted the rate of degradation to the severe and extreme classes, yielding 98.5×10^6 ha year^{-1}. Equal rates of growth of the light and moderate classes are assumed and degradation in tropical moist forest (TMF) subtracted, yielding $0.5(98.5 \times 10^6) - (14.2 \times 10^6)$ = 35.0×10^6 ha year^{-1}. † These rates assume that 100% of disturbed (but not clear-cut) TMF become lightly or moderately degraded, along with 84% of the clear-cut TMF [as 15% and 1% of the latter become severely and extremely degraded, assuming the same proportions as found in the soil degradation survey (17)]. Assuming equal partitioning between the light and moderate classes, the rate is $0.5[(15.4 \times 10^6)$ + $0.84 (15.4 \times 10^6)]$ ha year^{-1} for each class. ‡ Assuming 10×10^6 ha year^{-1} become severely or extremely degraded (see above) in the same relative proportions as reported for the soil degradation survey (17), namely, 15/16 and 1/16, respectively. § Fifteen percent of clear-cut forest $[0.15 (15.4 \times 10^6)$ ha year^{-1}]. ‖One percent of clear-cut TMF $[0.01 (15.4 \times 10^6)$ ha year^{-1}].

column 2) (54). These estimates are optimistic in that all assume the higher rate of recovery from ranges of possibilities and that rehabilitation will not be hindered by soil loss, lack of colonists, climate change, further human impact, or other important factors.

Three scenarios of future changes in global PDIV are considered (table 17.1) (55). In scenario A, degradation is arrested immediately. In 25 years, complete recovery occurs on 100% of land in the light class and on 50% of land in the moderate class; the other 50% in the moderate class moves into the light category; 0% of the land in the severe and extreme classes recover sufficiently to move up into another class. Scenario B assumes conservative rates of growth of each degradation class (derived in table 17.2). Scenario C assumes rates of degradation double those used in B; these accelerated rates approximate what could occur if vigorous measures to prevent and reverse land degradation are not taken.

The analysis suggests that ~10% of global PDIV of land has already been lost (table 17.3). From a biophysical perspective, recovery of half of this loss may be feasible in 25 years, provided that degradation is halted and strong rehabilitation measures are initiated immediately. In the absence of such measures, a very conservative extrapolation of present rates of degradation suggests that ~16% of global PDIV could be lost in 25 years. At more realistic, accelerated rates of degradation, this loss could reach ~20%. In the latter scenario (C), the land area irreversibly degraded from a socioeconomic perspective (in the severe and extreme classes) would increase by a factor of 1.8

Table 17.3 [orig. table 3] Global extent of land degradation and corresponding loss of PDIV. Scenarios as in table 17.1.

	Total degraded land (10^6 ha)	Vegetated land degraded (%)	Loss of PDIV on degraded lands (%)	Loss of PDIV on all vegetated land (%)
1995	5,000	43	24	10
Scenario A	1,950	17	28	5
Scenario B	7,771	68	23	16
Scenario C	10,543	92	23	21

over 1995 levels. These results are most useful for relative, rather than absolute, comparisons.

Costs and Benefits of Rehabilitation

Although a general lack of information on rehabilitation costs constitutes a serious shortcoming (56), the utter dependence of human well-being on productive land makes its continued degradation for short-term gain an unwise course. Moreover, the costs of off-site degradation may be substantial (57).

UNEP estimates the direct, on-site cost of failure to prevent desertification during the period 1978 to 1991 at between $300 billion and $600 billion (in U.S. dollars) (58). Currently, the total direct, on-site income foregone as a result of desertification is ~$42.3 billion year^{-1}. By contrast, UNEP's estimates of the direct annual cost of all preventive and rehabilitational measures range between $10.0 billion and $22.4 billion.

An enormous potential for recovery is inherent in most land types, but failure to realize this potential can result in rapid, essentially irreversible deterioration. Historically, land degradation has been implicated in the fall of great civilizations (59) and merits serious attention by this one (60).

References and Notes

1 The United Nations medium-range projection indicates that the world population may reach 10 billion by 2054 ultimately 11.6 billion before halting growth [United Nations Population Division, *Long-Range World Population Projections* (United Nations, ST/SEA/SER.A/125, NY, 1991); United Nations, *Population Newsletter*, 7 June 1994].

2 J. Perlin, *A Forest Journey* (Norton, New York, 1989); G. C. Daily and P. R. Ehrlich, *BioScience* 42, 761 (1992); D. O. Hall, F. Rosillo-Calle, R. H. Williams, in *Renewable Energy: Sources for Fuels and Electricity*, T. B. Johansson et al., Eds. (Island Press, Washington, DC, 1993), pp. 593–652; P. R. Ehrlich, A. H. Ehrlich, G. C. Daily, *The Stork and the Plow* (Putnam, New York, in press).

3 H. A. Mooney, P. M. Vitousek, P. A. Matson,

Science 238, 926 (1987); J. T. Houghton, G. J. Jenkins, J. J. Ephraums, Eds., *Climate Change: The IPCC Scientific Assessment* (Cambridge Univ. Press, Cambridge, 1990); R. A. Houghton, *Can. J. For. Res.* 21, 132 (1991); W. H. Schlesinger, *Biogeochemistry: An Analysis of Global Change* (Academic Press, San Diego, CA, 1991).

4 N. Meyers, *A Wealth of Wild Species* (Westview, Boulder, CO, 1983); P. R. Ehrlich and A. H. Ehrlich, *Ambio* 21, 219 (1992); E. Barbier, J. Burgess, C. Folke, *Paradise Lost? The Ecological Economics of Biodiversity* (Earthscan, London, 1994).

5 C. A. S. Hall, in *Ecosystem Rehabilitation*, M. K. Wali, Ed. (SPB, The Hague, 1992), vol. 1, pp. 101–126; M. Gadgil, *Ambio* 22, 167 (1993).

6 T. N. Khoshoo, in *Ecosystem Rehabilitation*, M. K. Wali, Ed. (SPB, The Hague, 1992), vol. 2, pp. 3–17.

7 T. Homer-Dixon, J. Boutwell, G. Rathjens, *Sci. Am.* 268, 38 (February 1993); N. Myers, *Ultimate Security: The Environmental Basis of Potential Stability* (Norton, New York, 1993).

8 P. R. Ehrlich and H. A. Mooney, *BioScience* 33, 248 (1983).

9 For a discussion of economic valuation, see D. W. Pearce and J. J. Warford, *World Without End* (Oxford Univ. Press, Oxford, 1993).

10 G. L. Atjay, P. Ketner, P. Duvigneaud, in *The Global Carbon Cycle*, B. Bolin, E. T. Degens, S. Kempe, P. Ketner, Eds. (Wiley, NY, 1979), p. 129; P. M. Vitousek, P. R. Ehrlich, A. H. Ehrlich, P. A. Matson, *BioScience* 36, 368 (1986).

11 A. Chisholm and R. Dumsday, Eds., *Land Degradation* (Cambridge Univ. Press, Cambridge, 1987); C. J. Barrow, *Land Degradation* (Cambridge Univ. Press, Cambridge, 1991); P. Dasgupta, *An Inquiry into Well-Being and Destitution* (Clarendon, Oxford, 1993).

12 *Biological Markers of Air-Pollution Stress and Damage in Forests* (National Academy of Sciences, Washington, DC, 1989); O. Loucks, in *Changing the Global Environment*, D. B. Botkin, M. F. Caswell, J. E. Estes, A. A. Orio, Eds. (Academic Press, London, 1989), p. 101; W. L. Chameides, P. S. Kasibhatla, J. Yienger, H. Levy II, *Science* 264, 74 (1994).

13 R. Worrest and L. Grant, in *Ozone Depletion: Health and Environmental Consequences*, R. Jones and T. Wigley, Eds. (Wiley, NY, 1989), p. 197.

14 G. C. Daily and P. R. Ehrlich, *Proc. R. Soc. London Ser. B* 241, 232 (1990); M. Parry, *Climate*

Change and World Agriculture (Earthscan, London, 1990).

15 Some valid criticisms of degradation assessments can be found in B. Forse [*New Sci.* 1650, 31 (1989)] and in F. Pearce [*ibid.* 1851, 38 (1992)]; for an overview, see J. L. Dodd, *BioScience* 44, 28 (1994). Improved, quantitative-methods of assessing land degradation, particularly by remote sensing, are available but remain little used, for example, A. K. Tiwari and J. S. Singh, *Environ. Conserv.* 14, 233 (1987); T. A. Stone and P. Schlesinger, in Proceedings of the IUFRO S4.02.05 Wacharakitti International Workshop, 13–17 January 1992, H. G. Lund, R. Päivinen, S. Thammincha, Eds. (IUFRO, Thailand, 1992), pp. 85–93; P. Aldhous, *Science* 259, 1390 (1993); B. V. Vinogradov, *Eurasian Soil Sci.* 25, 66 (1993).

16 W. H. Schlesinger et al., *Science* 247, 1043 (1990); D. Ponzi, *Desertification Control Bull.* 22, 36 (1993).

17 The UNEP study is by L. R. Oldeman, R. T. A. Hakkeling, W. G. Sombroek, *World Map of the Status of Human Induced Soil Degradation: An Explanatory Note*, rev. (International Soil Reference and Information Center, Wageningen, Netherlands, rev. ed. 2, 1990); see also D. Pimentel, Ed., *World Soil Erosion and Conservation* (Cambridge Univ. Press, Cambridge, 1993).

18 *Status of Desertification and Implementation of the United National Plan of Action to Combat Desertification* (United Nations Environment Program, Nairobi, Kenya, 1991); H. Dregne, M. Kassas, B. Rozanov, *Desertification Control Bull.* 20, 6 (1992).

19 C. J. Tucker, H. E. Dregne, W. W. Newcomb, *Science* 253, 299 (1991).

20 Vegetation degradation occurs in biomes other than drylands. For example, *Imperata* spp, is an unpalatable weed that occupies an estimated 40 to 100 million ha of potential cropland in Southeast Asia [references in (24); R. A. Houghton, *Ambio* 19, 204 (1990)]. It forms dense, rhizomatous mats that make cultivation impossible. Rehabilitation has only been achieved with complex, costly, and labor-intensive methods that have not been successfully applied on a wide scale [J. H. H. Eussen and W. DeGroot, *Wet Gent* 39, 451 (1974)].

21 H. E. Dregne, *Desertification of Arid Lands* (Harwood, Chur, Switzerland, 1983).

22 This estimate is calculated from (18); earlier estimates (53) were ~26 million ha year^{-1}.

23 A. Grainger, *The Threatening Desert* (Earthscan, London, 1990); W. Parham, P. Durana, A. Hess, Eds., *Improving Degraded Lands: Promising Experiences from South China* (Bishop Museum, Honolulu, 1993).

24 This area comprises ~137 million ha of logged tropical moist forest undergoing regeneration; ~203 million ha of rain forest in fallow state under shifting cultivation, much of which is no longer sustainable; and ~87 million ha of cleared montane forest regions [A. Grainger, *Int. Tree Crops J.* 5, 31 (1988); adapted in part from J. P. Lanly, Ed., *Tropical Forest Resource Assessment Project: Tropical Africa, Tropical Asia, Tropical America.* (FAO/UNEP, Rome, 1981)].

25 Food and Agriculture Organization of the United Nations, *Forest Resources Assessment 1990; Tropical Countries* (FAO, Rome, 1993); this estimate is subject to debate [N. Myers, in *The Causes of Tropical Deforestation*, K. Brown and D. W. Pearce, Eds. (University College Press, London, 1994), pp. 27–40; D. Skole and C. Tucker, *Science* 260, 1905 (1993)].

26 N. Myers, *Deforestation Rates in Tropical Forests and Their Climatic Implications* (Friends of the Earth, London, 1989).

27 ——, *Environ. Conserv.* 20, 9 (1993).

28 R. Buschbacher, C. Uhl, E. Serrao, in (6), pp. 257–274.

29 R. A. Houghton, *BioScience* 44, 305 (1994); increasingly, abandoned lands do not revert to forest (27).

30 This ignores all other land types.

31 S. Hardjowigeno, *Geojournal* 28, 131 (1992).

32 J. Skoupy, *Desertification Control Bull.* 22, 5 (1993); F. R. Troeh and L. M. Thompson, *Soils and Soil Fertility* (Oxford Univ. Press, New York, ed. 5, 1993).

33 Impediments include (in Amazonia, for example) the aerial extent of pastures, which frequently cover hundreds or even thousands of hectares and make re-establishment of forest problematic because ~90% of the tree species have animal seed dispersers, of which very few venture into open pasture (43). Seed and seedling predators, particularly some ant species, are abundant in pasture and will remove experimentally placed seeds within minutes of placement [C. Uhl, in *Biodiversity*, E. O. Wilson, Ed. (National Academy Press, Washington, DC, 1988), pp. 326–332]. Microclimatic conditions (air and soil temperatures and humidity) are harsh and further limit seedling survival [C. Uhl, *J. Eco.* 75, 377 (1987)].

34 M. B. Bush and R. J. Whitaker, *J. Biogeogr.* 18, 341 (1991); I. W. B. Thornton and T. R. New,

Geojournal 28, 219 (1992); but see R. W. Suatmadji, A. Coomans, F. Rashid, E. Geraert, D. A. McLaren, *Philos. Trans. R. Soc. London Ser. B* 322, 369 (1988).

35 I. W. B. Thornton, T. R. New, R. A. Zann, P. A. Rawlinson, *Philos. Trans. B. Soc. London Ser B* 328, 131 (1990).

36 S. Fridriksson, *Arct. Alp. Res.* 19, 425 (1987); *Environ. Conserv.* 16, 157 (1989).

37 D. R. Harris, *Geogr. Rev.* 61, 475 (1971); H. Ruthenberg, *Farming Systems in the Tropics* (Clarendon, Oxford, 1971).

38 Most experimental studies monitor the recovery of sites that were slashed and burned but not actually cultivated [but see O. P. Toky and P. S. Ramakrishnan, *Agro-Ecosystems* 7, 11 (1981)].

39 W. J. Peters and L. F. Neuenschwander, *Slash and Burn: Farming in the Third World Forest* (Univ. of Idaho Press, Moscow, ID, 1988).

40 F. E. Egler, *Vegetatio* 4, 412 (1954); E. S. D'Angela, J. M. Facelli, E. Jacobo, *ibid.* 74, 39 (1988); D. D. Steven, *Ecology* 72, 1076 (1991); R. W. Myster and S. T. A. Pickett, *ibid.* 75, 387 (1994).

41 C. Keever, *Ecol. Monogr.* 20, 229 (1950); F. A. Bazzaz, *Ecology* 49, 924 (1968).

42 P. M. Fearnside, *Ambio* 22, 537 (1993).

43 C. Uhl et al., *J. Ecol.* 76, 663 (1988).

44 R. Buschbacher, C. Uhl, E. Serrao, in (6), pp. 257–274.

45 C. Uhl and R. Buschbacher, *Biotropica* 17, 265 (1985).

46 Grasses often invade areas cleared of other types of vegetation, thereby establishing ecological conditions favoring their persistence [C. M. D'Antonio and P. M. Vitousek. *Annu. Rev. Ecol. Syst.* 23, 63 (1992)]. A state of arrested succession and lowered soil fertility often results, which sometimes strongly hinders efforts to establish crops or other vegetation (20).

47 J. Ewel, *Biotropica* 12, 2 (1980); M. Kellman, *ibid.*, p. 34.

48 E. G. Egler, *Rev. Bras. Geogr.* 23, 527 (1961).

49 P. S. Ramakrishnan, in (6), pp. 19–35.

50 J. Wallis, Ed, *Combating Desertification in China* (United Nations Environment Program, Nairobi, Kenya, 1982); J. Ewel, *Annu. Rev. Ecol. Syst.* 17, 245 (1986); E. B. Allen, Ed., *The Reconstruction of Disturbed Arid Lands* (Westview, Boulder, CO, 1988); J. Cairns, Ed., *Rehabilitating Damaged Ecosystems* (CRC, Boca Raton, FL, 1988); R. Lal and B. A. Stewart, Eds. *Advances in Soil Science: Soil Restoration* (Springer-Verlag, NY, 1992); D. Saunders,

R. Hobbs, P. Ehrlich, Eds., *Reconstruction of Fragmented Ecosystems* (Surrey Beatty, Chipping Norton, 1993); B. H. Walker, *Ambio* 22, 80 (1993).

51 V. M. Kline and E. A. Howell, in *Restoration Ecology*, W. R. Jordon III, M. E. Gilpin, J. D. Aber, Eds. (Cambridge Univ. Press, Cambridge, 1987), pp. 75–83; P. Singh, in (6), pp. 51–61.

52 P. M. Blaschke, N. A. Trustrum, R. C. DeRose, *Agric. Ecosyst. Environ.* 41, 153 (1992).

53 This value is conservative relative to other estimates that classify only 57% (as opposed to my assumption of 84%) of arid lands as retaining at least 75% of their productivity [H. E. Dregne, *Desertification of Arid Lands* (Harwood, Chur, Switzerland, 1983); J. A. Mabbutt, *Environ. Conser.* 11, 103 (1984); *Desertification Control Bull.* 12, 1 (1985)].

54 These are comparable to estimates given in N. -T. Chou and H. E. Dregne, *Desertification Control Bull.* 22, 20 (1993).

55 These projections assume that the average PDIV of degraded land is equivalent to that of nondegraded land. This may be an exaggeration, but the issue is complex and geographically varied; moreover, the assumed rates of recovery are sufficiently optimistic as to make the overall calculations conservative.

56 But see J. A. Dixon, D. E. James, P. B. Sherman, Eds., *Dryland Management: Economic Case Studies* (Earthscan, London, 1990); M. M. Mattos and C. Uhl, *World Dev.* 22, 145 (1994).

57 G. Upstill and T. Yapp in *Land Degradation Problems and Policies*, A. Chisholm and R. Dumsday, Eds, (University of Cambridge, Melbourne, Australia, 1987), chap. 5; E. H. Clark, J. A. Haverkamp, W. Chapman, *Eroding Soils: The Off-Farm Impacts* (The Conservation Foundation, Washington, DC, 1985).

58 This excludes other costs such as human suffering (18).

59 R. McC. Adams, *Heartland of Cities* (Univ. of Chicago Press, Chicago, 1981).

60 I gratefully acknowledge comments by R. Adams, A. Ehrlich, P. Ehrlich, R. Hanson, J. Holdren, K. Holl, P. Matson, S. Schneider, P. Vitousek, and two anonymous reviewers. Supported by the Winslow Foundation, the Heinz Foundation, and W. Alton Jones Foundation, an anonymous grant, and by P. Bing and H. Bing.

Part IV

Managing Water Resources

INTRODUCTION AND GUIDE TO FURTHER READING

18   Gleick, P. H. 1994: Water, war and peace in the Middle East. *Environment*, 36, 3, 6–15 and 35–42.

Gleick provides a useful article which describes the hydropolitical issues in the Middle East. He emphasizes the importance of water resources and their management on an international level. The paper is also useful because it provides recommendations for effective management that may help aid political stability within regions of conflict over water resources, such as the Middle East.

19   Anderson, D. M. 1994: Red tides. *Scientific American*, August, 52–8.

This is a very interesting paper which describes the nature of algal blooms. It describes how algal blooms can affect marine ecosystems and even lead to poisoning of fish, marine mammals and humans. The article discusses the possible connection between increased pollution and increased frequency of algal blooms. The paper also provides interesting examples of recent pollution and poisoning from the US Atlantic coast.

20   Mitchell, R. B. 1995: Lessons from international oil pollution. *Environment*, 37, 4, 10–15 and 36–41.

Mitchell provides a very comprehensive review of the nature of marine oil pollution and the effectiveness of international law on polluters. He discusses the methods of enforcing international law and provides suggestions for the best mechanisms to help enforce and develop these laws in the future.

21   Pearce, F. 1995: Dead in the water. *New Scientist*, 4 February, 26–31.

This article discusses the pollution of the Mediterranean Sea and it assesses the effectiveness of the Mediterranean Action Plan (MAP) which was agreed more than 20 years ago. It is a useful article because it illustrates the nature of pollution and the apathy regarding methods and legislation to reduce pollution.

22   Daly, D. 1992: Quaternary deposits and groundwater pollution. *Quaternary Proceedings*, 2, 79–89.

In this paper, Daly describes various ways that near-surface groundwater may become polluted and he discusses the vulnerability of various types of ground. The paper is useful because it provides information on how groundwater pollution can be assessed and prevented.

23   Brydges, T. G. and Wilson, R. B. 1991: Acid rain since 1985 – times are changing. In: Last, F. T. and Watling, R. 1991: *Acid Deposition: Its Nature and Impacts.* Edinburgh: The Royal Society of Edinburgh, pp. 1–16.

This paper provides an excellent review of the multi-national environmental issues surrounding acid deposition. It describes the development of scientific knowledge about acid deposition, and the public awareness of the problem, as well as the international action concerning the pollution control programmes.

24   Myers, M. F. and White, G. F. 1993: The challenge of the Mississippi floods. *Environment*, 38, 10, 5–9.

This is a useful review of the history of the management of the Mississippi river basin. It addresses the question of whether recent large floods, particularly the 1993 flood of the upper Mississippi and lower Missouri basins, are the result of canalization schemes or natural processes. The environmental and social impacts of flooding are considered as is the implementation of Federal policy to help alleviate the effects of flooding. In addition the paper provides a useful list of recommendations for floodplain management.

25   Pearce, F. 1995: The biggest dam in the world. *New Scientist*. 28 January, 25–9.

The environmental and social problems associated with the construction of the Three Gorges Dam on the Yangtse River in China are outlined in this article. In particular, the article assesses the likely geomorphological and ecological changes associated with flooding the Three Gorges. The paper is also useful because it illustrates the types of conflict that occur between environmental protection and development.

Clean water is essential for the well-being of humans, not only for domestic and industrial use, but also to ensure that contagious diseases are not transmitted between individuals and populations. Countries spend more money obtaining clean water than we do on petroleum, and the management of water resources is therefore not only a social, but also an economic and political issue. According to the WHO (World Resources Institute, 1990), an estimated 1,200 million people lack a satisfactory or safe supply of water. Water is an essential part of the global commons; rivers and oceans have no political boundaries and water must be managed on a national and international scale. In an attempt to help manage this essential resource, the International Conference on Water and the Environment (ICWE) was convened in Dublin in 1991. It prepared recommendations for the United Nations Conference on the Environment and Development (UNCED) in Rio de Janeiro in 1992. This became known as the Dublin Statement, and its recommendations were incorporated into the UNCED Agenda 21 document, the blueprint for action into the twenty-first century. The main principles from the Dublin Statement are summarized in Young, Dooge and Rodda (1994).

In many regions of the world, water resources have led to major political conflicts, exemplified by problems in the Middle East, particularly between Israel and Jordan. There is growing concern over potential future political conflicts between countries that may occur as a result of the over-exploitation of this essential resource, or the degradation of its quality by pollution. Such a resource must be shared fairly and efficiently managed. The chapter by Gleick (1994) illustrates this concern and provides management recommendations to help ensure political stability in regions such as the Middle East where there is a scarcity of water supplies.

There is growing concern over the pollution of water resources. Groundwaters and surface waters are polluted in four main ways: the direct release of effluents, seepage from groundwater sources, from surface waters, and from precipitation (acid rain/deposition). Pollution is greatest in the newly industrialized countries, where pollution protection measures have not been fully implemented and where money for remediation is not available. In developed countries, similar problems exist, but most polluters to date have escaped the legal and financial costs of cleaning up the mess. Slowly, too slowly, new legislation is being developed to help reduce the likelihood of pollution and to provide mechanisms for remediation. One of the most important legislative decrees was enforced by the US government in 1980 in response to pollution incidents. This is known as the Comprehensive Response, Compensation, and Liability Act (CERCLA), under which a Federal trust fund (Superfund) was established in response to contaminated land. CERCLA was later amended to the Superfunds Amendments and Reauthorisation Act (SARA) to enable the collection of funds for remedial action to pollution. These acts have proved vital for maintaining and improving water and land quality, and developing environmental strategies, as well as helping to provide models for other countries to consider in the fight against pollution.

On a global scale, developed countries are most important polluters of the oceans and seas. In recent years, this has been particularly noticeable in the Mediterranean, Adriatic and the east coast of North America, where the frequency and magnitude of algal blooms have increased markedly. There is much debate over this increased frequency, but there is a general consensus that it may be the result of increased nitrates, flushed from farmland which has been over-fertilized, and from increased discharges of effluents. These effects are described in the chapter by Anderson (1994).

Oil is another major pollutant in the oceans, with an estimated 3.6 million tonnes of oil spilt into the sea every year, mainly as the result of shipping accidents involving oil tankers and deliberate flushing of tanks and engines (World Resources Institute, 1990). The chapter by Mitchell (1995) provides a comprehensive review of the nature of marine oil pollution. It discusses the effectiveness of international law on polluters, and makes suggestions about the various ways of improving international law and enforcement in the future.

The chapter by Pearce (1995) emphasizes this by discussing the Mediterranean Action Plan which was agreed by countries that border the Mediterranean Sea. Pearce relates that little has been done to help reduce pollution problems.

Groundwater makes up 35 per cent of the total public water supply; it is also essential in helping to sustain the flow in rivers and as part of the hydrological cycle. There is growing concern, however, that groundwater reserves are rapidly being depleted for domestic and industrial uses, particularly in urban regions of high population density such as London, Venice and Bangkok (Wilkinson and Brassington, 1991; Pirazzoli, 1973; Rau and Nutalaya, 1982). The other major concern relates to the increased incidence of groundwater pollution resulting from the leaching of pollutants from contaminated lands. The chapter by Daly (1992) describes the various ways that near-surface groundwater may become polluted. This provides useful information on how pollution can be assessed and prevented.

Water pollution can also be the result of atmospheric pollution, noticeably the deposition of acids by rain, snow, fog or as dry particles. Acid deposition is primarily caused by the discharge of sulphur oxides and nitrogen oxides by industrial processes, particularly the burning of coal and oil from electricity-generation. Emissions of SO_2 and NO_x combine with water vapour in the atmosphere to form acids. During the late 1970s and early 1980s, it became increasingly apparent that soil and water bodies (mainly lakes) were becoming acidified by acid deposition and that this resulted in a reduction in the biological productivity of forest and lake ecosystems. During this period a number of important conferences were held to examine the links between emissions, acid precipitation and ecosystem degradation. A series of conventions was agreed as a result of these meetings, but it was not until the latter part of the 1980s that governments really began to accept the link between atmospheric pollution and acidification. The first major step toward control of emissions resulted from the Helsinki Conference on Security and Co-operation. 'The Convention on Long-Range Transboundary Air Pollution' arose from this and proposed controls on the levels of emissions of sulphur dioxide (SO_2) and other pollutants. This was followed by the 1982 Stockholm Convention which was later modified in 1983 with the aim of reducing SO_2 emissions by 30 per cent within a decade. In July 1985, the Executive Body of the 1982 Stockholm Convention met again and agreed a protocol binding its signatories to a 30 per cent reduction of SO_2 emissions by 1993. It was not until later that a protocol was formulated to fix levels on nitrous oxides emission. This was agreed at the Sofia meeting in 1988, setting targets at a 30 per cent reduction in NO_x by 1998. On 1 March 1993, the UN Economic Commission for Europe began negotiations for a new sulphur protocol, basing targets on critical loads of acid deposition that can be tolerated by sensitive ecosystems. Due to economic considerations, these levels could not be achieved, so a compromise was made to reduce levels to 50 per cent of the 1980 levels. On 14 June 1994, 29 countries supported by the EC signed a treaty in Oslo agreeing to reduce acidic deposition by 2010 to 80 per cent of the 1980 levels. This means that reductions of emissions in most signatory countries must fall by over 50 per cent of the 1994 levels. To achieve this there must be major commitments to introduce improved technologies for cleaner emission in power stations as well as a switch from coal-fired to gas-fired power stations, and/or the development of alternative energy sources. Conventions and protocols of this type are essential for the safeguarding of sensitive ecosystems and to sustain the global commons.

The chapter on acidification of water and soil, by Brydges and Wilson (1991), examines the nature and effects of acid deposition, and looks at the effects that legislation has had on reducing acid rain during the latter half of the 1980s.

Part IV also examines some of the issues surrounding the management of catchments. Catchment studies are important in helping to manage water resources as well as in helping to preserve and sustain the biological and physical value of a region. They are also important in predicting and mitigating natural hazards such as flooding. The management of catchments is a complex topic involving the understanding of climatology, geology, vegetation and hydrology, as well as human impact. To complicate matters further, no two catchments are identical, and the nature of catchments is continuously changing as humans, and possibly climate change, alter the physical characteristics of an area. The chapter by Myers and White (1993) provides an example of the issues surrounding the management of one of the world's largest catchments, the Mississippi river basin. The authors review the history of its management and question whether the recent large floods were a result of canalization schemes or natural processes. It also addresses the environmental and social impacts of flooding and assesses the US Federal policy on flooding and floodplain management. The chapter by Pearce (1995) examines the environmental issues surrounding the construction of the world's largest dam, the Three Gorges Dam on the Yangtse River in China. Pearce examines the effects of construction on the hydrology of the region and the geomorphological and ecological consequences of flooding the valley. It is a useful paper as it shows the conflict between environmental protection and development.

The management of water resources is complex, involving a variety of disciplines ranging from climatology, hydrology, geology and ecology, as well as a consideration of the economic, political and social aspects of the subject. The chapters in Part IV have provided a broad overview for environmental managers to consider in relation to water resources.

References

Pirazzoli, P. 1973: Inondations et niveaux marins à Venise. *Màm. Lab. Geomorph. Ecole Pratique Hautes*, 22, Dinard.

Rau, J. L. and Nutalaya, P. 1982: Geomorphology and land subsidence in Bangkok, Thailand. In: Craig, R. G. and Crafts, J. L. (eds) *Applied Geomorphology*. London: Mackeys and Chatham, pp. 181–201.

Wilkinson, W. B. and Brassington, F. C. 1991: Rising groundwater levels – an international problem. In: Downing, R. A. and Wilkinson, W. B. (eds) *Applied Groundwater Hydrology – a British Perspective*. Oxford: Clarendon Press, 35–53.

World Resources Institute 1990: *World Resources 1990–91*. Oxford: Oxford University Press.

Other references cited in text are listed in the selected readings below.

Suggested further reading

Bedient, P. B., Rifai, H. S. and Newell, C. J. 1993: *Groundwater Contamination: Transport and Remediation*. London: Prentice-Hall International.

In this very comprehensive text, Bedient et al. describe the scientific and engineering aspects of groundwater contamination transport and remediation. They examine waste site characterization and remedial design, including case studies that illustrate recent techniques. The text is particularly useful for hydrogeologists and geochemists, and others involved in waste disposal and groundwater resources.

Carter, F. W. and Turnock, D. (eds) 1996: *Environmental Problems in Eastern Europe*. London: Routledge.

This edited volume analyses the major forms of pollution and the resulting decline in the quality of life in each country in eastern Europe. Consideration is given to air, water, soil and vegetation pollution, dumping of waste, nuclear power and transboundary issues in relation to the roles of legislation, political movements, international co-operation, aid and education, all of which strive for a solution to environmental problems.

Cheremisinoff, P. N. 1993: *Water Management and Supply*. London: Prentice-Hall International.

This book examines the properties, sources and uses of water in industry, commerce and transportation. Sections include: water sources and uses; water quality and properties; chemical properties of water; water supply purification; groundwater; and industrial and commercial requirements.

Gleick, P. H. (ed.) 1993: *Water in Crisis: A Guide to the World's Fresh Water Resources*. Oxford: Oxford University Press.

This volume presents data on the status of water resources throughout the world. It comprises a collection of essays on: fresh water issues, providing a background to the dynamics of water systems; data on global and regional fresh water resources; rivers, lakes and waterfalls; sanitation and water-related diseases; water quality and contamination; water and agriculture; water and ecosystems; water and energy; water and human use; and water policy and politics.

Hinrichsen, D. 1990: *Our Common Seas: Coasts in Crisis*. London: Earthscan.

This book is based on UNEP data and examines the growing pressures on coastal ecosystems throughout the world. Case studies help illustrate the local successes in protecting marine and coastal environments.

Kamari, J., Brakke, D. F., Jenkins, A., Norton, S. A. and Wright, R. F. (eds) 1989: *Regional Acidification Models: Geographical Extent and Time Development*. London: Springer-Verlag.

This edited text reviews the development and use of mathematical models for the assessment of regional acidification and other effects of air pollutants on the environment. It comprises aspects of sensitivity distribution, time evolution of regional impacts, and uncertainty in model applications.

Kliot, N. 1993: *Water Resources and Conflict in the Middle East*. London: Routledge.

This book examines the hydrological, social, economic, political and legal issues in the Middle East. It shows how water shortages threaten the renewal of conflict and disruption in the Euphrates, Tigris, Nile and Jordan basins.

Mason, C. F. 1996: *Biology of Freshwater Pollution*. Harlow: Longman, 3rd edition.

This is an excellent and comprehensive text that provides an overview of the various biological aspects of freshwater pollution. It is a useful reference source for all who are interested in environmental pollution.

Open University 1991: *Case Studies in Oceanography and Marine Affairs*. Oxford: Pergamon Press.

This Open University text examines marine resources and activities, and the development and nature of international conventions on the sea. It emphasizes the complex interactions between the political, economic and environmental aspects of development within the marine realm.

Young, G. J., Dooge, J. C. I. and Rodda, J. C. 1994: *Global Water Resource Issues*. Cambridge: Cambridge University Press.

This is an excellent volume which examines the world's water resources, particularly the regional and global requirements for the management of water resources. Topics include: integrated water resources development and management; water resource assessment; protection of water resources, water quality and aquatic ecosystems; impact of climate change on water resources; water and sustainable development; water for sustainable food production and rural development; drinking water supply and sanitation; and mechanisms for implementation and co-ordination at global, national and local levels.

18

WATER, WAR AND PEACE IN THE MIDDLE EAST*

Peter H. Gleick

As the 21st century approaches, population pressures, irrigation demands, and growing resource needs throughout the world are increasing the competition for freshwater. Nowhere is this more evident than in the arid Middle East, where the scarcity of water has played a central role in defining the political relationships of the region for thousands of years. In the Middle East, ideological, religious, and geographical disputes go hand in hand with water-related tensions, and even those parts of the Middle East with relatively extensive water resources, such as the Nile, Tigris, and Euphrates river valleys, are coming under pressure. Competition for the limited water resources of the area is not new; people have been fighting over, and with, water since ancient times. The problem has become especially urgent in recent years, however, because of increasing demands for water, the limited options for improving overall supply and management, and the intense political conflicts in the region. At the same time, the need to manage jointly the shared water resources of the region may provide an unprecedented opportu-

nity to move toward an era of cooperation and peace.

During the last two years, water conflicts have become sufficiently important to merit separate explicit discussion in both the multilateral and bilateral Middle East peace talks now under way (see box 18.1). Among the issues that must be resolved are the allocation and control of water in, and the water rights to, the Jordan River and the three aquifers underlying the West Bank; a dispute between Syria and Jordan over the construction and operation of a number of Syrian dams on the Yarmuk River; the joint management of the Euphrates River between Turkey, Syria, and Iraq; and how to protect water quality for all those dependent on these resources.

Conflicts among nations are caused by many factors, including religious differences, ideological disputes, arguments over borders, and economic competition. Although it is difficult to disentangle the many intertwined causes of conflict, competition over natural resources and disputes over environmental factors are playing an increasing role in international relations.[1]

* Originally published in *Environment*, 1994, vol. 36, no. 3, pp. 6–15, 35–42.
[1] For a review of the principal points in the ongoing debate, see J. T. Mathews, "Redefining Security," *Foreign Affairs* 68, no. 2 (1989): 162–77; P. H. Gleick, "Environment, Resources, and International Security and Politics," in E. Arnett, ed., *Science and International Security: Responding to a Changing World* (Washington, D. C.: American Association for the Advancement of Science, 1990), 501–23;

Box 18.1   Water and the Middle East Peace Talks
By Peter Yolles and Peter H. Gleick

Water is such an important aspect of the international relationships in the Middle East that it has been made an explicit part of the ongoing peace talk. There are two tracks to these talks, the bilateral talks and the multilateral talks. The official goal of the bilateral negotiations is a "just, lasting, and comprehensive peace."[1] These talks are where the major political questions are being worked out in meetings between Israel and each of the other interests in the area. The major water issues in the bilateral talks are defining and securing appropriate shares of water rights. Discussion of the prime question of control of water and water rights was originally part of the multilateral talks but was recently moved to the bilateral talks. In the Israeli-Jordanian bilateral talks, a sub-committee on "Water, Energy, and Environment" was formed, and a sub-committee on "Land and Water" has been formed for the Israeli-Palestinian talks.

There are five separate working groups in the multilateral talks: Refugees, Arms Control and Regional Security, Economic Development, Environment, and Water. A steering committee oversees the work of these groups and provides links with the ongoing bilateral talks. In the water talks of the multilaterals, practical questions of regional cooperation are under discussion with all interested governmental parties. These questions include how to alleviate short-term and long-term water shortages, how to increase overall water supplies, and what institutions could enhance data sharing, conflict resolution, and river basin management. Four sets of multilateral water negotiations have already been held in Vienna (May 1992), Washington, D. C. (September 1992), Geneva (April 1993), and Beijing (October 1993). The next set is being held this month in Oman. The water track of the multilaterals is the only one to have successfully produced a signed agreement: to cooperate on a series of formal and informal "activities" around supply questions, data sharing, and institution building. These activities began in summer 1993 and are continuing.

In addition to the formal peace talks now under way, there is an informal track of separate independent, unofficial discussions. These are often academic meetings, workshops, and conferences. Among the recent meetings have been an Israeli-Palestinian conference in Zurich in December 1992; an academic workshop on the multilaterals held at the University of California at Los Angeles in April 1993, which included delegates from Jordan, Israel, and the Palestinians; a meeting in Champagne/Urbana, Illinois, sponsored by the International Water Resources Association in October 1993; and a Pugwash Conference on Middle East issues held outside of Stockholm in December 1993.

These meetings provide an unofficial forum for broaching ideas and exchanging information, and they are considered extremely fruitful both for the ideas that are raised and for the relationships that are formed. Several of the ideas that have made their way into the recent formal agreements between Israel and the Palestinians and Israel and Jordan originated at these unofficial meetings. These ideas include the goal of equitable utilization, the supply of minimum water requirements to existing inhabitants, and the need to examine certain new supply options.

1   Reuters, *Draft Agenda of the Israeli-Jordanian Delegation to the Peace Talks*, October 27, 1992 (1 November 1992).

T. Homer-Dixon, "On the Threshold: Environmental Changes as Causes of Acute Conflict," *International Security* 16, no. 2 (1991): 76–116; and P. H. Gleick, "Water and Conflict," *International Security* 18, no. 1 (1991): 79–112.

These conflicts can take several forms, including the use of resources or the environment as instruments of war or as goals of military conquest. History reveals that water has frequently provided a justification for going to war: It has been an object of military conquest, a source of economic or political strength, and both a tool and target of conflict. Also, on occasion, shortages of water have constrained a country's economic or political options.[2] No region has seen more water-related conflicts than the Middle East, and some of these go back more than 5,000 years to the earliest civilizations in Mesopotamia (see boxes 18.2 and 18.3).

Water can become a source of strategic rivalry because of its scarcity, the extent to which the supply is shared by more than one region or state, the relative power of the basin states, and the ease of access to alternative freshwater sources. In the Middle East, water is scarce and widely shared by countries with enormous economic, military, and political differences. Also, there are few economically or politically acceptable alternative sources of supply. Thus, the temptation to use water for political or military purposes has often proved irresistible. As water supplies and delivery systems become increasingly valuable in water-scarce regions, their value as military targets increases.

In modern times, the most pressing water conflicts in the Middle East have centered on control of the Jordan River basin, apportionment of the waters of the Euphrates and Nile Rivers, and management of the groundwater aquifers of the occupied territories.

The Water Resources

The water resources of the Middle East are unevenly distributed and used, and every major river in the region crosses international borders. Table 18.1 identifies the major river basins in the region and the countries that are part of those basins. The extent to which major rivers and groundwater basins are shared by two or more nations makes the allocation and sharing of water a striking political problem and greatly complicates the collection and dissemination of even the most basic data on water availability and use. In northeast Africa and the Middle East, more than 50 percent of the total population relies upon river water that flows across a political border. Two-thirds of all Arabic speaking people in the region depend upon water that originates in non-Arabic-speaking areas; two-thirds of Israel's freshwater comes from the occupied territories or the Jordan River basin; and one-quarter of the Arab people live in areas entirely dependent on non-renewable groundwater or on expensive, desalinized seawater.[3]

The major shared surface water supplies in the Middle East are the Jordan, Tigris, Euphrates, and Nile Rivers. Although the watershed of the Litani River lies entirely within Lebanon, control and allocation of its waters remain controversial. Several major groundwater aquifers are also heavily used and, in the occupied territories, strongly contested.

In the Middle East, actual water availability fluctuates dramatically both seasonally and from year to year. For many of the major rivers of the region, flows in dry years may be as low as one-half to one-third the volume of the average yearly flows, and there is a long history of persistent and severe droughts.[4]

[2] P. H. Gleick, ed., *Water in Crisis: A Guide to the World's Fresh Water Resources* (New York: Oxford University Press, 1993). See, also, M. Falkenmark, "Fresh Waters as a Factor in Strategic Policy and Action," in A. H. Westing, ed., *Global Resources and International Conflict* (New York: Oxford University Press, 1986), 85–113.

[3] J. Kolars, "The Future of the Euphrates River" (paper presented at the International Workshop on Comprehensive Water Resources Management Policy, World Bank, Washington, D. C., 24–28 June 1991).

[4] M. Shahin, "Hydrology of the Nile Basin," *Developments in Water Science*, vol. 21 (Amsterdam: Elsevier

Table 18.1 [*orig. table 1*]   International river basins in the Middle East.

River basin	Total area of basin (square kilometers)	Countries in basin	Area (square kilometers)	Percentage of total area
Tigris	378,850	Iran	220,000	58
		Iraq	110,000	29
		Turkey	48,000	13
		Syria	850	<1
Euphrates	444,000	Iraq	177,000	40
		Turkey	125,000	28
		Syria	76,000	17
		Saudi Arabia	66,000	15
Orontes	13,300	Syria	9,700	73
		Turkey	2,000	15
		Lebanon	1,600	12
Jordan	19,850	Jordan	7,650	39
		Syria	7,150	36
		Israel	4,100	21
		Lebanon	950	5
Nile	3,031,000	Sudan	1,900,000	63
		Ethiopia	368,000	12
		Egypt	300,000	10
		Uganda	233,000	8
		Tanzania	116,000	4
		Kenya	55,000	2
		Zaire	23,000	1
		Rwanda	21,500	1
		Burundi	14,500	<1

Water quality problems also affect the region. Heavy use of water for irrigation contaminates water with agricultural chemicals and salts and reduces the quality of water for downstream users. Overpumping from many underground aquifers is leading to the intrusion of saltwater and the contamination of remaining supplies – a problem especially evident in the coastal aquifers of the Gaza Strip.

The Jordan River

Despite its small size, the Jordan River is one of the most important in the region and the locus of intense international competition (see figure 18.1). Shared by Jordan, Syria, Israel, and Lebanon, the Jordan drains an area of slightly less than 20,000 square kilometers and flows 360 kilometers from its headwaters to the Dead Sea. Annual precipitation in the watershed ranges from less than 50 millimeters per year to more than

Science Publishers, 1985); and J. F. Kolars and W. A. Mitchell, *The Euphrates River and the South-east Anatolia Development Project* (Carbondale and Edwardsville, Ill.: Southern Illinois University Press, 1991).
[5] T. Naff, "The Jordan Basin: Political, Economic, and Institutional Issues," in G. LeMoigne, ed., Country Experiences with Water Resources Management, World Bank Technical Paper no. 175 (Washington, D.C.: World Bank, 1992), 115–18.

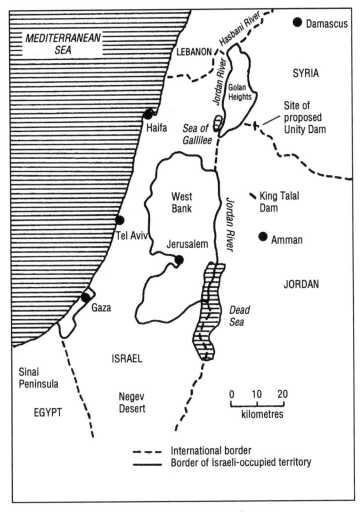

Figure 18.1 [*orig. figure 1*]   The Jordan River basin.
Source: Redrawn from a map of the General Staff Map Section, Director General of Military Survey, Ministry of Defence, United Kingdom, 1991.

1,000 millimeters per year and averages less than 200 millimeters per year,[5] which is insufficient for most rainfed agriculture. The upper Jordan is fed by three major springs, the Hasbani in Lebanon, the Banias in Syria, and the Dan in Israel. The major tributary of the Jordan, the Yarmuk River, originates in Syria and Jordan and constitutes part of the border between these countries and the Israeli-occupied Golan Heights before flowing into the Jordan River. The quality of Jordan River water is very good up to the point where it enters Lake Tiberias (also known as the Sea of Galilee); by the time it enters the Dead Sea, the water remaining in the Jordan has become too salty to use.

Total average unimpaired flow of the Jordan River is about 1,850 million cubic meters per year (m³/y). Israel normally uses 1,600 million to 1,800 million m³/y from all sources, including around 600 million m³/y from the Jordan River; about 800 million m³/y from groundwater aquifers; and 360 million m³/y from reuse of wastewater. Jordan has usually derived between 700 million and 900 million m³/y of usable water from all sources, including groundwater, the

Table 18.2 [*orig. table 2*]   Middle East population estimates and projections.

Country	1990	2000 (millions)	2025	Annual percentage rate of increase in 1990
West Bank[a]	0.90	1.12	2.37	3.40
Gaza Strip	0.62	0.76	1.23	1.98
Israel	4.66	6.34	8.15	1.67
Jordan	3.10	4.00	8.50	3.41
Lebanon	2.74	3.31	4.48	2.00
Syria	12.36	17.55	35.25	3.58
Saudi Arabia	14.87	20.67	40.43	3.28
Turkey	55.99	68.17	92.88	2.05
Iraq	18.08	24.78	46.26	3.21
Iran	58.27	77.93	144.63	2.71

[a] Population figures are for 1991 and 2020 instead of 1990 and 2025
Note: Growth rates in Jordan, Israel, the West Bank, and the Gaza Strip depend largely on immigration rates and thus are difficult to project.
Sources: United Nations, *World Population Prospects: The 1992 Revision, Annex Tables* (New York, 1993); and J. D. Priscoli and R. Brumbaugh, *Water in the Sand* (Washington, D.C.: U.S. Army Corps of Engineers, 1991).

Yarmuk, and a few other small surface sources.[6]

Additional population growth in this region is expected to be high: Even without immigration, Jordan's and Syria's populations are growing at around 3.5 percent per year, and Israel's is growing at about 1.7 percent per year (see table 18.2). Immigration of Soviet and other Jews and of Palestinians either displaced from other lands or returning to the region may add several million more people by the early 21st century.

Since the establishment of Israel in 1948, this basin has been the center of intense international conflict, and the dispute over the waters of the Jordan River is an integral part of the ongoing conflict. In the 1950s, when Syria tried to stop Israel from building its National Water Carrier, a system to provide water to southern Israel, fighting broke out across the demilitarized zone. When Syria tried to divert the headwaters of

the Jordan away from Israel in the mid-1960s, Israel used force, including air strikes against the diversion facilities.[7] These military actions contributed to the tensions that led to the 1967 Arab-Israeli War, the occupation of the West Bank, and control over much of the headwaters of the Jordan River by Israel.

Tensions also exist in the Jordan basin between Syria and Jordan over the construction and operation of a number of Syrian dams on the Yarmuk River. These dams were built to allow Syria to make use of the Yarmuk's flow, which would otherwise be available for use in Israel or Jordan.

Shared groundwater aquifers

A significant fraction of Israel's water comes from shared groundwater aquifers that underlie both the West Bank and the Gaza Strip (see figure 18.2). By some estimates, 40 percent of the groundwater that Israel uses – and more than one-third of its

[6] S. C. Lonergan and D. B. Brooks, *The Economic, Ecological and Geopolitical Dimensions of Water in Israel* (Victoria, B.C.: University of Victoria, Centre for Sustainable Regional Development, 1992); and J. D. Priscoli and R. Brumbaugh, *Water in the Sand: A Survey of Middle East Water Issues* (Washington, D.C.: U.S. Army Corps of Engineers, 1991).
[7] See, for example, T. Naff and R. C. Matson, eds., Water in the Middle East: Conflict or Cooperation? (Boulder, Colo.: Westview Press, 1984).

Box 18.2   Conflicts over Water in the Myths, Legends, and Ancient History of the Middle East

By Haleh Hatami and Peter H. Gleick

The history of water-related disputes in the Middle East goes back to antiquity and is described in the many myths, legends, and historical accounts that have survived from earlier times. These disputes range from conflicts over access to adequate water supplies to intentional attacks on water delivery systems during wars. A chronology of such water-related conflicts in the Tigris and Euphrates river valleys during the last 5,000 years appears in box 18.3.(1)

One of the earliest examples of the use of water as a weapon is the ancient Sumerian myth – which parallels the Biblical account of Noah and the deluge – recounting the deeds of the diety Ea, who punishes humanity's sins by inflicting the Earth with a great flood. According to the Sumerians, the patriarch Utu speaks with Ea, who warns him of the impending flood and orders him to build a large vessel filled with "all the seeds of life."(2)

A dispute between the city-states of Umma and Lagash over the fertile soils of Mesopotamia between the modern-day Tigris and Euphrates Rivers continued from 2500 to 2400 B.C. and included conflicts over irrigation systems and the intentional diversion of water supplies. Continuing disputes over water in the region later led Hammurabi of Babylon (around 1790 B.C.) to include several laws in the famous "Code of Hammurabi" pertaining to the negligence of irrigation systems and to water theft.

Many Biblical accounts include descriptions of the use of water as an instrument of conflict, including the banishment of Hagar and Ishmael to the wilderness with only a limited amount of water and their divine salvation when God leads them to a well (Genesis 21:1–23). According to Islam, Ishmael's offspring constitute the nation of Islam; a similar Qur'anic verse parallels this Biblical account. The well, called Zum Zum, is thought to be located at Mecca. Exodus recounts the miracle of Moses parting the Red Sea or, alternatively, damming a tributary of the Nile to prevent the Egyptians from reaching the Jews as they journeyed through the Sinai.(3) In Chronicles 32:3, Hezekiah digs a well outside the walls of Jerusalem and uses a conduit to bring in water to prepare for a siege by Sennacherib. By cutting off water supplies outside of the city walls, Jerusalem survives the attack.

Other historical accounts offer fascinating insights into the role of water in war and politics. Sargon II, the Assyrian king from 720 to 705 B.C., destroyed the intricate irrigation network of the Haldians after his successful campaign through Armenia. Sennacherib of Assyria devastated Babylon in 689 B.C. as retribution for the death of his son and intentionally destroyed the water supply canals to the city. Assurbanipal, King of Assyria from 669 to 626 B.C., seized water wells as part of his strategy of desert warfare against Arabia. According to inscriptions recorded during the reign of Esarhaddon (681–669 B.C.), the Assyrians besieged the city of Tyre, cutting off food and water. In another account, in 612 B.C., a coalition of Egyptian, Median (Persian), and Babylonian forces attacked and destroyed Ninevah, the capital of Assyria, by diverting the Khosr River to create a flood.

Nebuchadnezzar (605–562 B.C.) built immense walls around Babylon and used the Euphrates River and a series of canals as defensive moats surrounding the inner castle. Describing Nebuchadnezzar's plan to create an impregnable city, the ancient historian Berossus states, "He arranged it so that besiegers would no longer be able to divert the river against the city by surrounding the inner city with three circuits of walls."(4)

In one of the most intriguing legends, Herodotus describes how Cyrus the Great successfully invaded Babylon in 539 B.C. by diverting the Euphrates River above the city into the desert and marching his troops into the city along the dry riverbed.

1   This chronology comes from H. Hatami and P. H. Gleick, *Chronology of Conflict over Water in the Legends, Myths, and History of the Ancient Middle East* (Oakland, Calif.: Pacific Institute for Studies in Development, Environment, and Security, June 1993).
2   S. Burstein, "The Babyloniaca of Berossus," in *Sources from the Ancient Near East* (Malibu, Calif.: Undena Publications, 1978).
3   L. L. Honor, *Sennacherib's Invasion of Palestine* (New York: Columbia University Press, 1926).
4   Burstein, note 2 above.

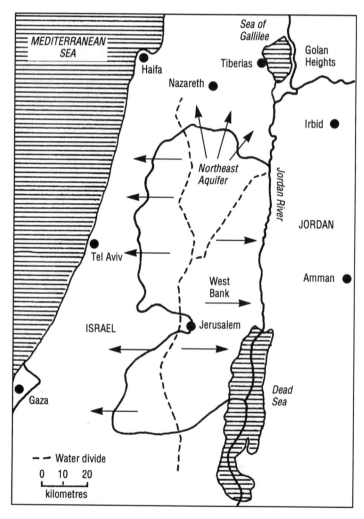

Figure 18.2 *[orig. figure 2]*   Groundwater aquifers underlying Isreal and the West Bank.
Source: Redrawn from a map in H. A. Amery, "Cooperative Water Management in the Middle East," in *Proceedings of the International Symposium on Water Resources in the Middle East: Policy and Institutional Aspects* (Urbana, Ill., 1993), 59–68.

sustainable annual water yield – comes from the occupied territories.[8] Though no accurate studies have been published, it is estimated that the long-term potential yield of the West Bank aquifers is just less than 700 million m^3/y, of which about 180 million is brackish water.[9] These aquifers are replenished almost entirely by rainfall on the West Bank. The largest of the aquifers, the Western (called the Yarkon-Taninim aquifer in Israel), flows west toward the Mediterranean Sea. This groundwater supply is tapped extensively by Israel, primarily from within the boundaries of pre-1967 Israel. The other aquifers also are largely controlled and heavily used by Israel, both within Israel proper and in the settlements in the occupied territories.

Box 18.3   A Partial Chronology of Conflict over Water in the Ancient mideast

3000 B.C. – The Flood

An ancient Sumerian legend recounts the deeds of the diety Ea, who punishes humanity for its sins by inflicting the Earth with a six-day storm. The Sumerian myth parallels the Biblical account of Noah and the deluge, although some details differ.

2500 B.C. – Lagash-Umma Border Dispute

The dispute over the "Gu'edena" (edge of paradise) region begins. Urlama, King of Lagash from 2450 to 2400 B.C., diverts water from this region to boundary canals, drying up boundary ditches to deprive Umma of water. His son cuts off the water supply to Girsu, a city in Umma.

1790 B.C. – Code of Hammurabi for the State of Sumer

Hammurabi lists several laws pertaining to irrigation that provide for possible negligence of irrigation systems and water theft.

1720–1684 B.C. – Abi-Eshuh v. Iluma-Ilum

A grandson of Hammurabi, Abish or Abi-Eshuh, dams the Tigris to prevent the retreat of rebels led by Iluma-Ilum, who declared independence of Babylon. This failed attempt marks the decline of the Sumerians who had reached their apex under Hammurabi.

1200 B.C. – Moses and the Parting of the Red Sea

When Moses and the retreating Jews find themselves trapped between the pharaoh's army and the Red Sea, Moses miraculously parts the waters of the Red Sea, allowing his followers to escape. The waters close behind them and cut off the Egyptians.

720–707 B.C. – Sargon II Destroys Armenian Waterworks

After a successful campaign against the Haldians of Armenia, Sargon II of Assyria destroys their intricate irrigation network and floods their land.

705–682 B.C. – Sennacherib and the Fall of Babylon

---

[8] These data come from M. R. Lowi, "The Politics of Water Under Conditions of Scarcity and Conflict: The Jordan River and Riparian States" (Ph.D. diss., Department of Politics, Princeton University, Princeton, N. J., 1990), 342; and Naff, note 5 above. But the unwillingness of all parties in the region to share water resources data makes a complete analysis difficult.

[9] H. I. Shuval, "Approaches to Finding an Equitable Solution to the Water Resources Problems Shared by Israel and the Palestinians over the Use of Mountain Aquifer," in G. Baskin, ed., *Israel/Palestine: Issues in Conflict, Issues for Cooperation*, vol. 1, no. 2 (Jerusalem: Israel/Palestine Center for Research and Information, 1992), 26–53.

In quelling rebellious Assyrians in 695 B.C., Sennacherib razes Babylon and diverts one of the principal irrigation canals so that its waters wash over the ruins.

Sennacherib and Hezekiah

As recounted in Chronicles 32:3, Hezekiah digs a well outside the walls of Jerusalem and uses a conduit to bring in water. Preparing for a possible siege by Sennacherib, he cuts off water supplies outside of the city walls, and Jerusalem survives the attack.

681–669 B.C. – Esarhaddon and the Siege of Tyre

Esarhaddon, an Assyrian, refers to an earlier period when gods, angered by insolent mortals, create a destructive flood. According to inscriptions recorded during his reign, Esarhaddon besieges Tyre, cutting off food and water.

669–626 B.C. – Assurbanipal, Siege of Tyre, Drying of Wells

Assurbanipal's inscriptions also refer to a siege against Tyre, although scholars attribute it to Esarhaddon. In campaigns against both Arabia and Elam in 645 B.C., Assurbanipal, son of Esarhaddon, dries up wells to deprive Elamite troops. He also guards wells from Arabian fugitives in an earlier Arabian war. On his return from victorious battle against Elam, Assurbanipal floods the city of Sapibel, an ally of Elam. According to inscriptions, he dams the Ulai River with the bodies of dead Elamite soldiers and deprives dead Elamite kings of their food and water offerings.

612 B.C. – Fall of Ninevah in Assyria and the Khosr River

A coalition of Egyptian, Median (Persian), and Babylonian forces attacks and destroys Ninevah, the capital of Assyria. Nebuchadnezzar's father, Nebopolassar, leads the Babylonians. The converging armies divert the Khosr River to create a flood, which allows them to elevate their siege engines on rafts.

605–562 B.C. – Nebuchadnezzar Uses Water to Defend Babylon

Nebuchadnezzar builds immense walls around Babylon, using the Euphrates and canals as defensive moats surrounding the inner castle.

558–528 B.C. – Cyrus the Great Digs 360 Canals

On his way from Sardis to defeat Nabonidus at Babylon, Cyrus faces a powerful tributary of the Tigris, probably the Diyalah. According to Herodotus's account, the river drowns his royal white horse and presents a formidable obstacle to his march. Cyrus, angered by the "insolence" of the river, halts his army and orders them to cut 360 canals to divert the river's flow. Other historians argue that Cyrus needed the water to maintain his troops on their southward journey, while another asserts that the construction was an attempt to win the confidence of the locals.

539 B.C. – Cyrus the Great Invades Babylon

According to Herodotus, Cyrus invades Babylon by diverting the Euphrates above the city and marching troops along the dry riverbed. This popular account describes a midnight attack that coincided with a Babylonian feast.

355–323 B.C. – Alexander the Great Destroys Persian Dams

Returning from the razing of Persepolis, Alexander proceeds to India. After the Indian campaigns, he heads back to Babylon via the Persian Gulf and Tigris, where he tears down defensive weirs that the Persians had constructed along the river. Arrian describes Alexander's disdain for the Persians' attempt to block navigation, which he saw as "unbecoming to men who are victorious in battle."

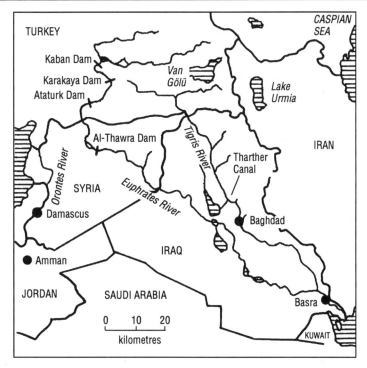

Figure 18.3 [*orig. figure 3*] Water supply projects in the Tigris and Euphrates river basins.
Sources: Redrawn from maps of J. Waterbury, "Dynamics of Basin-Wide Cooperation in the Utilization of the Euphrates" (paper prepared for the conference "The Economic Development of Syria: Problems, Progress, and Prospects," Damascus, 6–7 January 1990); and General Staff Map Section, Director General of Military Survey, Ministry of Defence, United Kingdom, 1991.

The control of the water from these aquifers is one of the major sources of tension between the Palestinians and the Israelis. Among the unresolved questions are the extent to which these three aquifers are used, disputes over their control and management, uncertainties about the effects of large withdrawals on water quality, and arguments over the yields that can be provided safely.

The Tigris and Euphrates

The Tigris and Euphrates Rivers are among the largest in the region. Both rivers originate in the mountains of Turkey, flow south through Syria and Iraq, and drain through the Shatt Al-Arab waterway into the Persian Gulf (see figure 18.3). Several tributaries of

the Tigris drain the Zagros Mountains between Iran and Iraq, and 15 percent of the Euphrates basin is in Saudi Arabia, though essentially none of its flow is generated there. Average annual runoff in these two rivers exceeds 80,000 million cubic meters, of which about 33,000 million are generated in the Euphrates and 47,000 in the Tigris.[10] Flows in both rivers are extremely variable. Minimum flows of the Euphrates have been reported as low as 180 cubic meters per second, while maximum flows as high as 5,200 cubic meters per second have occurred.[11] Half of the annual runoff of the Euphrates is generated during the brief spring (April and May) snowmelt, and runoff in dry years has amounted to as little as 30 percent of the annual average flow.

[10]   Kolars, note 3 above; J. Waterbury, "Dynamics of Basin-Wide Cooperation in the Utilization of the Euphrates" (paper prepared for the conference *Economic Development of Syria: Problems, Progress, and Prospects*, Damascus, 6–7 January 1990); and Kolars and Mitchell, note 4 above.
[11]   Kolars, note 3 above.

Ninety percent of the water in the Euphrates River originates in Turkey, though Turkey has only 28 percent of the area of the Euphrates basin. Almost all of the remainder of the flow originates in Syria. Turkey, Syria, and Iraq have large and rapidly growing populations (see table 18.2), and all three countries have ambitious plans to increase their withdrawals of water for irrigation.

Although Syria has other water resources, these are largely tapped or, like the Yarmuk, contested, and the Euphrates is the only major river crossing its territory with reliable annual flows. Iraq is the most heavily dependent upon the Euphrates at present, but it has an alternative source of water in the Tigris system, which currently is lightly used.

Recent developments on the Euphrates in southern Turkey, particularly the completion of the massive Ataturk Dam, are viewed by the other basin nations with mixed feelings. Such developments could help to reduce the extreme variations in flow and ensure predictable supplies in downstream countries, but they could lead to greater upstream withdrawals and a reduction in overall flows to Syria by as much as 40 percent and to Iraq by up to 80 percent, especially during dry years.[12] No formal agreement has yet been reached on minimum releases either by Turkey to Syria or by Syria to Iraq. Iraq believes that full development of the ambitious Turkish Southeast Anatolia Development Project (of which the Ataturk Dam is a part) and the more modest irrigation plans in Syria would deprive Iraq of sufficient water for its own irrigation plans.

Water quantity is not the only concern facing countries in the Euphrates basin. The quality of Euphrates water is being adversely affected by withdrawals and irrigation return flows. A large proportion of the water entering Iraq already consists of return flows containing high concentrations of both agricultural chemicals and salts. As a result, salinization of cropland and loss of agricultural productivity are growing concerns.

For 30 years, negotiations over the Euphrates among Turkey, Syria, and Iraq have produced no lasting agreement, in part because the three countries have long been at odds with each other. For example, Syria and Iraq have opposed Turkey over its membership in NATO, and Syria and Turkey opposed Iraqi military actions in the 1970s. In the 1980s, Turkey and Iraq tended to band together against Syrian military aggression, and Turkey and Syria sided with the allied forces against Iraq during the Persian Gulf War in the early 1990s.

Water-related disputes arose in the basin in the 1960s after both Turkey and Syria began to draw up plans for large-scale irrigation withdrawals. In 1965, tripartite talks were held in which each of the three countries put forth demands that, together, exceeded the natural yield of the river. Also in the mid-1960s, Syria and Iraq began bilateral negotiations over formal water allocations, but, by the end of the decade, no formal agreement had been reached. In the 1970s, an agreement was reached, though never signed, that allocated portions of overall flow to both Syria and Iraq. In the mid-1970s, dams at Keban, Turkey, and Tabqa, Syria, were completed, and their reservoirs had begun to fill, reducing flows to Iraq.[13]

In 1974, Iraq alleged that the flow of water in the Euphrates had been reduced by the Syrian dam, threatened to bomb it, and massed troops along the border. In spring of 1975, tensions between Iraq and Syria peaked as Iraq claimed that Syria was intentionally reducing flows to intolerably low levels. During April and May, the two countries traded hostile statements in which Iraq threatened to take any action necessary to ensure the Euphrates's flow. Iraq also issued a formal protest to the Arab League that Syria was intentionally depriving it of its rightful share. On 13 May, Syria closed its airspace to all Iraqi aircraft, suspended

---

[12] This estimate comes from Thomas Naff of the University of Pennsylvania and is cited in "Water Wars in the Middle East," *The Economist*, 12 May 1990, 54–59.
[13] Waterbury, note 10 above.

Syrian flights to Baghdad, and reportedly transferred troops from its front with Israel to the Iraqi border. The angry confrontation ended just short of military action after mediation by Saudi Arabia. Syria reportedly agreed to release additional water to Iraq, from "its own share" as a goodwill gesture.[14]

In the last few years, Turkey's new water supply projects have been the focus of new political concerns in the basin. Tensions arose in January 1990 when Turkey completed construction of the Ataturk Dam and closed the dam to begin filling the reservoir, interrupting the flow of the Euphrates for a month. Despite advance warning from Turkey of the temporary cutoff, Syria and Iraq both protested that Turkey now had a water weapon that could be used against them. Indeed, in October 1989, Turkish Prime Minister Turgut Ozal had threatened to restrict water flow to Syria to force it to withdraw support for Kurdish rebels operating in southern Turkey.[15] Thus, Turkish politicians' claims that the shutoff to fill the Ataturk's reservoir was entirely for technical, not political, reasons failed to appease Syrian and Iraqi officials, who argued that Turkey had already used its power over the headwaters of the Euphrates for political goals and could do so again.[16]

The ability of Turkey to shut off the flow of the Euphrates, even temporarily, was noted by political and military strategists at the beginning of the Persian Gulf conflict.[17] In the early days of the war, there were behind-the-scenes discussions at the United Nations about using Turkish dams on the Euphrates River to cut off water to Iraq in response to its invasion of Kuwait. Although no such action was taken, the threat of the "water weapon" was again made clear.[18]

The Nile River

The Nile River, the longest in the world, flows more than 6,800 kilometers from the highlands of central Africa and Ethiopia through nine nations to the Mediterranean. The nations that share the Nile are Egypt, Sudan, Ethiopia, Kenya, Tanzania, Zaire, Uganda, Rwanda, and Burundi, and the watershed covers nearly 10 percent of the African continent (see figure 18.4). Two major tributaries form the Nile: the White Nile, which starts in central Africa's Lake Plateau region, and the Blue Nile, which originates in the highlands of Ethiopia. More than 80 percent of the Nile's flow comes from the torrential seasonal flows of the Blue Nile. Like the other rivers in the Middle East, the Nile exhibits substantial hydrologic variability. Although the large Aswan Dam has buffered Egypt from some of this variability, a recent 10-year drought in the region has shown the limits of even major dams, and the Aswan has had a wide range of other impacts.[19]

The Nile River is also a shared water resource of tremendous regional importance, particularly for agriculture in Egypt and Sudan. Ninety-seven percent of Egypt's water comes from the Nile River, and more than 95 percent of the Nile's runoff originates outside of Egypt in the other eight nations of the basin. The Nile valley has sustained civilizations for more than 5 millennia, but historical evidence suggests

[14] Naff and Matson, note 7 above.
[15] *Middle East Economic Digest*, "Battle Lines Drawn for Euphrates," 13 October 1989, 4–5. See, also, Waterbury, note 10 above.
[16] A. Cowell, "Water Rights: Plenty of Mud to Sling," *New York Times*, 7 February 1990, A4.
[17] See P. Schweizer, "The Spigot Strategy," *New York Times*, 11 November 1990, op, ed.
[18] These closed-door discussions were described to the author by the ambassador of a member nation of the UN Security Council under the condition that he remain anonymous. See, also, the statement of the Minister of State of Turkey, Kamran Inan, at the Conference on Transboundary Waters in the Middle East: Prospects for Regional Cooperation, Ankara, 3 September 1991. At that meeting, Inan stated that Turkey would never use water as a means of political pressure and noted that it had declined to do so during the Gulf War.
[19] G. F. White, "The Environmental Effects of the High Dam at Aswan," *Environment*, September 1988, 4.

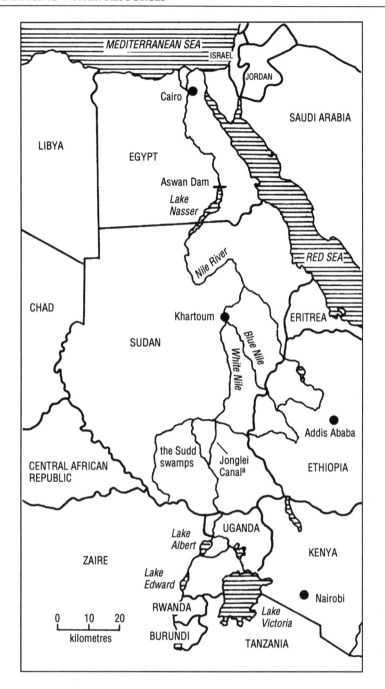

Figure 18.4 [*orig. figure 4*]   The Nile River basin.
[a] Construction of the canal has been stopped.
Source: Redrawn from a map of the General Staff Map Section, Director
General of Military Survey, Ministry of Defence, United Kingdom, 1991.

that the populations of ancient Egypt never exceeded 1.5 million to 2.5 million people.[20] Today, Egypt struggles to sustain a population rapidly approaching 60 million on the same limited base of natural resources. And Egypt's population grows by another million people every nine months.

A treaty was signed in 1959 allocating the water of the Nile between Egypt and Sudan.[21] Although this treaty has effectively reduced the risk of conflict between the two countries over water, none of the other seven nations of the basin is party to it, and several have expressed a desire to increase their use of Nile River water. Additional use of water by these other nations of the Nile basin, particularly by Ethiopia, could reduce water available to the downstream nations and greatly increase tensions over water. Concern over the security of Egypt's water supplies led President Anwar Sadat to say in 1979, "The only matter that could take Egypt to war again is water."[22] More recently, Egypt's foreign minister, Boutros Boutros-Ghali, now Secretary-General of the United Nations, was quoted as saying "The next war in our region will be over the waters of the Nile, not politics."[23] Although these statements partly reflect political rhetoric, they indicate the importance of the Nile to Egypt.

The Litani and Orontes Rivers

Two other important rivers flow through parts of the Middle East: the Orontes and the Litani. The Litani River is the only important river contained entirely within one country, Lebanon, and one of the few that have not been tapped to the limit. The Litani rises in the mountains surrounding the Bekaa Valley and flows 145 kilometers south and west into the Mediterranean Sea. Average annual flow in the basin is a little more than 900 million cubic meters, and two-thirds of this flow occurs between January and April.[24] Only about half of the average flow is currently used, primarily for irrigation. The waters of the Litani River also provide approximately 40 percent of Lebanon's total electricity supply and are of very high quality, although the effect of current agricultural water use in the lower Litani basin has not been documented.

The Orontes originates in central Lebanon and flows north through Syria and Turkey before emptying into the Mediterranean Sea. Three-quarters of the basin is in Syria, and the major use of Orontes River water is for irrigation in Syria's Ghab Valley. Although there may be some surplus water in the basin, additional developments in Syria and contamination of the water by sewage and industrial effluents limit any significant shared use of Orontes water by Turkey.

The Potential for Water Wars

The Persian Gulf War underscored the many connections between water and conflict. During this war, water and water supply systems were targets of attack, shared water supplies were used as instruments of politics, and water was considered a potential tool of warfare. The dams, desalination plants, and water conveyance systems of both sides were targeted for destruction. Most of Kuwait's extensive desalination capacity was destroyed by the retreating Iraqis. Oil spilled into the gulf threatened to contaminate desalination plants throughout the region. And the intentional destruction of Baghdad's modern water supply and sanitation system was so complete that the

[20] D. Hillel, *Out of the Earth: Civilization and the Life of the Soil* (Berkeley, Calif.: University of California Press, 1991).

[21] "Agreement Between the United Arab Republic and the Republic of Sudan for the Full Utilization of the Nile Waters," Cairo, 8 November 1959, *Egyptian Review of International Law* 15, 321.

[22] Cited by J. Starr in "Water Wars," *Foreign Policy* 82, (Spring 1991): 17–30.

[23] This statement has been widely cited. As one example of the widespread attention it has attracted, it appeared in the major newspaper of Nairobi. See T. Walker, "The Nile Struggles to Keep Up the Flow," *Sunday Nation*, 10 January 1988, 11.

[24] Priscoli and Brumbaugh, note 6 above.

Iraqis are still suffering severe problems as they rebuild them.[25]

Although water resources are only one source of tension in the Middle East, pressures over water are likely to grow in the future because of demographic trends, changing patterns of water use, and possible changes in supply caused by global climate change – the greenhouse effect. Few of the countries in the region believe that they have adequate water for their current populations; almost none believes that it can continue to provide adequate water as its population continues to grow and as industry and agriculture increase their demands for freshwater.

Population growth

In some of the most water-short parts of the Middle East, most notably the Jordan and Euphrates river basins, populations are expected to grow extremely rapidly (see table 18.2). At the same time, new demands for water are putting pressure on existing supplies. In Israel and Jordan, projected population growth could require the severe restriction or complete elimination of irrigated agriculture over the next several decades just to free up sufficient water to provide a reasonable minimum amount to their populations.

For example, the United Nations' medium projections show the population of Israel and the Gaza Strip reaching 10 million by 2025, not including the Palestinians presently included by the United Nations in Jordan's population.[26] Simply supplying this population with a minimum annual water requirement of 150 cubic meters per person

for drinking, sanitation, and all commercial and industrial activities would require 1,500 million m^3/y, which is approximately equal to Israel's entire long-term reliable supply. This level of use would leave only recycled wastewater for the agricultural sector and so would almost completely eliminate irrigated agriculture.

Table 18.3 shows how the per-capita availability of water in the countries of the Middle East and parts of the Persian Gulf is likely to decrease given the expected population growth between now and 2025. Most hydrologists believe that having less than 500 cubic meters available per person per year significantly limits the options available to a society.[27] Many countries in the region already fall into this category, and more will in the future as populations grow.

Climate change

All debates about regional water supplies assume that natural water availability in the future will not change and that flows will be subject only to natural variations. In fact, this assumption may no longer be true because of possible changes in the global climate.[28] Global climate change could affect water availability in many ways, though the precise nature of such changes is still obscure. Climate change could either increase or decrease overall water availability in different times and in different places.[29] Estimates of changes in temperature and precipitation patterns in the Middle East are mixed; average temperatures may rise between 3° and 6°C if the atmospheric concentration of carbon dioxide doubles, but precipitation projections show little con-

[25] *U.S. Water News*, "Iraq's Water Systems Still in Shambles," October 1992, 2.

[26] United Nations Department of Economic and Social Development, *World Population Prospects: The 1992 Revision, Annex Tables* (New York, 1993).

[27] Falkenmark, note 2 above. See, also, chapters 1, 7 and 9 in Gleick, note 2 above.

[28] This article is not the place for a discussion of climate change per se. For more detail on the science of this issue, see the report of the Intergovernmental Panel on Climate Change, *Climate Change: The IPCC Scientific Assessment* (Cambridge, England: Cambridge University Press, 1990); P. H. Gleick, "Effects of Climate Change on Shared Fresh Water Resources," in I. M. Mintzer, ed., *Confronting Climate Change: Risks, Implications and Responses* (Cambridge, England: Cambridge University Press, 1992), 127–40; and S. H. Schneider, P. H. Gleick, and L. O. Mearns, "Prospects for Climate Change," in P. E. Waggoner, ed., *Climate Change and U. S. Water Resources* (New York: John Wiley and Sons, 1990), 41–74.

[29] See the collection of articles in Waggoner, note 28 above; and P. H. Gleick, "Climate Change, Hydrology, and Water Resources," *Review of Geophysics 27*, no. 3 (1989): 329–44.

Table 18.3 [*orig. table 3*]   Per-capita water availability in 1990 and 2025.

Country	1990	2025
	(cubic meters per person per year)	
Kuwait	75	57
Saudi Arabia	306	113
United Arab Emirates	308	176
Jordan	327	121
Yemen	445	152
Israel	461	264
Qatar	1,171	684
Oman	1,266	410
Lebanon	1,818	1,113
Iran	2,025	816
Syria	2,914	1,021
Iraq	5,531	2,162

Source: P.H. Gleick, *Water in Crisis: A Guide to the World's Fresh Water Resources* (New York: Oxford University Press, 1993).

sistency across different climate models, reflecting the difficulty of accurately modeling precipitation and the uncertainty about regional model results. Hydrologists expect higher temperatures to lead to substantial increases in evaporation in the region, which would decrease overall water supply and increase demand. Despite the limited ability of the current models to project future conditions accurately, even slight decreases in long-term water availability would place severe political strains on the region, as was seen from 1979 to 1988, when a drought reduced the average runoff in the Nile by only 10 percent. Although the nature of future climate changes in the region cannot be predicted with confidence, there are indications that long-term

decreases in flow exceeding 10 percent are possible. Some preliminary modeling of the Nile basin suggests that Nile runoff would decrease by as much as 25 percent under some plausible conditions, and seasonal flows may experience even more significant changes.[30] Ironically, the possibility of increases in runoff during the snowmelt season raises the specter of increased frequency of severe flooding, as was experienced in Sudan in 1988.[31]

Future climate changes effectively make obsolete all old assumptions about the behavior of water supply. Perhaps the greatest certainty about future climate change is that the future will not look like the recent past. Changes are certainly coming, and, by the turn of the century, many of these changes may already be apparent. The challenge is to identify those cases in which conflicts are likely to be exacerbated and to reduce the probability and consequences of those conflicts.

Reducing Conflicts over Water

There is no single solution to the Middle East's water problems, and, ultimately, a combination of efforts and innovative ideas must be applied. Formal political agreements will have to be negotiated to apportion and manage the shared surface- and groundwater in the region, particularly in the Jordan and Euphrates river basins and the occupied territories. Unless all of the people who depend on the resources concerned are included in these agreements, conflicts will remain. In particular, definitions of *equitable utilization* of the existing water resources must be negotiated and applied.[32] Difficult decisions must also be made to prioritize water use within each

[30] P. H. Gleick, "The Vulnerability of Runoff in the Nile Basin to Climatic Changes," *The Environmental Professional* 13 (1991): 66–73.

[31] Gleick, "Water and Conflict," note 1 above; and S. Lonergan and B. Kavanagh, "Climate Change, Water Resources and Security in the Middle East," *Global Environmental Change*, September 1991, 272–90.

[32] For a discussion of the importance of the term *equitable utilization*, see, for example, P. H. Gleick, "Reducing the Risks of Conflict over Fresh Water Resources in the Middle East," in J. Isaac and H. Shuval, eds., *Water and Peace in the Middle East* (Dordrecht, the Netherlands: Elsevier Science Publishers, 1993); and S. C. McCaffrey, "Water, Politics, and International Law," in Gleick, note 2 above, pages 92–104.

country. Israel, like California and many other parts of the world, is wrestling with the conflicts between urban and rural water demands and between the agricultural and domestic sectors. Jordan is trying to improve its water-use efficiency so that it, like Israel, can make better use of its limited supplies. And all parties are exploring ways of increasing supply within serious economic and environmental constraints. Sharing of expertise, opening access to hydrologic data, and exploring joint water conservation and supply projects offer the best opportunities for reducing the risk of future tensions over water in the Middle East.

Water rights and control

The conflicts over water in the Middle East are not only about limited water availability; they also arise over the control of existing resources. In the Jordan basin, control over shared groundwater resources underlying the West Bank and the Gaza Strip are at the heart of the tension between Israelis and Palestinians. In 1967, Israel issued Military Order 92, which prohibits the drilling of new wells without permission from the military authorities, fixes quotas for pumping from existing wells, and expropriates wells in all occupied lands.[33] The Palestinians claim that these restrictions have, effectively, frozen Palestinian use of water in the occupied territories, resulted in insufficient water for Palestinian urban and industrial use, and stopped new agricultural development. At the same time, Israel has allowed the development of water wells for Jewish settlements in the occupied areas. One outcome of this situation is a gross discrepancy in per-capita water use by Israelis and Palestinians in the occupied territories. The perception that much of the Israeli water goes to nonessential uses, such as irrigation of lawns and the filling of swimming pools, has not helped the problem.[34]

This difference has played a direct role in framing the ongoing peace talks. Although many Israelis argue that efforts should focus on enlarging the "pie" so that existing uses can be maintained, Palestinians insist that a discussion of reallocation of water rights and control over the existing supply must precede any major efforts to enhance total availability. These distinct viewpoints have led to two distinct tracks in the peace talks: a discussion of water rights in the bilateral talks and a discussion of ways to enhance supply in the multilateral water talks (see box 18.1). Unless this disagreement is dealt with directly, the chances of resolving other water problems in the region are limited.

Once the issue of water rights is resolved, new options open up for reallocating existing water, including water marketing and sales or leasing of water rights. Recent experience in California with water marketing and "banking" suggests that short-term or long-term sales of water can be accomplished if appropriate institutions and incentives are developed. Under the right circumstances, those in possession of water rights may have an incentive to shift that water to more valued uses in return for economic benefits. Such transfers must be voluntary and equitable, but experience in the western United States suggests a variety of approaches that could be appropriate in the Middle East.[35] For example, the California Water Bank, created in 1991 during a long-term drought, bought nearly 1,000 million cubic meters of water from farmers at a price of $0.10 per cubic meter. This water was then sold to urban centers, and a small portion was set aside to aid threatened ecosystems.

The creation of a comparable water bank or market between the Palestinians and the Israelis could permit the Palestinians to sell or lease any unused West Bank water. If Israel were to "buy" groundwater from

---

[33] G. Baskin, "The West Bank and Israel's Water Crisis," in Baskin, note 9 above, pages 1–8.

[34] Shuval, note 9 above. See, also, Al Khatib, "Palestinian Water Rights," in G. Baskin, ed., *Water: Conflict or Cooperation*, rev. ed., vol. 2, no. 2 (Jerusalem: Israel/Palestine Center for Research and Information, 1993), 13–22.

[35] See, for example, R. H. Coppock and M. Kreith, eds., *California Water Transfers: Gainers and Losers in Two Northern Counties* (Davis, Calif.: University of California Press, 1993).

West Bank aquifers at $0.06 per cubic meter (about what agricultural users currently pay), the annual payment per 100 million cubic meters would be only $6 million. If the price were as high as $0.30 per cubic meter (about what urban users currently pay), this volume of water would cost about $30 million per year. At this higher price, there is a strong incentive to find alternative sources of supply or to increase investment in water-saving technologies. Creating this sort of market for water transfers would require some innovative institutional arrangements, but it could also offer benefits to all users in the region.

Supplying a "minimum water requirement"

Another new proposal that may begin to address the problem of water rights and the equitable distribution and utilization of existing water is to establish a minimum water requirement for the population in a region.[36] Such a minimum would be guaranteed to all residents and would provide for the minimum basic human needs of drinking water, sanitation, and domestic use and for moderate urban industrial and commercial uses. Although no minimum water requirement has been formally defined, present urban water uses suggest that an appropriate level may be between 75 and 150 m^3/y per person. Mechanisms must be developed to permit transfers from regions with water in excess of these amounts to regions without this minimum and to allocate the remaining water resources after the minimum is supplied. Supplying this minimum should be a higher priority than expanding overall supplies because it effectively defines an equitable water right.

Increased efficiency of use

Increasing the efficiency of water use in all the countries of the region may be the most economical and least controversial of all proposals. Even modest increases in the efficiency of agricultural water use and decreases in consumptive use[37] could dramatically increase overall availability in other sectors. A 10-percent reduction in agricultural water use in Israel, for example, would double the water available for urban users. Israel has pioneered many improvements in agricultural irrigation efficiency and the recycling and reuse of wastewater for certain uses, but, despite the fact that Israel is already one of the most water-efficient countries in the world, continued improvements in the efficiency with which water is used are possible. Jordan is now implementing similar measures to cope with its increasing water problems. In Jordan, the on-farm efficiency of water use is still low (only about 40 percent is productively used) and evaporation rates and seepage losses from open irrigation canals in the Jordan Valley are high. High subsidies for agricultural water by all of the countries in the region also contribute to continued wasteful, inefficient use.[38]

A major area for increasing the efficiency of water use is wastewater re-use and water

---

[36] This concept was raised at the December 1992 "First Israeli-Palestinian Joint International Academic Conference on Water" held in Zurich, Switzerland. For initial discussion of minimum water requirements, see H. Shuval, "Proposed Principles and Methodology for the Equitable Allocation of the Water Resources Shared by the Israelis, Palestinians, Jordanians, Lebanese, and Syrians," and P. H. Gleick, "Reducing the Risks of Conflict over Fresh Water Resources in the Middle East," in Isaac and Shuval, note 32 above.

[37] The "consumptive use" of water must be distinguished from water withdrawals. Consumptive use refers to water withdrawn and made unavailable for re-use through evaporative loss, percolation to deep groundwater layers, or contamination. In every country of the Middle East, the agricultural sector is responsible for the vast majority of the consumptive use of water. Reducing consumptive uses of water makes more water available for other uses.

[38] M. F. Abu-Taleb, J. P. Deason, E. Salameh, and B. Kefaya, "Water Resources Planning and Development in Jordan: Problems, Future Scenarios, and Recommendations," in G. LeMoigne, ed., *Country Experiences with Water Resources Management*, World Bank Technical Paper 175 (Washington, D. C.: World Bank, 1992), 119–27.

reclamation. In Israel, substantial advances have been made in water reclamation and reuse. Tel Aviv, for example, is reusing more than one-third of its wastewater for purposes other than drinking water.[39] Overall, approximately 5 percent of Israel's entire water use is recycled wastewater, and that percentage is increasing. By 2000, the Jordan Water Authority expects that one-quarter of east Jordan Valley irrigation water will come from recycled sewage water.[40] Although this use of water will be limited by religious, social, health, and environmental concerns, the technology exists to clean and recycle wastewater adequately for many purposes.

New supplies

The traditional reaction to resource pressures is to focus on how to increase supplies, and this is true in the Middle East as well. There are two principal ways to increase supplies: bring in outside sources of water and capture unused portions of the current supply by building reservoirs to store flows during wet periods for use during dry periods. Many ideas for developing new sources in the Middle East have been proposed, including building desalination plants to make freshwater out of seawater or brackish water; constructing enormous pipelines to divert underused rivers in Turkey or Pakistan to the parched regions of the Middle East and the Persian Gulf; tankering or towing enormous bags of freshwater to coastal areas; laying aqueducts from the Mediterranean Sea or the Red Sea to the Dead Sea to generate electricity and desalinate saltwater; and building new reservoirs on major rivers to increase storage for dry periods.

All of these proposals are controversial, and all have uncertain economic and environmental costs. In addition, political disputes over who would control the sources of some of these options make the construction of new facilities extremely unlikely in the absence of a lasting political settlement. On the other hand, some new sources of supply may eventually be developed as the economic value of water rises and as demands grow:

Desalination – Ninety-seven percent of the water on the planet is too salty to drink or to grow crops. This had led to great interest in devising ways of removing salt from water in the hope of providing unlimited supplies of freshwater. Indeed, by the beginning of 1990, there were more than 7,500 facilities worldwide producing more than 13.2 million cubic meters of freshwater per day. More than half of this desalination capacity is in the Persian Gulf region, where inexpensive fossil fuels provide the energy necessary to run the plants[41] (see table 18.4). For other regions, however, the high energy cost of desalination continues to make unlimited freshwater supplies an elusive goal. In the long run, the use of desalination will be limited by the amount and cost of the energy required to purify saltwater. Unless unanticipated major technical advances reduce overall energy requirements or the price of energy drops substantially, large-scale desalination will always be limited to extremely water-poor and energy-rich regions.

Peace Pipelines – Various proposals have been presented for pipelines to transfer water from Turkey to the Middle East and the countries around the Persian Gulf. Nicknamed the "Peace Pipeline," such a project would take water from the Seyhan and Ceyhan Rivers in southern Turkey as far south as Jidda and Mecca in Saudi Arabia and as far east as Sharjah in the United Arab Emirates. Along the way, water could be delivered to Damascus, Amman, Kuwait, and Israel. One version of the Peace Pipeline would deliver more than 1,000 million cubic meters of water per year, but little real progress has come of the various proposals. In part, the Arabs, particularly the Saudis, and the Israelis fear the political dominance

---

[39] Priscoli and Brumbaugh, note 6 above.

[40] A. Hindley, "Power and Water," *MEED Special Report*, 19 January 1990, v–xiv.

[41] K. Wangnick, *1990 IDA Worldwide Desalting Plants Inventory*, report no. 11 (Gnarrenburg, Germany: Wangnick Consulting, 1990); and Gleick, "Water and Energy," in Gleick, note 2 above, pages 67–79.

Table 18.4 [orig. table 4]   Desalination capacity in the Middle East as of 1990.

Country	Capacity (cubic meters per day)	Percentage of global capacity	Number of plants
Saudi Arabia	3,568,868	26.8	1,417
Kuwait	1,390,238	10.5	133
United Arab Emirates	1,332,477	10.0	290
Libya	619,354	4.7	386
Iraq	323,925	2.4	198
Qatar	308,611	2.3	59
Bahrain	275,767	2.1	126
Iran	260,609	2.0	218
Oman	186,741	1.4	79
Israel	70,062	0.5	32
Egypt	67,728	0.5	110
Jordan	8,445	0.1	13
Syria	5,743	<0.1	12
Lebanon	4,691	<0.1	10

Source: K. Wangnick, 1990 IDA Worldwide Desalting Plants Inventory, report no. 11 (Gnarrenburg, Germany, Wangnick Consulting, 1990).

of Turkey or the possible interference of other states across which the pipeline would pass.[42] The recent Turkish threats to cut off Euphrates River water to Syria have not helped to lessen this perception. Variants on the longer pipeline, such as a shorter version extending only as far as Amman, have also been proposed. Such variants may have fewer political constraints, but many environmental, economic, and political problems remain to be resolved before such a major transnational construction project could begin.

Other Out-of-Basin Transfers – There have been many other proposals to transfer water to the Middle East from basins where surplus water may be available. Such transfers could be accomplished via pipelines, aqueducts, tankers, floating bags, and even towing icebergs. Among the projects proposed have been pipelines from Baluchistan across the gulf to United Arab Emirates, from the Euphrates in Iraq to Jordan, and from the Nile through El Arish to the Gaza and Negev to alleviate the severe water crisis in the Gaza Strip.[43] Each of these projects

depends on the long-term availability of surplus water and the political, economic, and environmental feasibility of transferring that water. Similarly, it has been proposed that Israel and Jordan purchase water from the Litani River in Lebanon, build a short pipeline and set of pumping plants, and move water to northern Israel, the West Bank, and Jordan. While Litani River water is used for hydroelectricity, some surplus is currently thought to be available if the economic and political price is right.[44]

Moving water by tankers or by towing "trains" of bags filled with freshwater is also being explored for supplying coastal areas. For the Gaza Strip, where overpumping of limited groundwater supplies is leading to saltwater intrusion, such alternatives may prove feasible, though technical and political obstacles still must be removed.

Med-Dead or Red-Dead Canal – Another alternative that has been suggested in various forms is to bring large quantities of seawater from the Mediterranean Sea or the Red Sea to the Dead Sea, which lies well below sea level. The large elevation drop

[42] Kolars, note 3 above.

[43] Priscoli and Brumbaugh, note 6 above.

[44] Shuval, note 9 above.

would permit the generation of hydroelectricity, which in turn could be used to satisfy the energy requirements of a desalination plant. The freshwater provided by such a system could be allocated to Israel, the occupied territories, or western Jordan, where it would reduce pressures on the limited water supplies in those regions. Brine from the desalination process or additional seawater could be diverted into the Dead Sea to help raise its level, which has dropped nearly 20 meters over the last several decades because of the use of the Jordan River – its only inflow. Many different schemes and locations have been presented for such canals, and more work is needed to explore the best routes, the best allocation of water, and the many complicated environmental and economic uncertainties posed by such projects.

Politics and International Law

International water law and institutions have important roles to play despite the fact that no satisfactory water law has been developed that is acceptable to all nations. Developing such agreements is difficult because of the many intricacies of international politics, national practices, and other complicating political and social factors. For nations sharing river basins, factors affecting the successful negotiation and implementation of international agreements include whether a nation is upstream, downstream, or sharing a river as a border; the relative military and economic strength of the nation; and the availability of other sources of water supply.

In the last few decades, however, international organizations have attempted to derive more general principles and new concepts governing shared freshwater resources. The International Law Association's Helsinki Rules of 1966 (since

modified) and the work of the International Law Commission of the United Nations are among the most important examples. In 1991, the International Law Commission completed the drafting and provisional adoption of 32 articles on the law of the Non-Navigational Uses of International Watercourses.[45] Among the general principles set forth are those of equitable utilization, the obligation not to cause harm to other riparian nations, and the obligation to exchange hydrologic and other relevant data and information on a regular basis. Questions remain, however, about the principles' relative importance and means of enforcement.[46] In particular, defining *equitable utilization* of a shared water supply remains one of the most important and difficult problems facing many nations.

Until now, individual water treaties covering river basins have been more effective, albeit on a far more limited regional basis, than the broader principles described by the International Law Commission. International treaties concerning shared freshwater resources extend back centuries, and there are hundreds of international river treaties covering everything from navigation to water quality to water rights allocations. For example, freedom of navigation was granted to a monastery in Europe in the year 805, and a bilateral treaty on the Weser River, which today flows through Germany into the North Sea, was signed in 1221.[47] Such treaties have helped reduce the risk of water conflicts in many areas, but some of them are beginning to fail as changing levels of development alter the water needs of regions and nations. The 1959 treaty on the Nile River and some limited bilateral agreements on the Euphrates between Iraq and Syria and between Iraq and Turkey, for example, are now under pressure because of changes in the political and resource situations in the regions.

[45] UN International Law Commission, *Report of the International Law Commission on the Work of Its Forty-Third Session* (New York: United Nations, 1991).

[46] S. C. McCaffrey, "Water, Politics, and International Law," in Gleick, note 2 above, pages 92–104.

[47] Food and Agriculture Organization, *Systematic Index of International Water Resources Treaties, Declarations, Acts and Cases by Basin*, Legislative Study no. 15 (Rome, 1978). This index is irregularly updated.

To make both regional treaties and broader international agreements over water more flexible, detailed mechanisms for conflict resolution and negotiations must be developed, basic hydrologic data must be acquired and shared with all parties, flexible rather than fixed water allocations are needed, and strategies for sharing shortages and apportioning responsibilities for floods must be developed before shortages become an important factor. For example, both the 1944 Colorado River treaty between the United States and Mexico and the 1959 treaty on the Nile River between Egypt and Sudan allocate fixed quantities of water, which are based on assumptions about the total average flows of each river. However, mistaken estimates of average flows or future climate changes that could alter flows prove this type of allocation to be too rigid and prone to disputes. Proportional sharing agreements, if they include agreements for openly sharing all hydrologic data, can help to reduce the risk of conflicts over water, and modifications to these treaties should be undertaken by their signatories now, before such flow changes become evident.

Existing institutions appear sufficient to design and implement the kinds of conflict resolution mechanisms designed above, but some major improvements in them are needed. The United Nations has played an important role, through the International Law Commission, in developing guidelines and principles for internationally shared watercourses, but it should continue to press for the adoption and application of the principles in water-tense regions, such as the Jordan and Euphrates river basins. Similarly, bilateral or multilateral river treaties have been effective in the past, but they should consistently include all affected parties; establish joint management committees empowered to negotiate disputes; and be flexible enough to adapt to long-term changes in hydrologic conditions, such as those that may result from global climate change. Finally, disputes over shared groundwater resources are particularly important in the Middle East. However, international groundwater law and principles are poorly developed. Some recent progress has been made, but more attention should be given to this matter in the context of the Middle East.[48]

Toward Peace and Cooperation

For all of the countries of the Middle East, long-term sustainable economic development will depend in large part upon access to clean and dependable supplies of freshwater. Access to water, in turn, will depend upon regionwide comprehensive management of the shared major river and groundwater basins. Although new sources of water may eventually be developed, cooperation over the existing water resources is essential: Unless current water supplies are equitably and efficiently allocated and used, agreements to enlarge the overall pie will be stymied.

Enormous differences remain among the parties. Jordan still has a serious dispute with Syria over the damming of and withdrawals from the Yarmuk River; no formal agreements on water rights have been worked out between the Palestinians and Israelis; Turkey, Syria, and Iraq have no formal treaty allocating the waters of the Euphrates; and rapidly growing populations throughout the region are competing for an inadequate overall water supply, raising unanswered questions about the costs of alternative water sources.

At the broadest level, the Middle East needs a comprehensive framework for planning and managing shared water resources. If necessary, such a framework could be convened by third-party nations and institutions and include regional and national studies on water supply and demand, the development of standards for the collection and dissemination of data, the establishment of Jordan and Euphrates river basin authorities with representation from all of the

---

[48] Some progress has been made in this area with the Bellagio Draft Treaty of 1989; "Transboundary Groundwaters: The Bellagio Draft Treaty," revised and augmented by R. D. Hayton and A. E. Utton, *Natural Resources Journal* 29 (Summer 1989): 663–722.

people dependent on those water resources, and the identification of mechanisms for implementing joint projects. Some of the goals of a framework water convention would include identifying minimum water requirements and the equitable allocation of water; water-use efficiency capabilities and goals; means for shifting water use within and among sectors, such as through water "banks" or marketing; and objectives for providing new supplies. The opportunity for conflict over water in the Middle East is high, but peaceful, effective cooperation remains a goal worth striving for.

19

RED TIDES*

Donald M. Anderson

Late in 1987 scientists faced a baffling series of marine catastrophes. First, 14 humpback whales died in Cape Cod Bay, Mass., during a five-week period. This die-off, equivalent to 50 years of "natural" mortality, was not a stranding, in which healthy whales beach themselves. Instead the cetaceans died at sea – some rapidly – and then washed ashore. Postmortem examinations showed that the whales had been well immediately before their deaths and that many of them had abundant blubber and fish in their stomachs, evidence of recent feeding. Alarmed and saddened, the public and press blamed pollution or a chemical spill for the mysterious deaths.

Two more mass poisonings occurred that month, but the victims in these new cases were humans. Fishermen and beachgoers along the North Carolina coast started complaining of respiratory problems and eye irritation. Within days, residents and visitors who had eaten local shellfish experienced diarrhea, dizziness and other symptoms suggesting neurotoxic poisoning. The illnesses bewildered epidemiologists and even prompted public conjecture that a nearby sunken submarine was leaking poison gas.

Concurrently, hospitals in Canada began admitting patients suffering from disorientation, vomiting, diarrhea and abdominal cramps. All had eaten mussels from Prince Edward Island. Although Canadian authorities had dealt with shellfish poisoning outbreaks for decades, these symptoms were unfamiliar and disturbing: some patients exhibited permanent short-term memory loss. They could remember addresses but could not recall their most recent meal, for example. The officials quickly restricted the sale and distribution of mussels but eventually reported three deaths and 105 cases of acute poisoning in humans.

We now know that these seemingly unrelated events were all caused, either directly or indirectly, by toxic, single-celled algae called phytoplankton – vast blooms of which are commonly referred to as red tides. Although red tides have been recorded throughout history, the incidents mentioned above were entirely unexpected. As we shall see, they illustrate several major issues that have begun to challenge the scientific and regulatory communities.

Indeed, there is a conviction among many experts that the scale and complexity of this natural phenomenon are expanding. They note that the number of toxic blooms, the economic losses from them, the types of resources affected and the kinds of toxins and toxic species have all increased. Is this expansion real? Is it a global epidemic, as some claim? Is it related to human activities, such as rising coastal pollution? Or is it a

* Originally published in *Scientific American*, 1994, August, pp. 52–8.

result of increased scientific awareness and improved surveillance or analytical capabilities? To address these issues, we must understand the physiological, toxicological and ecological mechanisms underlying the growth and proliferation of red tide algae and the manner in which they cause harm.

Certain blooms of algae are termed red tides when the tiny pigmented plants grow in such abundance that they change the color of the seawater to red, brown or even green. The name is misleading, however, because many toxic events are called red tides even when the waters show no discoloration. Likewise, an accumulation of nontoxic, harmless algae can change the color of ocean water. The picture is even more complicated: some phytoplankton neither discolor the water nor produce toxins but kill marine animals in other ways. Many diverse phenomena thus fall under the "red tide" rubric.

Of the thousands of living phytoplankton species that make up the base of the marine food web, only a few dozen are known to be toxic. Most are dinoflagellates, prymnesiophytes or chloromonads. A bloom develops when these single-celled algae photosynthesize and multiply, converting dissolved nutrients and sunlight into plant biomass. The dominant mode of reproduction is simple asexual fission – one cell grows larger, then divides into two cells, the two split into four, and so on. Barring a shortage of nutrients or light, or heavy grazing by tiny zooplankton that consume the algae, the population's size can increase rapidly. In some cases, a milliliter of seawater can contain tens or hundreds of thousands of algal cells. Spread over large areas, the phenomenon can be both visually spectacular and catastrophic.

Some species switch to sexual reproduction when nutrients are scarce. They form thick-walled, dormant cells, called cysts, that settle on the seafloor and can survive there for years. When favorable growth conditions return, cysts germinate and reinoculate the water with swimming cells that can then bloom. Although not all red tide species form cysts, many do, and this transformation explains important aspects

of their ecology and biogeography. The timing and location of a bloom can depend on when the cysts germinate and where they were deposited, respectively. Cyst production facilitates species dispersal as well; blooms carried into new waters by currents or other means can deposit "seed" populations to colonize previously unaffected areas.

A dramatic example of natural dispersal occurred in 1972, when a massive red tide reaching from Maine to Massachusetts followed a September hurricane. The shellfish toxicity detected then for the first time has recurred in that region virtually every year now for two decades. The cyst stage has provided a very effective strategy for the survival and dispersal of many other red tide species as well.

How do algal blooms cause harm? One of the most serious impacts on human life occurs when clams, mussels, oysters or scallops ingest the algae as food and retain the toxins in their tissues. Typically the shellfish themselves are only marginally affected, but a single clam can sometimes accumulate enough toxin to kill a human being. These shellfish poisoning syndromes have been described as paralytic, diarrhetic and neurotoxic, shortened to PSP, DSP and NSP. The 1987 Canadian outbreak in which some patients suffered memory loss was appropriately characterized as amnesic shellfish poisoning, or ASP. The North Carolina episode was NSP.

A related problem, ciguatera fish poisoning, or CFP, causes more human illness than any other kind of toxicity originating in seafood. It occurs predominantly in tropical and subtropical islands, where from 10,000 to 50,000 individuals may be affected annually. Dinoflagellates that live attached to seaweeds produce the ciguatera toxins. Herbivorous fishes eat the seaweeds and the attached dinoflagellates as well. Because ciguatera toxin is soluble in fat, it is stored in the fishes' tissues and travels through the food web to carnivores. The most dangerous fish to eat are thus the largest and oldest, often considered the most desirable as well.

Symptoms do vary among the different

syndromes but are generally neurological or gastrointestinal, or both. DSP causes diarrhea, nausea and vomiting, whereas PSP symtoms include tingling and numbness of the mouth, lips and fingers, accompanied by general muscular weakness. Acute doses inhibit respiration, and death results from respiratory paralysis. NSP triggers diarrhea, vomiting and abdominal pain, followed by muscular aches, dizziness, anxiety, sweating and peripheral tingling. Ciguatera induces an intoxication syndrome nearly identical to NSP.

Illnesses and deaths from algal-derived shellfish poisons vary in number from year to year and from country to country. Environmental fluctuations profoundly influence the growth and accumulation of algae and thus their toxicity as well. Furthermore, countries differ in their ability to monitor shellfish and detect biotoxins before they reach the market. Developed countries typically operate monitoring programs that permit the timely closure of contaminated resources. Illnesses and deaths are thus rare, unless a new toxin appears (as in the ASP crisis in Canada) or an outbreak occurs in an area with no history of the problem (as in North Carolina). Developing countries, especially those having long coastlines or poor populations who rely primarily on the sea for food, are more likely to incur a higher incidence of sickness and death from algal blooms.

Phytoplankton can also kill marine animals directly. In the Gulf of Mexico, the dinoflagellate *Gymnodinium breve* frequently causes devastating fish kills. As the wild fish swim through *G. breve* blooms, the fragile algae rupture, releasing neurotoxins onto the gills of the fish. Within a short time, the animals asphyxiate. Tons of dead fish sometimes cover the beaches along Florida's Gulf Coast, causing several millions of dollars to be lost in tourism and other recreation-based businesses.

Farmed fish are especially vulnerable because the caged animals cannot avoid the blooms. Each year, farmed salmon, yellowtail and other economically important species fall victim to a variety of algal species. Blooms can wipe out entire fish farms within hours, killing fingerlings and large fish alike. Algal blooms thus pose a large threat to fish farms and their insurance providers. In Norway an extensive program is under way to minimize these impacts. Fish farmers make weekly observations of algal concentrations and water clarity. Other parameters are transmitted to shore from instruments on moored buoys. The Norwegian Ministry of Environment then combines this information with a five-day weather forecast to generate an "algal forecast" for fish farmers and authorities. Fish cages in peril are then towed to clear water.

Unfortunately, not much more can be done. The ways in which algae kill fish are poorly understood. Some phytoplankton species produce polyunsaturated fatty acids and galactolipids that destroy blood cells. Such an effect would explain the ruptured gills, hypoxia and edema in dying fish. Other algal species produce these hemolytic compounds and neurotoxins as well. The combination can significantly reduce a fish's heart rate, resulting in reduced blood flow and a deadly decrease in oxygen.

Moreover, nontoxic phytoplankton can kill fish. The diatom genus *Chaetoceros* has been linked to dying salmon in the Puget Sound area of Washington State, yet no toxin has ever been identified in this group. Instead species such as *C. convolutus* sport long, barbed spines that lodge between gill tissues and trigger the release of massive amounts of mucus. Continuous irritation exhausts the supply of mucus and mucous cells, causing lamellar degeneration and death from reduced oxygen exchange. These barbed spines probably did not evolve specifically to kill fish, since only caged fish succumb to the blooms. The problems faced by fish farmers are more likely the unfortunate result of an evolutionary strategy by certain *Chaetoceros* species to avoid predation or to stay afloat.

Algal toxins also cause mortalities as they move through the marine food web. Some years ago tons of herring died in the Bay of Fundy after consuming small planktonic snails that had eaten the PSP-producing dinoflagellate *Alexandrium*. From the human health standpoint, it is fortunate that herring, cod, salmon and other commercial fish are sensitive to these toxins and, unlike shellfish, die before toxins reach dangerous levels in their flesh. Some toxin, however, accumulates in the liver and other organs of certain fish, and so animals such as other fish, marine mammals and birds that consume whole fish, including the viscera, are at risk.

We now can reconstruct the events that killed the whales in 1987. A few weeks of intense investigations that year by marine pathologist Joseph R. Geraci of the Ontario Veterinary College, myself and many others revealed that the PSP toxins most likely caused these deaths. The dinoflagellate *Alexandrium tamarense* produced the toxins, which reached the whales via their food web. We analyzed mackerel that the whales had been eating and found saxitoxin, not in their flesh but concentrated in the liver and kidney. Presumably the mackerel ate zooplankton and small fish that had previously dined on *Alexandrium*.

The humpbacks were starting their southward migration and were feeding heavily. Assuming that they consumed 4 percent of their body weight daily, we calculated that they received a saxitoxin dosage of 3.2 micrograms per kilogram of body weight. But was this a fatal dose? Unfortunately, in 1987 we had no data that directly addressed how much toxin would kill a whale. We knew the minimum lethal dose of saxitoxin for humans is seven to 16 micrograms per kilogram of body weight, but that was two to five times more than what the whales had probably ingested.

Our calculations were initially disheartening, but as we thought about it we realized that whales might be more sensitive to the toxins than are humans. First, whales would have received continual doses of toxin as they fed, whereas human mortality statistics are based on single feedings. Second, during a dive, the mammalian diving reflex channels blood and oxygen predominantly to the heart and brain. The same mechanism sometimes protects young children who fall through thin ice and survive drowning, despite being underwater for half an hour or longer. For humans, cold water induces the reflex, but for whales, it is activated during every dive.

Each dive then would expose the most sensitive organs to the toxin, which would bypass the liver and kidney, where it could be metabolized and excreted. Finally, saxitoxin need not have killed the whales directly. Even a slightly incapacitated animal might have difficulty orienting to the water surface or breathing correctly. The whales may actually have drowned following a sub-lethal exposure to saxitoxin. The exact cause will never be known, but the evidence strongly suggests that these magnificent creatures died from a natural toxin originating in microscopic algae.

Other examples of toxins traveling up the food web appear nearly every year. In 1991 sick or dying brown pelicans and cormorants were found near Monterey Bay, California Wildlife experts could find no signs that pesticides, heavy metals or other pollutants were involved. The veterinarian in charge of the study telephoned Jeffrey Wright of the National Research Council laboratory in Halifax, Nova Scotia. Wright had directed the Canadian Mussel Toxin Crisis Team that identified the poison responsible for the mysterious ASP episode in 1987. His team had isolated a toxin from the Prince Edward Island mussels, called domoic acid, and traced it to its source – a diatom, *Pseudonitzschia pungens*, that had been considered harmless. Four years later members of the same Canadian team quickly ascertained that the sick and dying birds in California had eaten anchovies that contained domoic acid, again from *Pseudonitzschia* (but different species).

The toxins responsible for these syndromes are not single chemical entities but are families of compounds having similar chemical structures and effects. For example, the saxitoxins that cause PSP are a family of at least 18 different compounds with widely differing potencies. Most algal toxins cause human illness by disrupting electrical conduction, uncoupling communication between nerve and muscle, and impeding critical physiological processes. To do so, they bind to specific membrane receptors, leading to changes in the intracel-

Box 19.1   Algal Toxins
The structure of red tide toxins varies considerably. Saxitoxin compounds sport different combinations of H^+, OH^- and SO_3^- on the R1 to R4 sites, but all members of this family block the sodium channel and thus prevent communication between neurons and muscles. H. Robert Guy of the National Institutes of Health has proposed a structural model of this interaction. The carbon backbone of the sodium channel is colored gray, the carboxyls are red, nitrogen is blue, and hydrogen is white. Saxitoxin binds in the narrowest region of this channel. The brevetoxins that cause NSP are much larger molecules that also affect the sodium channel.

lular concentration of ions such as sodium or calcium.

The saxitoxins bind to sodium channels and block the flux of sodium in and out of nerve and muscle cells. Brevetoxins, the family of nine compounds responsible for NSP, bind to a different site on the sodium channel but cause the opposite effect from saxitoxin. Domoic acid disrupts normal neurochemical transmission in the brain. It binds to kainate receptors in the central nervous system, causing a sustained depolarization of the neurons and eventually cell degeneration and death. Memory loss in ASP victims apparently results from lesions in the hippocampus, where kainate receptors abound.

Why do algal species produce toxins? Some argue that toxins evolved as a defense mechanism against zooplankton and other grazers. Indeed, some zooplankton can become slowly incapacitated while feeding, as though they are being gradually paralyzed or otherwise impaired. (In one study, a tintinnid ciliate could swim only backward, away from its intended prey, after exposure to toxic dinoflagellates.) Sometimes grazing animals spit out the toxic algae as though they had an unpleasant taste. These responses would all reduce grazing and thus facilitate bloom formation.

All the same, nontoxic phytoplankton also form blooms, and so it is unlikely that toxins serve solely as self-defense. Scientists are looking within the algae for biochemical pathways that require the toxins, but the search thus far has been fruitless. The toxins are not proteins, and all are synthesized in a series of chemical steps requiring multiple genes. Investigators have proposed biosynthetic pathways, but they have not isolated chemical intermediates or enzymes used only in toxin production. It has thus been difficult to apply the powerful tools of molecular biology to these organisms, other than to study their genes or to develop detection tools.

We do have some tantalizing clues about toxin metabolism. For example, certain dinoflagellate strains produce different amounts of toxin and different sets of toxin derivatives when we vary their growth conditions. Metabolism of the toxins is a dynamic process, but we still do not know whether they have a specific biochemical role. As with the spiny diatoms that kill fish, the illnesses and mortalities caused by algal "toxins" may be the result of the accidental chemical affinity of those metabolites for receptor sites on ion channels in higher animals.

The potential role of bacteria or bacterial genes in phytoplankton toxin production is an area of active research. We wonder how a genetically diverse array of organisms, including phytoplankton, seaweeds, bacteria and cyanobacteria, could all have evolved the genes needed to produce saxitoxin [see "The Toxins of Cyanobacteria," by Wayne W. Carmichael; *Scientific American*, January]. Several years ago Masaaki Kodama of Kitasato University in Japan isolated intracellular bacteria from antibiotic-treated *A. tamarense* cultures and showed that the bacteria produced saxitoxin. This finding supported an old and long-ignored hypothesis that toxins might originate from bacteria living inside or on the dinoflagellate cell.

Despite considerable study, the jury is still out. Many scientists now accept that some bacteria produce saxitoxins, but they point out that dense bacterial cultures produce extremely small quantities. It is also not clear that those bacteria can be found inside dinoflagellates. That intracellular bacteria produce all of the toxin found in a dinoflagellate cell therefore seems unlikely, but perhaps some synergism occurs between a small number of symbionts and the host dinoflagellate that is lost when the bacteria are isolated in culture. Alternatively, a bacterial gene or plasmid might be involved.

Given the diverse array of algae that produce toxins or cause problems in a variety of oceanographic systems, attempts to generalize the dynamics of harmful algal blooms are doomed to fail. Many harmful species, however, share some mechanisms. Red tides often occur when heating or freshwater runoff creates a stratified surface layer above colder, nutrient-rich waters. Fast-growing algae quickly strip away nutrients in the upper layer, leaving nitrogen and phosphorus only below the interface of the layers, called the pycnocline. Nonmotile phytoplankton cannot easily get to this layer, whereas motile algae, including dinoflagellates, can thrive. Many swim at speeds in excess of 10 meters per day, and some undergo daily vertical migration: they reside in surface waters by day to harvest sunlight like sunbathers, then swim down to the pycnocline to take up nutrients at night. As a result, blooms can suddenly appear in surface waters that are devoid of nutrients and would seem incapable of supporting such prolific growth.

A similar sleight-of-hand can occur horizontally, though over much larger distances. The NSP outbreak in North Carolina illustrates how ocean currents can transport major toxic species from one area to another. Patricia A. Tester, a biologist at the National Oceanic and Atmospheric Administration's National Marine Fisheries Service laboratory in Beaufort, examined plankton from local waters under a microscope soon after the initial reports of human illnesses. She saw cells resembling the dinoflagellate *G. breve*, the cause of recurrent NSP along Florida's western coast. Experts quickly confirmed her tentative identification, and for the first time in state history, authorities closed shellfish beds because of algal toxins, resulting in a loss of $20 million.

Tester and her co-workers have since used satellite images of sea-surface temperatures to argue that the *G. breve* population in North Carolina originated off the southwestern coast of Florida, nearly 1,000 kilometers away. That bloom traveled from the Gulf of Mexico up the southeastern coast of the U.S., transported by several current systems culminating in the Gulf Stream. After 30 days of transport, a filament of water separated from the Gulf Stream and moved onto North Carolina's narrow continental shelf, carrying *G. breve* cells with it. The warm water mass remained in nearshore waters, identifiable in satellite images for three weeks. Fortunately, *G. breve* does not have a known cyst stage, so it could not establish a seedbed and colonize this new region.

This incident, taken together with many others like it throughout the world, speaks of an unsettling trend. Problems from harmful red tides have grown worse over the past two decades. The causes, however, are multiple, and only some relate to pollution or other human activities. For example, the global expansion in aquaculture means that more areas are monitored closely, and more fisheries' products that can be killed or take up toxins are in the water. Likewise, our discovery of toxins in algal species formerly considered nontoxic reflects the maturation of this field of science, now profiting from more investigators, better analytical techniques and chemical instrumentation, and more efficient communication among workers.

Long-term studies at the local or regional level do show that red tides (in the most general sense of the term) are increasing as coastal pollution worsens. Between 1976 and 1986, as the population around Tolo Harbor in Hong Kong grew sixfold, red tides increased eight-fold. Pollution presumably provided more nutrients to the algae. A similar pattern emerged in the Inland Sea of Japan, where visible red tides proliferated steadily from 44 per year in 1965 to more than 300 a decade later. Japanese authorities instituted rigorous effluent controls in the mid-1970s, and a 50 percent reduction in the number of red tides ensued.

These examples have been criticized, since both could be biased by changes in the numbers of observers through time, and both are tabulations of water discolorations from blooms, not just toxic or harmful episodes. Still, the data demonstrate what should be an obvious relationship: coastal waters receiving industrial, agricultural and domestic waste, frequently rich in plant nutrients, should experience a general increase in algal growth. These nutrients can enhance toxic or harmful episodes in several ways. Most simply, all phytoplankton species, toxic and nontoxic, benefit, but we notice the enrichment of toxic ones more. Fertilize your lawn, and you get more grass – and more dandelions.

Some scientists propose instead that pollution selectively stimulates harmful species. Theodore J. Smayda of the University of Rhode Island brings the nutrient ratio hypothesis, an old concept in the scientific literature, to bear on toxic bloom phenomena. He argues that human activities have altered the relative availability of specific nutrients in coastal waters in ways that favor toxic forms. For example, diatoms, most of which are harmless, require silicon in their cell walls, whereas other phytoplankton do not. Because silicon is not abundant in sewage, but nitrogen and phosphorus are, the ratio of nitrogen to silicon or of phosphorus to silicon in coastal waters has increased over the past several decades. Diatom growth ceases when silicon supplies are depleted, but other phytoplankton classes, which often include more toxic species, can proliferate using "excess" nitrogen and phosphorus. This idea is controversial but not unfounded. A 23-year time series from the German coast documents a fourfold rise in the nitrogen-silicon and phosphorus-silicon ratios, accompanied by a striking change in the composition of

the phytoplankton community: diatoms decreased, whereas flagellates increased more than 10-fold.

Another concern is the long-distance transport of algal species in cargo vessels. We have long recognized that ships carry marine organisms in their ballast water, but evidence is emerging that toxic algae have also been hitchhiking across the oceans. Gustaaf M. Hallegraeff of the University of Tasmania has frequently donned a miner's helmet and ventured into the bowels of massive cargo ships to sample sediments accumulated in ballast tanks. He found more than 300 million toxic dinoflagellate cysts in one vessel alone. Hallegraeff argues that one PSP-producing dinoflagellate species first appeared in Tasmanian waters during the past two decades, concurrent with the development of a local wood-chip industry. Empty vessels that begin a journey in a foreign harbor pump water and sediment into their tanks for ballast; when wood chips are loaded in Tasmania, the tanks are discharged. Cysts easily survive the transit cruise and colonize the new site. Australia has now issued strict guidelines for discharging ballast water in the country's ports. Unfortunately, most other nations do not have such restrictions.

The past decade may be remembered as the time that humankind's effect on the global environment caught the public eye in a powerful and ominous fashion. For some, signs of our neglect come with forecasts of global warming, deforestation or decreases in biodiversity. For me and my colleagues, this interval brought a bewildering expansion in the complexity and scale of the red tide phenomenon. The signs are clear that pollution has enhanced the abundance of algae, including harmful and toxic forms. This effect is obvious in Hong Kong and the Inland Sea of Japan and is perhaps real but less evident in regions where coastal pollution is more gradual and unobtrusive. But we cannot blame all new outbreaks and new problems on pollution. There are many other factors that contribute to the proliferation of toxic species; some involve human activities, and some do not. Nevertheless, we may well be witnessing a sign that should not be ignored. As a growing world population demands more and more of fisheries' resources, we must respect our coastal waters and minimize those activities that stimulate the spectacular and destructive outbreaks called red tides.

Further Reading

Primary production and the global epidemic of phytoplankton blooms in the sea: a linkage? Theodore J. Smayda in *Novel Phytoplankton Blooms: Causes and Impacts of Recurrent Brown Tide and Other Unusual Blooms*. Edited by E. M. Cosper, V. M. Bricelj and E. J. Carpenter. Springer-Verlag, 1989.

Marine biotoxins at the top of the food chain. Donald M. Anderson and Alan W. White in *Oceanus*, Vol. 35, No. 3, pages 55–61; Fall 1992.

Domoic acid and amnesic shellfish poisoning: a review. Ewen C. D. Todd in *Journal of Food Protection*, Vol. 56, No. 1, pages 69–83; January 1993.

A review of harmful algal blooms and their apparent global increase. Gustaaf M. Hallegraeff in *Phycologia*, Vol. 32, No. 2, pages 79–99; March 1993.

Marine toxins. Takeshi Yasumoto and Michio Murata in *Chemical Reviews*, Vol. 93, No. 5, pages 1897–1909; July/August 1993.

20

LESSONS FROM INTENTIONAL OIL POLLUTION*

Ronald B. Mitchell

When public concern about an international environmental problem is strong enough to require action, governments often respond by negotiating a treaty. More than 120 multilateral treaties have been negotiated to address such problems, along with hundreds of regional and bilateral ones.[1] This profusion of international environmental law will prove positive if it leads to substantive changes in behavior. Yet nations regularly violate even relatively low-cost

reporting requirements and ignore more substantive provisions in many environmental treaties.[2]

Why doesn't compliance follow automatically from a nation's signature? A recent study of compliance with international treaties regulating intentional oil pollution (discharges of waste oil at sea) by tankers may offer some insights into this discrepancy.[3] (The nature and extent of this problem are discussed in box 20.1). By comparing the

* Originally published in *Environment*, 1995, vol. 37, no. 4, pp. 10–15, 36–41.

[1] For the most extensive collection of environmental treaty texts, see B. Ruster and B. Simma, eds., *International Protection of the Environment: Treaties and Related Documents* (Dobbs Ferry, N.Y.: Oceana Publications, 1975), (hereafter cited as *IPE*).

[2] On reporting, see U.S. General Accounting Office, *International Environment: International Agreements Are Not Well Monitored*, GAO/RCED-92-43 (Washington, D.C., 1992); and on effectiveness, see P. H. Sands, ed., *The Effectiveness of International Environmental Agreements: A Survey of Existing Legal Instruments* (Cambridge, U.K.: Grotius Publications Limited, 1992).

[3] R. B. Mitchell, *Intentional Oil Pollution at Sea: Environmental Policy and Treaty Compliance* (Cambridge, Mass.: MIT Press, 1994), 9. Another excellent work on this issue is R. M. M'Gonigle and M. W. Zacher, *Pollution, Politics, and International Law: Tankers at Sea* (Berkeley, Calif.: University of California Press, 1979). Preliminary findings of several project teams working on the issues of treaty compliance and effectiveness are contained in A. Chayes and A. H. Chayes, *The New Sovereignty: Compliance with International Regulatory Agreements* (Cambridge, Mass.: Havard University Press, forthcoming); K. Hanf and A. Underdal (Erasmus University, Rotterdam, the Netherlands), "Domesticating International Commitments: Linking National and International Decisionmaking" (paper presented to the Helsinki meeting on Managing Foreign Policy Issues Under Conditions of Change, 1991); H. K. Jacobson and E. Brown Weiss, "Improving Compliance with International Environmental Accords," *Global Governance* 1, no. 2 (forthcoming); M. A. Levy, O. R. Young, and M. Zürn, *The Study of International Regimes*, IIASA Working Paper WP-94-113 (Laxenburg, Austria, 1994); J. Cameron, *Compliance and International Environmental Treaties* (London: Foundation for International Environmental Law and

different regulatory strategies within those treaties, it was possible to identify those that have had the most success in inducing compliance and causing the intended behavioral changes. This, in turn, suggested a number of policy prescriptions regarding the types of rules that should be imposed and the ways in which new environmental accords should be drafted (or existing ones redrafted) to elicit higher compliance. As Maurice Strong, secretary-general of the United Nations Conference on Environment and Development, has noted, "We have to push like hell to make sure implementation takes place."[4] These efforts will be most fruitful, however, if the correct direction in which to push is determined first.

Four decades of experience with the International Convention for Prevention of Pollution of the Sea by Oil (OILPOL) and the International Convention for Prevention of Pollution from Ships (MARPOL) show that treaties can – but do not always – lead powerful governments and corporations to behave in new ways, particularly if they opposed them initially.[5] (The provisions of these treaties and the technical means of preventing intentional oil pollution are summarized in box 20.2.) The study found reason to reject both the blind faith of international lawyers that international law "is a force with inherent strength of its own" and the cynical view of realist international

relations scholars that "considerations of power rather than of law determine compliance."[6] Even within this one problem area, compliance levels differed across treaty provisions. Higher compliance occurred when negotiators formulated those provisions to regulate the agents that had less power to violate them and/or greater interest in complying with them. The levels of environmental concern shown by governments and the resources they devoted to environmental protection clearly placed upper limits on compliance levels. But how policymakers chose to approach compliance, monitoring, and enforcement significantly influenced subsequent levels of compliance: Policymakers succeeded when they ensured that those responsible for complying, monitoring, or enforcing a treaty provision were placed in a "strategic triangle" in which they had the incentives, the ability, and the authority to perform those functions (see figure 20.1).

The Experience with Intentional Oil Pollution

International attempts to control intentional oil pollution include three cases in which those affected complied with the rules concerning one type of behavior while at the same time failing to comply with the rules concerning another, very similar type of

EarthScan, forthcoming); J. Wettestad and S. Andresen (Fridtjof Nansen Institute, Lysaker, Norway), "The Effectiveness of International Resource Cooperation: Some Preliminary Findings" (paper presented to the 32nd annual meeting of the International Studies Association, Vancouver, Canada, 1991); and O. Young (Dartmouth College, Hanover, New Hampshire), "On the Effectiveness of International Regimes: Defining Concepts and Identifying Variables" (working paper prepared for the Research Team Studying the Effectiveness of International Regimes, 1991).

[4] Quoted in J. Brooke, "U.N. Chief Closes Summit with an Appeal for Action," *New York Times*, 15 June 1992, A8.

[5] These agreements and their amendments are as follows: "International Convention for the Prevention of Pollution of the Sea by Oil," 12 May 1954, *IPE*, vol. 1, 332; "1962 Amendments to the International Convention for the Prevention of Pollution of the Sea by Oil," 11 April 1962, *IPE*, vol. 1, 346; "1969 Amendments to the International Convention for the Prevention of Pollution of the Sea by Oil," 21 October 1969, *IPE*, vol. 1, 366; "International Convention for the Prevention of Pollution from Ships," 2 November 1973, *International Legal Materials* 12 (1973): 1319; and "Protocol of 1978 Relating to the International Convention for the Prevention of Pollution from Ships," 17 February 1978, *International Legal Materials* 17 (1978): 1546.

[6] J. L. Brierly, *The Outlook for International Law* (Oxford, U.K.: The Clarendon Press, 1944), 1; and H. Morgenthau, *Politics Among Nations: The Struggle for Power and Peace* (New York: Alfred A. Knopf, 1978), 299.

Box 20.1   Intentional Oil Pollution

To most people, oil pollution conjures up images of massive oil spills due to tanker accidents like that of the *Exxon Valdez.*(1) Yet intentional oil discharges from tanker operations have consistently overshadowed accidents as the major source of ship-related oil pollution (see table 20.1).(2) After a tanker delivers its cargo, a small amount of oil remains clinging to the tank walls like the residue visible after a glass of milk is emptied. Traditionally, tanker captains filled empty cargo tanks with sea water to provide ballast on the return voyage and then used sea water in high-pressure cleaning machines to wash down their tanks before loading more oil. The resultant mixtures of waste oil and water were usually discharged at sea. Although clingage represents only about 0.4 percent of total cargo, this translates to 400 tons of oil discharged for each voyage of a typical tanker. Given the large volume of oil transported by sea, such discharges were a major pollution problem by the 1950s.

Discharges of crude oil do not persist indefinitely, but they can travel long distances before breaking up, thus posing environmental and aesthetic threats to coasts far from their release. Their most visible impact (and the most frequent source of public concern) has been on seabirds. Contact with oil destroys the insulating ability of the birds' feathers, and it can cause internal damage when they ingest it in attempting to clean themselves. Beyond this, however, scientists disagree over the extent of the environmental harm caused by oil discharges. Some contend that the low-concentration but

Table 20.1   Sources of oil in the sea.

Source	1971	1980	1989
		(Thousands of metric tons)	
Transportation:			
Tanker operations	1,080	700	159
Dry docking	250	30	4
Terminal operations	3	20	30
Pumping of bilges	500	300	253
Accidents	300	420	121
Scrappings of old tankers	*	*	3
Subtotal	2,133	1,470	570
Offshore production	80	50	*
Municipal and industrial wastes	2,700	1,180	*
Natural sources	600	250	*
Atmospheric emissions	600	300	*
Total	6,113	3,250	570
Selected detail:			
Discharge from tanker operations	1,080	700	159
Crude oil transported by sea	1,100,000	1,319,000	1,097,000
Discharge as a percentage of crude oil transported by sea	0.098	0.053	0.014

* Not available
Sources: National Academy of Sciences, *Petroleum in the Marine Environment* (Washington, D.C., 1975); National Academy of Sciences and National Research Council, *Oil in the Sea: Inputs, Fates and Effects* (Washington, D.C.: National Academy Press, 1985); and International Maritime Organization, *Petroleum in the Marine Environment*, MEPC 30/INF.13 (London, September 1990).

frequent oilings cause major long-term harm to fish, shellfish, and other marine life.(3) Others find no evidence that even major oil spills "have unalterably changed the world's oceans or marine resources."(4) Although less dramatic than major tanker spills, the appearance of small oil patches and tar balls on resort beaches has also prompted regular public complaints, especially in developed countries. These environmental and aesthetic concerns have provided the impetus for virtually all the international efforts to regulate oil pollution.(5)

1   The *Exxon Valdez* spilled 35,000 tons of oil into Prince William Sound, Alaska, on 24 March 1989.

2   See National Academy of Sciences, *Petroleum in the Marine Environment* (Washington, D.C., 1975); National Academy of Sciences and National Research Council, *Oil in the Sea: Inputs, Fates and Effects* (Washington, D.C.: National Academy Press, 1985); and International Maritime Organization, *Petroleum in the Marine Environment*, MEPC 30/Inf. 13 (London, September 1990).

3   On the impact of oil pollution on birds, see C. J. Camphuysen, *Beached Bird Surveys in the Netherlands, 1915–1988: Seabird Mortality in the Southern North Sea Since the Early Days of Oil Pollution* (Amsterdam: Werkgroep Noordzee, 1989). For estimates of the harm done to other creatures, see the sources in note 2 above.

4   National Academy of Sciences and National Research Council, note 2 above, 489. See also GESAMP [IMO/FAO/UNESCO/WMO/WHO/IAEA/UN/UNEP Joint Group of Experts on the Scientific Aspects of Marine Pollution], *The State of the Marine Environment*, GESAMP Reports and Studies No. 39 (New York: United Nations, 1990), 2; Second International Conference in the Protection of the North Sea, *Quality Status of the North Sea* (London: Her Majesty's Stationery Office, 1987), 72–73; United Kingdom Royal Commission On Environmental Pollution, *Eighth Report: Oil Pollution of the Sea* (London: Her Majesty's Stationery Office, 1981), 38, 46–49, and 266; and J. M. Baker, *Impact of Oil Pollution on Living Resources* (Gland, Switzerland: International Union for Conservation of Nature and Natural Resources, 1983), 40.

5   See United Kingdom Ministry of Transport, *Report of the Committee on the Prevention of Pollution of the Sea by Oil* (London: Her Majesty's Stationery Office, 1953), 1–2.

Figure 20.1 [*orig. figure 1*]   The strategic triangle of compliance.

behavior. The most striking contrast was between the almost universal compliance with MARPOL equipment standards requiring tankers to install segregated ballast tanks (SBTs), which remove a major source of oil pollution from ships, and the frequent violations of MARPOL discharge standards limiting the amount and location of discharges. Tanker owners installed SBTs by required dates, even though this entailed significant investments with no offsetting benefits and even though decreasing oil prices were increasing the pressure to cut costs. Even tankers registered in countries that had opposed adoption of the equipment rules complied. They did so not in response to the threat of enforcement by other countries but as a direct response to a "coerced compliance" approach that sought

Box 20.2   International Regulation of Intentional Oil Pollution

Intentional oil pollution was one of the first environmental problems to receive international attention, with draft treaties being negotiated (but never signed) in both 1926 and 1935. By 1954, growth in both the amount of crude oil being transported by sea and concern about birds and beaches led to the signing of the International Convention for Prevention of Pollution of the Sea by Oil (OILPOL). OILPOL required that all tankers keep the oil content of discharges below 100 parts per million when within 50 miles of land but left discharges outside these zones unregulated. Governments were required to impose fines large enough to deter violations, to ensure that reception facilities were provided for waste oil, and to report on various treaty-related matters.

Minor amendments in 1962 were followed by amendments in 1969 that required that discharges within these 50-mile zones leave no "visible sheen," that discharges outside these zones not exceed 60 liters of oil per mile, and that total discharges not exceed 1/15,000th of total cargo capacity. Tanker captains were expected to reduce their total discharges by consolidating ballast and tank-cleaning slops in a single tank, letting gravity separate the oil from the water, discharging the clean water from beneath the oil, and then loading the next cargo on top of the oil that remained (a system known as load-on-top).

In 1993, the International Convention for Prevention of Pollution from Ships (MARPOL) made the regulation of oil pollution one part of an agreement that addressed four other forms of ship-generated marine pollution. MARPOL retained the 1969 discharge standards essentially unchanged but added requirements that all large new tankers install segregated ballast tanks (SBTs), a system that outfits tanks and other ballast spaces equal to 30 percent of the tanker's capacity with a separate piping system that never carries oil. Difficulties in obtaining ratification of the initial MARPOL agreement prompted a protocol conference in 1978. The protocol, negotiated as an integral part of the earlier agreement, required all existing tankers to install either SBT or crude oil washing (COW) systems and required all new tankers, both large and small, to install both such systems. (COW systems use crude oil itself instead of water to remove oil clinging to tank walls for delivery with the current cargo, thereby significantly reducing the amount of oil wasted due to tank cleaning and ballasting.) MARPOL authorized countries to detain foreign tankers as a sanction, set specific dates by which reception facilities were required, and strengthened reporting requirements. Since 1978, major amendments to MARPOL have required accident mitigation equipment and limits on air pollutants produced by ships.

primarily to prevent violations.[7] The rules facilitated initial surveys and inspections by nongovernmental classification societies that made it hard for a tanker lacking the required equipment to receive the classification and insurance papers needed to trade internationally. The treaty reinforced this system by establishing the framework for more effective in-port monitoring and enforcement, the former of which has increased significantly.

Two facts confirm that the form of this compliance system, rather than some happy coincidence of economic or other factors, caused tanker owners to comply: Most tankers exempt from the requirements have

[7] Coerced compliance systems seek to reduce the opportunities for violations (by regulating easily observed and readily prevented activities) instead of deterring violations through the threat of detection and punishment. Domestic examples include banning handgun sales and requiring automobile manufacturers to install catalytic converters; in each case the potential violator has less opportunity to commit the violation.

not installed SBTs, and discharge limits failed to elicit high compliance levels even as equipment standards succeeded. Indeed, noncompliance with the economically more efficient discharge rules – even after many amendments – was a major impetus for adopting the equipment requirements. Clearly, tanker captains still have incentives to discharge waste oil at sea, but successful equipment regulations have prevented their acting on those incentives.

In the second case, 14 European countries signed a Memorandum of Understanding on Port State Control (MOU) in 1982 to increase regional enforcement of MARPOL and other maritime treaties.[8] Through 1990, all of those countries reported enforcement data to the MOU secretariat, while only half of them reported the same information to the secretariat of the International Maritime Organization (IMO), a United Nations body that is responsible for the regulation of international shipping. The MOU system succeeded because it had a daily reporting requirement (as opposed to IMO's annual requirement) that readily became incorporated into the enforcing bureaucracies' standard operating procedures. Port inspectors could query the MOU database to obtain otherwise unavailable information on which ships entering their harbors were most likely to be violating the law. In contrast, those responsible for reporting enforcement figures under the IMO system had no incentive to provide them.

Third, before MARPOL took effect in 1983, no government had ever detained a foreign-flagged tanker for violating oil pollution laws because this remedy was not available under international law. In contrast, at least seven countries have exercised their right to do this since that agreement took effect.[9] Although MARPOL did not transform unconcerned countries into rigorous enforcers (most countries have never detained any ships and others do so only rarely), it did create an effective enforcement tool for those with the practical ability and the political incentive to use it.

These three cases provide clear evidence that treaty rules independently influence behavior when other factors are controlled for (or absent). The cases include three successes – equipment standards, MOU reporting, and enforcement after 1983 – and three failures – discharge standards, IMO reporting, and enforcement before 1983. MARPOL presents two other examples of policy failure: The requirement that countries impose stiff penalties has been widely ignored and, despite MARPOL requirements, reception facilities for the waste oil that tankers would otherwise discharge at sea remain underprovided, especially in oil-loading ports where they are most needed.[10]

Lessons from the Oil Pollution Experience

These attempts to regulate intentional oil pollution offer several useful lessons regarding international efforts to protect the environment. First, despite the frequent assertions that most countries comply with most of their commitments most of the time,[11] significant noncompliance does occur and in some cases is quite common. The fact that treaty rules can influence behavior by no means implies that they always do. In many cases, however, noncompliance is not completely willful. Therefore, formulating treaty provisions along the proper lines can significantly increase overall compliance.

[8] G. Kasoulides, "Paris Memorandum of Understanding: A Regional Regime of Enforcement," *International Journal of Estuarine and Coastal Law* 5, no. 1–3 (1990): 180–92.

[9] Mitchell, note 3 above, table 5.5 on page 181.

[10] P. G. Sadler and J. King, "Study on Mechanisms for the Financing of Facilities in Ports for the Reception of Wastes from Ships" (discussion paper, College of Wales, Cardiff, Wales, March 1990).

[11] See L. Henkin, *How Nations Behave: Law and Foreign Policy* (New York: Columbia University Press, 1979), 47; Morgenthau, note 6 above, 299; and O. Young, *International Cooperation: Building Regimes for Natural Resources and the Environment* (Ithaca, N.Y.: Cornell University Press, 1989), 62.

Second, although nonreporting is common, it does not always indicate noncompliance. As many studies have noted, treaties frequently fail to induce countries to report as fully as required.[12] Largely unrecognized, however, is the fact that even developed countries that comply with the substantive provisions of a treaty sometimes fail to report that fact.

Third, achieving compliance with environmental treaties requires altering the behavior of corporations and individuals as well as governments. Although centralized enforcement is not possible, other means of exacting compliance have been very effective: As the right to detain tankers demonstrates, enforcement of treaty provisions by a few countries can induce compliance far beyond their borders. Similarly, private parties can greatly assist enforcement. In the oil pollution case, tanker classification societies, insurance companies, and tanker builders were instrumental in the success of the equipment requirements; their involvement not only prevented violations from occurring but also greatly reduced the monitoring demands on governments. Nongovernmental parties also played an important role in collecting, analyzing, and disseminating compliance and enforcement data on numerous aspects of oil pollution regulation.

Fourth, although reciprocal actions can be an effective means of inducing other countries to comply with international commitments, they are not necessarily appropriate in all cases.[13] In arms control or trade, retaliatory noncompliance may allow countries to recoup losses imposed by others' noncompliance. In environmental affairs, however, retaliation simply inflicts further damage on the planet, so countries have avoided it.

Fifth, in the case of intentional oil pollution, secretariats have been remarkably unwilling to analyze and disseminate reported data in ways that would allow interested parties to respond to noncompliance. IMO has rarely analyzed those enforcement reports it receives to identify nonreporting or noncompliance. Although the MOU secretariat publishes excellent aggregate statistics on an annual basis, they lack the detail needed to determine which countries have not met the 25 percent inspection rate that the agreement requires. The secretariat has also decided not to release a "black list" of the tankers violating MARPOL to private oil companies and chartering organizations, even though such information would provide a strong incentive for compliance. These failures have several origins: the inadequacy of secretariat resources, diplomatic deference to member governments, and fears of legal liability regarding the release of inaccurate information about specific corporations. Secretariats are also caught in the bind that national reports often provide the only basis for identifying noncompliance, but using them for that purpose may reduce the incentive to supply those reports.

Policy Prescriptions

These empirical lessons provide the foundation for several policy prescriptions. At the most general level, treaty provisions alter behavior when they create the conditions for a strategic triangle of compliance in which agents have the political and material incentives, the practical ability, and the legal authority to undertake (or refrain from) a specified activity.[14] In the case of intentional oil pollution, treaty provisions on compliance, monitoring, and enforcement only worked when they supplied all three components of this triangle. Tanker owners, for instance, always had the ability to install SBT equipment, but MARPOL gave them a new incentive to do so: Their ability to trade

---

[12] See A. Chayes and A. H. Chayes, note 3 above, chapter 8; and General Accounting Office, note 2 above.
[13] See, for example, R. O. Keohane, "Reciprocity in International Relations," *International Organization* 40, no. 1 (1986): 1–27; and R. Axelrod, *The Evolution of Cooperation* (New York: Basic Books, 1984).
[14] I am indebted to Robert Keohane for this notion.

internationally was now contingent on compliance with MARPOL equipment requirements, as determined by classification societies and insurance companies. Similarly, the 14 European countries had both the ability and the authority to conduct (and report) enforcement actions under the MARPOL and MOU agreements, but the MOU created greater incentives to do so because its computerized system enabled countries to deploy inspectors more effectively. Rules allowing detention also changed the enforcement tactics of those countries that already had the incentive and ability to impose stiff sanctions by giving them the legal authority.

In contrast, treaty provisions fail to improve compliance when they ignore this strategic triangle. Intentional discharges by older tankers have remained common because tanker captains, who are in the best position to prevent them, have little incentive to do so. Attempts to combat this problem by making detection of illegal discharges easier (countries may now regard any incoming tanker with an unusually low amount of waste oil as having discharged illegally at sea) have failed because oil-exporting countries have no desire to make their ports less attractive to tankers by rigorous enforcement. Similarly, attempts to increase fines for violations failed because countries already had the authority to increase fines and the rules did not create new incentives to do so. Lack of incentives also explains why many oil-loading states still lack reception facilities.

This triangle of incentives, ability, and authority is linked to the three major components of any compliance system: a primary rule system, a compliance information system, and a non-compliance response system. The primary rule system consists of the agents, rules, and procedures related to the substantive behavior targeted by the treaty. The compliance information system consists of the agents, rules, and procedures involved in the collection, analysis, and dissemination of data on

compliance and enforcement. The non-compliance response system consists of those factors that determine the type, likelihood, magnitude, and appropriateness of responses to noncompliance.

No treaty can induce perfect compliance. Even a good one requires refinement and adaptation to the particular problems and agents involved, as well as to the larger political and social context. Effective design also requires that all three components of the compliance system be closely integrated with one another: The choice of primary rules, for instance, influences the ease of monitoring and the likelihood that agents will respond to violations; similarly, a noncompliance response system that relies on funding mechanisms rather than sanctions may well induce greater self-reporting because parties seeking funding will have the incentive to report. Three general principles regarding the design of compliance systems emerge from this analysis, along with a number of specific policy recommendations:

Devise "opportunistic" primary rule systems

In general, negotiators can achieve the same environmental goal through quite different primary rules. In choosing among them, they must consider several important factors: the likelihood of compliance by the different parties involved, the ease of monitoring compliance, and the specificity of the rules themselves. This leads to three specific recommendations:

Match the compliance burden to the expected compliance by different parties. Defining primary rules to place the burden on those most likely to comply follows the path of least resistance. For instance, both the owners and the operators of oil tankers are in a position to curb intentional oil pollution; by placing the burden on the owners, however, MARPOL targeted those most likely to alter their behavior and most susceptible to monitoring and enforcement.[15]

Parties who are already regulated are

---

[15] A familiar domestic example of matching burdens to likely compliance is the policy of requiring

likely to comply more often because they are better informed about regulations, are subject to established monitoring and enforcement systems, and may have a "culture" or habit of compliance.[16] In the oil pollution case, well-established procedures to regulate safety and load lines for ships facilitated compliance with MARPOL equipment requirements: Although both the classification societies and the tanker owners initially opposed those requirements, both readily adhered to them once "those were the rules." International law was simply part and parcel of the specifications used to build new tankers.

Select primary rules that will ease the monitoring burden agents face. Regulatory limits defined to encourage self-monitoring and independent verification will elicit greater compliance than others.[17] For instance, the initial rules limited discharges to 100 parts oil per million parts water, but onboard instruments could not measure at that accuracy; and, after a discharge was made, independent surveillance could not measure oil content at all.[18] An amended criterion banning discharges that produced a "visible sheen" made detecting discharges easier but did not solve the problem of identifying the perpetrators.[19] Although conscientious oil transport companies would have had a strong incentive to identify and report tankers that were gaining a competitive edge by discharging illegally at sea, they were simply incapable of identifying the occurrence of discharge violations. The primary rules themselves thereby precluded self-policing, an often valuable means of inducing compliance with industrial regulations.

Primary rules also dictate the resources needed to collect compliance information effectively. Fewer resources are needed when the rules target behavior that involves few agents or actions; is inherently transparent; coincides with already monitored activities; and involves transactions between parties rather than actions over which a single party has control. For instance, equipment regulations improved monitoring because tanker construction occurs less often than tanker voyages, missing equipment is easier to detect than a discharge, classification societies already monitored tanker construction, and buyers, builders, and classification societies all control a tanker's construction. The Montreal Protocol's focus on the few chlorofluorocarbon producers rather than the myriad consumers is another important example.

Frame primary rules in specific terms. Rules that fail to specify the required actions and responsible parties provide a rationale for noncompliance. MARPOL's vague requirement that governments "ensure the provision" of "adequate" reception facilities at "major" ports has led to ongoing debates over whether governments or industry should provide them, what ports are "major," and whether existing facilities are adequate – while satisfactory facilities often remain unavailable.[20] Such ambiguity also makes sanctioning unlikely because one cannot identify what behavior to monitor or whether observed behavior is noncompliant. To take two examples from other regulatory areas, one cannot readily tell whether governments are meeting the UN Framework Convention on Climate

employers to withhold income taxes throughout the year rather than requiring employees to pay them in a lump sum at the end of the year.

[16] Young, note 11 above, pages 78–79.

[17] J. Ausubel and D. Victor, "Verification of International Environmental Agreements," *Annual Review of Energy and the Environment* 17, no. 1 (1992): 21.

[18] E. Somers, "The Role of the Courts in the Enforcement of Environmental Rules," *International Journal of Estuarine and Coastal Law* 5, no. 1–3 (1990), 196; and W. G. Waters, T. D. Heaver, and T. Verrier, *Oil Pollution from Tanker Operations: Causes, Costs, Controls* (Vancouver, Canada: Center for Transportation Studies, 1980), 121.

[19] T. Ijlstra, "Enforcement of MARPOL: Deficient or Impossible?" *Marine Pollution Bulletin* 20, no. 12 (1989): 596–97.

[20] Sadler and King, note 10 above.

Change's (FCCC) requirements to "coop-erate" in implementing policies "aimed at" stabilizing carbon dioxide emissions or the Ramsar Convention's requirements to make "wise use" of their wetlands.[21]

Devise useful compliance information systems

Once negotiators establish primary rules that facilitate monitoring, an effective compliance system needs self-reporting or independent verification systems (or both) as well as procedures to analyze and dissem-inate compliance information. The system must reduce the demands on those able to collect information and encourage them to report that information by facilitating their immediate goals. Three particular recom-mendations may be made in this regard:

Process and disseminate information to further the reporting parties' own goals. Of the five distinct reporting systems used in the oil pollution context,[22] the greatest response was to those that made the data available in readily usable ways. By incorporating daily inspection reports into a realtime database through which authorities could use other countries' recent inspections to identify likely violators, the MOU created incentives for each government to conscientiously report its own inspection results. Similarly, IMO's published reports on available recep-tion facilities promoted facility use, thereby increasing the reporting country's interest in reducing oil discharges off its coast. In contrast, IMO has made little use of the enforcement reports it receives, creating few incentives for parties to report. Of course, self-reporting systems face the paradox that to encourage honest reporting, they must reassure reporters that the information provided will not become the basis for sanc-tions. Such systems seem more likely to succeed when they seek simply to evaluate treaty effectiveness or to encourage compli-ance through positive inducements rather than sanctions.

Obtain information on noncompliance from parties with the incentives and capability to collect it. Self-reporting systems work best when they facilitate reporting by those who bear the costs of noncompliance. IMO has received regular, accurate information only when governments or other parties perceived themselves as directly harmed by treaty violations. Environmentally con-cerned governments have tended to report while others have not; shipping associations have reported on ports lacking reception facilities because this forces tankers to retain waste oil on board and thus wastes cargo space; classification societies and shipping consultants routinely collect data on tanker equipment; even oil companies have collected data on discharges when oil prices were high enough to raise their concerns regarding the loss of valuable cargo.[23] Surprisingly, however, IMO has not sought access to the latter two categories of infor-mation to evaluate treaty compliance or effectiveness.

Agenda 21, adopted at the United Nations Conference on Environment and Develop-ment in Rio de Janeiro in 1992, explicitly provided for reporting by nongovernmental parties with the incentive and capacity to detect violations. For example, the United Nations Commission on Sustainable Development has been structured to provide nongovernmental organizations

[21] See "United Nations Framework Convention on Climate Change," 15 May 1992, *International Legal Materials* 31 (1992): 849, article IV; and "Convention on Wetlands of International Importance Especially as Waterfowl Habitat," 2 February 1971, *International Legal Materials* 11 (1972): 969, article 3(1).

[22] Reporting systems under the International Maritime Organization have included an irregular, nonstandardized system for enforcement and compliance reporting (before 1985); a regular, standard-ized system for such reporting (from 1985); a periodic survey of countries regarding reception facilities in their own ports; and an ongoing requirement to report the lack of reception facilities in other coun-tries. The MOU requires daily reporting on the results of harbor inspections of all ships. For a detailed analysis of these systems, see Mitchell, note 3 above, pages 123–46.

[23] See, for example, Clarkson Research Studies, Ltd., *The Tanker Register* (London, 1990); Drewry Shipping Consultants, Ltd., *Tanker Regulations: Implications for the Market* (London, 1991); and Lloyd's Register of Shipping, *Register of Ships* (London, 1991).

considerable opportunities to report on the implementation (or lack of implementation) of Agenda 21 by member countries. In many instances, the incentives and capability to identify nonimplementation are greater among nongovernmental organizations than among governments. To convince parties to report, however, such systems must promise anonymity and demonstrate that the information will be used to press violators to comply.

Make monitoring and reporting both legal and easy. After the primary rules define the behavior to be monitored, the treaty must provide the legal authority for and otherwise facilitate such monitoring. MARPOL, for instance, legalized inspections by port countries and authorized them to delegate this authority to classification societies, which greatly reduced the burdens on governments. New inspection rights have not led countries previously uninterested in enforcement to begin inspection programs, but they have increased the size and effectiveness of existing programs. But, just as with arms control, new verification rights need to be created and old barriers removed before verification will occur.[24]

Both technical and political obstacles can make monitoring certain types of behavior quite difficult. All rules requiring that illegal discharges be detected at sea face the problem of "passive voice" violations, in which it is clear that a violation has occurred but it is not clear who committed it. Efforts to improve detection of violations by individual tankers have included improving radar detection capabilities along with proposals to "tag" oil cargos with traceable chemicals and to "fingerprint" oil based on the unique characteristics of each oilfield.[25] However, these technical improvements would still require large-scale aerial surveillance programs, as well as prompt sampling of the oil in the slick and inspection of all suspect tankers.

In addition to such practical problems, there are legal obstacles ranging from the renaming and reflagging of tankers to avoid prosecution to the variance in what courts in different countries deem sufficient legal evidence of a violation.[26] Even tankers caught redhanded often successfully avoid conviction.[27] Clearly, technological advances will only improve monitoring where the primary rules make monitoring easy.

Ease of reporting can also make a difference, however. IMO improved the number and quality of the annual reports it receives simply by clarifying and standardizing its reporting format. Although the MOU reporting system requires daily reporting, this demanding requirement has worked precisely because it has become a bureaucratic standard operating procedure and because the reports can be submitted by telex or computer.

Devise practical noncompliance response systems

International treaties have two options with respect to enforcement: They can either try to get reluctant parties to respond to noncompliance by others or they can remove the barriers that restrain those already prepared to respond. In the oil pollution case, the latter approach has proved far more successful; those lacking

---

[24] W. Fischer, "The Verification of a Greenhouse Gas Convention: A New Task for International Politics?" in J. B. Poole, ed., *Verification Report: 1991* (London VERTIC, 1991), 197–206.

[25] See International Maritime Organization, *Planned Operational Procedures for a Swedish Oil Tagging System*, MEPC I/Inf.5/Add.1 (London, March 1974); International Maritime Organization, *Review of Outstanding Items Undertaken by the Sub-committee on Marine Pollution*, MEPC I/6 (London, January 1974); International Maritime Organization, *Report of the MEPC on its 3rd Session*, MEPC II1/18 (London, July 1975); International Maritime Organization, *Identification of the Sources of Discharged Oil*, MEPC 23/14/2 (London, June 1986); and J. A. Butt, D. F. Duckworth, and S. G. Perry, eds., *Characterization of Spilled Oil Samples: Purpose, Sampling, Analysis and Interpretation* (London: John Wiley and Sons, 1985).

[26] "Shipowner Opposes Further Introduction of Pollution Rules Unless Clearly Justified," *International Environment Reporter*, 9 December 1987, 67; International Maritime Organization, *Visibility Limits of Oil Discharges*, MEPC 33/4/6 (London, July 1992); and T. Ijlstra, note 19 above.

[27] See A. J. O'Sullivan, "In Flagrante Delicto," *Marine Pollution Bulletin* 2, no. 10 (1971); 180–81.

independent incentives to sanction noncom-pliance by tankers or to fund reception facility construction have not been moved to do so by new treaty requirements. This suggests three recommendations for facili-tating enforcement:

Remove the obstacles to sanctioning faced by those with enforcement incentives. Countries do not appear to take legal obligations to enforce treaties seriously: Almost all govern-ments have failed to comply with the requirement to impose high penalties for discharge violations, and most governments have not complied with the requirement to detain ships for equipment violations. In contrast, countries do tend to observe inter-national prohibitions: Even activist governments refrained from detaining tankers until MARPOL entered into force. Therefore, removing legal barriers is crucial to effective governmental enforcement.

The same is true of enforcement by private parties. Tanker chartering companies have requested that the MOU secretariat release data on violations so they can use it to black-ball offending tankers. The secretariat, however, has refrained from doing so because of possible legal liability related to providing data on inspections rather than convictions.[28] Providing such information to (and authorizing sanctions by) nongovern-mental parties, could have a real effect on noncompliance, however: Violators would face the prospect of bad publicity, economic boycotts, or even stronger measures.

Reduce the number of potential violations and prevent initial violations to reduce the need for later enforcement. To succeed, a deterrence-based regulatory system must detect and respond forcefully to a large fraction of the violations that occur. Reducing the number of possible violations both decreases the resources needed for response and increases its likelihood. Equipment violations, for instance, can occur only once per tanker and represent a large, ongoing threat to the marine environment; it is thus more sensible to focus on them rather than on the smaller and more frequent discharge violations. Equipment standards have also relied on a coerced compliance strategy that prevents rather than punishes violations.[29]

Emphasize responses appropriate to the likely source of compliance. An effective non-compliance response system must take into account whether noncompliance is due to inadvertence and incapacity or to willful violation. Those advocating positive responses to noncompliance (which strive to promote future compliance through finan-cial assistance and technology transfers rather than to sanction previous violations) may be correct in assuming that much of it arises from the inability to comply.[30] Although such programs may be effective, disputes over the causes of noncompliance and the reluctance of developed countries to provide funding make their full implemen-tation unlikely; both the lack of such programs in the oil pollution case and the difficulty of negotiating relatively small financial mechanisms for the Montreal Protocol testify to this.

Sanctioning, too, is used only rarely, even though it seems to be an appropriate response to intentional noncompliance. Governments have tended to use sanctions only when they fit the crime and cost little to impose. Furthermore, the imposition of fines proportionate to the relatively small envi-ronmental harm of a discharge violation has had little effect as a deterrent. In contrast, tanker detentions are simultaneously appro-priate to the violation, have the effect of deterring others, and replace a cumbersome legal process with a far less costly adminis-trative one.

[28] H. E. Huibers, "Statement to the Seminar on Port State Control," *Report on the Joint IMIF/MOU Seminar on Port State Control* (London: International Maritime Industries Forum, 1991).

[29] On coerced compliance strategies for enforcement, see A. J. Reiss, Jr., "Consequences of Compliance and Deterrence Models of Law Enforcement for the Exercise of Police Discretion," *Law and Contemporary Problems* 47, no. 4 (1984): 83–122.

[30] A. Chayes and A. H. Chayes, "On Compliance," *International Organization*, 47, no. 2 (1993): 193–95.

Generalizing the Results

The findings and recommendations stemming from this analysis of intentional oil pollution are directly applicable to other environmental issues as well. First, this was a "hard case," in which it was highly unlikely that an international treaty would succeed: Well-organized and powerful industries would have to incur large costs to provide small, uncertain, and widely dispersed benefits to an unmobilized public.[31] Furthermore, enforcement would require collaboration among governments, each of which had an incentive to freeride, enjoying the benefits of enforcement by others without incurring the cost of it themselves. Second, all the major environmental treaties – on acid precipitation, ozone depletion, climate change, river pollution, and hazardous wastes – have faced the same choice as to whether to regulate operations directly or indirectly through equipment standards. Third, most environmental treaties have experienced problems with their reporting systems similar to those encountered in the area of oil pollution. Fourth, new enforcement obligations seem no more likely to get reluctant governments to enforce other agreements than they have with oil pollution. Fifth, the reluctance of developed countries to provide funds for waste oil reception facilities does not bode well for such treaties as the Montreal Protocol and the FCCC, which require far more costly transfers. Finally, and more positively, the increasingly international economy could give governments greater leverage against private offenders in many areas, including endangered species, hazardous wastes, tropical timber, and ozone depleting substances as well as intentional oil pollution.

At the same time, the recommendations made here have limits. For instance, they are probably more applicable to pollution problems than to wildlife and habitat preservation: Changes in production processes can often solve pollution problems without threatening the industry responsible, but preserving wildlife usually requires at least temporary bans on certain activities and involves direct conflict between opposing value systems. Then, too, compliance levels will always depend on compliance costs, which can vary considerably across problem areas.

Another important limit to these recommendations arises from the fact that strong pressure from the United Kingdom and the United States has been necessary both for the adoption of rules to increase compliance and for their enforcement. This pressure has not been sufficient to ensure implementation, however. In areas where concern is low, no efforts are likely to produce successful compliance systems; only where there is such concern can the recommendations made here contribute to a more effective approach.

Why have these suggestions not been implemented already? First, most environmental agreements have arrived on the international scene only in the last 20 years or so, and negotiators have been more interested in getting agreement on some rules than in ensuring that those rules would be implemented and complied with. Second, the implicit decision to approach different environmental problems through separate agreements and secretariats (rather than in a comprehensive, coordinated fashion) has made it more difficult for those in one area to learn from those in other areas. This problem is compounded by the fact that secretariats are notoriously understaffed, rarely having time to analyze data on their own treaties, let alone to take a more general perspective on compliance. Third, the nature of the treaty negotiation and amendment process makes it difficult to implement significant changes quickly (although this problem has recently been remedied in some treaties through framework-protocol arrangements and tacit acceptance procedures for amendments).[32] Fourth,

[31] For more on the theory underlying this argument, see M. Olson, *The Logic of Collective Action: Public Goods and the Theory of Groups* (Cambridge, Mass.: Harvard University Press, 1965).
[32] Framework-protocol arrangements entail initial framework agreements, that contain few behavioral

compliance with environmental treaties has only recently garnered scholarly attention because until recently there were few empirical data with which to evaluate success or failure.[33] Thus, the major obstacles to improving compliance with environmental treaties have lain not in the failure to implement previously identified solutions or in implementing "solutions" that do not work, but in the absence of solid evidence on what works and what does not.

So far, little has been said about the relationship of compliance to effectiveness and whether MARPOL has actually solved the oil pollution problem. Much current research has addressed the problem of evaluating the effectiveness of a treaty, and in the process several theoretical and empirical difficulties have been identified.[34] A high level of compliance is neither necessary nor sufficient for effectiveness: Positive behavioral change that falls short of full compliance may significantly mitigate some environmental problems, while strong compliance with inappropriate rules will leave environmental problems unresolved. Nevertheless, more compliance is usually bettern than less, and (other things being equal) will lead to greater effectiveness.

MARPOL's rules appear to have made significant – though not complete – progress toward eliminating intentional oil pollution, as most estimates show that discharges from tankers have declined during the last several decades.[35] Unfortunately, spill sightings and bird oilings have remained relatively constant, suggesting that there are still problems with other sources of marine oil pollution, including land-based sources.[36] As with most environmental problems, the necessary data are either unavailable for a long enough period of time or of sufficiently poor quality that reliable inferences as to increases or decreases in intentional oil discharges cannot be made. Available evidence does suggest, however, that intentional discharges would have continued to increase without MARPOL's equipment regulations.

Finally, it should be noted that rules that achieve high levels of compliance may not be the most cost-effective ones. If, for instance, it turns out that the benefits of reduced oil discharges are outweighed by the high cost of installing the required SBT equipment, then MARPOL's approach will have been a poor social choice. However, such options should not be ruled out simply because more efficient ones can conceivably achieve the desired environmental end. Negotiators must evaluate treaty provisions in terms of actual compliance, as well as in

prescriptions themselves but establish negotiating processes so that subsequent protocols can adopt such prescriptions when the political and scientific conditions arise. See T. Gehring, *Dynamic International Regimes: Institutions for International Environmental Governance* (Frankfurt, Germany: Peter Lang, GmbH, 1994). Tacit acceptance procedures provide for entry into force of treaty amendments unless some fraction of the member countries have objected before a designated date. R. Churchill, "Why Do Marine Pollution Conventions Take So Long to Enter into Force?" *Maritime Policy and Management* 4, no. 1 (1976): 41–49; and R. B. Mitchell, "The Impact of the Tacit Amendment Procedure under MARPOL" (unpublished paper, Eugene, Ore., November 1993).

[33] See note 3 above.

[34] See, for example, P. M. Haas, R. O. Keohane, and M. A. Levy, eds., *Institutions for the Earth: Sources of Effective International Environmental Protection* (Cambridge, Mass.: MIT Press, 1993); A Underdal, "The Concept of Regime 'Effectiveness'," *Cooperation and Conflict* 27, no. 3 (1992): 227–40; O. Young, "The Effectiveness of International Institutions: Hard Cases and Critical Variables," in J. N. Rosenau and E. O. Czempiel, eds., *Governance Without Government: Change and Order in World Politics* (New York: Cambridge University Press, 1991); and A. Nollkaemper, "On the Effectiveness of International Rules," *Acta Politica* 27, no. 1 (1992): 49–70.

[35] For a longer discussion of these problems and an evaluation of environmental improvement, see R. B. Mitchell, "Intentional Oil Pollution of the Oceans," in P. M. Haas, R. O. Keohane, and M. A. Levy, note 34 above, especially figure 5.1 on page 187.

[36] C. J. Camphuysen, *Beached Bird Surveys in the Netherlands, 1915–1988: Seabird Mortality in the Southern North Sea Since the Early Days of Oil Pollution* (Amsterdam: Werkgroep Noordzee, 1989), 39–42.

terms of efficiency, cost, and equity; the experience with discharge and equipment standards demonstrates that a nominally cheaper, more "efficient" policy simply could not achieve the desired level of compliance.

Conclusions

In international treaties to protect the environment, both the rules that are adopted and the degree of compliance they achieve will be strongly influenced by the distribution of power, economic interests, and environmental concern across countries. By acknowledging these limits, however, and recognizing that the same goal can often be achieved in quite different ways, policymakers can greatly improve compliance and benefit the environment. In concert with the proper primary rules, well-crafted compliance information and noncompliance response systems can increase detection and response through creating the right incentives for the right parties. All three elements of the compliance system must place agents in the strategic triangle of incentives, ability, and legal authority for undertaking the compliance, monitoring, and enforcement activities so essential to treaty effectiveness.

Whether the nations of the world avert the many environmental threats that loom on the horizon will depend not only on negotiating agreements to protect the air, land, and water but also on ensuring that those agreements get governments, industry, and individuals to change their behavior. The day may come when all nations are sufficiently concerned about the environment that there will be no need for international law to dictate proper behavior. Until then, however, careful crafting and recrafting of international treaties provides a valuable means of improving protection of the global environment.

21

DEAD IN THE WATER*

Fred Pearce

Last summer it was a plague of jellyfish, before that it was 50 million tonnes of mucus-like foam in the Adriatic, red tides in the Aegean, and a thousand dolphin corpses washed up on beaches from Morocco to Greece. Years of pollution, overfishing and overdevelopment have left the Mediterranean in crisis. Turtle nesting sites are under threat and monk seals are being driven to extinction. The problem is so severe that parts of the sea are devoid of life altogether.

But scientific understanding of what is happening to the sea that nurtured the birth of Western civilisation remains disjointed. A Mediterranean Action Plan (MAP) to clean up the sea and protect its ecosystems, agreed by the governments of the region 20 years ago this month, has failed. And a promised relaunch in June this year also seems doomed.

Case for Treatment

The problem is huge. More than 130 million people live along the 46 000 kilometres of Mediterranean coastline and their ranks are swelled each summer by 100 million tourists. Eighty per cent of their sewage – more than 500 million tonnes a year – pours into the sea without any treatment.

The sewage is a threat to tourist beaches from Cannes to Capri and from Rhodes to the Lido in Venice. Dozens of Italian beaches and 1 in 10 French beaches fail cleanliness standards laid down in the European Union's Bathing Waters Directive. The UN Environment Programme (UNEP) says that the dirty water can cause infections of the ear, nose and throat among bathers, not to mention hepatitis, dysentery, enteritis and occasional cases of cholera. The risks may be greater than at British beaches notorious for failing to meet EU standards since the warmth of the Mediterranean waters nurtures pathogens and encourages bathers to linger in polluted waters. Seafood is tainted, too. In 1990, UNEP reported that 93 per cent of shellfish taken from the Mediterranean contained more faecal bacteria than the maximum recommended by the WHO.

Sewage is only the start. Each year, according to UNEP, 120 000 tonnes of mineral oils, 60 000 tonnes of detergents, 100 tonnes of mercury, 3800 tonnes of lead and 3600 tonnes of phosphates enter the sea as a result of human activities on land. Plastic rubbish, which takes hundreds of years to disintegrate, continues to accumulate. And up to a million tonnes of crude oil are dumped from ships, often because harbours lack the facilities to collect waste oil or clean tanks.

* Originally published in *New Scientist*, 1995, February, pp. 26–31.

Three-quarters of the pollution comes from France, Spain and Italy. Mercury frequently reaches dangerous concentrations in seafood collected off their coasts. France and Italy have imposed statutory limits on the concentration of heavy metals in shellfish for human consumption of 500 and 700 parts per billion. But the laws are not enforced. According to a detailed report published in 1990 on the Mediterranean marine environment by Ljubomir Jeftic, deputy coordinator of the MAP, a large proportion of the French and Italian catches breach these limits.

The pollution has weakened the coastal ecosystem to the point where alien species are able to take over from native species. Monaco's oceanographic museum boasts an impressive collection of exotic weeds in its aquaria. One of the most striking is *Caulerpa taxifolia*, a bright green weed found in the tropics and, until recently, unknown in the Mediterranean. A decade ago, however, *Caulerpa* took root on the seabed outside the museum – perhaps after being flushed away accidentally when an aquarium was cleaned – and began spreading fast. Five years ago it turned up further along the French coast at Cap Martin. Now it has blanketed 1500 hectares of sea bed along a 300-kilometre stretch of the Riviera, from Toulon in France to Imperia in Italy. There have been isolated finds as far west as the Balearic Islands, and as far south as Italy's Elba and Sicily.

At the nearby University of Nice, Alexandre Meinesz, professor of marine biology and a leading expert on *Caulerpa*, says the weed poses a major risk to the Mediterranean. "We could be seeing the beginning of an ecological catastrophe," he says. His research students discovered the outbreak in Monaco in 1984. The Mediterranean version seems to have mutated from its tropical cousin, he says. It grows to more than twice the size, and contains more caulerpicin, a toxin that kills algae. He claims that the mutant invader is killing the Mediterranean's distinctive "sea meadows" of marine plants, known as posidonia beds, where hundreds of species of fish spawn and feed.

Collapsing Ecosystems

The weed has only been able to take hold because the coastal ecosystems of the Riviera are profoundly sick, says John Chisholm, an Australian marine biologist based in Monaco at the European Oceanographic Laboratory. The weed, he says, is an opportunist that has spread in response to a highly polluted environment. "It grows in sediments very rich in accumulated organic pollution – from sewage outfalls and so on. We are now seeing the long-term effects of pollution over two or three decades in the Côte d'Azur." Ecosystems are being turned on their head, Chisholm says. "Microbial populations [that support the ecosystem] can't cope any longer. And when they crash, large stands of posidonia can disappear within a few months. It is only then that the *Caulerpa* invades."

Caulerpa is not the only botanical interloper. According to Charles-François Boudouresque of the LBMEB, a marine biology lab in Marseilles, there are more than 300 alien species in the Mediterranean, two-thirds of them discovered since 1970. Most came from the Red Sea via the Suez Canal. This biological pollution, he says, is irreversible and may prove to be one of the major ecological problems of the coming century. Certainly, the Mediterranean seems increasingly vulnerable to plagues of all sorts. And few researchers doubt that sewage and chemical pollution are at the root of the problem.

It is now 20 years since all nations bordering the sea, except Albania, adopted the MAP, one of 11 conventions on international seas organised by UNEP. They promised to "take all appropriate measures to prevent, abate and combat pollution in the Mediterranean Sea area and to protect and improve the marine environment in the area."

Their actions have not matched their words, however. In 1985, after a decade of research but little action, the nations met again in Genoa to set themselves "ten priority objectives to be achieved by 1995". The priorities included sewage treatment works for all cities with a population over

100,000, new efforts to cut industrial pollution, protection of endangered marine species and 50 new marine reserves (see box 21.1 "Time's up"). But by the early 1990s the MAP was close to collapse, with major contributor European nations failing to pay their dues to its secretariat and many of its research activities cancelled. At the end of April 1994, unpaid contributions amounted to $3.7 million, about seven months' funds. There has been an improvement since, but "the situation is still very unstable", says the MAP's internal news bulletin. In November, with the 1995 deadline approaching, Jeftic told *New Scientist*: "I can't tell you what progress has been made in implementing the objectives. The countries are no good at giving us information. If they are implementing, they are not saying."

Perhaps the most serious problem that the relaunched MAP will have to tackle is eutrophication – the pollution of the sea by chemicals that fertilise its waters. This leads to the explosive growth of algae, especially in summer when the sea is warm and calm, allowing pollution to build up in the surface waters. Nitrates and phosphates – derived from sewage, detergents, farm fertilisers and slurry – enter the sea both directly from drains and sewage outfall pipes and down major rivers.

Eutrophication takes two main forms in the Mediterranean. The first is "redtides", caused by the build-up of dinoflagellates. These can release chemicals that are toxic to marine animals such as fish and anything eating fish. The French Oceanographic Institute, IFREMER, reported in 1993 that over the past 15 years dinoflagellate blooms on the Riviera had caused shellfish to accumulate so much toxin that they would be poisonous to humans if eaten in large quantities.

The second is the formation of huge quantities of mucus-like foam secreted by diatoms. The foam fouls beaches and removes oxygen from the water, killing many creatures on the seabed, such as heart urchins and shellfish. The pollution can last for several weeks and may only be cleared by autumn storms.

UNEP has records of frequent eutrophication from Juan-les Pins on the French Riviera to Castellón in Spain and from Split in Croatia to Alexandria on the Nile delta. During most summers, the Saronikós Gulf,

---

Box 21.1 Time's up

In 1985, Mediterranean nations set themselves 10 objectives for 1995 to clean up the sea and protect its environment. Ljubomir Jeftic, deputy coordinator of the Mediterranean Action Plan, says none of them is known to have been achieved. On the wish list were:

*   Measures to reduce the amount of contaminated water discharged by ships into the sea, including facilities at ports to offload dirty ballast waters and oily residues
*   Sewage treatment works for all cities with over 100 000 people, plus "appropriate" sewage outfall pipes into the sea or treatment plants for towns with over 10 000 people
*   Environmental impact assessments for development projects such as new ports, marinas and holiday resorts
*   Improved navigation safety to minimise the risk of collision, such as better shipping lanes in and out of ports, especially for ships carrying toxic cargo
*   Protection of endangered marine species such as monk seals and sea turtles
*   Reduced industrial pollution
*   Identification and protection of 100 historic coastal sites
*   Identification and protection of 50 marine and coastal conservation sites
*   More protection against forest fires and soil loss to minimise runoff into rivers
*   Reduction of acid rain

which receives pollution from Athens, becomes eutrophic. According to Basil Katsoulis of the University of Ioánnina, the bay around the port of Piraeus is almost dead. "The ecosystems show severe breakdown," he says. In 1960 there were 170 species of marine fauna in the gulf. Only 30 survive today.

In the Lac de Tunis, a large shallow saltwater lagoon and bird-watchers' paradise that receives the municipal waste of the Tunisian capital, a third of the lagoon is covered in algae that consume all the oxygen in the water each summer, killing any life below. Red tides also produce the foul-smelling gas hydrogen sulphide, while other algae release ammonia. Both pollutants further disrupt ecosystems.

On the Aegean coast of Turkey, half a million cubic metres of sewage and industrial waste from Izmir, a boom city which makes 15 per cent of Turkey's exports, pours into the Izmir Bay. A study for the MAP in 1993 found that eutrophication was a serious problem throughout the year and that red tides were becoming more frequent in the bay. Pollution-related illness among local swimmers and fishermen caused an estimated 10 000 lost working days a year.

The study suggested that a $1.5 billion cleanup programme over 30 years could boost the local economy by up to $10 billion, mainly from tourism. But plans for a waste treatment plant, first drawn up in 1969, remain on the drawing board. It also predicted that pollution would "reach a critical level by 1995, leading to a collapse in the ecosystem".

The worst eutrophication occurs in the northern Adriatic. Occasional outbreaks have been occurring for many centuries, but in a 1994 study for the MAP, Tarzan Legovic and Dubravko Justic of the University of Zagreb say that growths of algae are much more extensive today. They can cover up to 50 square kilometres and contain much more material than before – up to 50 million tonnes of mucus-like foam in one case. The reason, they say, is clearly pollution. The northern Adriatic's main source of freshwater is the River Po, which drains a large, heavily populated and intensively farmed region of northern Italy. The river is heavily polluted with nutrients. Each year some 5000 tonnes of phosphorus and 100 000 tonnes of nitrate and ammonia reach the sea from the Po – about ten times as much as 50 years ago.

The effect on the enclosed bays and lagoons of the Adriatic is devastating. The city authorities in Venice now dredge up to a million tonnes of algae from their lagoon each summer in order to stop it putrefying in the heat just as the tourists arrive. In the Gulf of Trieste, says Jeftic, large numbers of fish die as a result of algal blooms. He warns that about 250 square kilometres of the northern Adriatic is becoming devoid of life, with the posidonia sea meadows "on their way to extinction" in many areas. Italian plans to clean up the Po are proceeding only slowly.

A small tidal range and calm summer waters ensure that pollution in the Mediterranean's many bays does not disperse easily. And once out of bays it encounters a largely enclosed sea. Freshwater inputs to the Mediterranean are low, especially since virtually the entire flow of the Nile, the largest river, is now diverted onto farms. The Mediterranean exchanges water with the Black Sea, but if anything this is even more polluted: 90 per cent of it is anoxic, making the Black Sea the largest mass of water without dissolved oxygen on the planet. And renewal of water by exchange with the Atlantic, through the Strait of Gibraltar, takes around 150 years.

Scientists remain remarkably ignorant about what happens to pollution away from the shore, and what impact it has on fisheries and sea mammals such as dolphins. According to Jeftic, knowledge of a limited number of coastal areas is improving, but the effect of pollution in vast areas, including the entire deep Mediterranean basin, remains poorly understood.

Dead Dolphins

That is one reason nobody knows why several thousand striped dolphins, the most common cetacean in the Mediterranean, died between 1990 and 1992, struck down by a morbillivirus. First detected off Valencia in Spain, the epidemic spread east, with dead animals beached in Morocco, Algeria, France, Italy and eventually the Greek islands of the Aegean. At the height of the epidemic, schools of dolphins were only a third their normal size.

Some researchers believe that industrial chemicals had affected the dolphins' immune systems making them vulnerable to disease. Compounds such as polychlorinated biphenyls (PCBs), and other organochlorine compounds which do not mix easily in water, concentrate in plankton and the fish that feed on them, and end up in the body fat of predators that eat the fish. The cause of the epidemics has not been established. But Alex Aguilar, professor of animal biology at the University of Barcelona, who monitored the results of examinations of more than a thousand dolphin carcasses, reported in 1993 that the dolphins affected were found to carry high concentrations of pollutants. He says: "Striped dolphins in the Mediterranean have carried extremely high levels of PCBs for at least the last two decades."

Viruses and pollution are not the only threats to Mediterranean cetaceans. Greenpeace claims that dolphin hunting continues on a small scale from Italian ports such as San Stefano, near Imperia. Plastic rubbish – from ropes and nets to the plastic bands from six-packs of beer cans – chokes and entangles marine mammals. This deadly flotsam concentrates along the "fronts" between bodies of water, which is also where high concentrations of nutrients are found. The nutrients support abundant supplies of plankton which, in turn, attract large numbers of fish and cetaceans. One study of cetacean carcases found that 1 in 30 had choked on plastic debris.

Several thousand dolphins and whales die each year after being accidentally caught in driftnets in the Mediterranean, says Greenpeace. Driftnets hang like vast curtains from floating lines. About a thousand fishing vessels, most of them Italian, spread driftnets in the western Mediterranean. EU legislation banned nets longer than 2.5 kilometres after the

International Whaling Commission warned in 1990 that they threatened the survival of striped dolphins in the western Mediterranean. But Greenpeace claims the rules are widely flouted, especially by Italian driftnetters chasing the sea's declining stocks of swordfish.

Fleets of trawlers with drift nets and other modern fishing gear also threaten fish stocks. More than 100 of the 500 fish species found in the Mediterranean are harvested commercially. According to the World Conservation Union (IUCN) the hake, red mullet and sole fisheries off the European coast are badly depleted. Important fish nurseries on the seafloor such as posidonia and red coral beds are also being damaged by fishermen who drag weighted nets along the bottom to catch shellfish.

The Mediterranean currently yields around 2 million tonnes of fish a year. But this cannot last, says the IUCN, which believes the sustainable yield is between 1.1 and 1.4 million tonnes. Conservationists have called for a Mediterranean fisheries convention to protect fish in international waters. Top of the list for protection would be the Atlantic bluefin tuna, the largest and most valuable fish in the North Atlantic. Every summer, ships based in the Canary Islands follow the migrating tuna into the western Mediterranean, where it spawns. Their catch is destined for Japan, where a large tuna at auction can fetch enough money to pay for a new Porsche. But increasingly, juvenile bluefin tuna are being caught. With breeding disrupted, stocks in the western Atlantic have fallen by an estimated 90 per cent since 1975.

The destruction of fisheries has knock-on effects throughout the ecosystem. Marine mammals may starve. Legovic and Justic attribute the plagues of jelly-fish seen periodically over the past 20 years, including a major outbreak last summer, to overfishing of their predators.

Tourist Threat

Mediterranean wildlife is also threatened by human activity on land. Large beach-loving animals such as turtles and seals are especially vulnerable to tourism. The Mediterranean monk seals used to lie about on open beaches, and even sought human company, says Crassidas Zavras, secretary of the Hellenic Society for the Study and Protection of the Monk Seal. But the seal, which can grow up to 3 metres long, is one of the 12 most endangered species in the world, according to the IUCN. Once there were hundreds of seal colonies all over the Mediterranean, each containing thousands of animals. Today, estimates of the total number of animals range from 350 to 750. Most are in the Aegean, and (outside the Mediterranean itself) on the beaches of Madeira and the western Sahara. Tiny colonies of fewer than a dozen animals also persist in Morocco, Algeria, the Canaries and in the Ionian Sea.

Out on the remote islands of the Aegean Sea, between Greece and Turkey, conservationists fear that they are witnessing the demise of the monk seal. Once fish stocks began to decline in the Mediterranean, seals began to steal fish from nets. "They destroy the nets completely," says Zavras, "and fishermen have taken to destroying them, usually with guns." So now the seals live like fugitives, hiding in caves that can be entered only from the sea. The biggest colony is among the caves in the sheer cliffs of Pipéri, part of the northern Sporades islands in the Aegean. Even in midsummer this is a wild place, buffeted by winds and sea. In an effort to protect the seals, the islands have been declared Greece's first marine park. Fishing and boating are illegal here. The seals' only companions are falcons, wild mountain goats and automatic cameras, installed by conservationists to monitor them.

"We recorded five births last year," says Bill Johnson of the Bellerive Foundation, a Swiss-based conservation organisation that helped pay for the cameras. The foundation has also funded an emergency rescue service that has so far rehabilitated five abandoned baby seal pups. "The population is beginning to recover inside the park," says Johnson. "The animals are not necessarily doomed if we can preserve their last habitats."

A similar strategy is desperately needed for loggerhead turtles, which have also taken refuge in the Greek islands. An estimated 500 females lay their eggs in summer on the sandy beaches of Laganá Bay on the island of Zákinthos. There are other, smaller nesting sites at Lara beach in northern Cyprus, the Dalyan delta in Turkey, and in Egypt and Libya.

Laganá Bay has played host to loggerheads for thousands of years. Now it is also a major resort, catering for most of the planeloads of package holiday tourists that fly to the island. Last summer, for the first time, local authorities banned speedboats from the bay. This followed the deaths of nine turtles that were hit by boats the previous summer.

Futile Journey

The beach itself is also dangerous. The soft sands are cluttered with sun beds and umbrellas that shade incubating eggs, preventing the embryos from developing, and puncture eggshells. Half of Laganá beach is also brightly lit at night by bars, hotels and street lighting. When baby turtles hatch out in the sand, they turn to the nearest light and head for it. For millions of years that light was always the moon and stars reflected on the sea. Now it is often the Four Brothers' disco or Bob and Wendy's Bar, and many turtles expire during their futile journey or are picked off by birds the following day as they stumble through the dunes.

The Mediterranean's crisis arises from the accumulated impact of poisons and habitat destruction in the northwest sector of the sea, and from the fast-growing populations and even faster-growing economies of the east and south. Tourists can see the truth. They are deserting the beaches of Spain, Italy, the South of France and even the Greek islands for the less spoilt shores of Turkey and north Africa. But their search for clean waters is doomed unless the nations that border the sea can act together to meet past promises.

This summer, for the third time in 20 years, these nations will again declare their intent to clean up the sea and protect its wildlife. The countries, which since 1990 have included Albania, are due to present detailed reports on their progress at a meeting in Barcelona in June, where UNEP hopes to relaunch the MAP. But Jeftic warns: "Don't be too optimistic." When he asked nations two years ago to report on the implementation of antipollution legislation, most simply did not reply. Even if they can agree to act, the harm already done – the habitats destroyed and the pollution stored in sediments and the bodies of fish and marine mammals – will take many decades to undo. Right now, like much of the Mediterranean's wildlife, the MAP looks dead in the water.

22

Quaternary Deposits and Groundwater Pollution*

Donal Daly

Introduction

Never in the history of humanity has environmental concern and awareness been so great. This is reflected not only in the increasing media coverage of environmental issues, but also in environmental protection legislation. Never before has humankind had such an ability to damage and control the environment. This is due largely to our technological advancement and in the "developed" world, is reflected in an increased standard of living. How can these be balanced with the need to protect the environment? Land-use planning attempts to do this. But land-use planning is a dynamic process with social, economic and environmental aspects and impacts influencing to various degrees the use of land and water. In a rural area, housing, farming, industry, tourism, recreation, conservation, mining, waste disposal, groundwater supplies, surface water supplies, etc. are potentially interactive and may compete for priority. What information is needed for balancing the various interests in the planning process? It is argued in this paper that information on the geological materials is vital in land-use planning, pollution prevention and environmental protection. The

balancing of interests will always be difficult and complex, but at least it should be made in the knowledge of all aspects, including the geological materials. Quaternary geologists have a major role to play in providing this information.

This paper examines and describes the role of Quaternary deposits in assessments of the natural vulnerability of groundwater, the pollution loading that results from human activities, groundwater pollution investigation and prevention, and groundwater protection schemes.

This paper is written from the perspective of a hydrogeologist and not a Quaternary scientist. It is based on the experiences of working mainly in Ireland and for two years in Britain. The content and conclusions apply chiefly to the glaciated areas of Ireland and Britain that are underlain by bedrock of pre-Permian age. The main emphasis is on the situation in Ireland, however, most of the points also apply to Britain.

Groundwater Resources in Quaternary Deposits: A Summary

Sand and gravel aquifers, particularly in Ireland, are major potential sources of water supplies. In the Republic of Ireland they are

* Originally published in *Quaternary Proceedings*, 1992, no. 2, pp. 79–89.

the second most important aquifer, after the Carboniferous limestones, although they are largely undeveloped at present (Wright et al. 1982). In Northern Ireland, sands and gravel are the most important aquifer providing 40% of the groundwater supplies (Monkhouse & Richards 1982). In Scotland there has been little development of groundwater. However sand and gravel deposits are major potential sources of water which may be developed in the future, at least as emergency supplies. In England and Wales, sands and gravels provided only 1.5% of groundwater supplies in 1977 (Monkhouse & Richards 1982). In England they are not likely to be a major source of water in the future. In Wales they are one of the main alternative sources of water to surface streams, although seldom used at present. In the future they may have to be developed as a source of emergency water supplies.

In considering Quaternary deposits and groundwater pollution, not only are the groundwater resources in Quaternary deposits, as outlined in this section, an issue, but also the Quaternary deposits that overlie many of the bedrock aquifers. Both situations are considered in the following sections. But first let us consider groundwater pollution risk.

Groundwater Pollution Risk

Many human activities and developments produce potentially polluting wastes – for instance, agricultural activities, septic tank systems, waste disposal sites, piggeries, dairies, etc. Many of the wastes are disposed of directly into the geological environment.

Groundwater pollution risk depends on the interaction between:

(a)   the natural vulnerability of the groundwater; and
(b)   the pollution loading that is, or will be, applied to the subsurface environment as a result of human activity (Foster, 1987).

The main problems arise where there is a high groundwater vulnerability and a high

pollution loading. It should be noted that vulnerability is an intrinsic characteristic of an area whereas pollution loading can usually be controlled or modified by good land-use planning and monitoring.

Groundwater Vulnerability to Pollution

Groundwater vulnerability to pollution is defined as the sensitivity of its quality to human activities which may prove detrimental to the usage value of the resource (Bachmat & Collin, 1987).

In general groundwater is less prone to pollution than surface water as it is naturally protected against microbiological and chemical contaminants by the geological deposits above the aquifers or water table.

Factors influencing groundwater vulnerability to pollution

The vulnerability of groundwater varies greatly, depending on a number of interrelated geological and hydrogeological factors which are listed below and are summarised in table 22.1:

(a)   the type, permeability and thickness of the Quaternary deposits;
(b)   the potential for purification of polluted groundwater in the Quaternary deposits and in the underlying aquifers;
(c)   the type of rock permeability – whether intergranular or fissure;
(d)   the depth to the water table;
(e)   the thickness of aquifers;
(f)   the type of recharge.

Influence of Quaternary deposits

The Quaternary deposits act as a protecting layer over groundwater by both physical and chemical/biochemical means. Fine grained sediments such as clayey till, lacustrine clays and peats have a low permeability and consequently can act as a barrier to the vertical movement of pollutants. Even if the permeability is sufficiently high to allow slow intergranular movement of pollutants, for instance in sandy tills or silts, the sediments can strain out and absorb bacteria and viruses. In contrast, high permeability deposits – sands and gravels – allow easy access of pollutants to the water table although they provide opportunities for dispersion of the pollutants among the pore spaces.

Sorption, ion-exchange and precipitation are vital chemical processes in attenuating pollutants. The cation-exchange capacity of Quaternary sediments depends on the clay and/or organic content and ranges from essentially zero for sands to about 500 meq/100g for clay soils to over 100 meq/100g for peat. Consequently, clays and peats can attenuate bacteria, viruses and chemical pollutants such as cadmium, mercury, lead, potassium and ammonium whereas clean sand and gravel has little effect. Laboratory experiments conducted by Thorn (1987) show that the ratio of adsorption of potassium from landfill leachate is 36:11:9:1 per unit weight of peat, loam soil, glacial sand/silt/clay/gravel mix and acid washed sand, respectively.

In general the higher the clay content and the lower the permeability, the greater the

Table 22.1 [*orig. table 1*]   Geological/hydrogeological factors and degree of protection.

Good Protection	Poor Protection
1 High clay or peat content.	1 Low clay or peat content.
2 Low permeability subsoil e.g. clayey till, peat.	2 High permeability subsoil e.g. sand and gravel.
3 Thick (greater than 5m) subsoil.	3 Thin subsoil.
4 Intergranular flow.	4 Fissure or karstic flow.
5 Thick unsaturated zone.	5 Thin unsaturated zone.
6 Recharge by percolation through topsoil and subsoil.	6 Recharge by sinking streams.

protection of groundwater from pollution. However the corollary of this is that the pollutants may run off and pollute surface water.

Influence of permeability type

Permeability, which is a measure of the capacity of a rock to transmit water, can be subdivided into two types. Firstly, where the water moves between the grains, it is called primary or intergranular permeability. Secondly, where the water moves through fractures or fissures or joints and along bedding planes, it is called secondary or fissure permeability.

Intergranular flow is slower than fissure flow in rocks under most conditions, void or pore sizes are usually smaller and flow paths are more irregular. Also the amount of water stored in granular rocks is generally greater than in fissured rocks. These factors have an important bearing on pollutant attenuation. In contrast to rocks in which fissure flow dominates, the slow flow rate in rocks with an intergranular permeability delays the entry of pollutants into groundwater and, particularly in the unsaturated zone, allows time and opportunities for interactions between pollutants and rock grains. Also, the relatively small pore sizes allow filtration and absorption of bacteria and viruses. The irregular flow paths within a porous matrix causes hydrodynamic dispersion which decreases pollutant concentration. For pollutants that reach the water table and enter groundwater, dilution is much greater in rocks with an intergranular permeability and thus the resultant pollutant concentrations are much less. The worst situation is in karst limestone areas where flow rates are very high – over 100m/hr in some instances – due to widening of fissures by solution and there is little scope for attenuation other than by dilution. *Consequently there is generally far greater degradation and purification of pollutants in rocks with an intergranular permeability than in those with a fissure permeability.*

The presence of an intergranular permeability in rocks is related to the geological history. In general, the greater the age and the more forceful the tectonic history the greater the degree of fracturing, fissuring and infilling of intergranular pore spaces. So rocks of pre-Permian age (older than 290 million years) seldom have an intergranular permeability whereas Quaternary rocks seldom have a fissure permeability (certain tills are a minor exception). Rocks of intervening age usually have both, although fracturing is less common in the more clayey rocks. One of the Permian aquifers – the Magnesian Limestone – is dominated by fissure flow and can be considered similar to the pre-Permian rocks for the purposes of this paper.

Groundwater vulnerability in Britain and Ireland

Examination of the bedrock geology of Ireland shows that apart from the north-east where Jurassic and Triassic volcanics and sediments are present, most of the country is underlain by older rocks – rocks with a fissure permeability only. Similar rocks with a fissure permeability only are present in Scotland, much of Northern England apart from the Carlisle area and east Yorkshire, Wales, Cornwall and West Devon. Pollutants entering these rocks receive minimal attenuation apart from some dilution. The principal natural feature influencing the movement of pollutants into fissured rocks, particularly into fissured aquifers, is the protecting, filtering cover provided by the overlying Quaternary deposits.

Groundwater in Ireland and Britain is vulnerable to pollution where the Quaternary deposit cover is absent, thin or very permeable. This situation arises in Ireland particularly where karstified limestones in parts of Galway, Roscommon, Clare, Mayo, Sligo and Fermanagh are outcropping or are overlain by thin, permeable sandy tills, sands and gravels. Consequently, there are many polluted wells and springs in these areas – more than 50% of the wells are polluted at some time during their use. Groundwater is relatively well protected from pollution in both countries where pre-Permian rocks are overlain by thick, low permeability clayey tills, clays

and peats. However the corollary of this is that the pollutants may run off and pollute surface water.

This paper will deal firstly, with areas of sand and gravel aquifers and secondly, the areas in Ireland and Britain which are underlain by pre-Permian rocks, as these are the areas where the Quaternary deposits have a significant role to play in groundwater pollution assessment and prevention. In areas underlain by bedrock aquifers with both an intergranular and fissure permeability such as the Chalk and Sherwood Sandstone – the most highly productive aquifers in these islands – the Quaternary deposits play a useful but less significant role.

Groundwater vulnerability mapping

There are several approaches to vulnerability assessments, for instance using water quality data and numerical modelling. However one of the most cost-effective and useful methods is vulnerability mapping. This is the technique of assessing all the geological and hydrogeological factors and ranking the vulnerability of groundwater and displaying it on a map in a manner which is understandable and useful. Vulnerability maps can be prepared using existing geological, depth-to-bedrock and hydrogeological data, provided the existing information on the Quaternary deposits is adequate. This is not the case for most of Ireland and for large areas of Britain. Consequently reconnaissance surveys are necessary.

Vulnerability maps are increasingly becoming an important part of groundwater protection schemes in Ireland and their use is being advocated by the Geological Survey of Ireland. However the scarcity of both up-to-date Quaternary geology maps at useful scales of 1:100,000 or larger – and of Quaternary geologists is hindering the preparation of adequate, defensible groundwater protection schemes.

Vulnerability rating and hydrogeological setting

Vulnerability ratings for different typical Irish hydrogeological settings are given in

Table 22.2 [*orig. table 2*]   Vulnerability ratings for typical Irish hydrogeological settings.

Vulnerability Rating	Hydrogeological Setting
Extreme:	1 Outcropping bedrock aquifers (particularly karst limestone) or where overlain by shallow (less than 3m*) subsoil.
	2 Sand and gravel aquifers with a shallow (less than 3m*) unsaturated zone.
	3 Areas near karst features such as sink holes.
High:	1 Bedrock – major, minor and poor aquifers and non-aquifers – overlain by 3m+ sand and gravel or 3–10m sandy till or 3–5m low permeability clayey tills or clays.
	2 Unconfined sand and gravel aquifers with 3m+ unsaturated zone.
Moderate:	1 Bedrock – major, minor and poor aquifers and non-aquifers – overlain by 10m+ sandy till or 5–10m clayey till, clay or peat.
	2 Sand and gravel aquifers overlain by 10m+ sandy till or 5–10m clayey till, clay or peat.
Low:	1 Confined bedrock aquifers overlain by 10m+ clayey till or clay or low permeability bedrock such as shales.
	2 Non-aquifers and poor aquifers overlain by 10m+ clayey till or clay.
	3 Confined gravel aquifers overlain by 10m+ clayey till or clay.

* Note: Less than 1m subsoil or unsaturated zone beneath a development rather than 3m could be the cut-off depth for the "extreme" rating. However, taking a thickness of 3m rather than 1m is regarded as more practical and useful for the following reasons:

(a) the base of many developments – septic tank systems or farmyard effluent holding tanks for instance – are 1–3m b.g.l.;

(b) in preparing a vulnerability map the general rather than the site specific situation must be taken into account; and

(c) a 3m cut-off depth allows for lateral variations and often provides a safety margin. Obviously if the base of a potentially polluting development is more than 3m deep, the rating classification may be affected.

table 22.2, which clearly illustrates the crucial role of Quaternary deposits. These ratings are subjective and qualitative. This is inevitable as geological and hydrogeological knowledge (particularly of the Quaternary deposits) will normally be inadequate to enable quantitative and impartial ratings to be determined. However provided the limitations are appreciated, vulnerability assessments can be a powerful tool in protecting groundwater.

Sources of Groundwater Pollution

The potential sources of groundwater pollution can be classified into two groups: point sources and diffuse or non-point sources, and these are listed below.

Point Sources	Diffuse Sources
1 Farmyards	1 Organic and inorganic fertilizers
a) Manure and slurry	
b) Dirty water	
c) Silage effluent	2 Pesticides
2 Septic tank systems	3 Urban areas
3 Landfill sites	4 Rainfall
4 Spillages and leakages	
5 Contaminated surface water	

In Ireland the main contamination sources are farmyards, septic tank systems and urban areas. Minor sources include landfill sites, contaminated surface water, spillages, leakages and fertilizers. In central and southern Britain inorganic fertilizers are the main cause of concern due to rising nitrate levels in groundwater and surface water. Landfill sites, pesticides, and leakages and spillages of solvents and petroleum products have also been a focus of research and concern.

Two of these sources – landfill sites and septic tank systems are now considered. The points raised can also be applied to most of the other pollution sources.

Landfill Sites

A useful classification of landfill sites is one based on hydrogeological characteristics because these dictate the design and operational procedures used. This classification depends on the degree and type of permeability of the underlying and adjacent rock types and on the depth to the water table. The three classes are as follows:

(i) Containment sites.

(ii) Slow Migration sites.

    (a) With a thick unsaturated zone

    (b) With little or no unsaturated zone

(iii) Rapid Migration Sites.

Containment Sites: These sites are located on or in impermeable or relatively impermeable strata (less than 10^{-3}m/d) so that the leachate is contained on the site. Rock types such as lacustrine clay, clayey till, peat, mudstone, muddy limestones and certain igneous and metamorphic rocks may have sufficiently low permeabilities. However achieving these low permeabilities is more likely in Quaternary deposits than bedrock, as fissures in the upper layers of bedrock can allow migration of pollutants. Alternatively, containment can be created by using an artificial liner such as puddled clay, asphalt, plastic or butyl rubber. The advantages and disadvantages of containment sites are as follows:

Advantages	Disadvantages
1 Commonest safe geological situation in Ireland (and, in my view, those areas of Britain underlain by fissured rocks).	1 "Bathtub effect" causing the site to fill up with leachate and then overflow unless designed and operated properly.
2 If constructed and managed properly it gives full control over leachate.	2 Often difficult to achieve a sufficiently low permeability to ensure containment.
	3 Leachate has to be collected and disposed of. This can be expensive.

In certain situations, use of cut-over bog sites may be a relatively cost-effective way of disposing of refuse in an environmentally safe manner. The advantages and disadvantages of cut-over bog sites are listed below:

DH' type of sites that can be 'utilised'

Advantages	Disadvantages	Advantages	Disadvantages
1 Peat relatively impermeable, therefore can be used as a containment site.	1 Bogs invariably are wet and difficult to operate.	1 Allows natural degradation of wastes to less noxious compounds.	1 Difficult to locate because water table is usually high – within 10m of ground surface – in Ireland.
2 Peat has high cation exchange capacity (CEC) which would assist in attenuating leachate.	2 No clay cover material on site and may not be present nearby.	2 No leachate to collect and dispose of.	2 Difficult (perhaps impossible) to predict degree of attenuation in unsaturated zone. Pollution by trace organics likely.
3 Adjoining peatland could be used to treat the leachate.	3 Has low load-bearing capacity.	3 Cheaper site to run than containment sites.	3 If incorrectly located, contamination of groundwater and/or surface water could readily occur.
4 Low population density in area.	4 Peat is a fire hazard.		
5 Low land-use value, therefore cheaper to buy.	5 Further research and trials are necessary to test their suitability.		

Slow Migration Sites: These can be located on geological formations in which groundwater movement is by intergranular flow – gravels, sands, silts and free-draining sandy tills in most of Ireland, Scotland, Wales and Northern England. (In Central and Southern England they have been located on the Sherwood Sandstone and the Chalk aquifers). The pollution potential and pollutant attenuating capacity of these sites depends largely on the thickness of the unsaturated zone and the clay content of the geological materials.

Sites with an unsaturated zone: This is the "dilute and disperse" type site where the leachate as it moves through the unsaturated zone is attenuated by various chemical and biological processes and is diluted when it reaches the water table to a degree where it is hoped that significant contamination does not occur. The advantages and disadvantages are as follows:

Sites with little or no unsaturated zone: The absence of an unsaturated zone reduces the ability of the rock to attenuate the leachate and consequently there are many examples in the literature where contamination of aquifers and streams have occurred. Most gravel quarries in Ireland fall into this category. Consequently, waste disposal sites should not be located in this situation unless:

(1) the site is artificially sealed,
(2) the consequences of pollution of an aquifer or stream are acceptable to the regulatory authorities; or
(3) the wastes deposited are inert.

Sites Allowing Rapid Migration: About 30% of Ireland is underlain by rocks with a high fissure or karstic permeability which allows rapid migration of leachate without much natural degradation or dilution. These are the limestone, sandstone and volcanic bedrock aquifers which are very susceptible to pollution. Many limestone and sandstone quarries fall into this category. Waste disposal sites should not be allowed in this situation.

Optimum site characteristics

The optimum area of a site from a geological viewpoint is considered to be a greenfield site underlain by thick unfissured clayey till or clay which in turn is underlain by low permeability bedrock. The till can be excavated and used for intermediate and top cover and for building bunds around the waste. This type of site allows control of the leachate (the noxious liquid produced in the waste as water passes through it) and does not depend on the unpredictable attenuating processes beneath "dilute and disperse" sites which may not be adequate to prevent pollution. The intermediate cover (till spread daily on the waste) reduces problems with litter, fires and vermin. A possible alternative to this situation is a cutover bog area underlain by clays and adjoining an area of till, which can be used as cover material.

The use of sands/gravels either at greenfield sites or in disused pits as "dilute and disperse" sites has been criticised in Ireland (Daly 1987) and sites such as these are now seldom used.

A planned approach to the location of tip sites

Locating tip sites, whether for domestic waste alone or for co-disposal of domestic and industrial waste, can be time-consuming, difficult and expensive for local authorities. Difficulties arise particularly due to public resistance to sites in their area and to financial restrictions. Site investigations cost substantial sums and consequently it is important not to waste time and money on unsuitable sites. Problem sites can be engineered (e.g. an artificial liner might be used) and operated so that environmental effects are minimised, but this can be costly. Also, the pollution risks from highly engineered problem sites are usually greater than from good natural sites. The presence of cover material at or close to a site simplifies and reduces the cost of operating the site. Consequently, locating geologically and hydrogeologically optimum sites has many benefits for local authorities.

Optimum sites can be chosen by a local authority if there is sufficient geological and hydrogeological information compiled as maps. The following maps are required:

1   Bedrock geology;
2   Subsoil geology (Quaternary deposits);
3   Depth to bedrock;
4   Aquifer maps with a synopsis of hydrogeological data; and
5   Groundwater vulnerability and protection maps.

By superimposing various maps on one another, optimum areas can be located; for instance, areas with thick boulder clay overlying a low permeability bedrock or areas where the groundwater is saline and unpotable. By identifying these areas, inappropriate site investigations and expensive engineering solutions and operational methods can be avoided. Obviously there are also non-geological factors to be considered, such as distances from the waste source, and cost and difficulty of site acquisition, which are important. But this approach allows comparison of the different areas including consideration of non-geological factors, and allows the best potential site or sites to be chosen, thereby beneficial to both the environment and the local authority finances. This approach also has another major advantage for local authorities: when asked the difficult question "why this site? why not some other area?" the local authority can state that the site is in one of the optimum hydrogeological areas and that the whole county or a large area has been examined.

This planned approach can be part of groundwater protection schemes.

In Ireland, in the experience of this author, the main limiting factor in this approach is the lack of adequate Quaternary geology maps, information and expertise.

Septic Tank Systems

General

Behind many rural houses is a major environmental issue that has received little publicity – namely, the pollution of the environment by septic tank effluent. This is, obviously, a somewhat unpleasant topic – few people like to think or talk about the disposal of their own faeces and urine in their back garden. Yet it is a serious environmental and health issue for the hundreds of thousands of people who have septic tank systems and for the staff who deal with planning permissions for houses with septic tank systems.

There are over 300,000 septic tank systems in Ireland serving a rural population of about 1 million people (almost 25% of the population). Over 40 million gallons of effluent are produced daily, all of which enters the ground (usually Quaternary deposits) and a high proportion of which ultimately enters goundwater. These figures indicate a potential problem, a problem which is likely also to exist in the rural areas of Britain not served by public sewers.

A septic tank is a buried, watertight container which removes the gross solids from sewage by settlement. It provides only minor treatment of sewage and so the effluent leaving the tank is highly polluting if it enters water. The effluent contains faecal bacteria (up to 100 million/gallon of effluent) and viruses, nitrogen (40–80mg/l), and phosphorous (10–30mg/l), while the B.O.D. ranges from 20–450mg/l (Bouwer 1978).

Effluent treatment

The main treatment area for the effluent is not in the septic tank, but occurs after it has been discharged from a soakage pit or pipe distribution system into the ground. It is the Quaternary deposits that treat the effluent, therefore they are crucial in pollution prevention.

Water pollution problems

Septic tank systems cause water pollution either where:

(a) soakage is inadequate; or
(b) soakage is excessive.

Table 22.3 [*orig. table 3*] Suggested distance of a well from a percolation area.

Type of Subsoil	Minimum Depth Subsoil Above Rock (m)	Minimum Depth to Water Table (m)	Minimum Distance from percolation area (m)
Clay, Clayey Till (low percolation rate)	1.0	1.0	1.0
Sandy Till (medium percolation rate)	1.0	1.0	45
	2.0–5.0	1.0	30–45
	5.0	1.0	30
Sand and Gravel (high percolation rate)	1.0	1.0	60
	2.0–5.0	1.0	40–60
	5.0	5.0	30
			(from N.S.A.I., 1990)

Notes: (i)   Depths are from the level of the percolation pipe and not from ground level.
(ii) A distance less than 30m can be justified in certain circumstances, for instance (a) where the depth of clay is several metres and the well is upslope of the percolation area or (b) where both the thickness of sandy clay and the depth to the water table is greater than 5m.

Soakage is inadequate where the Quaternary deposits have a low permeability, usually due to the presence of clay or peat. This results in ponding of effluent at the surface, odour nuisance and surface water pollution. Where soakage is excessive, groundwater is at risk as the effluent flows rapidly through the ground with minimal attenuation. Studies carried out by Sligo Regional Technical College and the Geological Survey of Ireland suggest that septic tank systems are one of the major sources of pollution to wells and springs in Ireland (Henry 1990; Doyle et al. 1986; and Daly 1987). The main problems arise in areas where the subsoil consists of highly permeable sand and gravel, and where karstic or fissured rocks are present at or close to the ground surface.

Optimum site characteristics

The most suitable areas geologically for septic tank systems are those underlain by free-draining sandy tills or clayey gravels.

It has been suggested (Daly 1987) that there should be at least 1.0m of subsoil between the effluent distribution pipes of septic tank systems and fissured rocks or the water table if pollution is to be prevented. This is now becoming the recommended practice in Ireland (N.S.A.I. 1990).

Distances between percolation areas and wells

A minimum distance of 30m between a well and a septic tank system is considered to be "safe" by many regulatory authorities. However it has been shown (McGinnis & DeWalle 1983) that bacteria and viruses can travel more than 30m in various geological situations. Therefore the 30m distance is inadequate.

The level of effluent contamination of groundwater sources normally decreases with distance from the percolation area. This distance depends on a number of complex geological and hydrogeological factors, many of which are difficult to quantify. While it is not possible to specify with certainty the minimum safe distance between percolation areas and groundwater sources, those in table 22.3 are suggested as

a guide. These distances are based chiefly on the most important geological and hydrogeological factors – the type and thickness of the Quaternary deposits and also the depth to the water table, all of which can be assessed. If these distances are adhered to, the risk of pollution of wells should be minimised.

Quaternary Geology in Pollution Assessment and Prevention

General

According to Hageman (1987) "many Quaternary geologists are unaware of the decisive role this (i.e. applied Quaternary Studies) should play in the development, enrichment and survival of human society".

The main contention in this paper is that the Quaternary geology of any area is a crucial aspect in dictating the pollution risk from potentially polluting developments and activities. In considering groundwater pollution problems, hydrogeologists, engineers, planners, farm advisors and environmental health officers often have inadequate expertise on Quaternary geology and insufficient appreciation of the role of Quaternary deposits. This has arisen partly because of the inadequacy of the geological maps available and also, in my view, to the lack of input of Quaternary geologists in bedrock geology, hydrogeology, planning and engineering third-level courses.

Communicating with decision-makers

Quaternary geologists can seldom be classed as decision-makers. As Quaternary deposits are an important factor not only in the groundwater pollution area but also in many infrastructural developments and in land-use planning, it is essential that, as Hageman (1987) puts it, "to be constantly knocking at the doors of decision-makers". However in doing this effective communication is essential. Academic or applied research findings are of little use unless they can be translated and presented as facts and arguments that decision-makers can understand and be willing to act upon. Therefore,

Quaternary scientists must, in my view, present simplified interpretations in the form of user-friendly, thematic or interpretive maps and in reports that are readily intelligible to non-specialists. These maps are not only a means of communicating facts and interpretations in ways that can be understood and appreciated, they are also a means of marketing the importance of Quaternary deposits.

Quaternary geology and depth-to-bedrock maps

As emphasised already, the type, permeability and thickness of the Quaternary deposits are crucial factors in considering the location of potential pollution sources. Consequently Quaternary geology maps should be available in the offices of the relevant decision-makers, such as engineers and planners, at a useful scale – preferably 1:100,000 or larger. Depth-to-bedrock maps are also essential. In preparing Quaternary geology maps suitable for applied uses, thought must be given to the information provided and the language used.

Planners, engineers and hydrogeologists require information on the lithology, degree of sorting, grain size distribution and thickness. Details on the formation name or age have a low usage value.

The use of the term Quaternary in the title of maps aimed at decision-makers, such as planners and engineers, is not recommended as it is not readily meaningful. Terms such as "surficial", "superficial" or "drift" are also not helpful and the term "overburden" is ambiguous. The term "unconsolidated deposits" is an option, but the term used by this author, in the Irish situation, is "subsoil", which is the material between the topsoil and bedrock. This may not suit the situation in Britain but an effort should be made, in my view, to finding a suitable alternative term to "Quaternary".

Site investigations

Two of the main purposes of a site investigation are to give sufficient geological and hydrogeological information to enable firstly, prediction of the likely detrimental effects of a development on the environment and secondly, the installation of pollution control mechanisms. The level of investigation obviously depends on the magnitude of the risk posed to the environment. For landfill sites it is necessary to drill and test boreholes. For smaller developments such as septic tank systems and farmyards, a visual assessment of the site and a number of trial pits give useful information on the geology and consequently on the pollution risks at the site. This relatively simple and basic site investigation can also give useful preliminary information prior to more sophisticated and expensive investigations as they can help focus on the aspects that need checking by drilling and provide a qualitative and "common sense" assessment of the accuracy of quantitative determinations, for instance of permeability.

Quaternary geologists can have a role in carrying out these basic site investigations and perhaps more importantly in training other specialists in assessing sites.

Among the main factors relevant to engineers, planners, farm advisors etc., during a visual assessment are vegetation, composition of subsoil, iron pans and depth to bedrock. An outline of the basic knowledge relevant to these specialists and which can be taught in an expanded form by Quaternary geologists is given below.

Vegetation: Rushes, yellow flags (irises) and poaching all indicate poor percolation characteristics or high groundwater levels. Grasses, bracken fern and furze may indicate good percolation characteristics.

Composition of subsoil: Simple tests which assist in the recognition of the subsoil are as follows:

(a)   Take a moist sample of subsoil and rub it between the palms of the hands.

Subsoil	Result
1 Clean gravel	Leaves no residue on hands; friable
2 Sandy gravel, sand	Leaves some grains on hands; friable

| 3 Silt | Leaves a dirty film, can also be detected by rubbing between the fingers. |

| 4 Clay | Sticky on hands, tenacious. If rolled in the palm of the hand, a moist sediment will form a ball which can be rolled into a narrow string; the thinner the string that can be made without crumbling the higher the clay content. |

(b)   Take a dry sample of subsoil and rub it between the palms of the hands.

Subsoil	Result
1 Clean gravel, sandy gravel, sand	Clods (if formed) can be crushed readily.
2 Clay, silty clay	Clods will be formed which will be difficult to crush.

Colour of subsoil: Well drained and well aerated subsoil is usually brown, yellow or reddish in colour. Poorly drained or saturated subsoil is often grey or blue in colour. Brown and grey mottling usually indicates periodic saturation.

Quaternary Geology and Groundwater Protection Schemes

Groundwater protection schemes

"Prevention is better than cure". This is an often quoted phrase in many areas of living. In few areas is it a more appropriate or important aim and philosophy than in groundwater management and protection. Why? Because if groundwater gets polluted it takes months and sometimes years to clean up. This is in contrast with a surface stream which, if it gets polluted, usually cleans itself in a matter of days as the pollutants are flushed out. In the ground, water moves more slowly – sometimes as little as a few metres/year and quick flushing is not an option. Remedial measures such as attempting to pump out pollutants is usually very costly, slow and not always effective. So emphasis must be on "prevention" and not "cure".

There are several ways of preventing pollution – for instance, sanitary protection of wells. However, one of the main ways is through the use of groundwater (or aquifer) protection schemes as part of the planning process. Groundwater protection schemes are increasingly being used in developed countries, although they vary depending on local factors such as geology, hydrogeology, main pollution sources, statutory regulations, culture and attitudes. Most are based on information about the geological framework and, in glaciated countries, particularly about the Quaternary deposits. In Illinois, for instance, data on the composition, thickness and regional distribution of glacial deposits are used to construct maps of: geological materials to a depth of 20ft.; bedrock topography; thickness of glacial deposits and sand and gravel aquifers (Berg et al. 1984). On the basis of these maps, interpretative maps were then prepared that rate geological sequences in their capacity to protect aquifers and surface water from contamination by landfill sites, septic tank effluent and surface spreading of wastes and agricultural chemicals. These maps are available to planners in decision-making.

Groundwater protection schemes in Ireland

For several years the Geological Survey of Ireland has been advocating the use of groundwater protection schemes. The recommended approach, which is based on the natural conditions, the availability of resources, the main pollution sources and organisational framework, involves five stages:

(1)   Preparation of an *initial* groundwater protection scheme;
(2)   Vulnerability assessments;
(3)   Assessments of pollution loading;

(4) Site investigations and hydrogeological assessments; and

(5) Preparation of a *final* groundwater protection scheme (Daly, 1991).

Preparation of an Initial Groundwater Protection Scheme: This stage involves the production of a map which divides any area, such as a public (river, local, county) authority area, into a number of (usually four) groundwater protection zones according to the degree of protection required, Zone 1 requiring the highest degree of protection and Zone 4 the least.

Zone 1:   Source protection zone around public and major industrial supplies. Usually 1km radius around boreholes and divided into subzones.

Zone 2:   Major aquifers.

Zone 3:   Minor aquifers.

Zone 4:   Poor and non-aquifers.

A code of practice is then prepared which lists the generally acceptable and unacceptable activities for each zone and describes the recommended controls for developments. For instance in Zone 1 it is recommended that septic tank systems should not be allowed within 300m of public supply boreholes unless there is substantial information that they pose no risks to groundwater. Landfill sites and piggeries are not recommended in Zone 2. In Zone 4 there is no objection to any activities except where an existing ground source is being put at risk.

Several local authorities in Ireland have already carried out the first stage – Offaly, Galway, Mayo, Wexford, north Cork. However they are rather crude and basic as they are based on limited hydrogeological information. Information on the Quaternary deposits is not required at this stage, apart from the location of sand and gravel aquifers. Consequently in order to refine and improve the scheme further stages are required.

Vulnerability Assessments: Two main approaches are adopted:

(a) Vulnerability mapping using geological and hydrogeological data; and

(b) using water quality data.

Vulnerability mapping has already been described in this paper. It is a critical way of providing the information needed to improve the initial groundwater protection scheme and it requires information on the Quaternary deposits. A small number of local authorities have commenced this stage – Offaly, Roscommon, and Tipperary (South Riding) County Councils.

Assessment of Pollution Loading: Vulnerability assessments of any area by themselves do not provide a system of groundwater protection. They have to be linked to surveys and assessments of existing potentially polluting activities.

One method of taking account of the pollution loading is to survey the existing potentially polluting sources and to overlay the vulnerability map with a map showing the locations of these sources. An assessment of pollution risk is then possible. Also the relevant public authority can give priority to monitoring the developments placing groundwater at most risk.

Site Investigations and Hydrogeological Assessments: Depending on the assessments during stages 2 and 3 and on the available financial resources, it may be necessary to dig trial pits or drill boreholes to provide extra information.

Preparation of Final Groundwater Protection Scheme: The final stage is to refine the code of practice and the zonal boundaries prepared in stage 1 based on the information collected in stages 2, 3 and 4. The scheme should then be integrated into the planning process, which in Ireland would mean incorporating it into the County Development Plans.

Data requirements

In preparing a comprehensive groundwater

protection scheme the following maps and accompanying reports are needed:

Bedrock Geology

Subsoil (Quaternary) Geology

Depth-to-bedrock

Hydrogeological data

Aquifers Hydrochemical and bacterio-logical data

Groundwater Vulnerability

So a suite of maps should be available in the relevant public authority office for consultation by the engineers or planners or as a "back-up" to the final map – the ground-water protection map – when considering the location of a potentially polluting development.

Benefits of groundwater protection schemes

Groundwater protection schemes have the following benefits and uses:

(1) Groundwater is hidden underground and there is often insufficient aware-ness of its presence and importance. The schemes bring the groundwater interest to the attention of decision-makers and enable groundwater to be taken into account in balancing the various interests in land-use planning.

(2) They assist planners and engineers in locating and regulating potentially polluting activities:
    – by providing basic information and therefore assisting in the preparation of development, water quality and waste management plans and in assessing environmental impact statements,
    – by helping to show generally suitable and unsuitable locations for landfill sites and for the land spreading of wastes,
    – by acting as a guide for local authority staff and providing a "first-off" warning system which can be used before site visits are made. So for instance, an engineer assessing the environmental impact of blood or

slurry disposal or even a septic tank on an area could look at the Groundwater Protection Map and would have basic information on the groundwater resources in the area and the vulnera-bility of these resources before making the site visit.

(3) They enable more detailed and expen-sive investigations to be directed where the threat is greatest.

(4) They assist public authorities in preparing Emergency Plans.

(5) They enable an assessment of the full implications of the implementation of some E.C. directives – for instance, the proposed directive on nitrates in groundwater.

(6) They can assist in planning and under-taking regional and local groundwater monitoring schemes.

However, protection schemes are not a panacea for solving all problems. By their nature vulnerability maps generalise what in fact are often very variable geological conditions. In particular their value can be reduced by the absence of good quality Quaternary geology information.

Groundwater protection schemes are a means of not only protecting groundwater, but of bringing geological information into the planning process and therefore enabling the work of geologists to have a wide influ-ence. However, in Ireland the lack of Quaternary geology information and maps is hindering this process.

Conclusions

1 The Quaternary deposits are arguably the single most important natural feature in influencing groundwater vulnerability and groundwater pollu-tion prevention in areas underlain by pre-Permian bedrock. In areas under-lain by younger rocks, they have a less important but significant role. Where an appreciable thickness of low to medium permeability materials overlie groundwater, they act as a barrier to the vertical movement of pollutants or as a filtering, purifying medium.

Where they are absent or very thin or where they are very permeable, the groundwater is vulnerable to pollution and polluted wells are common. It follows that Quaternary scientists should have a major role to play in this area of environmental protection.

2   The use of groundwater protection schemes and interpretive maps are a practical means of preventing groundwater pollution. However, their value is dependent on the availability of accurate high quality Quaternary geology information.

3   It is suggested that a high priority within the Quaternary science area should be given to sedimentological and stratigraphical research and particularly to mapping.

4   The application of geological information to the human use and exploitation of the environment is by no means new, but the systematic collection, synthesis and application of geological and hydrogeological data to land-use planning and environmental protection are tasks of increasing urgency in today's society. Technological ("engineering") solutions alone are frequently expensive and risky. Sound engineering involves the proper assessment of the natural environment – the geological materials combined with good engineering design and practice based on this assessment. Geoscientists (including Quaternary geologists), engineers, planners and other decision-makers must work together to conserve our environment in a planned, cost-effective manner.

Acknowledgements

This paper is published with permission of the Assistant Director, Geological Survey of Ireland.

References

Bachmat, Y. & Collin, M. (1987). Mapping to access groundwater vulnerability to pollution. *In:* van Duijvenbooden, W. and H. G. van Waegeningh (eds). *Proceedings of the International Conference "Vulnerability of soil and groundwater to pollutants"* The Netherlands, TNO Committee on Hydrological Research. pp. 297–308.

Berg, R. C., Kempton, J. P & Stecyk, A. N. (1984). Geology for planning in Boone and Winnebago Counties. *Illinois State Geological Survey, Circular* 531.

Bouwer, H. (1978) *Groundwater Hydrology.* McGraw-Hill.

Daly, D. (1987). Septic Tanks and Groundwater. *Proceedings of International Association of Hydrogeologists (Irish Branch) Seminar, Portlaoise.*

Daly, D. (1991). Groundwater Protection Schemes. *Proceedings of Spring Show Conference of Local Authority Engineers, Dublin.* Department of the Environment.

Doyle, M., Henry, H. P. & Thorn, R. H. (1986). Septic Tanks: A Survey. *The GSI Groundwater Newsletter,* 2, 8.

Foster, S. S. D. (1987). Fundamental concepts in aquifer vulnerability, pollution risk and protection strategy. *In:* van Duijenbooden, W. and H. G. van Waegeningh (eds). *Proceedings of the International Conference "Vulnerability of soil and groundwater to pollutants"* The Netherlands TNO Committee on Hydrological Research, pp. 69–86.

Hageman, B. P. (1987). Applied Quaternary studies. *Episodes.* 10,1.

Henry, H. P. (1990). An evaluation of septic tank effluent movement in soil and groundwater systems. *Unpublished Ph.D. Thesis. Sligo Regional Technical College.*

McGinnis, J. A. & Dewalle, F. (1983). The movement of typhoid organisms in saturated permeable soil. *Journal of American Water Works Association.* pp. 266–271.

Monkhouse, R. A. & Richards, H. J. (1982). *Groundwater resources of the United Kingdom.* Directorate-general for the environment, consumer protection and nuclear safety, Commission of the European Communities. Brussels and Luxemburg.

N.S.A.I. (1990). *Septic tank systems.* Recommendations for Domestic Effluent Treatment and Disposal from a Single Dwelling House. In draft. National Standards Authority of Ireland.

Wright, G. R., Aldwell, C. R., Daly, D. & Daly, E. P. (1982). Groundwater resources of the Republic of Ireland. Directorate-general for the environment, consumer protection Communities. Brussels and Luxembourg.

23

ACID RAIN SINCE 1985 – TIMES ARE CHANGING*

T. G. Brydges and R. B. Wilson

Introduction

Last year, when Bob Wilson and I were asked by the Conference Organisers to prepare this introductory lecture for Glasgow '90, we felt both honoured and a little intimidated by the task. Acid rain has been one of the first and longest running of the multi-national environmental issues. Acid rain has been characterised by the trail-blazing and insightful studies of many scientists, by an intense and at times acrimonious scientific and public debate and also by a generous measure of success in building an international environmental science network and in implementing effective and innovative control programmes. Nearly every industrialised country is actively involved in emission control programmes. It is appropriate to reflect on some of the lessons we have learned, to review major areas of progress and to contemplate the challenges for the future.

Over 30 years ago, Drs Gorham and Gordon provided examples and warnings of things to come in their classical studies in the English Pennines and in Canada, near large sulphur dioxide sources at Sudbury, Ontario and in Nova Scotia (Gorham 1957; Gorham & Gordon 1960, 1963). As is often the case

with new scientific ideas, they did not generate much reaction in scientific or public domains.

Dr Harold Harvey also learnt firsthand of the lack of interest in acid rain. In the early 1960s, while attempting to find suitable habitat for stocking sockeye and pink salmon (*Onchorynchus nerka and O. gorbuscha*) he began work in the LaCloche Mountain Area near Sudbury. He and his graduate student, Richard Beamish, became aware of the collapse of fish populations in Lumsden and George Lakes and, under the influence of the preliminary results beginning to come from Sweden on the effects of acidification, they were able to determine that acidification was at the root of the failure of the fisheries.

The public reaction was definitely not supportive. For example, Dr Harvey relates that one sport camp operator who was aware that the fish populations exploited by his clients were declining, was planning to sell and recover his money before the problem became well known. A public petition was posted in one of the local stores seeking support for a resolution to have the University of Toronto researchers leave the area because the local perception was that there had been no problem in the lakes until

* Originally published in *Proceedings of the Royal Society of Edinburgh*, 1991, vol. 97B, pp. 1–16.

they had started their investigations (Brydges 1987).

In the late 1960s, the incomparable Svante Odén (1967) successfully alerted the Swedish public and the scientific community to the problem of long-range transport and deposition of acidic substances, a phenomenon now known in virtually every household in the industrialised world as 'acid rain'. The term is frequently used to describe all forms of deposition of acidic substances and for convenience, we will use the term in this way throughout our paper. Svante Odén's findings and hypotheses were part of Sweden's case study on acid rain which was developed by a group of scientists chaired by Professor Bert Bolin that provided a major input to the 1972 United Nations Conference on the Human Environment held in Stockholm. Years of environmental study have confirmed Odén's three main observations about the nature of the phenomenon of long-range transport of pollutants:

(1) in much of Europe, acid precipitation was a large-scale regional phenomenon with well-defined source and sink regions;

(2) both precipitation and surface waters were becoming more acidic; and

(3) long-distance (100–2000 km) transport

of both sulphur- and nitrogen-containing air pollutants was taking place among the nations of Europe (Cowling 1987).

In addition, Odén made 10 predictions of future environmental consequences (table 23.1): virtually all have been validated by subsequent studies. For example, he postulated acidification of rivers and lakes with resulting decline of fish populations. These phenomena were well documented fairly soon after Odén's predictions by such studies as the Norwegian Study on acid precipitation (Overrein et al. 1980). He also predicted the acidification of soils which was regarded by many scientists as highly unlikely to occur yet it has now been well documented (Tamm & Hallbacken 1989).

While Odén's predictions were not readily accepted by the scientific community of the day and, in some quarters, were even derided, they triggered international discussion. In 1975, the first large international conference on acid rain was held in Columbus, Ohio, and Odén gave the keynote lecture. The discussion is continuing and expanding to this day.

In the mid 1970s, Norway began one of the first large-scale integrated studies of the effects of long-range transport of acidic materials and the results were the focal point

Table 23.1 [orig. table 1] Predictions made by Svante Odén in 1986 about the probable ecological consequences of acid precipitation and other changes in the chemical climate (Cowling, 1987).

(1) Increased acidification of lakes, streams, and rivers

(2) Decline of fish populations

(3) Acidification of soil systems

(4) Displacement of nutrient cations from soils

(5) Displacement of toxic metals from soils to surface and ground waters

(6) Acidification of ground waters

(7) Decreased biological fixation of atmospheric nitrogen

(8) Decreased growth of forest trees

(9) Increased disease in plants

(10) Accelerated corrosion and weathering of engineering and cultural materials

of the 1980 Sandefjord Conference (Overrein et al. 1980). By that time, acidification of surface waters by acidic deposition and the resulting death of fish populations had become a well established fact. At the 1980 Sandefjord Conference, Dr Bernard Ulrich expressed concern about forest decline, although the possibilities of acidification damage to forests was considered little more than a curiosity.

After the Sandefjord Conference it was obvious that the acid rain issue would be with us for at least a few more years and in 1981 Canada proposed the Muskoka 1985 conference. Some of us thought that the issue would have lost a considerable amount of public and scientific interest by 1985, but there would still be sufficient concern that we would attract as many as 500 participants. Registration was finally closed at 750 at which time the facilities were taxed beyond their limit as many participants who shared overcrowded rooms will unfortunately remember. By 1985, the accelerating, rather than diminishing interest was mainly due to forest decline having become a dominant issue. Dying trees were observed by the public in widespread areas in central Europe, in eastern Canada and in the U.S.A. (Bruck 1985; Matzner & Murach 1985; McLaughlin et al. 1985). Ulrich's warnings from five years before seemed all too true and when asked how he felt about the new findings he said "In 1980 I had hoped that I was wrong, but it appears that I was right".

In 1985, however, we were able to take considerable pride in the fact that the flow of scientific and public information was resulting in substantial sulphur dioxide control activity. Within the Economic Commission for Europe (ECE) the Convention on Long Range Transboundary Air Pollution was signed in 1979. This led in turn to the sulphur dioxide protocol of July, 1985 whereby the signatory nations agreed to implement a 30% reduction in sulphur dioxide emissions or in transboundary fluxes. We were even further encouraged by the fact that many countries had recognised their national and international environmental responsibilities prior to the final signing of the protocol and were already reducing sulphur dioxide emissions. Moreover, many had made agreements to reduce emissions by more than 30%. For example, in 1984, Canada committed itself to a 50% reduction in SO_2 emissions from eastern Canada to be implemented by the end of 1994. The European countries are committed to reductions from large combustion plants of 20, 40 and 60% SO_2 by the years 1993, 1998 and 2003. NO_x reductions of 30% have also been agreed upon for the years 1993 and 1998.

Our improving scientific understanding of the consequences and control of acid rain has been documented by the many national and international acid rain assessment programmes that are coming to fruition. In addition, numerous countries have progressed well on sulphur dioxide emission reductions. There are international negotiations in Europe with regard to the post-1993 extension of the ECE Sulphur Dioxide Protocol, and in North America, to a Canada/U.S.A. agreement on acid rain control. Developments such as these since Muskoka '85 have led inevitably to the identification of a need for an international audit of global acid deposition and here we are, in Glasgow in 1990, to do just that!

We will begin the audit with some thoughts on the lessons and progress of the past, but most importantly, on how the public and industrial perceptions have changed and how these changes may influence scientific developments.

Scientific Progress

The scientific evaluation of the causes of acid rain and of present, and projected, environmental damage has provided the basis for public concern and consequent social action for nearly two decades. However, defining the causes and making these projections has been a long and difficult task with the scientific community itself frequently – and we believe correctly – being the strongest critic of the information. Scientific debate of complex issues is a necessary component of progress. Communication of scientific facts and uncertainties to the public has been, and remains, a major challenge that we will discuss later.

Under the influence of the pioneering work of Gorham, Harvey, Odén and others, studies in the 1970s were dominated by atmospheric transport, deposition and the resulting aquatic effects. These studies led to the documentation of acidification of lakes and rivers by sulphate deposition and the resulting loss of fish and other aquatic species so well reported at the Sandefjord Conference and confirmed at Muskoka '85.

In the early 1980s, unexplained forest decline reached the headlines in Europe and North America and this led directly to an expansion of acid rain studies to include forests and soils.

Natural factors such as diseases and insects certainly contribute to forest decline. However, in addition, air pollution has become a factor. The wax coatings of leaves and needles are known to be affected by acid mists (Percy & Baker 1987, 1988). High elevation forests and forests in some coastal regions seem to be particularly affected by acid rain. For example, in New Brunswick, Canada, near the Bay of Fundy, birch (*Betula* spp.) are being killed by acidic mists (Cox et al. 1990). It is also known that many trees are in decline attributable to nutrient deficiencies thought to be induced or aggravated by acid rain (Zoettl & Huettl 1986).

Damage to buildings and materials has also been verified and will be given further corroboration by the UNECE International Cooperative Program on Buildings and Materials, including Cultural Heritage, led by Dr Kucera of Sweden.

Decreasing sulphate deposition, in response to reduced sulphur dioxide emissions, has shifted some of the emphasis on soil and aquatic studies toward understanding the role of nitrogen (oxides of nitrogen (NO_x) and ammonium (NH_4)) and toward monitoring and predicting the rate and extent of chemical and biological recovery. Perhaps one of the most satisfying results of scientific endeavour and emission control has been the slow but steady documentation of recovery of aquatic ecosystems in Europe and North America (Hauhs & Wright 1989). Gunn & Keller (1990) have

published very encouraging responses in lakes near Sudbury where 25 have returned to an acceptable pH range for lake trout (*Salvelinus namoycush*). This is mainly in response to large-scale reductions in sulphur dioxide emissions at local smelters.

There were many arguments put forward in the 1970s and early 1980s that decreasing sulphur dioxide emissions would not result in improved water quality. Some claimed that acidic deposition was not the cause of the acidification and/or that emission control would not change downwind deposition significantly. These extraneous and distorted interpretations of the science of acid rain have been totally resolved by the documentation of the predicted benefits to surface waters from emission controls.

It appears quite likely that, by removing the acid rain component of the total stress applied to forests, a major percentage of the decline might be halted in some locations. Although this has yet to be proven or observed in the field in areas affected by long-range transport, there are encouraging signs of forest re-growth in the severely damaged soils near Sudbury (Winterhalder 1988). Perhaps this is an 'early warning' of forest improvements to come in other areas!

Since 1985, scientific developments in relation to acid rain have been rapid and significant and we will be brought up to date during the next few days. But what of the future?

The concepts of 'critical loads' and 'target loads' for environmental protection have become central themes in control activities and they will take on greater importance as costs of control increase. Target and critical loads directly link environmental protection to deposition and consequent emission controls. The wish of society to maximise environmental protection and minimise control costs places great demands on our understanding of the present environmental effects and on our ability to predict the future reactions to deposition changes.

In the early 1980s, in North America, Canada developed a target loading of 20 kg of wet sulphate deposition per hectare per annum as being protective of all but the most sensitive aquatic ecosystems (Canada/U.S.

Memorandum of Intent, 1983). The 20 kg ha^{-1} a^{-1} was based on the limited empirical and modelling information available in the early 1980s. Canada has effectively used this target as a guiding principle in designing and implementing the eastern Canadian control programme although both the concept and numerical values were rejected by the U.S. members of the Work Group.

Nitrogen was not deemed to be a problem with respect to long-term acidification based on the data at the time. However, the possibility that nitrogen leaching from watersheds might offset the benefits of sulphur dioxide control (Henriksen 1987) is a matter of considerable concern.

In 1986, the Nordic Council revitalised the concept of establishing environmentally defined deposition values and introduced the definition of critical load, i.e. "the highest load which will not cause chemical changes leading to long-term harmful effects on the most sensitive ecological systems" (Nilsson 1986). It was recognised that, in some cases, the critical load would be unattainable in the short term, so the concept of a 'target loading' was introduced as a loading which was achievable when taking into account socio-economic considerations. It was expected that the target loading would be less stringent than the critical loading and would be associated with some degree of environmental damage that must be defined or estimated by the scientific community.

The ECE countries are discussing the use of critical and target loads for application to the post-1993 sulphur dioxide protocol and the ECE NO$_x$ Protocol specifically calls for critical loads for nitrogen to be developed by 1994.

Both sulphur and nitrogen are essential elements and the problem of acid rain is caused more by their amount, and chemical form, than some inherently 'toxic' feature. Therefore, it is quite valid to seek an acceptable, or critical loading. This is in contrast to substances such as DDT or lead where a zero concentration is a valid objective. The great variability of ecosystem sensitivities and the ecological and social values to be placed on measured 'effects', represents an increasing

challenge for the scientific community over the next few years.

Society and industry expect answers to these difficult questions and scientists must respond if we are to maintain credibility and sustain the successes of the past. However, in addition to the basic research and monitoring programmes being conducted by most countries, we think there are two particularly important scientific tools to deal with the challenge.

The first tool is the use of ecosystem manipulation experiments. In Europe, in the early 1960s Bleasdale studied the effects of air pollution on crops by the use of exclusion chambers. The concept of deliberately altering ecosystems to evaluate changes in different physical chemical and biological processes was pioneered by Dr David Schindler at the Experimental Lakes area in Kenora, Canada (Schindler 1987). This is a fairly remote area located on Precambrian rock with no local source of pollution. In the 1960s he began by testing different fertiliser treatments on different (whole) lakes (Schindler et al. 1971): the results of these experiments had a direct influence on Canada adopting phosphorus control programmes to solve the eutrophication problem.

Dr Schindler began deliberate acidification of lakes in the mid-1970s using both sulphuric and nitric acids. The records of biological changes in Lake 223, the main study lake, have given direction to field surveys for many years and provided convincing and surprising evidence of the loss of species at pHs of no more than 5.8 (Schindler et al. 1985).

At the time of the Muskoka Conference (1985) the Reversing Acidification in Norway (RAIN) project was just getting underway: it has provided key data on the recovery rates and mechanisms for watersheds (Wright et al. 1988). The project design also served as the prototype for 'Fertigation experiments dealing with the effects of acid deposition and forest soils. There are now at least fifteen major projects evaluating effects of stresses on ecosystem responses with many being discussed here in Glasgow.

However, it seems that man's ability to

stress natural ecosystems is almost outstripping his ability to conduct studies and apply appropriate emission controls. In spite of progress, we are still three to five years away from seeing the full implementation of sulphur dioxide controls in Western Europe and Canada and five to ten years away from seeing controls in Eastern Europe and the U.S.A. Meanwhile, the global environment is being faced with additional stresses such as climate change (IPCC 1990) and the alarming global loss of species diversity (Barney 1982).

We think that the techniques of ecosystem manipulation offer great potential in addressing the outstanding and emerging environmental issues of forest decline and ecosystem response to deposited-nitrogen, CO_2, methane production in wetlands and changing patterns of temperatures and rainfall.

The second tool is an important result of the acid rain study, namely the exceptional international cooperation and linkage of scientific communities. It is very well demonstrated by the attendance here in Glasgow of nearly seven hundred participants from more than twenty countries. There are many outstanding examples of international cooperative projects: EMEP, the European air pollution monitoring programme, the development and calibration of the Eulerain Model involving Canada, U.S.A. and West Germany; the RAIN project in Norway involving Norway, Sweden, U.K., Canada, F.R.G. and U.S.A.; the forest 'Fertigation' experiments that involve Sweden, Denmark, F.R.G. and The Netherlands; the Northern American sugar maple decline study involving Canada and the U.S.A.; the SWAP study, involving the U.K., Norway and Sweden, a range of UNECE International Cooperative Programs and the EEC COST Programmes. All of these initiatives have added greatly to our understanding of the mechanisms and processes related to acid deposition and its effects. The successes are, in large part, attributable to the mobilisation of the international scientific community.

When summing-up the results of Muskoka '85, Hans Martin (1985) wrote that

"Acidification may well be a training ground for the scientific community in preparation for more complex interdisciplinary environmental problems". It is essential that this training be put to good use and that we maintain and build on this international network in tackling the myriad of environmental issues facing society.

Industry Response

Since the 1985 conference, the attitudes of industry to environmental protection have changed greatly with a major shift towards a positive approach to solving environmental problems. For many years we have been accustomed to industry, and its defenders, putting forward facts and interpretations of science that have not helped the legitimate scientific debate. For example, Katzenstein (1981) stated in a booklet for public distribution that "most people are surprised to learn that a delicious pear can be more acidic than a tomato or that bananas and carrots are nearly as acidic. All of these have pH values well in the range of the rain that is the subject of scare headlines in the popular media". The covering letter indicated the booklet was to promote "a clear and common understanding of the facts". However, the booklet does not point out that aquatic ecosystems could not survive in tomato juice!

The economics of pollution control are also subject to some misunderstanding. Table 23.2 shows the projected percentage increases in electricity costs for consumers in four regions of the U.S.A. (Regens & Rycroft 1988) that would result from implementing a 10 million ton sulphur dioxide control programme (a measure included in the Clean Air Act Amendments now being discussed by the U.S. Congress). The percentage increases projected for Ontario Hydro's 45% sulphur reduction programme are also included (Ontario Ministry of the Environment 1985; Ontario Hydro 1989). While absolute costs of pollution control may run into several billion dollars, the percentage change in costs to consumers is relatively small. These percentage changes should also be considered in relation to the

Table 23.2 [*orig. table 2*]   Estimated percentage increases in consumer electricity charges attributable to the implementation of acidic rain controls. (Based on the Utilities Specific Control Plan, Ontario Ministry of the Environment, 1989.)

A.	U.S.A. (10 million ton reduction in sulphur dioxide emissions)	
	New England States	4.7%
	Mid-Atlantic States	3.6%
	Midwest States	7.4%
	Southern States	4.5%
B.	Ontario Hydro (60% reduction in sulphur emissions)	2.9%

overall variation in electricity charges. Figure 23.1 shows the average costs for the ten utility districts in the U.S.A. (National Association of Regulatory Utility Commissioners 1985) plotted against the number of scrubbers operating in each of the districts (Edison Electric Institute 1985). For a large number of reasons, the average utility charges varied by 200% and even appear to be *negatively* related to numbers of scrubbers. While it is clear that pollution control costs money, the associated cost increases, relative to other factors governing costs of services to society, are relatively small.

In the case of acid rain controls in North America, the economic debate has created

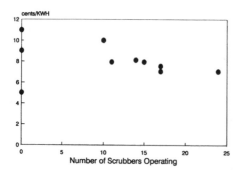

Figure 23.1 [*orig. figure 1*]   Average costs of electricity (cents per kilowatt hour) in the 10 U.S. Utility Districts plotted against the numbers of scrubbers for emission control operating in each district (National Association of Regulatory Utility Commissioners 1985).

considerable delay in building public support for controls. In addition, public concern over excessive cost increases, loss of jobs and severe social disruption caused by implementing control programmes has seriously detracted from the scientific discussion of the problem.

It is also a true, but often overlooked, fact that it is cheaper in the long run to remove or reduce pollution at source than to clean up afterwards – 'prevention is better than cure'. We must become more aware of the economic factors and possible distortions of economic and scientific information if we are to continue to extend pollution control successfully in the future.

In the Canadian sulphur dioxide control programme, control orders have been issued to the larger polluters. One of these, Inco Limited, must make provision to retain 90% of the sulphur contained in its ore by 1994 and while it will spend in the neighbourhood of $500 million constructing new facilities, there seems to be no indication that the company will close. In fact, quite the opposite is true. The president, Mr Roy Aitkin (1989), recently stated "There is no question that we have had a free ride on the environment. Now we have got to spend some money to make sure it is restored". This sort of positive approach is most encouraging to all who have worked for environmental improvements in the Sudbury area.

We must not lose sight of the fact that when Inco Ltd reaches the 90% control level in 1994, it will be 26 years since the first control orders were issued in 1968. It has been a long and difficult process and indeed for Inco, it has been expensive. However, the company has survived, and subject to recovery taking place, so has the environment.

Other Canadian companies such as Falconbridge Ltd., Algoma Steel, Noranda Limited, Kidd Creek Mines and Ontario Hydro have, and will be, implementing major sulphur dioxide pollution control measures without serious economic consequences either to themselves or society.

The entire city of Sudbury, Canada is a rather positive example of industrial and social reactions to the need for pollution control and industrial efficiency. The mining employment in the Sudbury area has dropped from approximately 20 000 workers (40% of the work force) to less than 12 000 workers (20%) since 1971 (Richardson et al. 1990). This has been mainly brought about by modernisation of the plants and mining operations rather than by losses attributable to the implementation of environmental protection measures. Major changes in employment created serious difficulties for the local economy in the short term and special measures were needed to assist unemployed miners (Ontario Ministry of Municipal Affairs and Housing 1985). However, in the long term, the economy has been diversified by an effective combination of planning and actions taken by industry, local government, provincial and federal governments. Consequently, the economy of Sudbury is now much more varied than previously. For example, employment in the accommodation and food service industries increased from 2325 to 4750 between 1971 and 1986 – the cleaner environment did not *cause* the diversified economy; instead it allowed the diversification to take place as a result of generating an attractive environment (Richardson et al. 1990).

Other industries are also taking a more positive attitude. The North American Motor Vehicle Manufacturers Association (MVMA) advertises programmes concerned with vehicle emission control – "The members of the MVMA are committed to continuing the search for cleaner emissions

from cars and trucks and from our manufacturing processes". Under EEC regulations, all new petrol-engined vehicles will be fitted with catalytic converters from 1992; this implies a move towards the exclusive use of lead-free petrol shortly afterwards.

Industries are still faced with the reality of having to implement pollution abatement technology and to meet the costs which are far from trivial. For example, Mr G. Capobianco, president of the Coal Association of Canada, recently warned "When we consider the magnitude of economic and social costs we face in the battle against environmental degradation, it is clear we cannot attack the wrong foe" (*Calgary Herald*, 13 June 1990). He is making the valid point that we had better have our science right! It is essential that industry be given time and clear direction as to their emission requirements. This places a great responsibility on the scientific community to provide the best understanding and effective predictions of 'environmental requirements'. Consequently, the need to understand complex environmental issues in order to design the best possible control programme and the more positive attitude of industry are likely to make the 1990s a decade in which discussions and negotiations with the industrial community may be more constructive than in the past. For example the Canadian Council of Environment Ministers has requested a plan to control ozone in Canada and the plan is being developed with full participation by scientists, industrialists and the public at open meetings.

We need look no farther than our host city, Glasgow, for an outstanding example of environmental recovery. The reality and image of enshrouding pollution from coal and iron ore mines and smelters has given way to a 'renewed' urban environment marked by international conferences and garden festivals.

These examples show that the economics of environmental protection can be managed to minimise costs and detrimental effects. However, if we are to continue to make progress on large and complex environmental issues, scientists must be fully prepared to incorporate their work more effectively into the socio-economic discussion and, indeed, to be in a position to challenge economic studies in the same way as economists/industrialists change the validity of the work of scientists.

Public Response

In the past, segments of society lived with pollution for so long that it was almost accepted as a fact of life, and it was even regarded as a sign of progress and prosperity. However, this view of progress is rapidly changing. Public response is a necessary component of acid rain research and control. Whether we like it or not, it is public reaction to environmental issues that provides most of the social priority for funding research and implementing pollution control programmes. Svante Odén understood this fundamental truth and so introduced the acid rain problem to the Swedish public via the Stockholm newspapers in 1968. As a result, acid rain quickly became a very important issue in Sweden.

National reactions to acid rain have been varied in nature and time. Indifference has been characteristic of many European countries, especially the most industrialised. After several years of apathy in North America the issue became of major importance in 1978 following a Canadian newspaper article linking acid rain with damage to the socially and environmentally important Muskoka/Haliburton recreational area in central Canada (Brydges 1987).

Further, in 1978, in response to a proposal to build a coal-fired power plant in central Canada, the U.S. Congress became very concerned about transboundary aspects of 'acid rain'. On October 7, 1978, public law 95–426 was passed by the 95th U.S. Congress which stated *inter alia* that "the United States and Canada are both becoming increasingly concerned about the effects of pollution, particularly that resulting from power generating facilities since the facilities of each country affect the environment of the other ... It is the sense of the Congress that the President should make every effort

to negotiate a cooperative agreement with the Government of Canada aimed at preserving the mutual airshed of the United States and Canada so as to protect and enhance air resources and ensure the attainment and maintenance of air quality protective of public health and welfare ... It is further the sense of Congress that the President, through the Secretary of State working in concert with interested federal agencies and affected states, should take whatever diplomatic action appears necessary to reduce or eliminate any undesirable impact upon the United States and Canada resulting from air pollution from any source". The United States thus became one of the first governments to take such a strong position with regard to the need for controls. Regrettably, when the full extent of the necessary control programmes became known, the public response to sulphur dioxide controls in the U.S.A. was quite negative, encouraged by the outdated environmental attitudes of the Administration of the day. However, increasing public anxiety, encouraged by scientific information about national and international damage caused by acid rain has prompted President Bush to introduce a proposal for acid control legislation. Although much progress has been made the U.S. Congress has not yet passed the Clean Air Act Amendments; it is expected that they will become law this year.

Through the Info Globe Information Service, we have examined the way in which one Canadian National Newspaper, the *Globe and Mail*, has reacted to acid rain issues. Assuming that the number of newspaper articles is some measure of public interest, then it is apparent that interest increased dramatically from 1978; numbers of articles peaking in 1985 (figure 23.2). While the 1985 Muskoka conference generated considerable media coverage, other events of environmental importance also occurred in Canada that year. The Federal government announced the availability of funds to ensure implementation of the Canadian control programme, while the Province of Ontario, in relation to its large sources of emissions, launched a major

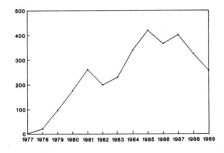

Figure 23.2 [*orig. figure 3*] Number of articles related to acid rain published annually in the Canadian national newspaper *Globe and Mail* between 1977 and 1989.

Table 23.4 [*orig. table 4*] The increasing public concern in the U.K. between 1986 and 1989 for environmental issues as compared with others (U.K. Department of Environment 1989). The figures indicate the percentage of respondents who indicated that an issue was important and required government attention.

Issues*	1986	1989
Environment/Pollution	8	30
Health/Social services	22	32
Unemployment	75	26
Pensions/Social Security	15	18
Crime/Law and order	17	17
Rising prices/Inflation	4	17
Housing/Mortgage rates	8	15
Education	14	13

* Respondents were able to indicate that more than one answer was of importance; therefore the sum of the percentages exceeds 100. All other issues were mentioned by 10% or less of respondents.

control plan called the 'Countdown Acid Rain Program'. However, the issue of acid rain is still frequently given media coverage and doubtless 1990 will see an upsurge due to Acidic Deposition '90 and the U.S. Clean Air Act deliberations. We look forward to the day when there are no articles; the problem having been solved!

In many countries, there has been a continued increase in both public awareness and distress about environmental issues. This is demonstrated by a recent public opinion poll published in one of Canada's national magazines (*Maclean's* January 1, 1990). Prior to 1987, an insignificant number of people rated the environment as the most important issue facing the country. This is surprising because, even before 1987, it was

widespread public understanding of, and concern about, acid rain that established the basis of the political actions that were taken in Canada. In 1984, a decision had been reached to implement a unilateral 50% reduction in sulphur dioxide emissions in the absence of appropriate controls by the U.S.A. However, we can see in table 23.3 that Canadians have placed increasing importance on the environment since 1985 and it now ranks as the top issue for the 1990s.

Table 23.3 [*orig. table 3*] Trends in answers of respondents to a poll assessing public opinions in Canada of the most important issues (Maclean's Magazine, January 1, 1990).

Issues	1985	1986	1987	1988	1989	1990
Environment	*	*	2	10	18	29
GST	*	*	*	*	15	5
Inflation/Economy	16	12	12	5	10	9
Deficit/Government	6	10	10	6	10	8
National Unity	*	*	*	*	7	4
Free Trade	2	5	26	42	7	5
Abortion	*	*	*	*	6	3
Employment	45	39	20	10	6	6

* Not cited by a significant number of poll respondents.

The high priority now given to the environment should provide ample drive to tackle the oustanding issues. Public responses have been similar in the U.K. A survey of public attitudes to the environment (U.K. Department of the Environment, 1989) has shown that 30% of the people rate the environment as one of the most important issues (table 23.4).

Environmental issues are taking a greater place within the political arena with ministerial meetings, such as Bergen 1990, becoming more frequent. Twenty-four Heads of State signed the Hague Declaration in 1989 calling for more international action on environmental issues. The 'G-7' Nations took a strong position on the need for environmental protection (Environmental Statement of the G-7 Summit, 1989). Mrs Thatcher recently stated that protection of the environment and the balance of nature was "one of the great challenges of the twentieth century". In practice, public concern and worry over the consequences of environmental damage are likely to make scientific debate of the issues more difficult. However, we think it is the responsibility of scientists to accept the challenge and help the public to put each environmental issue in its proper context to enable public and politicians to reach considered decisions. In the past, scientists have not always been the best communicators: all too often they have abrogated this responsibility, sometimes with disastrous results. We think that the communication of scientific results and environmental consequences, uncertainties and risks must be seen as a necessary and integral part of environmental research.

Conclusions

Substantial progress towards the elimination of acid rain has been made due in large part to the support which grew from the communication of objective knowledge of environmental processes and the alterations (damage) made to them. This process must continue at this conference.

Sulphur dioxide controls are being implemented or planned in virtually all industrialised countries. While full imple-

mentation of sulphur dioxide controls is still at least 10 years away, the recovery of some lakes in some areas has proven the effectiveness of the actions for which many scientists have been pressing since the 1950s. Industrial attitudes toward the need for environmental protection and hence emission reductions have become much more positive over the past few years.

Concern over the economic consequences of pollution control, mainly centred on increased costs of electricity and possible job losses, has, for ten years in both Europe and North America, been a major impediment to the implementation of pollution controls. Nevertheless, while costs to industry and society can be large, it has been shown that the costs of a clean, acid-free environment are not prohibitive and that a clean environment aids economic development.

There are many large and complex environmental problems that must be faced including the consequences of nitrogen emissions to the atmosphere, the alarming global loss of species and climate change. The associated socio-economic issues are many and varied but they must be tackled during the 1990s. However their move into the public arena poses problems of integrity for research scientists with the responsibility for generating objective science.

In summary, the more positive industrial attitude, the increased public concern and the need to understand the science of complex issues in order to design the best possible programmes of 'control' together necessitate the involvement of environmental scientists in public debates, if rational decisions are to be taken. We have assembled an effective international scientific network to tackle the issue, and believe that *Acidic Deposition '90* will be a milestone in meeting the challenge.

Acknowledgements
We greatly appreciate the efforts of Ms Ruth Tung and Ms Josie Slack and Ms Melanie Chown for their patient typing and retyping of drafts.

We wish to thank the conference organizers for providing us the opportunity to participate in *Acidic Deposition '90*.

References

Aitken, R. 1989. The Greening of the Boardroom. *Toronto Globe and Mail*, Report on Business, July.

Bruck, R. I. 1985. Observations of Boreal Montane forests decline in the southern Appalachian mountains, International Symposium on Acidic Precipitation, Muskoka, Canada.

Brydges, T. G. 1987. Some Observations on the Public Response to Acid Rain in Canada, Svante Oden Commemorative Symposium, Skokloster, Sweden.

Canada/U.S. Memorandum of Intent. 1983. Work Group on Effects Report.

Cowling, E. B. 1987. A Tribute to Svante Oden – Svante Oden Commemorative Symposium, Skokloster, Sweden.

Cox, R. M., Spavold-Tims, J. & Hughes, R. N. 1990. Acidic fog and ozone: their possible role in birch deterioration around the Bay of Fundy, Canada, *Water, Air and Soil Pollution*, (in press).

Edison Electric Institute. 1985. Location of FYD systems in North America, September 1983. Map showing locations.

Gorham, E. 1957. The chemical composition of lake waters in Halifax County, Nova Scotia. *Limnology and Oceanography* 2, 12.

—— & Gordon, A. G. 1960. The influence of smelter fumes upon the chemical composition of lake waters near Sudbury, Ontario and upon the surrounding vegetation. *Canadian Journal of Botany* 38, 477.

—— & Gordon, A. G. 1963. Some effects of smelter pollution upon aquatic vegetation near Sudbury, Ontario. *Canadian Journal of Botany* 41, 371.

Gunn, J. M. & Keller, W. 1990. Biological recovery of an acid lake after reductions in industrial emissions of sulphur. *Nature*, 345, 431–3.

Hauhs, M. & Wright, R. F. 1989. Acid Deposition Reversibility of Soil and Water Acidification – A Review. Commission of the European Communities, Air Pollution Research Report II.

Henriksen, A. 1987. 1000 Lake Survey, The Norwegian State Pollution Control Authority.

IPCC (Intergovernmental Panel on Climate Change) 1990 – Work Group I. Scientific Assessment of Climate Change. World Meteorological and United Nations Environmental Programs, Barney, N. O. 1982. The Global 2000 Report to the President. A Report to the Council on Environmental Liability and the Department of State, Washington, DC, U.S.A.

Katzenstein, A. W. 1981. An updated perspective on acid rain. Edison Electric Institute, Washington, D.C.

Martin, H. 1985. Introduction. International Symposium on Acidic Precipitation, Muskoka, Canada.

Matzner, E. & Murach, D. 1985. Soil acidification and its relationship to root growth in declining forest stands in Germany. International Symposium on Acidic Precipitation, Muskoka Canada.

McLaughlin, D. L., Linzon, S. N., Dimma, D. E. & McIveen, W. D. 1985. Sugar maple decline in Ontario: Is acidic precipitation a contributing factor? International Symposium on Acidic Precipitation, Muskoka, Canada.

National Association of Regulatory Utility Commissioners, Washington, D.C. 1985. 1984–85 Winter Survey of Residential Electric Bills.

Nilsson, J. (Ed.) 1986. Critical Loads for Sulphur and Nitrogen. Report from a Nordic working group. Norwegian Ministry of Environment Report 11.

Oden, S. 1967, Stockholm Newspaper, *Dagens Nyheter*, October 24, 1967.

—— 1976. The Acidity Problem – An Outline of Concepts. Proc. First Int. Symp. Acid Precipitation For. Ecosys. U.S.D.A. Forest Service, Gen. Tech. Rept. NE-23 1–35.

Ontario Hydro. 1989. Options available to meet acid gas limits and selection of preferred options. Ontario Hydro Report to the Lieutenant Governor in Council.

Ontario Ministry of the Environment. 1985. Countdown Acid Rain: Ontario's Acid Gas Control Program for 1986–1994.

Ontario Ministry of Municipal Affairs and Housing. 1985. Project Group on Urban Development, Case Study Report: Sudbury, Ontario, Canada. Community Planning Programs Division. Report to the O.E.C.D.

Overrein, L. N., Seip, H. M. & Tollan, A. 1980. Acid Precipitation – Effects on Forest and Fish, Final Report of the SNSF – Project – 1972–1980.

Percy, K. E. & Baker, E. A. 1987. Effects of simulated acid rain on production, morphology and composition of epicuticular wax and on cavicular membrane development. *New Phytologist* 107, 577.

—— & Baker, E. A. 1988. Effects of simulated acidic rain and leaf wettability, rain retention and uptake of some inorganic ions. *New Phytologist* 108, 75.

Regens, J. L. & Rycroft, R. W. 1988. *The Acid Rain Controversy*, p. 110. Pittsburgh: University of Pittsburgh Press.

Richardson, N. H., Savan, B. I. & Bodnar, L. 1990.

Economic Benefits of a Clean Environment. Sudbury Case Study. Consultants Report to Environment Canada.

Schindler, D. W. 1987. Detecting Ecosystem Responses to Anthropogenic Stress. *Canadian Journal of Fisheries and Aquatic Sciences* 44, 6–25.

——, Mills, K. H., Malley, D. F., Findlay, D. L., Shearer, J. A., Davies, I. J., Turner, M. A., Linsey, G. A. & Cruikshank, D. R. 1985. Long Term Ecosystem Stress: The effects of years of experimental acidification on a small lake. *Science* 228, 1395–1401.

——, Armstrong, F. A. J., Holmgren, S. K. & Brunshill, G. J. 1971. Eutrophication of lake 227. Experimental Lakes Area, northwestern Ontario, by addition of phosphate and nitrate. *Journal of the Fish Resources Board, Canada* 28, 1763–82.

Tamm, C. O. & Hallbacken, L. 1988. Changes in soil acidity in two forest areas with different acid deposition: 1920's to 1980's. *Ambio* 17, 49–55.

U.K. Department of Environment, Digest of Environmental Protection and Water Statistics, No. 12, 1989, HMSO ISBN.0.11.752284.8.

Winterhalder, K. 1988. Trigger factors initiating natural revegetation processes on barren acid, metal-toxic soils near Sudbury Ontario Smelters. Paper to the 1988 Mine Drainage and Surface Mine Reclamation Conference, Pittsburgh, P. A.

Wright, R. F., Lotse, E & Semb, A. 1988. Reversibility of Acidification shown by whole-catchment experiments. *Nature* 334, 670–5.

Zoettl, H. & Huettl, R. 1986. Nutrient supply and forest decline in southwest Germany. *Water, Air and Soil Pollution* 31, 449.

24

THE CHALLENGE OF THE MISSISSIPPI FLOODS*

Mary Fran Myers and Gilbert F. White

Massive flooding in the upper Mississippi and lower Missouri river basins last summer caused widespread human distress and provoked sober public questioning of the wisdom of the policies and programs that had contributed to that disaster. The slowly rising muddy water floated a cluster of tough issues into the national arena.

As the waters recede and recovery efforts pass their peak, it is appropriate to ask how the United States came to mount its long, giant struggle to master the flows of the Mississippi. What were the dimensions of the 1993 floods in terms of rainfall, streamflow, and control works – both surviving and damaged? Why did so many levees fail? What were the consequences for people, buildings, and the natural landscape? How effective has the prevailing patchwork of federal, state, and local efforts been in dealing with flood losses? Did the Mississippi overflow waters carry significant messages for citizens and administrators who have sought and still seek lasting harmony with extreme forces in nature? Can these messages motivate fundamental changes in policies on disaster

response, recovery, and mitigation and on long-term management of the nation's waters and associated lands?

Central Issues

Frequently in U.S. history, one dramatic event that stirs public concern and shakes up legislators and administrators has led to basic changes in the course of action on natural resources. The record of flood policy illustrates this reaction handsomely. The immediate window of opportunity for change is likely to last a few months or as much as a year or two. Then, the opening may be expected to contract.

It seems possible that, within the current window of opportunity, the nation could resolve three major issues. First and most pressing, decisions must be made about whether to rebuild, strengthen, raise, lower, or abandon the levees along the upper Mississippi and lower Missouri rivers. How those decisions are to be made is now the subject of searching and testing by concerned citizens and public officials. Will that review and further study result in

* Originally published in *Environment*, 1993, vol. 35, no. 10.

leaving the rivers very much as they were last June, or will the debate trigger radical revisions?

Second, it remains to be seen what effect these early decisions, in focusing on the emergency, will have on the long-term quality of the natural landscapes and human communities in the region. They could optimize the distinctive floodplain values without degrading basic resources, or they could exacerbate problems in the long run. Of course, both the problems and the solutions are likely to differ greatly from place to place within the basin. Can the knowledge and skills of the region be mobilized to achieve optimal solutions?

Third, it is apparent that the actions taken to cope with the floods may stimulate measures that could have large significance for water management in the entire nation. Some of these possible measures are improving methods for comparative evaluation of levees with other adjustments to floods; reshaping policy for dealing with substantially damaged structures after flooding; revising the strategy for protecting vulnerable public facilities, such as water treatment plants; changing the policy for extension of federal assistance to property owners who have not elected to purchase flood insurance; increasing the degree to which the availability of federal crop or flood insurance is tied to mitigation of flood, erosion, or drought vulnerability; and expanding the federal government's capacity to assist local communities in drawing up and carrying out plans to address jointly residential, commercial, recreational, agricultural, and wildlife aims in adjusting to the flood hazard. Certain of these measures, such as improvements in flood insurance practice, might have been undertaken in any case, but most are currently receiving attention only because of last summer's floods.

How We Arrived Here

The federal government's involvement in coping with flood hazards has grown in sporadic jumps ever since 1825, when a unit of the Army Corps of Engineers was authorized to make waterway improvements (see box 24.1). Although navigation initially was the principal aim, the channel works had incidental benefits for flood management. Floods were then considered by all concerned to be the province of local and state levee districts. By the major flood of 1850, however, it was recognized in the lower Mississippi basin that some broader kind of program was needed to cope with recurrent flood losses, even though the primary cost was to be nonfederal.[1]

After discussions of various possible approaches, a report by two army engineers, Andrew A. Humphreys and Henry L. Abbott, was received in 1861 as a basis for guiding investment, chiefly by local agencies, in further protection works.[2] The approach was limited to levees only. Farmers and town dwellers who had invaded the alluvial lands of the Mississippi called for control by levees. They also wanted assurance that the great barriers of earth being erected along the stream courses would not cause undue damage to areas across the channel or downstream or to the essential maintenance of a navigable channel.

As the levee system was extended and strengthened and as the channel was improved, the river flow was greatly confined. Flood stages (the heights of water in the channels) for given discharges (volumes of water) increased. Meander belts were

---

[1] J. W. Moore and D. P. Moore, *The Army Corps of Engineers and the Evolution of Federal Flood Plain Management Policy* (Boulder, Colo.: Institute of Behavioral Science, 1989), 1–2.

[2] A. A. Humphreys and H. L. Abbott, *Report upon the Physics and Hydraulics of the Mississippi River: Upon the Protection of the Alluvial Region Against Overflow* (Washington, D.C.: Professional Papers of the Corps of Topographical Engineers, U.S. Army, 1861). See, also, C. S. Ellet, Jr., *The Mississippi and Ohio Rivers: Containing Plans for the Protection of the Delta from Inundation* (Philadelphia: Lippincott, Grambo and Co., 1853).

Box 24.1   Significant Events in the Development of U.S. Flood Control Policy

1825 Board of Engineers for Internal Improvements is authorized to undertake waterway improvements; levees are incidental. Early settlers already active in building local levees.

1850 Because of destructive flooding in the Mississippi basin, Congress authorizes surveys of the Mississippi Delta's flood problem.

1853 C. S. Ellet, Jr., an engineer, proposes comprehensive flood control for the Mississippi, including levees below Cairo, Illinois, and reservoirs in the Ohio River basin.

1861 Two army engineers, Andrew A. Humphreys and Henry L. Abbott, propose reliance on levees only, and their proposal is adopted.

1862
1865 } Flooding in the Mississippi River basin.
1869
1874

1879 The Mississippi River Commission is created.

1913 The Ohio Valley floods.
      The Board of Officers on River Flooding is created.

1917 Federal Flood Control Act of 1917 is passed and establishes a policy of federal funding in restricted areas.

1927 The great flood of 1927 overwhelms the lower Mississippi basin. Comprehensive "308" basin surveys are authorized by the Rivers and Harbors Act.

1928 The Flood Control Act of 1928 is passed, ending the "levees only" policy.

1935–36   Heavy flooding in New England.

1936 The Ohio Valley floods.
      The Flood Control Act of 1936 establishes a national structural program with local cost sharing, including soil conservation and watershed protection.

1937 The Ohio Valley floods.

1938 The Flood Control Act of 1938 reduces local cost sharing for reservoirs.

1951 The Missouri Valley floods.

1961 The Senate Select Committee on Water Resources recommends expanding the scope of water planning and confirms recreational benefits as appropriate for federal funding.

1966 The Bureau of the Budget Task Force on Federal Flood Control Policy recommends a Unified National Program for Managing Flood Losses, including flood insurance, to reduce flood losses.

1968 The National Flood Insurance Act establishes the National Flood Insurance Program.

1979 The Federal Emergency Management Agency is established.

1988 The Stafford Act guides flood recovery and mitigation practices in damaged areas.

1992 The Federal Interagency Floodplain Management Task Force publishes *Floodplain Management in the United States: An Assessment Report.*

1993 The Mississippi-Missouri Valley floods.

curbed as a levee system was completed from Cairo, Illinois, to the delta. When the great floods of 1927 poured into the valley below Cairo, they exceeded the designed channel capacity. Levees ruptured, and the river breached its confined course. Water spread over about 20,000 square miles, displaced more than 700,000 people, damaged at least 135,000 buildings, and took at least 200 lives.[3] As the water receded, it was clear that the strategy of "levees only" was no longer sufficient. Attention extended to other, supplemental structural measures, such as reservoirs, fuse-plug levees, floodways, and channel improvements. Upstream forest land improvement was discussed but did not figure in plans adopted by Congress in the Flood Control Act of 1928.

Support for the other types of structures was fully incorporated into the policies set in 1936 and 1938 through which the federal government took on a larger burden of the cost and full responsibility for reservoir projects. Parallel support was authorized for soil conservation and watershed protection. This policy did not again change significantly until after a 1966 report from a Bureau of the Budget Task Force on Federal Flood Control Policy recommended an expanded approach to floodplain management, comprising a variety of measures, including flood insurance.[4] As a result, the Federal Insurance Administration and its National Flood Insurance Program were instituted in 1968, providing a basis for communities to examine how a wider range of activities, including land-use planning, flood-proofing of buildings, and integration of warning systems, might be encouraged. Progress along these lines was slow, however.

In 1991 and 1992, the Federal Interagency Floodplain Management Task Force reviewed the status of floodplain management throughout the country and published an assessment report that provided a comprehensive view of where the nation stood as of 1990.[5] That report did not lead to any immediate changes in policy or practice. The Reagan administration had abolished the Water Resources Council as a coordinating agency; there was little subsequent interest on the part of the Executive Office of the President in the assessment report. Congressional attention at the time was focused on correcting deficiencies in the emergency reponse program that were revealed by Hurricane Andrew and on the failure of the flood insurance program to deal adequately with mitigation. Then, the flood of 1993 gave new impetus for criticism and change.

The Flood of '93

The flooding in the upper Mississippi and lower Missouri basins from mid June through early August 1993 was caused by intense rainstorms in late June and July that came on the heels of six months of heavy and persistent rainfall. Precipitation between January and July in the affected area was 1.5 to 2 times the normal for that period.[6] In June, a stalled weather pattern caused by a strong low pressure system in the western U.S. and a large high pressure system in the southeast resulted in large amounts of rain in the upper Midwest.[7] By late June, flood storage reservoirs were at or near capacity and soils throughout the area were saturated.

Flood peak discharges exceeding the estimated 10-year recurrence interval were recorded at 154 gauging stations in the upper Mississippi River basin, and the

[3] Moore and Moore, note 1 above, page 6.
[4] U.S. Congress, House of Representatives, *A Unified National Program for Managing Flood Losses: Communication for the President of the United States*, 89th Cong., 2d sess., H. Doc. 465, 1966.
[5] Federal Interagency Floodplain Management Task Force, *Floodplain Management in the United States: An Assessment Report* (Washington, D.C.: FEMA, 1992).
[6] C. Parrett, N. B. Melcher, and R. W. James, Jr., *Flood Discharges in the Upper Mississippi River Basin, 1993*, circular no. 1120–A (Washington, D.C.: U.S. Geological Survey, 1993), 4.
[7] K. L. Wahl, K. C. Vining, and G. J. Wiche, *Precipitation in the Upper Mississippi Basin, January 1 Through July 31, 1993*, circular no. 1120–B (Washington, D.C.: U.S. Geological Survey, 1993), 2–3.

Table 24.1 [orig. table 1]   Preliminary estimates of damage to levees in the Mississippi River basin in 1993.

Designation	Eligible for federal assistance[a]	Total number of levees	Number of levees damaged[b]	Percentage of levees damaged
Federally constructed and maintained	Yes	15	3	20
Federally constructed and locally maintained	Yes	214	36	16.8
Subtotal for federally constructed		229	39	17
Not federally constructed and locally maintained	Yes	268[c]	164	61.2
	No	1,079[d]	879	81.5
Subtotal for not federally constructed		1,347	1,043	77.4
Total		1,576	1,082	68.7

[a] Eligible levees meet the requirements for assistance under Public Law 84-99. Ineligible levees either do not meet those requirements or did not have an application for assistance submitted for their repair. However, some levees that are ineligible under Public Law 84-99 may be eligible for assistance from the Soil Conservation Service.
[b] *Damaged* means that the levees were breached or overtopped. There may be other levees that were damaged without being breached or overtopped.
[c] These levees together stretch for some 1,800 miles.
[d] These levees together stretch for some 4,000 miles.
Source: Department of the Army, U.S. Army Corps of Engineers, Civil Works, "Tabulation of Levees" (Washington, D.C., 12 August 1993).

maximum known peak discharge was exceeded at 56 gauging stations.[8] The discharges measured at some stations were reduced by the operation of reservoirs upstream. Peak discharges occurred as late as the first week in August and flooding in some areas continued into October.

Peak flood stages were higher than the previously recorded maximum stages at 73 of these 154 sites.[9] The fact that previous maximum peak discharges were not exceeded at 22 of these 73 sites is often attributable to changes in landscape, such as those imposed by the construction of levees and control structures, that allow a smaller volume of water to produce a higher flood stage. This effect is illustrated by historical discharge and stage data at selected points. Because of the short data record and changing observation methods, however, it is difficult to assign precise recurrence intervals and elevations to past flows.[10]

The flooding caused significant damage in nine states: Illinois, Iowa, Kansas, Minnesota, Missouri, North and South Dakota, Nebraska, and Wisconsin. More than 1,000 levees stretching nearly 6,000 miles in length were breached or overtopped (see table 24.1). Many others were significantly damaged. Taking into account the excessive soil moisture as well as surface overflow, a total of 487 counties (including

[8] Parrett, Melcher, and James, note 6 above.
[9] Ibid.
[10] Problems of estimating volume and fequency are illustrated in C. Belt, Jr., "The 1973 Flood and Man's Constriction of the Mississippi River," *Science* 189 (1975): 681–84.

all 99 in Iowa) were included in a Presidential Disaster Declaration.[11]

The number of people affected by this event is not certain. The American Red Cross has estimated that 56,295 family dwellings were affected in some way. Alone, it spent more than $30 million in flood relief efforts, sheltered 14,502 people in 145 shelters in the region, and served more than 2.5 million meals.[12]

Property damages from the flood have been estimated at $12 billion.[13] Disaster relief from the Federal Emergency Management Agency (FEMA) will cover about $650 million, including $250 million for individual assistance and $400 million for public assistance.[14] By 27 September, the Small Business Administration (SBA) had received more than 16,200 applications for low-interest disaster-assistance loans from individuals and business owners and had approved $277 million worth. SBA continues to process the applications and will accept applications to provide working capital to make up for economic hardship caused by the flood for up to nine months after the peak of the disaster in July.[15]

The Federal Insurance Administration estimates that, there are 88,400 policies in force in the nine states, but it is difficult to ascertain how many of these are on structures damaged by the flood. As of 30 September, the administration had received more than 10,500 claims and had processed about one-third of them for more than $72.4 million.[16]

The Federal Crop Insurance Corporation estimates that of 122.9 million insurable acres in the affected states (statewide, not just the flooded areas), 69.7 million acres, or 56.7 per cent, were insured. As of September, payments of claims for crop damages resulting from flooding and excess moisture in the upper Mississippi basin exceeded $51.7 million. More than $600 million in additional claims are expected.[17] Some losses resulted from heavy rainfall rather than flooding, and many farmers could not purchase crop insurance because they were unable to plant a crop.

Comprehensive data on the number of bridges, roads, water and wastewater treatment facilities, and other public infrastructure that were destroyed or damaged by the flood are not available. The suspension of the Des Moines, Iowa, water treatment plant received major media attention, but many smaller operations were also seriously affected.

The Impact on Individuals

In looking at these aggregate numbers, it is important to remember that the flood did not affect the Midwest per se; rather, it affected thousands of individuals. Even though television news no longer brings news of the flood into people's homes every night, it is still a very big story. Take, for example, the plight of an individual family of townspeople who were victims of this flood. How were they affected?

An urban family

When the levee broke, the family's house was inundated up to eight feet deep.

[11] Larry Zensinger, Federal Emergency Management Agency, personal communication with the authors, 12 October 1993.

[12] American Red Cross, "Midwest Flooding Operations Summary" (Arlington, Va., 10 September 1993).

[13] B. D. Ayers, Jr., "What's left from the Great Flood of '93," New York Times, 10 August 1993, A-1.

[14] Federal Emergency Management Agency, "Mid-west Flooding Disaster Damage Estimates" (Washington, D.C., 28 September 1993).

[15] Alfred Judd, Small Business Administration, personal communication with the authors, 27 September 1993.

[16] Federal Emergency Management Agency, "National Flood Insurance Program: Update on the Mississippi River Flooding" (Washington, D.C., September 1993); and John Gambel, FEMA, personal communication with the authors, 5 October 1993.

[17] Federal Crop Insurance Corporation, "Midwest Flood/Excess Moisture Loss Projections" (Washington, D.C., 22 September 1993).

Initially, the family went to a public shelter, although others went to stay with a luckier neighbor whose house sat on dry land. When the flood receded, the family returned to discover their home a sodden, uninhabitable mess. They had no potable water because the community's water treatment plant was put out of service by the flood. The local health inspector issued notices that flooded buildings had to be tested for molds and fungus and alerted residents to be aware that the waters that had swept through their town carried pesticides and other toxic substances in concentrations that approached permissible contaminant levels.

The local building official told the family that their house was "substantially damaged" and that, before they moved back in, it would have to be elevated. The family had not purchased flood insurance (none of their neighbors had ever thought the levee would break), so they looked to disaster assistance to help them recover. They were interested in moving out of the area, but, for the time being, the only assistance was a check from FEMA for temporary housing to stay in a hotel for a while and an Individual and Family Grant for $11,900. They were told they could use the grant to flood-proof some items in their house – for example, they could put the furnace and water heater on the second floor – but this was not enough money to elevate the structure plus clean or replace the siding, furniture, and carpeting. So, they cleaned up as well as they could and received permission to move back into their house on a temporary basis until the city decides whether it will buy out the properties in this neighborhood.

The city is puzzling about how to do this. City officials do not know if the levee will be repaired. The city does not have funds to match the federal grant that might be available for a buy-out. Even if it found the money, however, it likely would not be enough to compensate the family adequately to find a new home; property values in other parts of town are higher. Equally important, the city lacks an employee with the expertise to carry out a program to acquire and relocate flood-damaged properties.

The full, traumatic effects of the fight against the levee break, the evacuation, and the tiring search for recovery are impossible to calculate in the short run. Some consequences will only be recognized as solutions of some kind evolve over months of searching.

It is possible that, at some point in the future, a review of alternative flood protection strategies may make adequate funds and expertise available to the community to buy out this family's house and move them to a new area. That could be months or, more likely, years from now. By that time, the family probably will be re-established in its former location and may have little desire to uproot and move. It will be tempted to argue that a flood like the one in 1993 probably will not happen for another 100 or so years.

Alongside the experience of towns that were flooded by levee breaks must be placed the arguments of officials in Davenport, Iowa. They believe their city wisely refused to build large levees, in contrast to Rock Island, Illinois, across the river. Davenport had adjusted its land use with a riverside park and waterfront gambling boats and had designed buildings to minimize potential flood losses. Thus, the flood's costs to the city were lower than what the city would have paid for levee construction and for restricted riverfront access.

A rural family

Now, consider the situation of a farm family that, for decades, cultivated 250 to 300 acres of rich alluvial soil in bottoms along the lower Missouri and now operates a consolidated farm several times that size. The waters overtopped a levee and inundated the land for the first time since the embankment was strengthened 50 years ago.

The increased high-velocity flow washed soil out of some land, deposited coarse material across one low area, and left a further layer of fertile silt in another stretch. The corn crop was ruined. Drainage ditches were clogged. Debris was deposited in a few low places and along fence and tree rows. The family had worked night and day to strengthen levees as the waters had risen in the stream channel and then to evacuate livestock, people, and especially valuable equipment and goods when it seemed likely that their defenses would fail. The dwelling and major barns were inundated to depths of 3 to 10 feet, and it was 30 days before they again were accessible by trucks on muddy roads.

The family collected $70 per acre in crop insurance for which it paid an annual premium of $10 per acre. The neighbor's claims varied according to the duration of and damage done by the flood. The initial costs of flood fighting, evacuation, temporary shelter elsewhere, restoring operations, repairing damaged facilities, replacing lost property, and applying for indemnification or assistance were painfully apparent and difficult to estimate precisely.

Deciding what the family should do beyond recovery raises troublesome questions. Should they seek, with other members of the levee district, to have the levee rebuilt at the previous height and accept the risk of its failing again after another unusual precipitation event? That judgment would be affected by whether levees across the channel or up- or downstream are being raised, lowered, or abandoned. Whatever the levee protection, the family must also decide whether it would be desirable to elevate or otherwise flood-proof any of the farm structures to make them less vulnerable to damage in the next flood. Could some or all of the area be left unprotected from floods and dedicated to hunting and fishing in the natural habitat, thereby increasing to some unspecified degree the flow capacity of the channel?

There are no easy, general answers to these questions. The floodplain differs in soils, vegetation, flood vulnerability, and drainage from one reach to the next. It is affected by improvements to the navigation channel and by the height and configuration of the levees. Stream discharge is influenced in a number of ways by upstream precipitation, land use, and reservoir storage. Farm management in the floodplain also varies greatly.

The Repair/Rebuild Dilemma

When response activities to any disaster turn to long-term recovery operations, conventional wisdom suggests that people should go about rebuilding and reconstructing in a way that reduces the likelihood of significant damage from future similar events. In this case, two major dilemmas have arisen that serve as poignant examples of why conventional wisdom is not always followed: how to deal with the reconstruction of thousands of damaged buildings and what to do with all of the broken levees that have left hundreds of square miles of floodplain without structural flood control protection.

In regard to damaged buildings, many local codes require structures damaged beyond 50 percent of their value to be rebuilt in compliance with the minimum standards of the National Flood Insurance Program, which state that the lowest floor must be at or above the level of the 1-percent chance flood. This requirement often presents an overwhelming economic burden on victims who must struggle just to replace what they had prior to a flood, much less to absorb the cost of elevation or relocation.

As for the more than 1,000 damaged levees, many of them were locally owned and operated and are not eligible for federal reconstruction assistance. Many were designed and built to provide protection only from frequent flood events. That they were overtopped or failed last summer should have surprised no one.

The opportunities

There are opportunities, however, to rebuild in a safer manner. For example, reports from post disaster mitigation teams indicate that as many as 200 communities may be interested in acquiring and relocating or demolishing their structures that were substantially damaged.[18] FEMA is focusing the use of its Section 404 Hazard Mitigation Program grant funds – an estimated $45 million, which must be matched by the

[18] Federal Emergency Management Agency, "Potential Interest in Acquisition Projects" (Washington, D.C., 28 September 1993).

recipients on a 50/50 cost-sharing basis – on acquisition and relocation projects and also has funds available to provide technical expertise to communities wishing to undertake such projects.[19] Lessons from past experience with acquisition and relocation projects for substantially damaged structures provide much guidance on how to ensure the success of such efforts. If these lessons are applied, the wheel does not have to be reinvented.[20] Several other financial assistance programs exist that can be used to help fund mitigation projects for substantially damaged structures, including SBA disaster assistance loans, various FEMA disaster assistance programs, Housing and Urban Development grant and loan programs, and state housing and development finance authorities.

Opportunities for changing "business as usual" for levee reconstruction also exist. To facilitate the search for appropriate alternatives, the Office of Management and Budget (OMB) issued guidance on 23 August for the establishment of an unprecedented review procedure to assess strategies for levee reconstruction.[21] The agencies involved in this review include FEMA, the Army Corps of Engineers, the Soil Conservation Service, the Fish and Wildlife Service, and the

Environmental Protection Agency as well as state and local government agencies and other interested organizations. OMB's short-term policy guidance calls for the agencies' representatives to consider nonstructural alternatives to levee repair that would benefit both flood control and natural resource protection. The implication for FEMA's disaster field offices is that the interagency team is reviewing each application for levee repair and trying to determine whether a nonstructural alternative is appropriate for that levee.

The White House later called for a broader review of alternatives – one that looks at the entire upper Mississippi and lower Missouri watershed with a view to preparing, by February 1994, recommendations for further action.[22] Recently, the House Committee on Public Works and Transportation began hearings on legislation to strengthen the cost-sharing in support of hazard mitigation projects such as relocation.

The Association of State Floodplain Managers and the Association of State Wetland Managers have also furthered the debate by sponsoring two major conferences among interested citizens and government officials in August and September in St. Louis, Missouri.[23] On 30 and 31 August,

[19] M. F. Myers and F. Wetmore, *Meeting Summary Record: Post Flood Recovery in the Mississippi Basin, August 30–31, 1993* (Madison, Wisc.: Association of State Floodplain Managers, 1993).
[20] See, for example, Minnesota Department of Natural Resources, *Reducing Flood Damages by Acquisition and Relocation: The Experiences of Four Minnesota Communities* (St. Paul, Minn.: Department of Natural Resources, 1980); R. M. Field, *State and Local Acquisition of Floodplains and Wetlands: A Handbook on the Use of Acquisition in Floodplain Management* (Washington, D.C.: U.S. Water Resources Council, 1981); R. J. Burby and E. J. Kaiser, *An Assessment of Urban Floodplain Management in the United States: The Case for Land Acquisition in Comprehensive Floodplain Management* (Madison, Wisc.: Association of State Floodplain Mangers, 1987); and U.S. Department of the Interior, *A Casebook in Managing Rivers for Multiple Uses* (Washington, D.C.: National Park Service, Association of State Wetland Managers and Association of State Floodplain Managers, 1991).
[21] Executive Office of the President, Office of Management and Budget, "Procedures for Evaluation and Review of Repair and Restoration Projects for Levees," memorandum (Washington, D.C., 23 August 1993).
[22] At press time, there was no detailed public statement as to the composition and mission of the task force. There was some discussion of a possible programmatic environmental impact statement for recovery measures.
[23] The Association of State Floodplain Mangers (P.O. Box 2051, Madison, Wisc., 53701–2051) is distributing a summary of the first meeting's discussions. A summary of the second meeting is being distributed by the Association of State Wetland Managers (P.O. Box 2463, Berne, N.Y., 12023). A great deal of information was made available at these meetings, including D. Hey, "Wetlands: A Strategic National Resource," *National Wetlands Newsletter* 7, no. 1 (January–February 1985): 1–2; J. Kusler, "Post

representatives of the nine states affected by the flood, federal agencies, environmental organizations, the White House, and Congress gathered to share information on post-flood recovery activities as well as to identify opportunities to help communities and states recover in a way that would prevent a future similar event from causing such extensive damage. The three key issues addressed at this meeting were substantially damaged structures, levee reconstruction, and community recovery planning. The second, follow-up meeting was held on 27 through 29 September. Discussions at this meeting focused primarily on agricultural concerns related to the flood and on opportunities for the restoration of wetlands. Although much of the background information and issues discussed during the conferences had already been appraised by the leaders of the associations in the year before the floods,[24] the body politic had adopted no plan or policy as to what to do next.

Overcoming the dilemma

Although interest in the acquisition and relocation of flood-damaged structures is unprecedented and the consideration of alternatives to levee reconstruction – under the leadership of the White House – is a pioneering effort, problems do remain. For instance, there is a continuing, critical need for better public awareness, training and education about disaster relief programs and mitigation options. Local officials who are caught up in day-to-day immediate response and recovery activities are often unfamiliar with long-term mitigation assistance programs. These programs take considerable planning effort, and most

communities have not learned a lesson from other communities hit by disaster – the lesson that pre-event planning for recovery clearly pays off afterwards.

Mitigation work is also hampered by the conflicting goals of the short and long-term disaster assistance programs. Initial disaster assistance enables people to get back on their feet and thus makes them less likely to support more comprehensive programs, such as acquisition or relocation, at a later date. If interest does exist, slow processing of applications for projects eligible for hazard mitigation funds from FEMA often causes people to lose interest and become unwilling to participate.

As for the levees, the interagency teams considering their reconstruction or repair are having difficulty striking a balance between the need to restore flood protection quickly and the need for long-term planning for alternative flood protection that incorporates the concerns of sound ecosystem management. This is particularly problematic because the review procedure outlined in the Office of Management and Budget's directive[25] on levee repair does not match well with existing statutes and regulations under which the Army Corps of Engineers and the Department of Agriculture carry out such efforts. Because the current flood control strategy in the country is biased toward structural solutions, programs for viable alternatives and guidance on how to finance and implement them are not well formulated. Policy options suggested by federal agency headquarters or other executive offices may not be carried out in the field simply because adequate information to do so does not exist. Nor is the appropriate level of rigor for the cost-benefit analyses of alter-

---

Disaster Response and Recovery in the Mississippi Basin and the Restoration of Floodplain and Wetland Areas," in *Proceeding of the Second Workshop on Post-Flood Recovery and the Restoration of Mississippi Basin Floodplains Including Riparian Habitat and Wetlands* (St. Louis, Mo.: Association of State Wetland Managers, September 1993), 1–7; and A. Riley, "Background on National Efforts to Manage Flood Losses and Proposed Strategies for the Future Management of the Mississippi River Basin," in *Proceeding of the Second Workshop on Post-Flood Recovery and the Restoration of Mississippi Basin Floodplains Including Riparian Habitat and Wetlands* (St. Louis, Mo.: Association of State Wetland Managers, September 1993), 23–28.

[24] J. Kusler and L. Larson, "Beyond the Ark: A New Approach to U.S. Floodplain Management," *Environment*, June 1993, 6.

[25] Executive Office of the President, Office of Management and Budget, note 21 above.

natives clear. How are downstream impacts considered; how is future loss accounted for; how are other government expenditures, such as subsidies to farmers to forgo planting crops, incorporated into the equation; and how are environmental benefits quantified?

Recognizing Floodplain Values

Over the past three decades, several streams of thought have converged to shift both popular and technical views of floodplains. The limitations of engineering measures to restrict and channel river flow have been recognized more clearly. The uncertainties inherent in water engineering have been demonstrated along the Mississippi by the ways in which levees have increased flood stages. Levee failures have begun to be viewed as the river reclaiming its natural terrain.

During the same period, the significance of wetlands and floodplains in providing distinctive habitat, soil conditions, water storage, and ground- and surface-water quality was established in the scientific and political arenas. This significance was appraised in *Restoration of Aquatic Systems: Science, Technology, and Public Policy*, a 1992 report by a committee of the National Research Council.[26] That document summed up the state of knowledge of the interrelations of physical, biological, and social factors in wetlands, including floodplains. The overall opportunities for wetland restoration had already been brought to national attention by the controversy over the federal policy of "no net loss" of wetlands and the accompanying issue of how to define what is a wetland.[27] For the first time,

questions of habitat diversity and function came to be widely considered in association with questions of watershed hydrology and sedimentation, multiobjective river management, and land-use plans along stream channels.

These various aims now are reflected in an uncoordinated set of federal and loosely related state and local initiatives. The National Park Service assists in planning rivers and trails projects. The Environmental Protection Agency supports a Watershed Planning Program and seeks to link it with provisions of the Clean Water Act. Under the Soil Conservation Service, there is a Wetlands Reserve Program. Within the Department of the Interior, there are several programs dealing, at least in part, with waterfowl management, including Partners for Wildlife, the Waterbank Program, and the Land and Water Conservation Fund. The Army Corps of Engineers also has authority to support environmental improvements and is undertaking major studies in that direction.[28] Some of these agencies work closely with nongovernmental organizations such as The Nature Conservancy. The program's effectiveness would be greatly enhanced by early prioritization, in terms of location and timing, of wetlands for restoration. Such prioritization would facilitate attainment of the programs' goals and eliminate much of the ambiguity as to which wetlands – mainstream floodplains, tributary floodplains, upland marsh areas, potholes, and the like – would be targeted for early treatment.

These deepened scientific understandings and broadened views of the ecological linkages have fostered an unprecedented growth of concerned citizen organizations at

---

[26] U.S. National Research Council, Committee on Restoration of Aquatic Ecosystems, *Restoration of Aquatic Ecosystems: Science, Technology, and Public Policy* (Washington, D.C.: National Academy Press, 1992).

[27] J. Kusler, "Wetlands Delineation: An Issue of Science or Politics?" *Environment*, March 1992, 6.

[28] The Army Corps of Engineers is authorized to repair certain damaged levees under Public Law 84–99, but it has broader authorization to assist in a number of ways in wetlands preservation and restoration. Major authority for environmental measures was provided by the Water Resources Development Act of 1990 and by Public Law 99–662 of 1986 (the Upper Mississippi River Management Act of 1986). The Army Corps of Engineers has been more vigorous in undertaking appropriate studies, such as on wetland mitigation banking, than Congress has been in appropriating funds.

the local, state, interstate, and national levels. Appeals for action have been strong and pointed from a variety of traditional environmental organizations: American Farmland Trust, American Rivers, Environmental Defense Fund, Environmental Law Institute, Izaak Walton League, National Fish and Wildlife Federation, Sierra Club, and World Wildlife Fund. Newer groups, such as the Coalition for the Restoration of Coastal Louisiana and the Coalition to Restore Urban Waters,[29] are addressing related problems of wetlands' values in less conventional ways.

A call for action along broader lines, incorporating agricultural and conservation goals, was issued in August by the Upper Mississippi River Conservation Committee.[30] Its call, representing the states in the upper basin, was for an unprecedented level of environmental restoration. Many other citizen groups are also becoming involved. The major significance of these multiple initiatives is that they are moving in roughly converging directions and, for the first time, show signs of harmonious action, with urban and rural groups involved in addressing other values in floodplain management. Thus, the alternative of wetland management has emerged as practicable.

The Need for Mitigation

In a similar vein, the concept of mitigating damages from disasters has gained full currency over the past few decades. There is now a well-established consensus that the disaster cycle of preparedness, response, and recovery must include the fourth component of mitigation. This component ensures that damaged facilities are not automatically repaired and replaced to their pre-event status; rather, they are removed, so far as is practicable, from harm's way or reconstructed in such a way as to avoid future damages.

Federal policy has done much to promote the concept of mitigation, especially immediately after a disaster, when awareness of the problems that hazards present is high and political support to correct deficiencies in existing programs is strong. For example, since 1988 when the Stafford Act[31] was passed, the federal disaster relief assistance available to communities has included funds needed to repair or replace damaged public structures in a safer (and often more expensive) manner. Since the early 1980s, there has been a concerted effort made by federal interagency hazard mitigation teams to identify opportunities for mitigation in communities that have just experienced a major disaster.[32]

Unfortunately, no systematic assessment has been made to track the effectiveness of the teams or to determine the outcome of the recommendations they make immediately after disasters. Also unfortunately, the push for mitigation seems to be at its peak only when disasters occur. Although some agencies such as FEMA work consistently during both disaster and nondisaster times to promote mitigation (and James Lee Witt, the new director of FEMA, has announced that mitigation will be his central concern),[33] other local, state, and federal agencies, many citizens, and Congress appear content to let hazards and the need for mitigation fade into the background soon after an event occurs.

[29] Mark Davies, executive director of the Coalition for the Restoration of Coastal Louisiana, statement made at a meeting of the Association of State Wetland Managers, St. Louis, Mo., 27–29 September 1993.

[30] Upper Mississippi River Conservation Committee, "The Upper Mississippi River: America's River Faces Its Greatest Threat" (St. Louis, Mo., August 1993).

[31] The Robert T. Stafford Disaster Relief and Emergency Assistance Act, Pub. L. 100–707, 23 November 1988.

[32] Federal Emergency Management Agency, *Post-Disaster Hazard Mitigation Planning Guidance for State and Local Governments*, DAP-12 (Washington, D.C.: FEMA, September 1990), 24.

[33] J. L. Witt, memorandum for participants of the Natural Hazards Workshop, Boulder, Colo., 28 July 1993.

Marshaling Government

In light of the decisions that have been made regarding the human occupation and development of the Mississippi watershed that led to the summer's flood and the dilemmas it has presented, the challenge that lies ahead is clear. Disasters such as the flood of '93 cannot be considered impossible or highly unlikely; rather, they must be viewed as a reality.

Although the early studies by Humphreys and Abbott looked to a system of levees for which the local landowners would carry the financial burden, the evolution of commitments and activities has blurred public concepts of what is an effective national strategy for dealing with floods and where responsibilities should rest among property owners, local governments, states, and the federal agencies. The "levees only" policy has been demonstrated to have severe limitations. Some floodways and reservoirs have also proven inadequate in coping with great floods. Although the concept of unified floodplain management has received endorsement in theory,[34] it is far from being realized in practice. One striking example of this problem may be seen in how FEMA's National Flood Insurance Program, which was intended to promote such management, has instead encouraged uneconomic reconstruction and new building in the floodplain with insurance coverage and how disaster assistance has in some instances abetted it. Another example is how the federal government has provided grants for building public facilities, such as water treatment plants, in flood-prone areas. Natural values of floodplain environments are given lip service but little concrete support. For example, landowners receive little technical advice about how they can use wetlands instead of engineered protection.

A key consideration in the near future is timing. Can the concern to recover promptly be reconciled with the need for further scientific and technical studies and with the need to forge stronger private and government instruments to carry out the desired actions at the community level? One possible strategy relies on the recognition that the next potentially damaging flood might come next year or might not occur for decades. Its recurrence is certain, but its timing is uncertain. Under these circumstances, the federal government might institute new provisions of crop and flood insurance to indemnify property owners behind unrepaired levees against any damage they might suffer from a flood until such time as a decision is made whether to rebuild the levees to previous levels. Thus, the breathing space required for developing alternative programs could be supplied at little or no cost to the taxpayer and at little or no financial risk to the floodplain occupant.

The Months Ahead

If the work of the White House's new task force is to lead to truly constructive action, its appraisal of options to use and preserve the natural resources of the area must be supplemented with a set of hard-headed suggestions as to how current procedures and programs can be revised to reach those aims. In developing these suggestions, the task force must also address the interrelationship between hazards and the myriad decisions made on a daily basis at the individual, local, regional, and national levels. The nation's strategy for mitigating damages from disasters should not be driven by its disaster relief policy.

In the next year or two, the national approach to coping with the flooding may not change at all; disaster assistance, insurance offerings, and the current practices of federal agencies may have altered little. That is one extreme. At the other, a vision of a new

[34] Natural Hazards Research and Applications Information Center, *Action Agenda for Managing the Nation's Floodplains*, special publication no. 25 (Boulder, Colo.: Natural Hazards Research and Applications Information Center, 1992), 2. This agenda was compiled by a committee of independent experts who reviewed a draft of the Federal Interagency Floodplain Management Task Force's assessment report (see note 5 above) at the task force's request.

harmony in the aspirations for the Mississippi basin and for other basins around the country might emerge. In addition to sharing that vision, the task force might find practical ways to help put it into effect. Such a vision might regard hazards not as single, rare events but, rather, as part of a continuum wherein the resiliency of natural and social resources is tested.

In between these two extremes, there might well be a few major improvements that fall short of a full reorientation of policy and programs. The current congressional review of legislation affecting property insurance is likely to lead to helpful changes in procedures affecting requirements for federal financial assistance. For example, if lending institutions were penalized for not requiring property buyers to purchase flood insurance for flood-prone property as a condition for receiving a mortgage, more properties would be insured. Also, more buyers would be forewarned of the risk of flooding and so could choose not to buy. In either case, the need for federal disaster relief assistance would likely decrease. In addition, FEMA's commitment to mitigation may be expressed in operating rules that will reduce vulnerability before the next disaster strikes. Other steps outlined in the Federal Interagency Floodplain Management Task Force's assessment report and its companion, *Action Agenda for Managing the Nation's Floodplains*,[35] might be adopted (see box 24.2). Those several actions would together mark a significant advance.

Box 24.2   Recommendations for Floodplain Management

In 1992, the Federal Interagency Floodplain Management Task Force published an assessment report on the status of the United States' floodplain management. Before the report was published, the task force invited a panel of independent experts to review the report. The following is a list of recommendations put forward by the task force and the panel.

To integrate flood loss vulnerability and protection of floodplains' natural values into broader state and community development and resource management processes,

- vigorously foster the preparation of state floodplain management plans, involving both public and private interests and, where appropriate, interstate agreements;
- require states to prepare comprehensive floodplain management plans as a condition for continued participation in the National Flood Insurance Program;
- prepare an executive order requiring that new federal investments, regulations, and grants-in-aid be consistent with state and local floodplain management plans that conform to federal standards; and
- adjust tax codes to discourage building in flood-prone areas.

To improve the database for floodplain management,

- remap the nation's floodplains, where appropriate, to take into account potential changes in hydraulic conditions associated with full development of the drainage areas under existing land-use plans;
- establish a cooperative, jointly funded program by the National Science Foundation and interested federal agencies to develop methods for mapping, regulating, and identifying natural values in areas with special flood hazards;
- develop an accurate, affordable, national system for gathering flood loss data; and
- fund research to examine, in a selected sample of communities, the full benefits and costs of floodplain management measures.

---

[35] Ibid.

To give weight to local conditions,

- without loosening the limits on permissible vulnerability, examine the practicability of using performance standards for the preservation, use, and development of floodplains;
- further experiment with the National Flood Insurance Program's community-rating system to encourage communities to adopt a variety of flood hazard mitigation measures, such as zoning, particularly suited to their local circumstances; and
- redesign local zoning and subdivision regulations and building codes to contribute to the management of floodplains' natural resources.

To minimize conflicts and gaps among federal programs,

- establish an independent task force to continue to review and recommend changes in the current Unified National Program for Floodplain Management.

To reduce vulnerabilities,

- prepare assessments of the vulnerability to flooding of a sample of facilities built with federal aid;
- fund research on techniques for estimating benefits and costs to encourage local agencies to improve flood preparedness and retrofitting;
- assess existing aging flood control structures and make recommendations for sustainable improvements or replacements;
- improve land-use regulations to reduce encroachment onto floodplains downstream of dams;
- assess alternative designs to channel modification that include less straightening of channels, employ more gradual slopes, and use natural vegetation or rip-rap rather than concrete-lined channels;
- assess pre-existing storm water networks and make suggestions for alternatives such as on-site retention, natural drainage systems, and zero-increment runoff for new development;
- establish positive tax incentives and technical and financial assistance for land treatment measures, such as maintaining trees, shrubbery, and vegetative cover; terracing; slope stabilization; grass waterways; contour plowing; conservation tillage; and strip farming.

To improve professional skills and public education,

- develop training programs and conduct regional training, at an affordable rate, for appropriate government personnel;
- expand and evaluate efforts to inform and educate the public about flood hazards and flood management; and
- improve documentation and quantification of the values of natural floodplains to improve public understanding of possible needs for protecting those values.

Sources: Federal Interagency Floodplain Management Task Force, *Floodplain Management in the United States: An Assessment Report* (Washington, D.C.: FEMA, 1992); and Natural Hazards Research and Applications Information Center, *Action Agenda for Managing the Nation's Floodplains*, special publication no. 25 (Boulder, Colo.: Natural Hazards Research and Applications Information Center, 1992).

It is not unduly sanguine to expect that scientific and public support may be marshaled for a series of executive orders or administrative agreements that would promote comprehensive state floodplain management planning; steer federal activities in harmony with that planning; establish broad priorities for wetland restoration; strengthen coordination of federal agencies' efforts to assess the vulnerability of public facilities and preparedness and flood-proofing measures; and step up education and training relating to the nature of flood hazards and natural values of floodplains. These goals could be achieved without major new legislation, but their achievement would be accelerated if federal agencies were allowed more flexibility in how they allocate available disaster relief funds. The long-term benefits would extend far beyond the Mississippi basin.

In less than two centuries, the nation has moved slowly, often by trial and error, from a belief that levees could surely protect humans and their built environment to a recognition that the built and managed environment must be more open to accommodating its natural components. The need to integrate the nation's disaster response and relief policies more closely with broader environmental and economic policies is now accepted. Events like the flood of '93 offer an opportunity to speed the formation of sounder policy. This opportunity should not be squandered.

Acknowledgments
The authors are indebted to Sharon Gabel, David Morton, and Eve Passerini for their assistance in preparing this article.

25

THE BIGGEST DAM IN THE WORLD*

Fred Pearce

China will have seen nothing like it since the building of the Great Wall. Last month, premier Li Peng inaugurated work on the world's largest hydroelectric dam project, the Three Gorges scheme on the Yangtze river. Its dam will be nearly 2 kilometres long and some 100 metres high, and it will consume enough material to build 44 Great Pyramids.

More than a million people will have to move to make way for the reservoir, which will stretch 600 kilometres upstream, longer than any other. Its water will drive turbines with a generating capacity of almost 180 000 megawatts, eight times that of the Aswan Dam on the Nile, four times greater than any power station in Europe, and 50 per cent more than the world's largest existing hydroelectric dam, the Itaipu Dam in Paraguay.

Itaipu powers Brazil's two great megacities, São Paulo and Rio de Janeiro. Three Gorges will supply power to Shanghai, the world's fifth largest city, and fuel a $175 billion programme of economic development of the upper Yangtze River basin around Chongqing, a city of three million people, with airports and heavy industry. A consortium of international investors, headed by the American merchant bank Merrill Lynch, has been given access to markets in the region in return for its help in arranging loans for the project.

The dam, sited at Sandouping, will take between 15 and 20 years to build, but nobody is sure how much it will cost. The Chinese government estimates $12 billion at 1990 prices. But independent estimates range from $22 to 70 billion, including interest charges and inflation. China expects to raise 90 per cent of the money itself, and has already levied a 2 per cent tax on electricity to help fund the project.

Besides powering China's industrialisation, the Three Gorges project is designed to protect some 10 million people downstream from floods that have killed 300 000 people this century, and to open the river above the gorges to shipping. The Qutang, Wu and Xiling gorges form 180 kilometres of rapids and whirlpools, bounded by sheer cliffs rising into fog-shrouded mountains. The dam will plug the Xiling Gorge, raise water levels in all three gorges by around 90 metres and turn the churning waters into a placid slow-moving lake.

The Yangtze pours from Tibet, through the mountains of southwest China and the

* Originally published in *New Scientist*, 1995, January, pp. 25–9.

gorges, onto a wide floodplain and to the sea at Shanghai. It drains 1.8 million square kilometres of China and discharges 700 cubic kilometres of water to the sea each year. The floodplain provides two-thirds of China's rice, and is home to 400 million people, or one person in 13 on the planet. Any flood is therefore disastrous. To hold back the waters, Chinese peasants under both Imperial and Communist rule, have built and raised dykes that stretch for thousands of kilometres and rise up to 16 metres above the fields. But their leaders have increasingly looked to augment the dykes with large dams to control floods before they reach the floodplain.

Sun Yat-Sen, leader of the 1911 Chinese revolution, first proposed damming the Yangtze at Three Gorges in 1919. Later, after floods killed 200 000 people in 1931 and 1935, nationalist leaders surveyed the area with American engineers. After floods in 1954, which left 30 000 dead, Mao Tse-tung called in Soviet engineers to design a dam that would combine flood protection and power generation. Over the next three decades, Three Gorges went in and out of favour, until Li Peng, a former hydroengineer, forced the project onto the current 10-year plan.

China's engineers calculate that the country has a theoretical hydroelectric capacity of 380 000 megawatts, the world's largest, but that only 10 per cent of this is so far tapped. They regard Sandouping as one of the best sites in engineering terms, and note that the dam will flood out only half as many people per megawatt of generating capacity as the average for Chinese dams (fewer than 70 people compared with 140).

Sandouping is also in a good position. Most of the mountain gorges that can generate large amounts of electricity are in the west of China, while the large factories that consume three-quarters of the electricity are in the east. Three Gorges is the most easterly gorge site in China. The dam will help China hold back from burning its vast coal reserves, thus reducing the country's chronic smogs and its contribution to the greenhouse effect, which has doubled in the past 15 years.

But critics say China uses its power inefficiently. Its heavy industries typically consume twice as much energy for every tonne of output as their Western rivals. If China brought its industrial efficiency up to existing Western levels, it could cut demand by a third, making Three Gorges redundant. And if these measures failed to end the country's long blackouts and power rationing, they say smaller hydroelectric dams on the tributaries of the Yangtze could be brought on stream much more quickly than one giant scheme.

One of the big questions to be answered about the Three Gorges project, as with other large multipurpose dams, is what the "normal" operating level of the reservoir should be. Should it be kept as full as possible to maximise hydroelectric power? Or should it be empty in readiness for capturing floods? The situation is complicated by the need to prevent the reservoir from filling with silt brought down by the river during the annual flood. To do this, the water level should be kept low during the flood season, to allow the silty floodwaters to flow through as quickly as possible. However, if heavy floods occur, requiring downstream regions to be protected, then the reservoir would have to be filled.

Power Struggle

The debate about water levels has rumbled on for many years. In 1990, the plan was to maintain the reservoir level at 160 metres above sea level, which is some 25 metres below the maximum safe height. This was to have been reduced to 140 metres during the flood season when the reservoir might suddenly have to store floodwaters. The arrangement was a compromise between flood prevention managers, who wanted a normal level of 150 metres throughout the year, and power and navigation authorities who wanted a higher level – they complained that a level of 140 metres would reduce power production substantially, and that several rapids would reappear at the top end of the reservoir near Chongqing.

In 1992 Li Peng ordered that the normal level be raised to 175 metres. As power and navigation authorities rejoiced, flood prevention managers warned that more people would need to be resettled and the risk of flooding would rise.

But even if the dam were operated primarily for flood protection, it could only shield a relatively short stretch of the river immediately downstream. This is because most floodwaters in recent disasters have come from tributaries that enter the Yangtze below the gorges. That was the case in both 1954 and, most recently, in 1991 when more than 2000 people died. Indeed the Yangtze could well become more dangerous with the dam in place. With part of its burden of silt left behind on the bottom of the reservoir, the river would be able to pick up more material and thus its currents would have greater power to destroy dykes. A detailed evaluation of the scheme, the Three Gorges Water Control Project Feasibility Study funded in the late 1980s by the Canadian government's international aid agency and the World Bank at a cost of $14 million, warned that the silt-free river would tend to alter its course, eating at banks and dykes and increasing the risk of disastrous floods. The likely death toll from a major breach of the main JingJiang dyke is put by Chinese officials at 100 000.

There are other fears. The region of the Three Gorges is seismically active and landslides are frequent. An earthquake or a landslide overtopping or breaking the dam could submerge downstream cities such as Wuhan, with a population of 4 million. And dams do fail. In 1975 the Banqiao dam in Hunan province, which held just half of a cubic kilometre of water compared with the 40 cubic kilometres to be impounded by the Three Gorges dam, failed killing an estimated 10 000 people.

American engineers who visited the Three Gorges site in the early 1980s concluded that the project would not prevent flooding. According to one, John Morris, former chief of the US Army Corps of Engineers, landslides, earthquakes or military attack could all breach the dam. They backed alternative plans drawn up by Chinese engineers for a series of smaller dams on the river's tributaries. Philip Williams, an hydrologist and

partner of the Californian consultant, Philip Williams Associates, argues that large flood-protection dams have an inbuilt tendency to create the disasters that they are designed to prevent. Williams, who is also president of the International Rivers Network, a nongovernmental organisation that often campaigns against large dams, says that large dams encourage undue confidence downstream and building in flood-prone areas. If the dam were unable to hold back a future flood, "the loss of life would be greater than if the dam had never been built", he notes. And he believes that "the consequences of failure at Three Gorges would rank as history's worst man-made disaster", on a par with the estimated loss of between 300 000 and more than one million people when Chinese generals deliberately broke dykes on the Yellow River to halt an enemy advance at the height of the Sino-Japanese War in 1938.

The dam should allow large ships unimpeded passage up the Yangtze, through ship locks at the dam and along the tranquil reservoir as far as Chongqing. The Ministry of Communications predicts that shipping traffic will increase fivefold to 50 million tonnes per year once the dam is built. But again there are complications. At the 140-metre operating level, rapids would reappear and, during the two decades of construction, existing boat traffic would be severely disrupted.

And there are doubts about whether Chongqing port, the intended destination for the shipping, can survive the creation of the reservoir which will have its head close by. As river waters enter the reservoir and the speed of flow drops, much of its silt will settle onto the reservoir floor, forming a fan of sediment that may destroy the city's port, without massive dredging. And it may also increase the risk of flooding in the city.

Against the problematic benefits of the dam are some all-too tangible disbenefits. Up to 1.2 million people will have to move to make way for the dam and reservoir. The precise number depends on final decisions about the normal water level in the reservoir. The reservoir will inundate dozens of towns including the city of Wanxian, an unlovely industrial town of 140 000 people, and Fuling, with a population of 80 000 people. And it will lap at the banks of the homes of Chongqing's 3 million people. Building new houses for these people will be the largest dam resettlement programme ever attempted anywhere in the world. Already tens of thousands of people have been moved.

The Chinese government says it will spend up to $4 billion on resettlement, mostly on rebuilding towns on higher ground. The original idea was for the 300 000 farmers who will lose their land to move farther up the hillsides. But, says the Canadian study, more than half the land available for resettling farmers is more than 800 metres above sea level. These remote hillsides are colder with poorer soils and will grow many fewer crops. A study by soil scientists in 1991 at the Chinese Academy of Sciences concluded that five times as much new land would be needed to replace the fields in the valley bottom than would be lost to the reservoir.

How long will the reservoir so expensively created actually last? Engineers round the world are increasingly worried about the accumulation of silt in large reservoirs, with many lasting only a few decades ("A dammed fine mess", *New Scientist*, 4 May 1991). On the Yellow River in China, the reservoir behind the Sanmenxia Dam filled with silt within four years of opening in 1960, and was emptied, dredged and rebuilt. The reservoir today has less than half of its original capacity.

The Yangtze is not as yellow as the Yellow River, but it carries 530 million tonnes of silt through the gorges each year. The first dam on the river, the Gezhouba Dam, regarded as a test run for the Three Gorges, lost more than a third of its capacity within seven years of opening. Lin Bingan, a silt expert on the project, warned earlier this year that to avoid a similar fate at Three Gorges: "During the high-water season the reservoir should be kept at a lower level, allowing water and silt to flow into the lower reaches of the river as much as possible." If operated in this way, with water levels kept to 140 metres during the flood, Chinese

government engineers estimate that only 10 per cent of the reservoir's capacity will be lost.

Nevertheless, there are considerable uncertainties in these calculations. They assume the river will erode many of the sand bars it creates during the flood season, a process known as retrogressive scouring. But Shiu-hung Luk, a Chinese soil scientist now at the University of Toronto, says retrogressive scouring occurs only near the dam. It will not occur 600 kilometres away at the head of the reservoir, where the majority of the silt at Three Gorges will accumulate. If it does not, the risks of flooding in Chongqing would increase greatly, he says.

To reduce the river's silt load, China's Ministry of Forestry announced in 1993 that it plans to double the upper Yangtze valley's tree cover to 40 per cent within 40 years in order to bind the soil and reduce erosion. But there are no examples anywhere in the world of reforestation projects having an effect on the flow of river silt on this scale. Moreover, other factors could increase the silt load. The thousands of people forced to farm steep hillsides above the flooded river valley are likely to increase soil erosion. And Shiu-hung Luk warns that the Yangtze's silt load is currently reduced by small dams in tributaries upstream that collect silt. "Within a few decades, these reservoirs will become clogged. They will have to be flushed out, causing a significant increase in the river's total sediment load."

Naturally, the full impact of the dam on the environment is as yet unknown. Some warnings may be exaggerated. The dam will undoubtedly interrupt fish swimming upstream. However, their way is already blocked by the much smaller Gezhouba Dam downstream, says Philip Fearnside of Brazil's National Institute for Research in the Amazon, who has investigated the impact of large dam projects worldwide. Existing disruption to the river's hydrology and silt flows will be drastically worsened by the Three Gorges Dam, however. Much of the floodplain of the lower Yangtze is farmed using traditional methods that are dependent on the seasonal fluctuation of the river to provide free irrigation and wetland grazing for animals. But with the Three Gorges dam in place, the flood regime will drastically change.

Among endangered species downstream of the dam are the 300 surviving White Flag dolphin and the 500 surviving Chinese alligator, as well as Chinese sturgeon and finless porpoise. Most of these species are in serious decline. But the dam could be the final straw. Most of the world's Siberian white cranes winter on Poyang Lake, where they eat weeds rooted to the lake bottom. Changes to water levels in the lake during that season could eliminate this food source, and threaten the species' survival. But China is unlikely to abandon its plans to save a few species that are, arguably, already doomed.

Browned Relics

Internationally, there may be greater concern about the destruction of archaeological treasures in the Yangtze valley. More than a hundred major cultural and historic sites will be inundated. Already there are calls for an effort to save the best of them, along the lines of the efforts in Egypt during construction of the Aswan Dam. China's National Bureau of Cultural Relics says it will move several major temples, such as Zhang Fei, to higher ground. But the ancient city of Fengdu (the City of Ghosts), one of the capitals of the Ba Kingdom, will be largely lost, as will much of the 2000-year-old plank road through Wuhan and Wuqi and hundreds of lesser archaeological sites, many not yet explored.

China's decision to go ahead with Three Gorges is a major shot in the arm for the world's dam builders. After the investment of hundreds of billions of dollars over the past five decades or so in one of the great postwar engineering enterprises, dam-building has seemed to founder in the past five years. A history of cost overruns and corruption scandals, growing concern about the environmental impact, civil unrest arising from failed resettlement programmes and the campaigning of an international network of anti-dam pro-

testers, have all undermined the case for giant dams.

Last May, Daniel Beard, the commissioner of the US government's Bureau of Reclamation, which built the Hoover and Grand Coulee dams and dozens of others, said that "the dam-building era in the US is over". He told delegates to a meeting of the International Commission on Irrigation and Drainage that most projects cost 50 per cent more than predicted and that "often, project benefits were never realised". In Canada, the grand and controversial James Bay hydro-electric project was halted last year. And in Australia last year, scientists were pushing a plan to drain the giant reservoir created 20 years ago when Lake Pedder in Tasmania was converted into a huge hydroelectric reservoir.

In Africa, the money for giant dams has run out as Western aid agencies withdraw funds. The World Bank, which had lent $60 billion for dams over 50 years, has got cold feet, with funding more than halved in the 1990s. Britain's Overseas Development administration stopped funding large dams even before the Pergau Dam scandal broke, concluding they were usually bad value for money and had hidden environmental and social costs.

Only in Asia are governments determined to continue to build dams, with or without international aid. Within days of the public inauguration of Three Gorges, the countries along the Mekong, which include Vietnam, Thailand, Laos and Cambodia, announced the formation of a new organisation to begin dam-building on the river. India has persisted with the controversial Sardar Sarovar Dam on the Narmada river, despite losing World Bank support. And late last year, it resumed construction of the 250-metre high Tehri Dam in the headwaters of the Ganges. Malaysia, fresh from its battles with Britain over aid funding for the Pergau Dam, announced the go-ahead for a 230-metre high dam in the rainforest of Borneo.

Now the Three Gorges project has created a honey pot that few construction companies or governments can resist. At the height of environmental concern about large dams at the end of the 1980s, Canada's Liberal Party said the project would "impoverish and dislocate millions of people, spread debilitating diseases [and] endanger rare species". But last November, Jean Chretian, the country's premier and Liberal Party member, urged Canadian companies to capitalise on the detailed feasibility study completed earlier. The US government, which two years ago expressed concern about the project's impact and refused to get involved, is now also considering renewing its backing to help ensure American companies win contracts. The signs are that, with environmental concerns on the retreat worldwide, the drive to build large dams may be resuming.

Today, there are more than 100 "super-dams" with heights greater than 150 metres. Their reservoirs cover 600 000 square kilometres, an area greater than the North Sea, and have a capacity of 6000 cubic kilometres, equivalent to 15 per cent of the annual runoff of the world's rivers.

Of the world's 30 largest rivers, apart from those flowing into the Arctic, most have dams across them, holding back substantial parts of their flow. They include the Ganges, Parana, Tocantins (a major tributary of the Amazon), Columbia, Zambezi, Niger, Danube, Nile and Indus. After the Yangtze has been dammed, engineers will be left with only two major rivers to plug: the Zaire and the main stem of the Amazon.

Part V

Environmental Economics

INTRODUCTION AND GUIDE TO FURTHER READING

26 Pearce, D. W. and Turner, R. K. 1990: The historical development of environmental economics. In: Pearce, D. W. and Turner, R. K. (eds) *Economics of Natural Resources and the Environment*. New York and London: Harvester Wheatsheaf, pp. 3–28.

The introductory chapter of the authors' key text on the economics of the environment provides a comprehensive overview of the historical development of the different approaches encompassed within environmental economics. Most of the chapter is reproduced here, although the sections on the Marxist paradigm, the property rights approach, policy analysis: fixed standard versus cost-benefit framework, economic and environmental values, and the ecological and co-evolutionary economic paradigm have been abbreviated.

27 Daly, H. E. 1991: Towards an environmental macroeconomics. *Land Economics*, 67, 2, 255.

In this brief paper, Daly advocates the need for a macroeconomic approach to environmental economics, and addresses the optimal scale of the economy.

28 Tietenberg, T. H. 1990: Economic instruments for environmental regulation. *Oxford Review of Economic Policy*, 6, 1, 17–33.

This paper provides a review of emissions trading and charges. It is reprinted as in the original, but omits a fifth section entitled 'Lessons from Implementation'.

29 Hahn, R. W. 1989: Economic prescriptions for environmental problems: how the patient followed the doctor's orders. *Journal of Economic Perspectives*, 3, 2, 95–114.

An economic analysis of the use of marketable permits and emissions charges in environmental management is presented. The following material has been omitted from the original text: a case study of the Wisconsin Fox River Water Permits programme, some of the section on emissions trading, a section on new directions for marketable permits, the case study of charges in Germany which the author argues are very similar to those in France, and some of the concluding discussion on the implementation of market-based environmental programmes.

30 Turner, R. K. 1993: Sustainability: principles and practice. In Turner, R. K. (ed.) *Sustainable Environmental Economics and Management: Principles and Practice*, London: Belhaven.

This article is part of the first chapter of Kerry Turner's edited volume on sustainable environmental economics and management. The parts reproduced here include the introduction and most of the section on principles of sustainable development. The remainder of the original chapter, which is not published here, covers the practice of sustainable development.

Parts I–IV have focused largely on the technical and scientific aspects of environmental management, addressing for example the physical processes of atmospheric change and land degradation alongside the technical responses that have been developed to 'manage' these processes. However, the very word 'management' implies human intervention in such processes, and such human involvement in turn involves financial costs. Moreover, it is important to recognize that there are also key issues which need to be addressed concerning the 'management' of human systems involved in influencing physical processes. There are fundamental economic implications associated with all environmental management, and it is with these that Part V is concerned.

Michael Redclift (1993, p. 115) has suggested that the attention that has been paid in recent years to environmental policy 'can be interpreted in one of two ways. Either it is evidence of our concern and the priority we have chosen to give to environmental factors. This, broadly, is the position of most governments in the industrialized world. Alternatively, the current preoccupation with environmental policy is evidence of environmental neglect, pointing to our failure to examine the underlying assumptions on which our economic growth has been purchased.' This second alternative implies that the apparent successes of the capitalist economic system have been purchased with a direct cost to the environment. Redclift (1993, p. 115) goes on to argue that if the second hypothesis is true we are faced with the enormous challenge of regarding 'the management of a consensus over environmental issues as part of the problem, rather than part of the solution. By seeking to maintain the social authority of science, and assure the public that governments are able to use science to redress environmental problems successfully, we are effectively disabling ourselves.' Redclift here makes management itself a problematic issue, and in doing so he emphasizes that we cannot consider economic matters alone without the social and political contexts within which economics is practised. Parts V and VI are therefore fundamentally inter-

linked. While Part V concentrates specifically on environmental economics and management, Part VI explores the social and political implications of such management.

As well as providing a broad context within which to interpret the selected readings on economic management that have been included here, Part V also seeks to provide a brief historical interpretation of the emergence of environmental economics and a summary of key themes of contemporary concern in the economics of environmental management.

The Emergence of Environmental Economics

The chapter by David Pearce and Kerry Turner (1990) is a comprehensive overview of the historical development of environmental economics. This chapter provides a highly readable introduction to the central themes at the core of environmental economics since it emerged as a distinct field of research and teaching in the 1970s. As the authors emphasize, from the end of the nineteenth century until around 1970 the majority of mainstream economists appeared to believe that it was possible for economic growth to continue unabated into the foreseeable future. Over the last 25 years, though, there has been a growing realization that such growth can have serious environmental implications.

Classical economics

As with all disciplines, the concerns and methods of economics have closely reflected the interests of those societies in which it has been practised. Classical economics is widely considered to have emerged in the late eighteenth and early nineteenth centuries, beginning with the publication of Adam Smith's (1723–90) *Wealth of Nations* in 1776 and ending with John Stuart Mill's (1806–73) *Principles of Political Economy* published in 1848. This was a period of considerable change, expansion and increasing global integration of economic activity, and the interests of classical economics therefore came to be concerned primarily with the origins and growth of economic surpluses. The dominant figure of classical economics was David Ricardo (1772–1823), whose theory of relative prices was based primarily on the costs of production, which he saw as being dominated by labour costs. Adam Smith's so-called 'invisible hand' of market competition was seen as being the main mechanism through which conflicting self-interest could be resolved, and through which 'self-interested rational behaviour by individuals could satisfy individual wants but also serve the interests of society as a whole' (Pearce and Turner, 1990, p. 6).

Although both Ricardo and Smith considered that a lack of natural resources meant that the period of economic growth through which they were living could not be sustained, it is with the work of Thomas Malthus (1766–1834) that the linkages between population growth, economic expansion and the physical environment have become most widely associated. In his *Essay on the Principle of Population* (1798), Malthus systematized many of the views then current, and argued that the natural tendency of populations to increase faster than the means of subsistence would lead to a reduction in per capita food supply, a lowering of living standards, and an eventual cessation of population growth. For Malthus, land was the ultimate source of material wealth, and given the inability of people to increase food production at a rate commensurate with the potential rate of growth of population, both preventive and positive checks on population growth were required. Among preventive checks, Malthus included abstinence and delays in marriage, and typical of the positive checks he envisaged were disease and warfare, resulting from competition for increasingly limited resources.

For classical economists, therefore, the environment, in the broadest sense, provided a constraint on economic growth, and Malthus's arguments concerning abstinence and warfare can be interpreted as ways in which populations sought to manage their affairs within such environmental constraints. The continued economic growth of the nineteenth century, however, and the apparent ability of technological change to outpace population growth, provided the context within which such formulations were increasingly challenged. Two contrasting positions, those of Marxist and neo-classical economics, came to dominate economic thinking over the next century.

Marxist economics

Karl Marx (1818–83) built on the classical economists' labour theory of value, but shifted his focus of concern from the origin and production of surplus to the way in which that surplus was distributed (for a useful introduction to Marx's political economy see Howard and King, 1985). For Marx (1976, p. 283), labour was 'first of all, a process between man and nature, a process by which man, through his own actions, mediates, regulates and controls the metabolism between himself and nature'. Moreover, Marx saw labour not only as acting upon and changing external nature, but he also argued that through this process human nature itself was changed. Under capitalism, it is the class of capitalists who, through their ownership of the means of production, vested essentially in land, are able to expropriate the surplus value produced through this application of labour power to the products of nature. Significantly, Marx (1976, p. 649) also argued that 'It is not the absolute fertility of the soil but its degree of differentiation, the variety of its natural products, which forms the natural basis for the social division of labour, and which, by changes in the natural surroundings, spurs man on to the multiplication of his needs, his capacities, and the instruments and modes of his labour. It is the necessity of bringing a natural force under the control of society, of economising on its energy, of appropriating or subduing it on a large scale by the work of human hand, that plays the most decisive role in the history of industry.'

While there is therefore some direct discussion of the interaction between human society and nature in Marx's writing, Deleage (1989) argues that his primary emphasis on the relationship between capital and labour meant that these environmental linkages generally remained implicit and largely undeveloped. Pepper (1993, p. 60), moreover, asserts that this relationship 'remains largely unexplored by Marxists to this day. Nevertheless, both Marx and Engels (1820–95) were only too aware of the environmental changes taking place in England as a result of industrial capitalism. In *The Condition of the Working Class of England*, written in 1844 and 1845, Engels thus described in graphic detail how the centralizing forces of capitalism led to mass deprivation and squalor in the poorer quarters of England's industrial cities. In describing London, he observed that 'After roaming the streets of the capital a day or two, making headway with difficulty through the human turmoil and the endless lines of vehicles, after visiting the slums of the metropolis, one realises for the first time that these Londoners have been forced to sacrifice the best qualities of their human nature, to bring to pass all the marvels of civilisation which crowd their city . . . The very turmoil of the streets has something repulsive, something against which human nature rebels' (Engels, 1975, pp. 328–9). In commenting on London's slums, he goes on to note that 'The streets are generally unpaved, rough, dirty, filled with vegetable and animal refuse, without sewers or gutters, but supplied with foul, stagnant pools instead. Moreover, ventilation is impeded by the bad, confused method of building of the whole quarter, and since many human beings here lived crowded into a small space, the atmosphere that prevails in these working-men's quarters may readily be imagined' (Engels, 1975, p. 331).

Despite such evident awareness of the environmental implications of capitalist 'development', the political economy of Marx and Engels was above all concerned with a critique of capitalism as an economic system; they wrote little about environmental management, and most subsequent Marxists have until recently likewise paid scant attention to practical concerns of environmental management. As Emel and Peet (1989, p. 59) have commented, 'Generally,

Marxists have avoided decision-making models relative to environmental management for political and epistemological reasons'. If the citicisms of Marxists about environmental degradation have been largely focused on the capitalist system, it is also important to emphasize here that the practical expression of Marxist ideology in the former Soviet Union itself led to widespread adverse environmental change. In Pearce and Turner's (1990, p. 3) words: 'Despite the impression given in some of the environmental literature, environmental degradation is not the unique attribute of advanced Western industrial capitalism . . . The Soviet environment has suffered from a catalogue of pollution abuses over a long period of heavy industrialisation.' While it is possible to argue from a Soviet viewpoint that such pollution was essential if the Soviet Union was to attempt to compete with capitalist industrialization, it is salient to recall that one of the factors leading to the collapse of the Soviet Union in the late 1980s and early 1990s, particularly in the Baltic States such as Estonia, was an increasingly vocal movement expressing its outrage at the damaging Soviet exploitation of the environment.

Neoclassical economics

The alternative economic position that emerged during the 1870s was that of neoclassical economics, as expressed in the work of William Jevons's (1835–82) *The Theory of Political Economy* and Karl Menger's (1840–1921) *Grundsatze der Volkswirtschaftslehre*. This has remained the dominant discourse in economics throughout the twentieth century, and has also been extensively applied in economic research on environmental management. As Emel and Peet (1989, p. 51) argue, 'Neoclassical economic theories of allocation and development are the dominant social science perspective in natural resource issues'.

The main aim of neoclassical economics has been to explain activity through the use of laws governing economic activity, and the key methodological achievement of early neoclassical economists was the introduction of marginal analysis, based on the study of relationships between small or incremental changes (Pearce and Turner, 1990). Neoclassical theories make important assumptions about economic behaviour, notable among which are that the market is neutral and value free, that the market place allocates resources and distributes income, that people behave rationally, and above all that the system is regulated by supply and demand in the market place. This, therefore, represents a marked departure from both classical and Marxist economics where the labour theory of value was the key to an understanding of economic activity.

With respect to environmental issues, neoclassical economic approaches have assumed 'the objective of maximising economic welfare from resource use' (Emel and Peet, 1989, p. 51), and have built on the theorems of welfare economics, which seek 'to legitimise rational behaviour as being socially desirable and also to justify some government intervention to improve the conditions under which individuals make choices' (Pearce and Turner, 1990, p. 11). The central criterion by which neoclassical economists judge the social welfare of environmental use is that of Pareto optimality, or the point at which no person is made better off without someone else being made worse off. Within neoclassical economics, renewable resources should therefore ideally be used at the maximum and most efficient rate of continuable usage, and non-renewable resources should be exploited at a rate which maximizes all future benefits.

While neoclassical economics provides an elegant, comprehensive and apparently objective scientific approach to economic activity, it has not been without its critics. Within mainstream economics, the problems of negative externalities of economic growth, the changed nature of business organizations, and the difficulties of central assumptions, such as rational economic behaviour, have all been addressed (for a brief review see Smith, 1994). From a more radical perspective, as Emel and Peet (1989, p. 52) comment, 'More recent work from a political-economy perspective takes issue with nearly all neoclassical assumptions, particularly the ability and opportunity of consumers and producers to enter into market decisions, the perfect competitiveness of the market, the mobility of factors of production, the absence of unpriced goods and services, and the absence of political intervention.'

The emergence of a specifically 'environmental' economics

During the 1960s, the central belief of economists over the previous century that economic growth was sustainable, came to be challenged as a result of a growing awareness of the environmental implications of such economic activity. In particular, increased awareness of the damaging effects of industrial pollution, recognition of the rapidly increasing levels of global population, a growing concern about the dangers of nuclear energy provision, and greater anxiety about the effects of deforestation in regions such as Amazonia, all led to a situation where the conventional economic approaches to growth came to be challenged. Interestingly, these challenges first emerged, and have achieved the greatest success, in the most advanced capitalist economies, such as the United States and Germany, in which there has been sufficient surplus to enable people to be altruistically concerned with environmental issues, and where economic growth has not provided the social benefits and welfare that neoclassical economics had predicted.

Environmental economics thus emerged primarily as an alternative to the neoclassical approach which had dominated the discipline over the previous century. Pearce and Turner (1990) suggest that there have been four main 'world views' which have provided the basic theoretical frameworks for the emerging sub-discipline of environmental economics since 1970:

- *extreme cornucopian* – based on resource exploitation and continued economic growth
- *accommodating* – managerialist in conceptualization, and adopting a resource conservationist approach

- *communalist* – ecocentric, and based on a resource preservationist approach
- *deep ecology* – derived from an extreme preservationist position

All of these approaches have different views with respect to the continued possibility of economic growth, ranging from the extreme cornucopian position which adheres to the belief that market mechanisms will ensure technical innovation and continued substitution thus enabling economic growth to be maintained, to the deep ecological position which seeks a radical socio-economic transformation to ensure a minimum resource-take system in which moral rights are conferred on non-human species.

Selected Themes in Environmental Economics

Given the great diversity of approaches to environmental economics, any collection of readings can only be highly selective. Those chosen from Part V have been grouped into three broad headings. First there are two introductory essays which provide a historical and conceptual framework to the economics of environmental management. Most environmental economics can be considered to fall within micro-economics, focusing in particular on the internalization of environmental costs and their equivalents in monetary terms that reflect their social marginal opportunity costs. The second section of Part V therefore includes two key essays on the micro-economics of environmental management. As this introduction has stressed, one of the main issues underlying the emergence of environmental economics has been whether or not economic growth is sustainable. This raises central questions concerned with the sustainable economic management of the environment, and Part V therefore concludes with an overview by Turner on the economics of sustainability.

Conceptual frameworks of environmental economics

The chapter by Pearce and Turner builds on their historical overview of the emergence of

environmental economics, and explores five main conceptual frameworks which have emerged within the sub-discipline: the market model of environmental management; the policy analysis framework; concerns with economic and environmental value; issues surrounding sustainable development; and the ecological and co-evolutionary framework.

Within conventional economics, two main *market oriented approaches* to environmental management have been developed: the property rights approach, and the materials balance approach. In the former, environmental costs such as pollution are seen as being a type of market failure, resulting from inadequately specified property rights. It suggests that where such property rights are well defined, natural resources will instead be used as efficiently as possible, both by individuals and by groups such as companies. In contrast, the materials balance approach suggests that pollution is inevitable, and therefore requires government intervention in order to regulate it. Society, therefore, through its elected representatives, defines acceptable levels of environmental deterioration, and the task of environmental economics is to identify the least-cost policy package which meets these levels.

The *policy analysis framework* can again be divided into two approaches: the cost-benefit framework, and the fixed standard approach. Within the former, while financial valuations continue to be placed on environmental changes, some of these are recognized to be irreversible. Furthermore, cost benefit analysis generally assumes that some environmental change can be desirable in circumstances where exploitation exceeds the environmental costs, however these are defined. In the latter, government intervention is once again seen as being necessary in order to impose basic environmental standards which would operate as binding constraints. In part, demand for such standards results from the difficulties of calculating damage functions, and also from the essentially political nature of the determination of appropriate levels of environmental quality. This gives rise to the need to specify the relationships between *economic value and environmental value*. Pearce and Turner suggest that there have been three main approaches here: value as determined by individual preferences; value reflecting public preferences and derived from social norms; and value resulting from the characteristics of functional physical ecosystems. Following an overview of the *sustainable development* debate, Pearce and Turner finally summarize some of the central characteristics of the *ecological and co-evolutionary approach*, which they see as addressing the complex two-way interactions between humans and the physical environment.

The chapter by Daly seeks to redress the overwhelmingly micro-economic focus of environmental economics by exploring the connections between the environment and macro-economics. He suggests that the macro-economy needs to be seen as an open sub-system of the finite environment, and this then forces him to examine the optimal scale of the human economy. Once again, he thus addresses issues concerned with the global level of economic activity, and in contrast to Goeller and Zucker, he reaches a much more pessimistic conclusion, suggesting that we have already passed a prudent size for the global macro-economy.

The micro-economics of environmental policy: emissions trading and emissions charging

The two chapters in this section examine central micro-economic issues of environmental management, and focus in particular on two of the most important policies used to control environmental pollution: emissions trading and emissions charging. The chapter by Tietenberg incorporates both a theoretical approach and empirical examples to explore the role of the market in pollution control. He argues that the use of economic incentives offers a flexible and relatively low-cost solution to environmental control, and that emissions trading works effectively in scenarios with uniformly mixed pollutants, whereas emissions charging works best when the transaction costs of bargaining are high.

Hahn provides further case studies of marketable permits (emissions trading) and emissions charging practices, drawn from the United States and Europe, and concludes that it is important to ensure a satisfactory balance between different kinds of policy in the context of both economic and political realities. As increased public pressure leads to the introduction of tighter environmental controls, Hahn argues that the experiences of those governments which have adopted these measures, at a variety of scales, will stimulate wider use of such market-based environmental management policy instruments.

The economics of sustainable development

Much has recently been written on sustainable development, but, as Turner's comprehensive overview of the economics of sustainability suggests, the term has a great variety of meanings. In the most general sense, the term sustainable development has been used to refer to economic development that continues into the future. However, the most widely accepted definition of sustainable development, that generated by the World Commission on Environment and Development (1987, p. 43), which defined it as 'development that meets the needs of the present without compromising the ability of future generations to meet their own needs', introduces both a social and a political dimension to the issue of sustainability. More formally, economic sustainability has most usually been seen as non-declining consumption in relation to some specific welfare indicator, such as per capita or per unit of GNP. Beginning with this broad definition, Turner then provides a detailed assessment of the different economic characteristics of four kinds of sustainability: very weak sustainability, weak sustainability, strong sustainability, and very strong sustainability. In using a systems approach, he emphasizes that economic systems are underlain by ecological systems, and that in the final analysis the biological and physical environment provides constraints within which economic activity takes place.

These concerns with the environmental

constraints on economic growth are not dissimilar to those that challenged the classical economists two centuries ago. Once again, economic growth does not appear inevitable, and some environmental economists are therefore again seeking to understand and explain the grounds upon which such growth can be sustained. Others are returning to the Marxist critiques of capitalism that challenge the very basis of neoclassical economics. However, as several of the readings selected for Part V illustrate, all these approaches are themselves situated within particular social and political contexts. It is thus to a more detailed examination of these agendas that Part VI will turn.

References

Deleage, J P. 1989: Eco-Marxist critique of political ecology, *Capitalism, Nature, Socialism*, 3, 15–31.

Emel, J. and Peet, R. 1989: Resource management and natural hazards. In Peet, R. and Thrift, N. (eds) *New Models in Geography: the Political-Economy Perspective. Volume 1*. London: Unwin Hyman, pp. 49–76.

Engels, F. 1975: The condition of the working-class in England: from personal observation and authentic sources. In: *Karl Marx and Friedrich Engels Collected Works. Volume 4. Marx and Engels: 1844–45*. London: Lawrence & Wishart, pp. 295–583.

Howard, M. C. and King, J. E. 1985: *The Political Economy of Marx*. Harlow: Longman, 2nd edition.

Marx, K. 1976: *Capital. Volume 1*. Harmondsworth: Penguin in association with New Left Review.

Pearce, D. W. and Turner, R. K. 1990: The historical development of environmental economics. In Pearce, D. W. and Turner, R. K. (eds) *Economics of Natural Resources and the Environment*. New York and London: Harvester, pp. 3–28.

Pepper, D. 1993: *Eco-socialism: from Deep Ecology to Social Justice*. London: Routledge.

Redclift, M. 1993: Environmental economics, policy consensus and political empowerment. In: Turner, R. K. (ed.) *Sustainable Environmental Economics and Management*. London: Belhaven, pp. 106–19.

Smith, D. M. 1994: Neoclassical economics. In Johnston, R. J., Gregory, D. and Smith, D. M. (eds) *The Dictionary of Human Geography*. Oxford: Blackwell Publishers, pp. 410–15.

Suggested further reading

Adams W. M. 1992: *Green Development: Environment and Sustainability in the Third World*. London: Routledge.

This is an excellent introduction to issues of green development, which seeks to bridge the gap between ecology and development studies. It argues that the 'greenness' of development planning is to be found in issues of Third World self-determination rather than in concerns for ecology and the environment.

Barbier E. B. (ed.) 1993: *Economics and Ecology: New Frontiers and Sustainable Development*. London: Chapman and Hall.

This edited collection aims at integrating the concerns of social development, ecology, environmental policy and economic development. In exploring issues associated with sustainable development, it examines ways in which economic management can be combined with ecological policies.

Elkington, J. and Burke, T. 1989: *The Green Capitalists: How Industry can Make Money and Protect the Environment*. London: Victor Gollancz.

This populist book argues that it is only by harnessing the commercial and innovative energies of industry that we can hope to solve the growing environmental problems facing society. It calls for green growth, and suggests that this can only be achieved by environmentalists and industrialists working together.

Hopfenbeck, W. 1992: *The Green Management Revolution: Lessons in Environmental Excellence*. New York: Prentice Hall.

This well-illustrated book aims to raise awareness among business managers of the challenge of ecology, and seeks to demonstrate that environmentally sensitive approaches to management can be cost effective.

Jacobs, M. 1991: *The Green Economy: Environment, Sustainable Development and the Politics of the Future*. London: Pluto Press.

This book rejects both the traditional green demand for zero growth and the economic arguments for giving the environment an economic value. Instead, the author argues for sustainable planning in which economic activity is considered within environmental limits.

Markandya, A. and Richardson, J. (eds) 1992: *The*

Earthscan Reader in Environmental Economics. London: Earthscan.

This is a comprehensive collection of papers on environmental economics divided into sections on basic concepts, validation methods and applications, instruments for environmental control and applications, environment and sustainable development, and international and global environmental problems.

May, P. (ed.) 1996: *Environmental Management and Governance.* London: Routledge.

This book examines aspects and problems of environmental management. It considers the role of governments, both at a local and national level, and the strengths and weaknesses of cooperative versus coercive environmental management, through a focus on the management of natural hazards. It presents new and innovative environmental management and planning programmes, with particular focus on North America and Australia.

Mikesell, R. F. 1995: *Economic Development and the Environment.* London: Cassell.

This book examines how the environment and sustainability can be integrated with development programmes and strategies. It outlines the conceptual and theoretical issues involved in sustainable development, as well as providing case studies that compare the successfulness of various types of development projects.

Omara-Ojungu, P. H. 1992: *Resource Management in Developing Countries.* Harlow: Longman.

This text examines the problems of resource management in developing countries, outlining the basic ecological, economic, technological and ethnological aspects of resource management. It emphasizes that poverty is the critical problem facing resource management and development. Examples are provided from Africa, South-East Asia and Latin America.

Pearce, D. W. and Turner, R. K. 1990: *The Economics of Natural Resources and the Environment.* Hemel Hempstead: Harvester.

A thorough and comprehensive introduction to the economics of natural resources and the environment. It is essential reading for both students and practitioners.

Smith, L. G. 1993: *Impact Assessment and Sustainable Resource Management.* Harlow: Longman.

This book provides an integrated approach to environmental planning. It balances both academic and practical considerations with regard to impact assessment and sustainability. Various aspects of environmental planning include: decision making, dispute resolution, environmental law, public policy, administration, the nature of planning, impact assessment and methodology. This is a useful text for planners, managers and policy-makers.

Turner, R. K. (ed.) 1993: *Sustainable Environmental Economics and Management: Principles and Practice.* London: Belhaven.

This is one of the definitive texts on environmental economics and management. It is divided into two main sections covering the principles and then the practice of environmental economics. It is concerned particularly with issues of sustainability, in both the developed and in developing countries.

Turner, R. K., Pearce, D. and Bateman, I. 1994: *Environmental Economics: An Elementary Introduction.* New York: Harvester Wheatsheaf.

This useful book provides a highly readable introduction for students and non-specialists alike on environmental problems and their economic influences. Particular attention is paid to issues of environmental degradation and economic incentives which can be designed to slow, halt or reverse them.

World Commission on Environment and Development 1987: *Our Common Future.* Oxford: Oxford University Press.

Our Common Future is the report of the World Commission on Environment and Development which seeks to examine the critical environmental and development problems facing the world at the end of the twentieth century. It includes a wealth of empirical data, and is an important reference source for college and university teachers and students, as well as policy-makers concerned with environmental and development issues.

26

THE HISTORICAL
DEVELOPMENT OF
ENVIRONMENTAL ECONOMICS*

D. W. Pearce and R. K. Turner

Knock on effect

1.1 Introduction

Environmental stresses and strains are now ubiquitous phenomena appearing in all economic systems, regardless of political ideology, from the very poorest to the very rich. Despite the impression given in some of the environmental literature, environmental degradation is not the unique attribute of advanced Western industrial capitalism. Eastern bloc economies face acute water and air pollution threats, notable examples being river-water pollution in many industrial areas of Poland and declining urban air quality levels in industrial Czechoslovakia. The Soviet environment has suffered from a catalogue of pollution abuses over a long period of heavy industrialisation. Pollution there now threatens even the most precious of biospherical assets such as Lake Baikal. Among the developing economies, air pollution in cities such as Caracas, Mexico City and Sao Paulo is extremely severe and poses a significant health hazard. For the group of thirty-six (the poorest countries on earth) their very poverty is a major cause and effect of environmental problems. Poverty, which

denies poor people the means to act in their own long-term interest, creates environmental stress (such as overgrazing of rangeland, soil erosion and eventual desertification) leading to resource degradation and growing population pressures.

Uncertainty still surrounds the exact nature and extent of the global interdependencies between economic growth and the supporting environmental systems. We still cannot fully quantify the risks to future human well-being posed by acid rain, ozone depletion and the greenhouse effect. Even so, half of the net natural output produced by environmental systems is now utilised by humans. Future necessary global economic growth will further diminish that sector of nature in which self-regulating natural systems can regenerate free of human intervention. The margin for error in economic planning that has the capacity to inflict irreversible change on the natural resource base is, in the view of many, narrowing.

Environmental issues, on the boundaries of economic and natural systems, are undoubtedly complex and in many cases contain inherently uncertain outcomes. The

* Originally published in D. W. Pearce and R. K. Turner (eds) *Economics of Natural Resources and the Environment*. New York and London: Harvester Wheatsheaf, pp. 3–28.

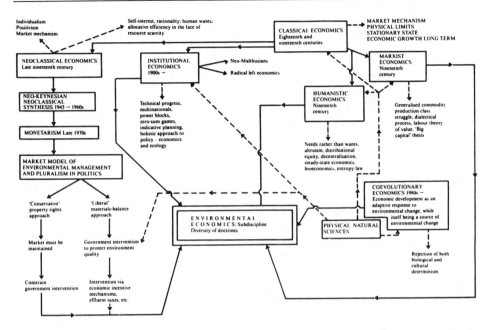

Figure 26.1 [*orig. figure 1*]   Economic paradigms and the environment. Some caveats are in order. The figure is meant to be descriptive rather than analytical. It is probably not correct to view changing economic doctrines over time in terms of Kuhnian 'scientific revolutions'. Rather it is more fruitful to think of clusters of interconnected theories or 'scientific research programmes' which compete against each other.

sub-discipline of environmental economics which seeks to analyse such issues consequently sits on the boundaries of a range of social science and natural science disciplines. The current position in the economics profession is one in which mainstream analyses are firmly anchored to earlier neoclassical foundations (neoclassicism is explained on pp. 369–70). Since any economics needs value judgements, and in the absence of agreed meta-ethical criteria for choosing between value judgements, it cannot be argued that neoclassical economics and its Paretian value judgements are 'worse' or 'better' than any other economic doctrine.

Particularly in its formative years (1960s), environmental economics encompassed a diversity of economic doctrines. A pluralistic view (i.e. one that recognises more than a single tradition in the development of economic thought) of the contribution that economics can make would guard against narrowness in economics, as well as

fostering more interdisciplinary analytical linkages. We devote most of the chapters in this book to an analysis which expands the horizons of modern conventional economic thought. 'Alternative' economic paradigms are surveyed in this first chapter.

1.2   Early Economic Paradigms and the Environment

In order to appreciate the modern arguments (roughly over the period 1960 into the 1980s) both between economists themselves and between economists and other environmental analysts, the historical roots of environmental economics must be explored.

Figure 26.1 summarises some of the more important concepts and ideas that have influenced environmental economists and traces their origins back to past doctrines. Economic theories ought to be appraised within the context of their wider framework ('paradigm'). There is a complex interaction taking place as both scientific (natural, physical and social) theory and social order evolve. The ways in which scientific research asks its questions of the human and natural worlds it seeks to explain will at times be influenced by social, cultural and political factors. Thus attitudes toward nature and preservation/conservation will change as humanity and nature evolve.

The classical economic paradigm

The classical economists left a legacy of ideas many of which are relevant to, and have been re-introduced into, contemporary environmental debates. Classical political economy stressed the power of the market to stimulate both growth and innovation, but remained essentially pessimistic about long-run growth prospects. The growth economy was thought to be merely a temporary phase between two stable equilibrium positions, with the final position representing a barren subsistence level existence – the *stationary state*.

Adam Smith (1723–1790), through what became known as the doctrine of the *invisible hand*, argued that there were circumstances in which self-interested rational behaviour by individuals could satisfy individual wants but also serve the interests of society as a whole. Governments were important only in the sense of providing 'night-watchman' services (law and order, national defence, education). What was vital to economic and social progress was that economic transactions should be allowed to operate on the basis of freely competitive markets.

Thomas Malthus (1766–1834) and David Ricardo (1772–1823) were, like Smith, pessimistic about the prospects of long-term economic growth. They expressed their 'environmental limits thinking' in terms of the limits on the supply of good quality agricultural land and therefore diminishing returns in agricultural production. For Malthus, the fixed amount of land available (absolute scarcity limit) meant that as population grew, diminishing returns would reduce the per capita food supply. Standards of living would be forced down to subsistence level and the population would cease to grow.

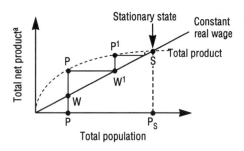

Figure 26.2 [*orig. figure 2*]   Ricardo's simple commodity production model. (*Source*: M. Blaug, 1978.) Simple commodity production, i.e. production for sale by independent producers in which labour power is a commodity (a measure and a source of value); but capitalist institutions and activities treated as external to the model.

Note: Model assumes the entire economy operates like a giant farm producing a crop by applying homogeneous doses of capital and labour to a fixed supply of land of varying quality. With population at level OP, the wage bill = PW and farmer profits = WP, which induces investment and consequent rise in market wage rate. Population growth serves to force down wages back towards subsistence. Over time profits are squeezed W^1P^1 etc., until all investment and growth ceases at the stationary state point, S.

[a] Total product subject to the axiom of diminishing returns minus rent accruing to non-productive landlord class; workers are paid wages, farmers gain profits and undertake productive investment. No technical progress assumed.

In Ricardo's more complex economic model, economic growth again peters out in the long run because of a scarcity of natural resources. Diminishing returns set in not so much because of absolute scarcity but because the available land varies in quality and society is forced to move on to successively less productive land. Figure 26.2 summarises the main Ricardian analytics. Note that the lack of technical progress in the model means that the total product curve (subject to diminishing returns) remains fixed. Technical innovation (e.g. artificial fertilisers, irrigation and deep drainage, etc.) would shift the total product curve upwards, increasing output per unit of input and offsetting, but not eliminating, the tendency towards diminishing returns.

John Stuart Mill (1806–1873) conceived of economic progress in terms of a race between technical change and diminishing returns in agriculture. But unlike the other classical economists he viewed the distant prospect of the stationary state with some optimism. By then, he reasoned, technical progress would have provided for much of mankind's individualistic material wants and society would be free to pursue educational, aesthetic and other social goals.

During the nineteenth century fundamental changes in these traditional classical patterns of thought were established with Marxism, neoclassicism and humanism.

The Marxist paradigm

Karl Marx (1818–1883) adopted the *labour theory of value* from the classical economists (workers were the sole source of net economic product) and was equally pessimistic about the future standard of living for the majority of people (working class) in capitalist society. According to Marx, the classical economists had failed to place capitalist economic organisation in its historical context. He sought to formulate a generalised commodity production model which characterised commodity production as a social relationship. History was to be interpreted as a dialectical phenomenon, a process of conflicting material or economic forces out of which a synthesis, a resolution of the conflict, would emerge. Capitalist society would inevitably be beset by a class struggle (workers versus capitalist entrepreneurs) for social power. Power would be gained via control of economic resources. Marx predicted that the capitalist economic system would be faced with a falling rate of profit over time, increasing destitution for the majority working class and increased monopoly. Ultimately, the majority would overthrow the small capitalist class (existing via the exploitation of surplus value produced by workers) and seize power to create a socialist society.

Marx believed that progress was a process of natural development, inherent in human history. Progress itself was to be defined in terms of material and technological advance made possible by the exploitation ('humanising') of nature. He saw the political state as apart from nature, created as an alternative to a 'natural' environment. Nature was there to be humanised via science so that inherent value could be turned into use value. But some modern Marxian writers have pointed out that Marx does emphasise the *process* of production and the fact that a viable basis for any society can only be provided if the system of production is capable of reproducing itself. There is a strong hint in this analysis that natural systems could be a limit to reproduction, as well as the economic and political make-up of society. In this sense, it can be argued that Marx took what we call today a 'materials balance' approach to the process of production over time. In terms of modern environmental issues this reproductive economic system analysis raises questions about the sources and nature of technological change. Does such change alleviate or aggravate the environmental constraints on an economy's capacity to reproduce itself? Further, is the process of reproduction consistent with reasonably stable social systems?

According to Marxian analysis, modern capitalist economic systems fail the reproduction test, i.e. capitalist systems are not sustainable, and *one* source of non-sustainability is environmental destruction. Questions of economic power, exploitation and the dialectical process encompassing the two classes in society are at the root of an inevitable environmental despoliation process, which in turn contributes to the failure of capitalism.

In the international economy the exploitation process manifests itself in terms of the operations of the transnational corporations. There are structural linkages between economic development in the Northern economies and the South, and these linkages radically affect the environment in the South. Changes in the environments of the South need, according to Marxians, to be understood in terms of the international redivision of labour.

Neoclassical and humanistic paradigms

Starting around 1870, neoclassical economic thought began to be developed by analysts within the mainstream of the economics profession. The labour theory of value was abandoned and a commodity's price was seen not as a measure of its labour cost but of its scarcity. The concentration on scarcity value allowed both sides of the market to be analysed simultaneously. Analysts compared the amount of a commodity that was available (supply) with the amount required (demand). The interaction of supply and demand then determined the equilibrium market price for the commodity. The economic activity that was observed in the real world was seen as the result of the interaction between productive activity (determined by technological progress) and the preferences of individual buyers constrained by the feasible range of choice and income.

The neoclassical economists also introduced a new methodology, *marginal analysis*, i.e. the study of the relationships between small or incremental changes. This type of approach was well suited to the investigation of price determination and market structures. Consequently, the classical concern with long-term growth patterns was sidelined almost completely over the period 1870–1950.

The neoclassical theory of the market was supposed to be neutral and value-free. The basic aim had been to define a set of economic laws which governs economic activity (in much the same way as physicists had done following Newton's discoveries). Rational individuals were seen in terms of seeking to satisfy substitutable wants (or preferences) and this pursuit of individual self-interest was also believed to be improving societal welfare; thus, within the 'hard core' of the neoclassical system was a particular model of human nature – the 'rational and egotistic person'. In its modern version the model has economic person holding the preference structure of indifference and operating on the basis of constrained satisfaction (utility) maximisation. The economic (in-strumental) value of marketable commodities, unpriced environmental goods and services, or sympathy for future generations, is determined according to the amount of personal utility yielded. Economic person makes trade-offs at the margin to identify positions of equal personal satisfaction. The preferences of individuals are revealed by the choices they make, and efficiency and consistency of choice reflect rational behaviour.

The criterion of social desirability is usually expressed in terms of the so-called *Pareto criterion*. A Pareto optimum situation is one in which it is impossible to make any individual better off without making someone else worse off, where 'better off' means 'more preferred' and 'worse off' means 'less preferred'. Every competitive market equilibrium is a Pareto optimum and every Pareto optimum is a competitive equilibrium, as long as a set of restrictive assumptions (e.g. perfect information, absence of externalities, etc.) hold true. The 'basic theorem of welfare economics' seeks to legitimise rational behaviour as being socially desirable and also to justify some government intervention to improve the conditions under which individuals make choices. Intervention would be especially justified whenever so-called *market failures* exist, i.e. when it is clear that markets are not maximising collective welfare. The basic neoclassical view sees government as an essentially ethical agent only intervening in the market in the public interest to ease the inevitable tension between individual rationality and collective ethics. Ethical or moral obligations are not recognised at the level of the individual.

Supporters of the minority *humanistic paradigm* reject the 'rational economic person' model and instead adopt a behavioural psychology approach which emphasises a hierarchy of *needs* in place of a flat plane of substitutable wants. Humanistic analysts emphasise that preferences (tastes) are not static, independent, and determined by genetics (some of this critique is misplaced in that neoclassics does not say wants are genetically determined; it does not ask the question at all). Instead they are interdependent and can and do change over time because they are at least partially learned via the culture.

In the absence of a theory of how tastes are determined, how they differ between individuals and how they change over time, neoclassical theory treats tastes as 'exogenous'. Wants and needs are therefore not separable in the conventional analysis. Quite recently supporters of the 'human capital theory' movement have argued that all economic agents do hold exactly the same set of 'stable preferences'. It is then possible to interpret these particular preferences as beliefs about how to meet basic human needs. Needs cannot be traded off against each other or against market commodities without threatening survival. As we shall see later in this chapter, environmentalists would argue that 'high' levels of environmental quality are human needs.

Humanists, among others, have also been critical of the neoclassical theory of self-interested rationality. They argue that individuals are capable of truly altruistic acts and that an extended notion of rationality is required. *Extended rationality* could be analysed in terms of multiple preference rankings within a single individual – one self-interested and the other altruistic (group-interested). Moral consideration will then determine a 'meta-ranking' of alternative motivations, e.g. altruistic motives might be judged morally superior to ones based on self-interest. Individuals possess a sense of community which is reflected in a willingness to view assets as a common pool. This extended rationality also generates a strong obligation to abide by particular laws which are seen by the individual as promoting his/her meta-preferences, despite a potential tension between the law and narrow self-interest. The law is seen as personally beneficial not just because of its restrictive effects on others, but also because of its direct effects on the individual concerned. Given these sorts of assumptions the humanistic economy, especially in its transition phase, would be subject to central planning and direction. The government's role would not be restricted merely to correcting market failures.

The humanistic economics viewpoint would not seek to abolish the market mechanism but would seek to restrain and supplement it to a significant degree. In order to facilitate greater social system stability over the long term, increased government intervention would be required in order both to decentralise economic activity and to promote a more deliberately egalitarian distribution of income.

1.3   Post-war Economics and the Rise of Environmentalism

Neoclassical economics contained the basic assumption that the economy had an in-built tendency to operate at an overall level of activity fixed by the full employment of labour. Full employment would be the norm because of the further assumption of flexible wage rates: wages would simply vary up or down until full employment is achieved. The experiences of the inter-war years (1920s and 1930s), when mass unemployment became the norm, led to the formulation of Keynesian economics with its emphasis on government intervention and deficit spending. Thus during the 1950s economic growth got back onto both the economic and political agendas. Economic growth driven by technical innovation appeared to offer limitless progress.

During the 1960s environmental pollution intensified and became more widespread. Environmental awareness was consequently heightened in some sections of industrialised societies, spawning new environmental ideologies. A number of these ideologies were basically anti-economic growth.

These events caused economists to look afresh at a central economic idea: resource scarcity in relation to possible uses. Between 1870 and 1970, mainstream economists (with some notable exceptions) appeared to believe that economic growth was sustainable indefinitely. After 1970 a majority of economists continued to argue that economic growth remained both feasible (a growing economy need not run out of natural resources) and desirable (economic growth need not reduce the

Table 26.1 [*orig. figure 1*] Environmental ideologies.

TECHNOCENTRIC		ECOCENTRIC	
Extreme 'Cornucopian'	'Accommodating'	'Communalist'	'Deep Ecology'
Resource exploitative, growth-orientated position	Resource conservationist and 'managerial' position	Resource preservationist position	Extreme preservationist position
Economic growth ethic in material value terms Maximise Gross National Product It is taken as axiomatic that unfettered market mechanisms or central planning (depending on the ruling political ideology) in conjunction with technological innovation will ensure infinite substitution possibilities capable of mitigating long-run physical resource scarcity	Infinite substitution is not thought realistic but sustainable growth is a practicable option as long as certain resource management rules (e.g. for renewable resource sustainable yield management) are followed	Pre-emptive macro-environmental constraints on economic growth are required, because of physical and social limits Decentralised socio-economic system is necessary for sustainability	Minimum 'resource-take' socio-economic system (e.g. based on organic agriculture and de-industrialisation) Acceptance of bioethics (i.e. non-conventional ethical thinking which confers moral rights or interests on non-human species)
Instrumental value (i.e. of recognised value to humans) in nature	Instrumental value in nature	Instrumental and intrinsic value in nature (i.e. valuable in its own right regardless of human experience)	Intrinsic value in nature

Source: Adapted from O'Riordan and Turner, 1983.

overall quality of life). What was required, however, was an efficiently functioning price system. Such a system was capable of accommodating higher levels of economic activity while still preserving an acceptable level of ambient environmental quality. The 'depletion effect' of resource exhaustion would be countered by technical change (including recycling) and substitutions which would augment the quality of labour and capital, and allow for, among other things, the continued extraction of lower quality non-renewable resources.

Since 1970 a number of 'world views' have crystallised within environmentalism, providing the background to the emerging environmental economics sub-discipline. Four basic world views can be distinguished, ranging from support for a market and technology-driven growth process which is environmentally damaging, through a position favouring managed resource conservation and growth, to 'eco-preservationist' positions which explicitly reject economic growth. Table 26.1 outlines some of the main features of these different positions.

It was against this backdrop of emerging environmental ideologies that environmental economics became established as a sub-discipline. Its development within the economics profession was in one sense a reaction to the prevailing conventional paradigm. A minority of revisionists wished to alter the 'hard core' of the conventional economic research programme, in order to speed up the evolution of economics towards a paradigm that was 'relevant' to the coming zero-growth society. Others merely saw an opportunity to accommodate better the environmental systems implications of the growth economy and society within a modified, but not radically different, set of economic models. The majority mainstream view remained optimistic about future growth prospects, with 'Ricardian scarcity' being offset by technology and compensatory market processes.

From outside economics, *ecocentrists* tried to bring to the forefront of public debate profound questions relating to the 'acceptability' of conventional growth objectives, strategies and policies. The influential Meadows Report (Meadows et al., 1972) adopted a distinctive Malthusian position which implied that environmental protection policies and the promotion of economic growth objectives were incompatible (i.e. that long-run economic growth objectives were not feasible). This line of thinking led eventually to calls for steady-state (zero growth) economies, and even more radical bio-economic communities based on organic agriculture which in some people's minds ought to be guided by the ethical principles of 'deep ecology' (see table 26.1).

The anti-growth argument was buttressed by economic analyses which sought to highlight the social costs, especially the environmental costs, of living in a 'growth society' (i.e. desirability of economic growth and social system stability). Easterlin's 'paradox' (i.e. survey data indicating that material affluence and human happiness were not closely correlated), Hirsch's 'positional goods' concept (i.e. that the enjoyment of a range of commodities is necessarily restricted to a small group of high income earners, despite the illusion given that all sections of society might one day participate in such consumption), and Scitovsky's 'joyless economy' analysis (again emphasising human need for more than mere material affluence) are representative of 'social limits' thinking (Boskin, 1979; Hirsch, 1977; Scitovsky, 1976).

1.4   Institutional Economics Paradigm

This minority economic doctrine began to emerge around the beginning of the twentieth century, although it has remained a somewhat diverse collection of views. Institutionalists have adopted what they call a 'processual paradigm' which encompasses the concept of the economy as a dynamic process. Their explanation of socio-economic change is one based on *cultural determinism*. Culture is an on-going complex of ideas, attitudes and beliefs that is absorbed by individuals ('cultural person' not 'rational economic person') in a habitual manner through institutional arrangements. Special significance is attached to scientific and technological change as factors that provide for dynamic change in the structure and functioning of the economic system.

Individual preferences are *learned* preferences which change over time and any one individual will hold both private and public preferences. The latter are thought to be important and justify an active public sector in the economy, and some analysts go as far as advocating indicative planning. Environmental problems are judged to be an inevitable result of economic growth in advanced industrial economies. Institutionalists have long accepted an approach which encompasses the notion of social costs of pollution and stresses the importance of the ecological foundations of any economic system. State intervention is required to control, as far as possible, the activities of transnational corporations and also to mediate between the interest groups (power blocks) that have emerged in modern economies. Institutionalists remain divided over the extent of intervention required in order to reach a social consensus. Some 'neo-Malthusians' believe that only an authoritarian system would be capable of bringing about the necessary changes to protect the environment, while others put their faith in decentralised socialist systems.

1.5   The Market Model of Environmental Management: Property Rights Paradigm versus Materials Balance Analysis

The conventional approach has generated two variants of an environmental resource management model, one more revisionist than the other in terms of the required modifications to the neoclassical blueprint. These approaches are the property rights approach and the materials balance approach.

The property rights approach

Some analysts at first maintained that pollution cost problems were non-pervasive and could be adequately mitigated via a process of re-defining the existing structure of property rights. A particular interpretation of the 'Coase theorem' (Coase, 1960) was used as the theoretical basis for this non-interventionist pollution control policy. According to Coase, given certain assumptions the most efficient solution to pollution damage situations is a bargaining process between polluter and sufferer. Each could compensate the other according to who possesses property rights: if the polluter has the right, the sufferer can 'compensate' him *not* to pollute; if the sufferer has the right, the

polluter can compensate him to tolerate damage.

The non-revisionist 'property rights paradigm' approach to environmental economics has become more sophisticated. Key neoclassical assumptions about human behaviour in the marketplace (i.e. self-interested utility maximisation) have been extended to cover the activities of bureaucrats in the public sector (drawing on public choice economics literature) and notions of extended rationality (i.e. the possession of motivation other than self-interest alone) have been resisted.

Sociobiological explanations for 'rational economic man' have also been advanced. Self-interested behaviour, it is argued, is genetically programmed into humans and is therefore inevitable. At the level of the individual, economic person is still seen as making trade-offs at the margin to identify positions of equal satisfaction. The idea of 'rational ignorance' has, however, been added to the model, i.e. it is rational for individuals to obtain less than complete information before making a decision, because information is scarce and something must be given up – time, effort or money – to obtain more of it. Exactly how much information is rationally required is not specified.

It is argued that in an economy with well-defined and transferable property rights, individuals and firms have every incentive to use natural resources as efficiently as possible. Markets and prices emerge from collective economic behaviour provided exclusion is possible – i.e. any individual consuming a good can exclude other individuals from consuming the same good – and property rights exist. Environmental pollution is a form of market failure, usually because of the over-exploitation of resources held as common property or not owned at all. The market fails therefore when property rights are inadequately specified or are not controlled by those who can benefit personally by putting the resources to their most highly valued use.

According to the property rights approach, increased government intervention should be resisted because public ownership of many natural resources lies at the root of resource control conflicts: there is 'government failure'. It is assumed that the theory of the public sector should be based on the same motivational assumptions (self-interest) employed in the analysis of private individual behaviour. Thus the decision-maker will seek to maximise his own utility, not that of some institution or state, in whatever situation he finds himself. The public sector, it is argued, provides no incentives for politicians or bureaucrats to resist pressures from special interest groups. Gains to such groups often come only at a net cost to society.

The misallocation of environmental resources is not, therefore, just a question of market failure. A range of government intervention policies have themselves been the cause of environmental disruption (government failures). For example, non-integrative government policy and inefficient government intervention have resulted in 'created' land-use conflicts in wetland ecosystems and consequent sub-optimal wetland protection levels in industrialised and developing countries.

Overall, property rights paradigm supporters would probably concede that markets are imperfect but equally they would emphasise that their failings do not automatically imply that collective action is superior. The market mechanism is then judged to be superior to any other practical alternative. Any further relinquishing of private rights and the rule of willing consent in favour of collective action will create rather than resolve environmental problems.

The materials balance approach

Revisionists have sought to incorporate materials balance models and to a more limited extent entropy limits into economic analysis. While pollution is seen as a symptom of market failure, it is also recognised that it is a pervasive and inevitable phenomenon (because of the laws of thermodynamics) requiring government intervention via a package of regulatory and incentive instruments.

In principle, an economic optimum (effi-cient) level of pollution can be defined, given certain simplifying assumptions. It is that level of pollution at which the marginal net private benefits of the polluting firm are just equal to the marginal external damage costs. Because of data deficiencies and the limitations of this static approach, the optimum situtation is not a practicable policy objective. Instead society sets 'acceptable' levels of ambient environmental quality, and policy instruments are directed at these standards. The analytical task is to seek out the least-cost policy package sufficient to meet acceptable ambient quality standards. Many economists favour the use of effluent taxes (per unit of pollution) but actual pollution control policy has been based on a regulatory approach often involving uniform reductions in pollution emissions across classes of industry. Because of the uncertainties involved, pollution control policy should be seen as an iterative search process based on a 'satisfising' (extended rationality) rather than an optimising principle.

1.6 Policy Analysis: Fixed Standard versus Cost-benefit Framework

In the face of the complexity of ecological interdependence and uncertainties surrounding resource management, two alternative approaches have been suggested. Some analysts have argued for the adoption of a *cost-benefit framework*, utilising monetary valuations but also incorporating explicit recognition of uncertainty and irreversibilities. Others urge the adoption of a *fixed standard approach*, either in selected cases or as a way of implementing a general 'macroenvironmental policy'. Macroenvironmental standards could encompass land-use zoning policy, ambient environmental quality standards for air and water, etc. In this form, such standards would operate as binding constraints and, although perhaps flexible over time (as knowledge increases), would limit the scope of cost-benefit analysis to cost effectiveness analysis.

1.7 Economic and Environmental Values

There are various interpretations of the term 'value', but economists have concentrated on monetary value as expressed via individual consumer preferences. On this basis, value only occurs because of the interaction between a subject and an object and, in terms of this explanation, is not an intrinsic quality of anything. A given object can then have a number of assigned values because of differences in the perception of held values of human valuators and different valuation contexts. Economic assigned values are expressed in terms of individual willingness to pay (WTP) and willingness to accept compensation (WTA).

The environmental literature has identified three basic value relationships which seem to underlie the policy and ethics adopted in society: values expressed via individual preferences; public preference value which finds expression via social norms; and functional physical ecosystem value. Some writers argue that economic value measures are context-specific, assigned values and may therefore be inappropriate as the *sole* value measures for public resource allocation. Ecocentric ideologies seek to base policy on social norms that individuals accept as members of a community (public preferences) and that are operationalised via 'social' legislation. Deep ecology advocates place primary emphasis on a distinction between instrumental value (expressed via human-held values) and intrinsic, non-preference-related value. They lay particular stress on the argument that functions and potentials of ecosystems themselves are a rich source of intrinsic value. This value would, it is argued, exist even if humans and their experiences were extinct. For now we merely raise the possibility that the intrinsic value and object-subject value distinction is not clear-cut. Humans may capture part of the intrinsic value in their preferences, e.g. valuing 'on behalf of' other species. Economists use the term 'existence value' to encompass these notions.

1.8 Sustainable Economic Growth and Development

The re-birth of environmentalism in the 1960s was confined to the industrialised countries of the North. In the developing countries of the South, environmental policies, over and above a concern for basic necessities, were regarded as unaffordable luxuries. It was not until 1972, with the Stockholm Conference on the Human Environment, that a milestone was reached in the development of international environmental policy. It resulted in the establishment of the United Nations Environment Programme and the creation of national environmental protection agencies in the economies of the North. In the years that followed, developing countries, while pressing for the establishment of a new 'International Economic Order', also came to realise that the health of the environment should concern them as much as it did the industrialised countries.

In 1980, the US *Global 200 Report* (Barney, 1980) appeared to confirm environmental prophesies about the consequences of the neglect of the global 'common interest' and the over-exploitation of open-access resources. But in a re-run of the original *Limits to Growth* debate based around the Meadows report, *Global 200* stimulated a 'cornucopian technocentrist' backlash and the publication of *The Resourceful Earth* report in 1984 (Simon and Kahn, 1984).

The rejection of the physical limits to growth thesis, the appropriate role of market forces in the development process, the role of poverty in natural resource degradation and the need to recognise and build on common interests, are all themes that reappear in highlighted form in reports such as *Our Common Future* (WCED, 1987) and *The Global Possible* (Repetto, 1985). In these documents it is accepted, in principle, that the world's resources are sufficient to meet long-term human needs. The critical issues under debate, therefore, concern the uneven spatial distribution of population relative to natural carrying capacities, together with the extent and degree of inefficient and irrational uses of natural resources.

The 1980s have also seen a re-orientation of some environmental thinking. The term *sustainability* has appeared in a range of contexts and probably most prominently in the *World Conservation Strategy* (IUCN, 1980). Underlying some sustainability thinking is an increased recognition that knowledge accumulated in the natural sciences ought to be applied to economic processes. For instance, the scale and rate of throughput (matter and energy) passing through the economic system is subject to an entropy constraint. Intervention is required because the market by itself is unable to reflect accurately this constraint. Modern economics lacks what we call an *existence theorem*: a guarantee that any economic optimum is associated with a stable ecological equilibrium. The Pareto optimality of allocation, for example, is independent of whether or not the scale of physical throughput is ecologically sustainable. There is a risk that some 'ecologically relevant' externality situations may involve damage to the ecosystem itself. Such externalities can cause false signals to regulatory authorities in such a way that sustainability conditions are not fulfilled.

A working definition of sustainable development might be as follows: it involves maximising the net benefits of economic development, subject to maintaining the services and quality of natural resources over time. Economic development is broadly construed to include not just increases in real per capita incomes but also other elements in social welfare. Development will necessarily involve structural change within the economy and in society. Maintaining the services and quality of the stock of resources over time implies, as far as is practicable, acceptance of the following rules:

(a) Utilise renewable resources at rates less than or equal to the natural rate at which they can regenerate.
(b) Optimise the efficiency with which non-renewable resources are used, subject to substitutability between resources and technological progress.

Economic development and natural resource maintenance are related in the following two broad ways:

1   Up to some level of resource base utilisation there is likely to be a trade-off between development and the services of the resource base (complementary relationship).
2   Beyond this level, economic development is likely to involve reductions in one or more of the functions of natural environments – as inputs to economic production, a waste assimilation service and recreation/amenity provision. In this trade-off context, the multifunctionality of natural resources is a critical concept.

1.9   Ecological and Co-evolutionary Economic Paradigm

Thinkers within this paradigm (Norgaard, 1984) have seriously questioned the validity of either biological determinism or cultural determinism as explanations of development and change. Reality is more complex and dynamic. There is a constant and active interaction of the organism with its environment. Organisms (especially humans) do not merely receive a given environment but actively seek alternatives or change what they find. Organisms are not simply the results but are also the causes of their own environments.

Economic development can therefore be viewed as a process of adaptation to a changing environment, while itself being a source of environmental change. From this perspective there are three distinct sources of change – the breakdown of ecological equilibrium (i.e. any combination of a method and a rate of resource use which the environment can sustain for long periods), the demands of technical consistency, and the development of new forms of need as the real costs of living are changed – none of these alone explain all change. Development is then a process of moving through a succession of ecological niches. Niche occupancy is variable, and a niche may be destroyed by means external to a society's own development process.

The *co-evolutionary perspective* has been designed to provide a link between ecological and economic analysis. Co-evolution refers to any on-going feedback process between two evolving systems. During co-evolution, energy surpluses are generated within systems, and these are then available for stimulating new interactions between systems. If the interactions prove favourable to society the development process continues. But co-evolutionary development feedback systems frequently shift from the ecosystem to the sociosystem, i.e. production systems become more round-about and complex and environmental exploitation increases. Since learning, knowledge and evolution are interrelated, additional co-evolutionary development potential remains untapped. However, the magnitude and extent of this development potential, which will determine how tolerable survival will be, remain uncertain.

The physical limits to growth are manifestations of the increasing complexity of the productive system. Individual subsistence requirements depend on the technology and culture of contemporary society. As complexity in the social system increases so do subsistence requirements. Preferences change because the context in which individuals learn their preferences is changed by development. This explains the 'Easterlin paradox' (i.e. survey data indicating no close correlation between material affluence and human happiness), along with explanations couched in terms of the interdependency of preferences (relative income hypothesis) and 'positional goods'. Steady-state models can be seen to be advocating the deliberate selection of a new niche of a particular kind, one which is thought would offer the prospect of prolonged occupancy (sustainability).

1.10 Conclusions

A brief overview of alternative ways of thinking about natural environments can do no more than touch on salient issues. Three in particular stand out.

First, while economists typically acknowledge the fact that there are several varied objectives possessed by individuals and by societies generally, they tend to work with only one – economic efficiency. One possibly powerful reason for this is that it may well dominate in many of the contexts that economists have typically analysed – markets in goods and services, for example. Translating that objective to the supply and demand for non-marketed goods, which is the typical context for environmental assets, permits the application of the economist's tools of the trade – optimising, marginal analysis and so on. But it may well be that finding the economically 'optimal' provision of environmental assets is not the dominant concern of individuals or societies. Issues such as fairness of access to those assets, both within a time period and over time, could be of equal and possibly greater significance. Similarly, uncertainty about the role which natural environments play in providing for the quality of our lives, and even for its very existence, might make us wary of engaging in a standard comparison of costs and benefits. Making allowance for the different objectives that people have with respect to environments could therefore alter the perspective that neoclassical economics places on those environments.

Second, some alternative views raise the issue of whether preference-based systems are relevant *at all* to the analysis of environmental issues. If habitats and their non-human occupants have some form of 'intrinsic' value, unrelated to the *act* of preferring or dispreferring by humans, then we face the problem of how to account for those values. That is, we may have values that are *of* people *for* a given environmental asset, and values that somehow reside *in* that environment and that are not *of* people. The extent to which this distinction poses a real problem depends on several factors. One is the extent to which people actually do capture some of this intrinsic value by expressing preference *on behalf* of species and habitats. If this is a feature of human preference expression, then the distinction between intrinsic and subject–object values may well remain valid but not be of great importance for real-world decisions. Another problem is the extent to which any prescription for social action can be based on 'rights' not possessed by humans. An animal rights campaigner would have no difficulty in affirming that human action should be guided in part by the rights of non-human beings. Others would query whether it makes sense to speak of 'rights' outside of the province of human attributes.

Third, much of the literature that questions the role of economic analysis and the environment does so, we believe, because it is not convinced that economics has come to terms with ecological conditions for sustainability. Put another way, economics does not appear to have an 'existence theorem' which enables us to be sure that whatever economy we devise will be sustainable ecologically. The only way to be sure of this sustainability is to ensure that economic models have sustainability conditions built into them. Beliefs about what would happen if we did this have clearly varied markedly over time. The 'limits to growth' school of thought would argue that such an extended economic-ecological model would show that economic growth, as traditionally understood, would not be sustainable. Others would argue that growth may be perfectly feasible but that the configuration of growth, the way in which it is achieved, would differ from the patterns we observe today. As an example, it might be growth based more on the sustainable use of renewable natural resources and less on exploitative use of exhaustible resources. But it might also be a 'hi-tech' economy in which growth is based on very low resource inputs and high technical progress: the *presumption* in much of the literature that an ecologically constrained economy is a low growth, austerity economy need not be true at all.

Our belief is that only by improving substantially our understanding of economy–environment interactions will we get a better grasp of these wider issues.

References and Further Reading

The standard work in the **history of economic thought** is Marc Blaug's *Economic Theory in Retrospect*, Cambridge University Press, Cambridge, 1978. A very readable and different survey of the same history can be found in Ken

Cole, John Cameron and Chris Edwards, *Why Economists Disagree: The Political Economy of Economics*, Longman, London, 1983. M. Lutz and K. Lux's *The Challenge of Humanistic Economics*, Benjamin Cummings, New York, 1979, surveys the humanistic economics paradigm. See also R. D. Hamrin: *Managing Growth in the 1980s: Toward a New Economics*, Praeger, New York, 1980. Some of the wider aspects of **environmentalism** are explored in Tim O'Riordan and R. Kerry Turner (eds). *An Annotated Reader in Environmental Planning and Management*, Pergamon Press, Oxford, 1983. The **'limits to growth' debate** has spawned an immense literature; the volume that acted as a catalyst for much of the modern contribution was D. H. Meadows et al., *The Limits to Growth*, Universe Books, New York, 1972. The 'social limits to growth' thesis is explored in the following texts: Michael J. Boskin, *Economics and Human Welfare*, Academic Press, New York, 1979; Fred Hirsch, *Social Limits To Growth*, Routledge and Kegan Paul, London 1977; Tibor Scitovsky, *The Joyless Economy*, Oxford University Press, London, 1976. The **sustainability** debate of the 1980s has been characterised by definitional confusion and a lack of precision – see R. Kerry Turner (ed.), *Sustainable Environmental Management: Principles and Practice*, Belhaven Press, London and Westview, Boulder, 1988; G. O. Barney *The Global 2000 Report to the President of the US* (2 vols), Pergamon Press, Oxford, 1980; Julian Simon and Herman Kahn, *The Resourceful Earth: A Response to Global 2000*, Basil Blackwell, Oxford, 1984; World Commission on Environment and Development, *Our Common Future*, Oxford University Press, London, 1987; Robert Repetto (ed.), *The Global Possible*, Yale University Press, New Haven, 1985; and International Union for the Conservation of Nature, *World Conservation Strategy*, IUCN-UNEP-WWF, Gland, Switzerland, 1980. Possible **ecological models** for economic development have been explored by: Michael Common, 'Poverty and progress revisited', in David Collard, David Pearce, and David Ulph (eds), *Economic Growth and Sustainable Environments*, Macmillan, London, 1988; and by Richard Norgaard, 'Coevolutionary development potential', *Land Economics*, 60 (2), 1984. The view that markets in environmental damage will emerge and secure the optimal level of damage was first elucidated in R. Coase, 'The problem of social cost', *Journal of Law and Economics*, 3, 1960.

27

TOWARDS AN ENVIRONMENTAL MACROECONOMICS*

Herman E. Daly

I Introduction

Environmental economics, as it is taught in universities and practiced in government agencies and development banks, is overwhelmingly microeconomics. The theoretical focus is on prices, and the big issue is how to internalize external environmental costs so as to arrive at prices that reflect full social marginal opportunity costs. Once prices are right the environmental problem is "solved" – there is no macroeconomic dimension. There are, of course, very good reasons for environmental economics to be closely tied to microeconomics, and it is not my intention to argue against that connection. Rather I want to ask if there is not a neglected connection between the environment and macroeconomics.

A search through the indexes of three leading textbooks in macroeconomics (Barro 1987; Dornbusch and Fischer 1987; Hall and Taylor 1988) reveals no entries under any of the following subjects: *environment, natural resources, pollution, depletion*. Is it really the case, as prominent textbook writers seem to think, that macroeconomics has nothing to do with the environment? What historically

has impeded the development of an environmental macroeconomics? If there is no such thing as environmental macroeconomics, should there be? What might it look like?

The reason that environmental macroeconomics is an empty box[1] lies in what Thomas Kuhn calls a "paradigm," and what Joseph Schumpeter more descriptively called a "preanalytic vision." As Schumpeter emphasized, analysis has to start somewhere – there has to be something to analyze. That something is given by a preanalytic cognitive act that Schumpeter called "vision." One might say that vision is what the "right brain" supplies to the "left brain" for analysis. Whatever is omitted from the preanalytic vision cannot be recaptured by subsequent analysis. Schumpeter is worth quoting at length on this point:

In practice we all start our own research from the work of our predecessors, that is, we hardly ever start from scratch. But suppose we did start from scratch, what are the steps we should have to take? Obviously, in order to be able to posit to ourselves any problems at all, we should

* Originally published in *Land Economics*, 1991, vol. 67, no. 2, pp. 255–9.
[1] The box is not entirely empty. Recent work on correcting national income accounts, along with applications of input–output models to environmental problems, should be noted.

first have to visualize a distinct set of coherent phenomena as a worthwhile object of our analytic effort. In other words, analytic effort is of necessity preceded by a preanalytic cognitive act that supplies the raw material for the analytic effort. In this book, this preanalytic cognitive act will be called Vision. It is interesting to note that vision of this kind not only must precede historically the emergence of analytic effort in any field, but also may re-enter the history of every established science each time somebody teaches us to *see* things in a light of which the source is not to be found in the facts, methods, and results of the pre-existing state of the science. (Schumpeter 1954, 41)

The vision of modern economics in general, and especially of macroeconomics, is the familiar circular flow diagram. The macroeconomy is seen as an isolated system (i.e., no exchanges of matter or energy with its environment) in which exchange value circulates between firms and households in a closed loop. What is "flowing in a circle" is variously referred to as production or consumption, but these have physical dimensions, and the circular flow does not refer to materials recycling, which in any case could not be a completely closed loop, and of course would require energy which cannot be recycled at all. What is truly flowing in a circle can only be abstract exchange value – exchange value abstracted from the physical dimensions of the goods and factors that are exchanged. Since an isolated system of abstract exchange value flowing in a circle has no dependence on an environment, there can be no problem of natural resource depletion, nor environmental pollution, nor any dependence of the macroeconomy on natural services, or indeed on anything at all outside itself (Georgescu-Roegen 1971; Daly 1985).

Since analysis cannot supply what the preanalytic vision omits, it is only to be expected that macroeconomics texts would be silent on environment, natural resources, depletion, and pollution. It is as if the preanalytic vision that biologists had of animals recognized only the circulatory system and

abstracted completely from the digestive tract. A biology textbook's index would then contain no entry under "assimilation" or "liver." The dependence of the animal on its environment would not be evident. It would appear as a perpetual motion machine.

Things are no better when we turn to the advanced chapters at the end of most macroeconomics texts, where the topic is growth theory. True to the preanalytic vision the aggregate production is written as Y = f(K,L), i.e. output is a function of capital and labor stocks. Resource flows (R) do not even enter! Neither is any waste output flow noted. And if occasionally R is stuck in the function along with K and L it makes little difference since the production function is almost always a multiplicative form, such as Cobb-Douglas, in which R can approach zero with Y constant if only we increase K or L in a compensatory fashion. Resources are seen as "necessary" for production, but the amount required can be as little as one likes!

What is needed is not ever more refined analysis of a faulty vision, but a new vision. This does not mean that everything built on the old vision will necessarily have to be scrapped – but fundamental changes are likely when the preanalytic vision is altered. The necessary change in vision is to picture the macroeconomy as an open subsystem of the finite natural ecosystem (environment) and not as an isolated circular flow of abstract exchange value, unconstrained by mass balance, entropy, and finitude. The circular flow of exchange value is a useful abstraction for some purposes. It highlights issues of aggregate demand, unemployment, and inflation that were of interest to Keynes in his analysis of the Great Depression. But it casts an impenetrable shadow on all physical relationships between the macroeconomy and the environment. For Keynes this shadow was not very important, but for us it is. Once the macroeconomy is viewed as an open subsystem, rather than an isolated system, then the issue of its relation to its parent system (the environment) cannot be avoided. And the most obvious question is how big should the subsystem be relative to the overall system?

II   The Macro-Economics of Optimal Scale

Just as the micro unit of the economy (firm or household) operates as part of a larger system (the aggregate or macroeconomy), so the aggregate economy is likewise a part of a larger system, the natural ecosystem. The macroeconomy is an open subsystem of the ecosystem and is totally dependent upon it, both as a source for inputs of low-entropy matter-energy and as a sink for outputs of high-entropy matter-energy. *The physical exchanges crossing the boundary between system and subsystem constitute the subject matter of environmental macroeconomics.* These flows are considered in terms of their scale or total volume relative to the ecosystem, not in terms of the price of one component of the total flow relative to another. Just as standard macroeconomics focuses on the volume of transactions rather than the relative prices of different items traded, so environmental macroeconomics focuses on the volume of exchanges that cross the boundary between system and subsystem, rather than the pricing and allocation of each part of the total flow within the human economy or even within the nonhuman part of the ecosystem.

The term "scale" is shorthand for "the physical scale or size of the human presence in the ecosystem, as measured by population times per capita resource use." Optimal *allocation* of a given scale of resource flow within the economy is one thing (a microeconomic problem). Optimal *scale* of the whole economy relative to the ecosystem is an entirely different problem (a macro-macro problem). The micro allocation problem is analogous to allocating optimally a given amount of weight in a boat. But once the best relative location of weight has been determined, there is still the question of the absolute amount of weight the boat should carry, even when optimally allocated. This absolute optimal scale of load is recognized in the maritime institution of the Plimsoll line. When the watermark hits the Plimsoll line the boat is full, it has reached its safe *carrying capacity.* Of course if the weight is badly allocated the waterline will touch the Plimsoll mark sooner. But eventually as the absolute load is increased the watermark will reach the Plimsoll line even for a boat whose load is optimally allocated. Optimally loaded boats will still sink under too much weight – even though they may sink optimally! It should be clear that optimal allocation and optimal scale are quite distinct problems. The major task of environmental macroeconomics is to design an economic institution analogous to the Plimsoll mark – to keep the weight, the absolute scale, of the economy from sinking our biospheric ark.

The market of course functions only within the economic subsystem, where it does only one thing: solves the allocation problem by providing the necessary information and incentive. It does that one thing very well. What it does not do is to solve the problems of optimal scale or of optimal distribution. The market's inability to solve the problem of just distribution is widely recognized, but its similar inability to solve the problem of optimal or even sustainable scale is not as widely appreciated.[2]

An example of the confusion that can result from the nonrecognition of the independence of the scale issue from the question of allocation is provided by the following dilemma (Pearce et al. 1989, 135). Which puts more pressure on the environment, a high or a low discount rate? The usual answer is that a high discount rate is

[2] This can be illustrated in terms of the familiar microeconomic tool of the Edgeworth box. Moving to the contract curve is an improvement in efficiency of *allocation.* Moving along the contract curve is a change in *distribution* which may be deemed just or unjust on ethical grounds. The *scale* is represented by the dimensions of the box, which are taken as given. Consequently the issue of optimal scale of the box itself escapes the limits of the analytical tool. A microeconomic tool cannot be expected to answer a macroeconomic question. But so far macroeconomics has not answered the question either – indeed, has not even asked it. The tacit answer to the implicit question seems to be that a bigger Edgeworth box is always better than a small one!

worse for the environment because it speeds the rate of depletion of nonrenewable resources and shortens the turnover and fallow periods in the exploitation of renewables. It shifts the allocation of capital and labor toward projects that exploit natural resources more intensively. But it restricts the total number of projects undertaken. A low discount rate will permit more projects to be undertaken even while encouraging less intensive resource use for each project. The allocation effect of a high discount rate is to increase throughput, but the scale effect is to lower throughput. Which effect is stronger is hard to say, although one suspects that over the long run the scale effect will dominate. The resolution to the dilemma is to recognize that two independent policy goals require two independent policy instruments – we cannot serve both optimal scale and optimal allocation with the single policy instrument of the discount rate (Tinbergen 1952). The discount rate should be allowed to solve the allocation problem, within the confines of a solution to the scale problem provided by a presently nonexistent policy instrument that we may for now call an "economic Plimsoll line" that limits the scale of the throughput.

Economists have recognized the independence of the goals of efficient allocation and just distribution and are in general agreement that it is better to let prices serve efficiency, and to serve equity with income redistribution policies. Proper scale is a third independent policy goal and requires a third policy instrument. This latter point has not yet been accepted by economists, but its logic is parallel to the logic underlying the separation of allocation and distribution.

Microeconomics has not discovered in the price system any built-in tendency to grow only up to the scale of aggregate resource use that is optimal (or even merely sustainable) in its demands on the biosphere. *Optimal scale, like distributive justice, full employment, or price level stability, is a macro-*

economic goal. And it is a goal that is likely to conflict with the other macroeconomic goals. The traditional solution to unemployment is growth in production, which means a larger scale. Frequently the solution to inflation is also thought to be growth in real output and a larger scale. And most of all the issue of distributive justice is "finessed" by the claim that aggregate growth will do more for the poor than redistributive measures. Macroeconomic goals tend to conflict, and certainly optimal scale conflicts with any goal that requires further growth, once the optimum has been reached.

III How Big is the Economy?

Probably the best index of the scale of the human economy as a part of the biosphere is the percentage of human appropriation of the total world product of photosynthesis. Net primary production (NPP) is the amount of solar energy captured in photosythesis by primary producers, less the energy used in their own growth and reproduction. NPP is thus the basic food resource for everything on earth not capable of photosynthesis. Vitousek et al. (1986) calculate that 25 percent of potential global (terrestrial and aquatic) NPP is now appropriated by human beings. If only terrestrial NPP is considered, the fraction rises to 40 percent.[3] Taking the 25 percent figure for the entire world it is apparent that two more doublings of the human scale will give 100 percent. Since this would mean zero energy left for all nonhuman and nondomesticated species, and since humans cannot survive without the services of ecosystems, which are made up of other species, it is clear that two more doublings of the human scale is an ecological impossibility, although arithmetically possible. Furthermore, the terrestrial figure of 40 percent is probably more relevant since we are unlikely to increase our take from the oceans very much. Total appropriation of the terrestrial NPP is only a bit over one

[3] The definition of human appropriation underlying the figures quoted includes direct use by human beings (food, fuel, fiber, timber), plus the reduction from the potential due to degradation of ecosystems caused by humans. The latter reflects deforestation, desertification, paving over, and human conversion to less productive systems (such as agriculture).

doubling time in the future. Perhaps it is theoretically possible to increase the earth's total photosynthetic capacity somewhat, but the actual trend of past economic growth is decidedly in the opposite direction. If the above figures are approximately correct, then expansion of the world economy by a factor of four (two doublings) is not possible. Yet the Brundtland Commission calls for economic expansion by a factor of five to ten. And the greenhouse effect, ozone layer depletion, and acid rain all constitute evidence that we have *already* gone beyond a prudent Plimsoll line for the scale of the macroeconomy.

IV   How Big Should the Economy Be?

Optimal scale of a single activity is not a strange concept to economists. Indeed microeconomics is about little else. An activity is identified, be it producing shoes or consuming ice cream. A cost function and a benefit function for the activity in question are defined. Good reasons are given for believing that marginal costs increase and marginal benefits decline as the scale of the activity grows. The message of microeconomics is to expand the scale of the activity in question up to the point where marginal costs equal marginal benefits, a condition which defines the optimal scale. All of microeconomics is an extended variation of this theme.

When we move to macroeconomics, however, we never again hear about optimal scale. There is apparently no optimal scale for the macro economy. There are no cost and benefit functions defined for growth in scale of the economy as a whole. It just doesn't matter how many people there are, or how much they each consume, as long as the proportions and relative prices are right! But if every micro activity has an optimal scale then why does not the aggregate of all micro activities have an optimal scale? If I am told in reply that the reason is that the constraint on any one activity is the fixity of all the others and that when all economic activities increase proportionally the restraints cancel out, then I will invite the

economist to increase the scale of the carbon cycle and the hydrologic cycle in proportion to the growth of industry and agriculture. I will admit that if the ecosystem can grow indefinitely then so can the aggregate economy. But, until the surface of the earth begins to grow at a rate equal to the rate of interest, one should not take this answer too seriously. The indifference to scale of the macroeconomy is due to the preanalytic vision of the economy as an isolated system – a view the inappropriateness of which has already been discussed.

Two concepts of optimal scale can be distinguished, both formalisms at this stage, but important for clarity.

1   *The anthropocentric optimum.* The rule is to expand scale, i.e., grow, to the point at which the marginal benefit to human beings of additional manmade physical capital is just equal to the marginal cost to human beings of sacrificed natural capital. All nonhuman species and their habitats are valued only instrumentally according to their capacity to satisfy human wants. Their intrinsic value (capacity to enjoy their own lives) is assumed to be zero.

2   *The biocentric optimum.* Other species and their habitats are preserved beyond the point of maximum instrumental convenience, out of a recognition that other species have intrinsic value independent of their instrumental value to human beings. The biocentric optimal scale of the human niche would therefore be smaller than the anthropocentric optimum.

The notion of sustainable development does not specify which concept of optimal scale to use. Sustainability is a necessary, but not sufficient condition for optimal scale and the further elaboration of an environmental macroeconomics.

References

Barro, Robert J. 1987. *Macroeconomics.* 2d ed. New York: John Wiley and Sons.

Daly, H. E. 1985 "The Circular Flow of Exchange Value and the Linear Throughput of Matter-Energy: A Case of Misplaced Concreteness." *Review of Social Economy* 43(3): 279–97.

Dornbusch, Rudiger, and Stanley Fischer. 1987. *Macroeconomics,* 4th ed. New York: McGraw-Hill.

Georgescu-Roegen, Nicholas. 1971. *The Entropy Law and the Economic Process*. Cambridge: Harvard University Press.

Hall, Robert E., and John B. Taylor. 1988. *Macroeconomics*. 2d ed. New York: W. W. Norton.

Pearce, David, et al. 1989. *Blueprint for a Green Economy*. London: Earthscan, Ltd., p. 135.

Schumpeter, Joseph. 1954. *History of Economic Analysis*. New York: Oxford University Press, p. 41.

Tinbergen, Jan. 1952. *On the Theory of Economic Policy*. Amsterdam: North-Holland Press.

Vitousek, Peter M., Paul R. Ehrlich, Anne H. Ehrlich, and Pamela A. Matson. 1986. "Human Appropriation of the Products of Photosynthesis." *BioScience* 34 (May): 368–73.

28

ECONOMIC INSTRUMENTS FOR ENVIRONMENTAL REGULATION*

T. H. Tietenberg

"BARGANING
- Concept"

1 Introduction

As recently as a decade ago environmental regulators and lobbying goups with a special interest in environmental protection looked upon the market system as a powerful adversary. That the market unleashed powerful forces was widely recognized and that those forces clearly acted to degrade the environment was widely lamented. Conflict and confrontation became the battle cry for those groups seeking to protect the environment as they set out to block market forces whenever possible.

Among the more enlightened participants in the environmental policy process the air of confrontation and conflict has now begun to recede in many parts of the world. Leading environmental groups and regulators have come to realize that the power of the market can be harnessed and channelled toward the achievement of environmental goals, through an economic incentives approach to regulation. Forward-looking business people have come to appreciate the fact that cost-effective regulation can make them more competitive in the global market-place than regulations which impose higher-than-necessary contol costs.

The change in attitude has been triggered by a recognition that this former adversary, the market, can be turned into a powerful ally. In contrast to the traditional regulatory approach, which makes mandatory particular forms of behaviour or specific technological choices, the economic incentive approach allows more flexibility in how the environmental goal is reached. By changing the incentives an individual agent faces, the best private choice can be made to coincide with the best social choice. Rather than relying on the regulatory authority to identify the best course of action, the individual agent can use his or her typically superior information to select the best means of meeting an assigned emission reduction responsibility. This flexibility achieves environmental goals at lower cost, which, in turn, makes the goals easier to achieve and easier to establish.

One indicator of the growing support for the use of economic incentive approaches for environmental control in the United States is the favourable treatment it has recently received both in the popular business,[1] and environmental[2] press. Some

* Originally published in *Oxford Review of Economic Policy*, 1990, vol. 6, no. 1, pp. 17–33.
[1] See, for example, Main (1988).
[2] See, for example, Stavins (1989).

public interest environmental organizations have now even adopted economic incentive approaches as a core part of their strategy for protecting the environment.[3]

In response to this support the emissions trading concept has recently been applied to reducing the lead content in gasoline, to controlling both ozone depletion and non-point sources of water pollution, and was also prominently featured in the Bush administration proposals for reducing acid rain and smog unveiled in June 1989.

Our knowledge about economic incentive approaches has grown rapidly in the two decades in which they have received serious analytical attention. Not only have the theoretical models become more focused and the empirical work more detailed, but we have now had over a decade of experience with emissions trading in the US and emission charges in Europe.

As the world community becomes increasingly conscious of both the need to tighten environmental controls and the local economic perils associated with tighter controls in a highly competitive global market-place, it seems a propitious time to stand back and to organize what we have learned about this practical and promising approach to pollution control that may be especially relevant to current circumstances. In this paper I will draw upon economic theory, empirical studies, and actual experience with implementation to provide a brief overview of some of the major lessons we have learned about two economic incentive approaches – emissions trading and emission charges – as well as their relationships to the more traditional regulatory policy.[4]

II   The Policy Context

(i) Emissions trading

Stripped to its bare essentials, the US Clean Air Act[5] relies upon a *command-and-control* approach to controlling pollution. Ambient standards establish the highest allowable concentration of the pollutant in the ambient air for each conventional pollutant. To reach these prescribed ambient standards, emission standards (legal emission ceilings) are imposed on a large number of specific emission points such as stacks, vents, or storage tanks. Following a survey of the technological options of control, the control authority selects a favoured control technology and calculates the amount of emission reduction achievable by that technology as the basis for setting the emission standard. Technologies yielding larger amounts of control (and, hence, supporting more stringent emission standards) are selected for new emitters and for existing emitters in areas where it is very difficult to meet the ambient standard. The responsibility for defining and enforcing these standards is shared in legislatively specified ways between the national government and the various state governments.

The emissions trading programme attempts to inject more flexibility into the manner in which the objectives of the Clean Air Act are met by allowing sources a much wider range of choice in how they satisfy their legal pollution control responsibilities than possible in the command-and-control approach. Any source choosing to reduce emissions at any discharge point more than required by its emission standard can apply to the control authority for certification of

---

[3] See the various issues in Volume XX of the EDF Letter, a report to members of the Environmental Defense Fund.

[4] In the limited space permitted by this paper only a few highlights can be illustrated. All of the details of the proofs and the empirical work can be found in the references listed at the end of the paper. For a comprehensive summary of this work see Tietenberg (1980), Liroff (1980), Bohm and Russell (1985), Tietenberg (1985), Liroff (1986), Dudek and Palmisano (1988), Hahn (1989), Hahn and Hester (1989a and 1989b), Tietenberg (1989b).

[5] The US Clean Air Act (42 U.S.C. 7401–642) was first passed in 1955. The central thrust of the approach described in this paragraph was initiated by the Clean Air Act Amendments of 1970 with mid-course corrections provided by the Clean Air Act Amendments of 1977.

the excess control as an 'emission reduction credit' (ERC). Defined in terms of a specific amount of a particular pollutant, the certified emissions reduction credit can be used to satisfy emission standards at other (presumably more expensive to control) discharge points controlled by the creating source or it can be sold to other sources. By making these credits transferable, the US Environmental Protection Agency (EPA) has allowed sources to find the cheapest means of satisfying their requirements, even if the cheapest means are under the control of another firm. The ERC is the currency used in emissions trading, while the offset, bubble, emissions banking, and netting policies govern how this currency can be stored and spent.[6]

The *offset policy* requires major new or expanding sources in 'non-attainment' areas (those areas with air quality worse than the ambient standards) to secure sufficient offsetting emission reductions (by acquiring ERCs) from existing firms so that the air is cleaner after their entry or expansion than before.[7] Prior to this policy no new firms were allowed to enter non-attainment areas on the grounds they would interfere with attaining the ambient standards. By introducing the offset policy EPA allowed economic growth to continue while assuring progress toward attainment.

The *bubble policy* receives its unusual name from the fact that it treats multiple emission points controlled by existing emitters (as opposed to those expanding or entering an area for the first time) as if they were enclosed in a bubble. Under this policy only the total emissions of each pollutant leaving the bubble are regulated. While the total leaving the bubble must be not larger than the total permitted by adding up all the corresponding emission standards within the bubble (and in some cases the total must be 20 per cent lower), emitters are free to control some discharge points less than dictated by the corresponding emission standard as long as sufficient compensating ERCs are obtained from other discharge points within the bubble. In essence sources are free to choose the mix of control among the discharge points as long as the overall emission reduction requirements are satisfied. Multi-plant bubbles are allowed, opening the possibility for trading ERCs among very different kinds of emitters.

Netting allows modifying or expanding sources (but not new sources) to escape from the need to meet the requirements of the rather stringent new source review process (including the need to acquire offsets) so long as any net increase in emissions (counting any ERCs earned elsewhere in the plant) is below an established threshold. In so far as it allows firms to escape particular regulatory requirements by using ERCs to remain under the threshold which triggers applicability, netting is more properly considered regulatory relief than regulatory reform.

Emissions banking allows firms to store certified ERCs for subsequent use in the offset, bubble, or netting programmes or for sale to others.

Although comprehensive data on the effects of the programme do not exist because substantial proportions of it are administered by local areas and no one collects information in a systematic way, some of the major aspects of the experience are clear.[8]

* The programme has unquestionably and substantially reduced the costs of complying with the requirements of the Clean Air Act. Most estimates place the accumulated capital savings for all components of the programme at over $10 billion. This does not include the recurring savings in operating cost. On

---

[6] The details of this policy can be found in 'Emissions Trading Policy Statement' 51 *Federal Register* 43829 (4 December 1986).

[7] Offsets are also required for major modifications in areas which have attained the standards if the modifications jeopardize attainment.

[8] See, for example, Tietenberg (1985), Hahn and Hester (1989a and 1989b), and Dudek and Palmisano (1988).

SAVES Money

the other hand the programme has not produced the magnitude of cost savings that was anticipated by its strongest proponents at its inception.

- The level of compliance with the basic provisions of the Clean Air Act has increased. The emissions trading programme increased the possible means for compliance and sources have responded.

- Somewhere between 7,000 and 12,000 trading transactions have been consummated. Each of these transactions was voluntary and for the participants represented an improvement over the traditional regulatory approach. Several of these transactions involved the introduction of innovative control technologies.

- The vast majority of emissions trading transactions have involved large pollution sources trading emissions reduction credits either created by excess control of uniformly mixed pollutants (those for which the location of emission is not an important policy concern) or involving facilities in close proximity to one another.

- Though air quality has certainly improved for most of the covered pollutants, it is virtually impossible to say how much of the improvement can be attributed to the emissions trading programme. The emissions trading programme complements the traditional regulatory approach, rather than replaces it. Therefore, while it can claim to have hastened compliance with the basic provisions of the act and in some cases to have encouraged improvements beyond the act, improved air quality resulted from the package taken together, rather than from any specific component.

(ii) Emissions charges

Emission charges are used in both Europe and Japan, though more commonly to control water pollution than air pollution.[9] Currently effluent charges are being used to control water pollution in France, Italy, Germany, and the Netherlands. In both France and the Netherlands the charges are designed to raise revenue for the purpose of funding activities specifically designed to improve water quality. *"GOOD"*

In Germany dischargers are required to meet minimum standards of waste water treatment for a number of defined pollutants. Simultaneously a fee is levied on every unit of discharge depending on the quantity and noxiousness of the effluent. Dischargers meeting or exceeding state-of-the-art effluent standards have to pay only half the normal rate.

The Italian effluent charge system was mainly designed to encourage polluters to achieve provisional effluent standards as soon as possible. The charge is nine times higher for firms that do not meet the prescribed standards than for firms that do meet them. This charge system was designed only to facilitate the transition to the prescribed standards so it is scheduled to expire once full compliance has been achieved.[10]

Air pollution emission charges have been implemented by France and Japan. The French air pollution charge was designed to encourage the early adoption of pollution control equipment with the revenues returned to those paying the charge as a subsidy for installing the equipment. In Japan the emission charge is designed to raise revenue to compensate victims of air pollution. The charge rate is determined primarily by the cost of the compensation programme in the previous year and the amount of remaining emissions over which this cost can be applied *pro rata*.

Charges have also been used in Sweden to increase the rate at which consumers would purchase cars equipped with a catalytic converter. Cars not equipped with a catalytic converter were taxed, while new cars equipped with a catalytic converter were subsidized. *GOOD*

---

[9] See Anderson (1977), Brown and Johnson (1984), Bressers (1988), Vos (1989), Opschoor and Vos (1989), and Sprenger (1989).

[10] The initial deadline for expiration was 1986, but it has since been postponed.

A 'personal' way to make the polluter pay

While data are limited a few highlights seem clear:

- Economists typically envisage two types of effluent or emissions charges. The first, an efficiency charge, is designed to produce an efficient outcome by forcing the polluter to compensate completely for all damage caused. The second, a cost-effective charge, is designed to achieve a predefined ambient standard at the lowest possible control cost. In practice, few, if any, implemented programmes fit either of these designs.
- Despite being designed mainly to raise revenue, effluent charges have typically improved water quality. Though the improvements in most cases have been small, apparently due to the low level at which the effluent charge rate is set, the Netherlands, with its higher effective rates, reports rather large improvements. Air pollution charges typically have not had much effect on air quality because the rates are too low and, in the case of France, most of the revenue is returned to the polluting sources.
- The revenue from charges is typically earmarked for specific environmental purposes rather than contributed to the general revenue as a means of reducing the reliance on taxes that produce more distortions in resource allocation.
- The Swedish tax on heavily polluting vehicles and subsidy for new low polluting vehicles was very successful in introducing low polluting vehicles into the automobile population at a much faster than normal rate. The policy was not revenue neutral, however; owing to the success of the programme in altering vehicle choices, the subsidy payments greatly exceeded the tax revenue.

III First Principles

Theory can help us understand the characteristics of these economic approaches in the most favourable circumstances for their use and assist in the process of designing the instruments for maximum effectiveness. Because of the dualistic nature of emission charges and emission reduction credits,[11] implications about emission charges and emissions trading flow from the same body of theory.

Drawing conclusions about either of these approaches from this type of analysis, however, must be done with care because operational versions typically differ considerably from the idealized versions modelled by the theory. For example, not all trades that would be allowed in an ideal emissions trading programme are allowed in the current US emissions trading programme. Similarly the types of emissions charges actually imposed differ considerably from their ideal versions, particularly in the design of the rate structure and the process for adjusting rates over time.

Assuming all participants are cost-minimizers, a 'well-defined' emissions trading or emission charge system could cost-effectively allocate the control responsibility for meeting a predefined pollution target among the various pollution sources despite incomplete information on the control possibilities by the regulatory authorities.[12]

The intuition behind this powerful proposition is not difficult to grasp. Cost-minimizing firms seek to minimize the sum of (a) either ERC acquisition costs or payments of emission charges and (b) control costs. Minimization will occur when the marginal cost of control is set equal to the emission reduction credit price or the emission charge. Since all cost-minimizing sources would choose to control until their marginal control costs were equal to the same price or charge, marginal control costs

[11] Under fairly general conditions any allocation of control responsibility achieved by an emissions trading programme could also be achieved by a suitably designed system of emission charges and vice versa.

[12] For the formal demonstration of this proposition see Baumol and Oates (1975), Montgomery (1972), and Tietenberg (1985).

would be equalized across all discharge points, precisely the condition required for cost-effectiveness.[13]

Emission charges could also sustain a cost effective allocation of the control responsibility for meeting a predefined pollution target, but only if the control authority knew the correct level of the charge to impose or was willing to engage in an iterative trial-and-error process over time to find the correct level. Emissions trading does not face this problem because the price level is established by the market, not the control authority.[14]

Though derived in the rarified world of theory, the practical importance of this theorem should not be underestimated. Economic incentive approaches offer unique opportunity for regulators to solve a fundamental dilemma. The control authorities' desire to allocate the responsibility for control cost-effectively is inevitably frustrated by a lack of information sufficient to achieve this objective. Economic incentive approaches create a system of incentives in which those who have the best knowledge about control opportunities, the environmental managers for the industries, are encouraged to use that knowledge to achieve environmental objectives at minimum cost. Information barriers do not preclude effective regulation.

What constitutes a 'well-defined' emissions trading or emission charge system depends crucially on the attributes of the pollutant being controlled.[15]

To be consistent with a cost-effective allocation of the control responsibility, the policy instruments would have to be defined in different ways for different types of pollutants. Two differentiating characteristics are of particular relevance. Approaches designed to control pollutants which are uniformly mixed in the atmosphere (such as volatile organic compounds, one type of precursor for ozone formation) can be defined simply in terms of a rate of emissions flow per unit time. Economic incentive approaches sharing this design characteristic are called *emission trades* or *emission charges*.

Instrument design is somewhat more difficult when the pollution target being pursued is defined in terms of concentrations measured at a number of specific receptor locations (such as particulates). In this case the cost-effective trade or charge design must take into account the *location* of the emissions (including injection height) as well as the *magnitude* of emissions. As long as the control authorities can define for each emitter a vector of transfer coefficients, which translates the effect of a unit increase of emissions by that emitter into an increase in concentration at each of the affected receptors, receptor-specific trades or charges can be defined which will allocate the responsibility cost-effectively. The design which is consistent with cost-effectiveness in this context is called an *ambient trade* or an *ambient charge*.

Unfortunately, while the design of the ambient ERC is not very complicated,[16] implementing the markets within which these ERCs would be traded is rather complicated. In particular for each unit of planned emissions an emitter would have to acquire separate ERCs for each affected receptor. When the number of receptors is large, the result is a rather complicated set of transactions. Similarly, establishing the correct rate structure for the charges in this context is particularly difficult because the set of charges which will satisfy the ambient air quality constraints is not unique; even a trial-and-error system would not necessarily result in the correct matrix of ambient charges being put into effect.

---

[13] It should be noted that while the allocation is cost-effective, it is not necessarily efficient (the amount of pollution indicated by a benefit-cost comparison). It would only be efficient if the predetermined target happened to coincide with the efficient amount of pollution. Nothing guarantees this outcome.

[14] See Tietenberg (1988) for a more detailed explanation of this point.

[15] For the technical details supporting this proposition see Montgomery (1972), and Tietenberg (1985).

[16] Each permit allows the holder to degrade the concentration level at the corresponding receptor by one unit.

As long as markets are competitive and transactions costs are low, the trading benchmark in an emissions trading approach does not affect the ultimate cost-effective allocation of control responsibility. When markets are non-competitive or transactions costs are high, however, the final allocation of control responsibility is affected.[17] Emission charge approaches do not face this problem.

Once the control authority has decided how much pollution of each type will be allowed, it must then decide how to allocate the operating permits among the sources. In theory emission reduction credits could either be auctioned off, with the sources purchasing them from the control authority at the market-clearing price, or (as in the US programme) created by the sources as surplus reductions over and above a predetermined set of emission standards. (Because this latter approach favours older sources over newer sources, it is known as 'grandfathering'.) The proposition suggests that either approach will ultimately result in a cost-effective allocation of the control responsibility among the various polluters as long as they are all price-takers, transactions costs are low, and ERCs are fully transferable. Any allocation of emission standards in a grandfathered approach is compatible with cost-effectiveness because the after-market in which firms can buy or sell ERCs corrects any problems with the initial allocation. This is a significant finding because it implies that under the right conditions the control authority can use this initial allocation of emissions standards to pursue distributional goals without interfering with cost-effectiveness.

When firms are price-setters rather than price-takers, however, cost-effectiveness will only be achieved if the control authority initially allocates the emission standards so a cost-effective allocation would be achieved even in the absence of any trading. (Implementing this particular allocation would, of course, require regulators to have complete information on control costs for all sources, an unlikely prospect.) In this special case cost-effectiveness would be achieved even in the presence of one or more price-setting firms because no trading would take place, eliminating the possibility of exploiting any market power.

For all other emission standard assignments an active market would exist, offering the opportunity for price-setting behaviour. The larger is the deviation of the price-setting source's emission standard from its cost-effective allocation, the larger is the deviation of ultimate control costs from the least-cost allocation. When the price-setting source is initially allocated an insufficiently stringent emission standard, it can inflict higher control costs on others by withholding some ERCs from the market. When an excessively stringent emission standard is imposed on a price-setting source, however, it necessarily bears a higher control cost as the means of reducing demand (and, hence, prices) for the ERCs.

Similar problems exist when transactions costs are high. High transactions costs preclude or reduce trading activity by diminishing the gains from trade. When the costs of consummating a transaction exceed its potential gains, the incentive to participate in emissions trading is lost.

IV   Lessons from Empirical Research

A vast majority, though not all, of the relevant empirical studies have found the control costs to be substantially higher with the regulatory command-and-control system than the least cost means of allocating the control responsibility.

While theory tells us unambiguously that the command-and-control system will not be cost-effective except by coincidence, it cannot tell us the magnitude of the excess costs. The empirical work cited in table 28.1 adds the important information that the excess costs are typically very large.[18] This is an important finding because it provides the

---

[17] See Hahn (1984) for the mathematical treatment of this point. Further discussions can be found in Tietenberg (1985) and Misiolek and Elder (1989).

[18] A value of 1.0 in the last column of table 28.1 would indicate that the traditional regulatory approach

Table 28.1 [*orig. table 1*]   Empirical studies of air pollution control.

Study	Pollutants Covered	Geographic Area	CAC Benchmark	Ration of CAC Cost to Least Cost
Atkinson and Lewis	Particulates	St Louis	SIP regulations	6.00[a]
Roach et al.	Sulphur dioxide	Four corners in Utah	SIP regulations Colorado, Arizona and New Mexico	4.25
Hahn and Noll	Sulphates	Los Angeles	California emission standards	1.07
Krupnick	Nitrogen dioxide	Baltimore	Proposed RACT	5.96[b]
Seskin et al.	Nitrogen dioxide regulations	Chicago	Proposed RACT	14.40[b]
McGartland	Particulates	Baltimore	SIP regulations	4.18
Spofford	Sulphur Dioxide	Lower Delaware Valley	Uniform percentage regulations	1.78
	Particulates	Lower Delaware Valley	Uniform percentage regulations	22.00
Harrison	Airport noise	United States	Mandatory retrofit	1.72[c]
Maloney and Yandle	Hydrocarbons	All domestic DuPont plants	Uniform percentage reduction	4.15[d]
Palmer et al.	CFC emissions from non-aerosol applications	United States	Proposed emission standards	1.96

Notes:
CAC = command and control, the traditional regulatory approach.
SIP = state implementation plan.
RACT = reasonably available control technologies, a set of standards imposed on existing sources in non-attainment areas.
[a]   Based on a 40 µg/m^3 at worst receptor.
[b]   Based on a short-term, one-hour average of 250 µg/m^3.
[c]   Because it is a benefit–cost study instead of a cost-effectiveness study, the Harrison comparison of the command-and-control approach with the least-cost allocation involves different benefit levels. Specifically, the benefit levels associated with the least-cost allocation are only 82 per cent of those associated with the command-and-control allocation. To produce cost estimates based on more comparable benefits, as a first approximation the least-cost allocation was divided by 0.82 and the resulting number was compared with the command-and-control cost.
[d]   Based on 85 per cent reduction of emissions from all sources.

motivation for introducing a reform programme; the potential social gains (in terms of reduced control cost) from breaking away from the status quo are sufficient to justify the trouble. Although the estimates of the excess costs attributable to a command and control presented in table 28.1 overstate the cost savings that would be achieved by even an ideal economic incentive approach (a point discussed in more detail below), the

was cost-effective. A value of 4.0 would indicate that the traditional regulatory approach results in an allocation of the control responsibility which is four times as expensive as necessary to reach the stipulated pollution target.

general conclusion that the potential cost savings from adopting economic incentive approaches are large seems accurate even after correcting for overstatement.

Economic incentive approaches which raise revenue (charges or auction ERC markets) offer an additional benefit – they allow the revenue raised from these policies to substitute for revenue raised in more traditional ways. Whereas it is well known that traditional revenue-raising approaches distort resource allocation, producing inefficiency, economic incentive approaches enhance efficiency. Some empirical work based on the US economy suggests that substituting economic incentive means of raising revenue for more traditional means could produce significant efficiency gains.[19]

When high degrees of control are necessary, ERC prices or charge levels would be correspondingly high. The financial outlays associated with acquiring ERCs in an auction market or paying charges on uncontrolled emissions would be sufficiently large that sources would typically have lower financial burdens with the traditional command-and-control approach than with these particular economic incentive approaches. Only a 'grandfathered' trading system would guarantee that sources would be no worse off than under the command-and-control system.[20]

Financial burden is a significant concern in a highly competitive global market-place. Firms bearing large financial burdens would be placed at a competitive disadvantage when forced to compete with firms not bearing those burdens. Their costs would be higher.

From the point of view of the source required to control its emissions, two components of financial burden are significant (a) control costs and (b) expenditures on permits or emission charges. While only the former represent real resource costs to society as a whole (the latter are merely transferred from one group in society to another), both represent a financial burden to the source. The empirical evidence suggests that when an auction market is used to distribute ERCs (or, equivalently, when all uncontrolled emissions are subject to an emissions charge), the ERC expenditures (charge outlays) would frequently be larger in magnitude than the control costs; the sources would spend more on ERCs (or pay more in charges) than they would on the control equipment. Under the traditional command-and-control system firms make no financial outlays to the government. Although control costs are necessarily higher with the command-and-control system than with an economic incentive approach, they are not so high as to outweigh the additional financial outlays required in an auction market permit system (or an emissions tax system). For this reason existing sources could be expected vehemently to oppose an auction market or emission charges despite their social appeal, unless the revenue derived is used in a manner which is approved by the sources, and the sources with which it competes are required to absorb similar expenses. When environmental policies are not co-ordinated across national boundaries, this latter condition would be particularly difficult to meet.

In the absence of either a politically popular way to use the revenue or assurances that competitors will face similar financial burdens, this political opposition could be substantially reduced by grandfathering. Under grandfathering, sources have only to purchase any additional ERCs they may need to meet their assigned emission standard (as opposed to purchasing sufficient ERCs or paying charges to cover all uncontrolled emissions in an auction market). Grandfathering is *de facto* the approach taken in the US emissions trading programme.

Grandfathering has its disadvantages. Because ERCs become very valuable,

---

[19] See Terkla (1984).
[20] See Atkinson and Tietenberg (1982, 1984), Hahn (1984), Harrison (1983), Krupnick (1986), Lyon (1982), Palmer et al. (1980), Roach et al. (1981), Seskin et al. (1983), and Shapiro and Warhit (1983) for the individual studies, and Tietenberg (1985) for a summary of the evidence.

especially in the face of stringent air quality regulations, sources selling emission reduction credits would be able to command very high prices. By placing heavy restrictions on the amount of emissions, the control authority is creating wealth for existing firms *vis-à-vis* new firms.

Although reserving some ERCs for new firms is possible (by assigning more stringent emission standards than needed to reach attainment and using the 'surplus' air quality to create government-held ERCs), this option is rarely exercised in practice. In the United States under the offset policy firms typically have to purchase sufficient ERCs to more than cover all uncontrolled emissions, while existing firms only have to purchase enough to comply with their assigned emission standard. Thus grandfathering imposes a bias against new sources in the sense that their financial burden is greater than that of an otherwise identical existing source, even if the two sources install exactly the same emission control devices. This new source bias could retard the introduction of new facilities and new technologies by reducing the cost advantage of building new facilities which embody the latest innovations.

While it is clear from theory that larger trading areas offer the opportunities for larger potential cost savings in an emissions trading programme, some empirical work suggests that substantial savings can be achieved in emissions trading even when the trading areas are rather small.

The point of this finding is *not* that small trading areas are fine; they do retard progress toward the standard. Rather, when political considerations allow only small trading areas or nothing, emissions trading still can play a significant role.

Sometimes political considerations demand a trading area which is smaller than the ideal design. Whether large trading areas are essential for the effective use of this

policy is therefore of some relevance. In general, the larger the trading area, the larger would be the potential cost savings due to a wider set of cost reduction opportunities that would become available. The empirical question is how sensitive the cost estimates are to the size of the trading areas.

One study of utilities found that even allowing a plant to trade among discharge points within that plant could save from 30 to 60 per cent of the costs of complying with new sulphur oxide reduction regulations, compared to a situation where no trading whatsoever was permitted.[21] Expanding the trading possibilities to other utilities within the same state permitted a further reduction of 20 per cent, while allowing interstate trading permitted another 15 per cent reduction in costs. If this study is replicated in other circumstances, it would appear that even small trading areas offer the opportunity for significant cost reduction.[22]

Although only a few studies of the empirical impact of market power on emissions trading have been accomplished, their results are consistent with a finding that market power does not seem to have a large effect on regional control costs in most realistic situations.[23]

Even in areas having especially stringent controls, the available evidence suggests that price manipulation is not a serious problem. In an auction market the price-setting source reduces its financial burden by purchasing fewer ERCs in order to drive the price down. To compensate for the smaller number of ERCs purchased, the price-setting source must spend more on controlling its own pollution, limiting the gains from price manipulation. Although these actions could have a rather large impact on *regional financial burden*, they would under normal circumstances have a rather small effect on *regional control costs*. Estimates typically suggest that control costs would rise by less than 1 per cent if market

---

[21] ICF, Inc. (1989).

[22] As indicated below, the fact that so many emissions trades have actually taken place within the same plant or among contiguous plants provides some confirmation for this result.

[23] For individual studies see de Lucia (1974), Hahn (1984), Stahl, Bergman and Mäler (1988), and Maloney and Yandle (1984). For a survey of the evidence see Tietenberg (1985).

power were exercised by one or more firms.

It should not be surprising that price manipulation could have rather dramatic effects on regional financial burden in an auction market, since the cost of *all* ERCs is affected, not merely those purchased by the price-setting source. The perhaps more surprising result is that control costs are quite insensitive to price-setting behaviour. This is due to the fact that the only control cost change is the net difference between the new larger control burden borne by the price searcher and the correspondingly smaller burden borne by the sources having larger-than-normal allocations of permits. Only the costs of the marginal units are affected.

Within the class of grandfathered distribution rules, some emission standard allocations create a larger potential for strategic price behaviour than others. In general the larger the divergence between the control responsibility assigned to the price-searching source by the emission standards and the cost-effective allocation of control responsibility, the larger the potential for market power. When allocated too little responsibility by the control authority, price-searching firms can exercise power on the selling side of the market, and when allocated too much, they can exercise power on the buying side of the market.

According to the existing studies it takes a rather considerable divergence from the cost-effective allocation of control responsibility to produce much difference in regional control costs. In practice the deviations from the least cost allocation caused by market power pale in comparison to the much larger potential cost reductions achievable by implementing emissions trading.[24]

V   Concluding Comments

Our experience with economic incentive programmes has demonstrated that they have had, and can continue to have, a positive role in environmental policy in the future. I would submit the issue is no longer *whether* they have a role to play, but rather *what kind* of role they should play. The available experience with operating versions of these programmes allows us to draw some specific conclusions which facilitate defining the boundaries for the optimal use of economic incentive approaches in general and for distinguishing the emissions trading and emission charges approaches in particular.

Emissions trading integrates particularly smoothly into any policy structure which is based either directly (through emission standards) or indirectly (through mandated technology or input limitations) on regulating emissions. In this case emission limitations embedded in the operating licences can serve as the trading benchmark if grandfathering is adopted.

Emissions charges work particularly well when transaction costs associated with bargaining are high. It appears that much of the trading activity in the United States has involved large corporations. Emissions trading is probably not equally applicable to large and small pollution sources. The transaction costs are sufficiently high that only large trades can absorb them without jeopardizing the gains from trade. For this reason charges seem a more appropriate instrument when sources are individually small, but numerous (such as residences or automobiles). Charges also work well as a device for increasing the rate of adoption of new technologies and for raising revenue to subsidize environmentally benign projects.

Emissions trading seems to work especially well for uniformly mixed pollutants. No diffusion modelling is necessary and regulators do not have to worry about trades creating 'hot spots' or localized areas of high pollution concentration. Trades can be on a one-to-one basis.

Because emissions trading allows the issue of who will pay for the control to be separated from who will install the control,

[24] Strategic price behaviour is not the only potential source of market power problems. Firms could conceivably use permit markets to drive competitors out of business. See Misiolek and Elder (1989). For an analysis which concludes that this problem is relatively rare and can be dealt with on a case-by-case basis should it arise, see Tietenberg (1985).

it introduces an additional degree of flexibility. This flexibility is particularly important in non-attainment areas since marginal control costs are so high. Sources which would not normally be controlled because they could not afford to implement the controls without going out of business, can be controlled with emissions trading. The revenue derived from the sale of emission reduction credits can be used to finance the controls, effectively preventing bankruptcy.

Because it is quantity based, emissions trading also offers a unique possibility for leasing. Leasing is particularly valuable when the temporal pattern of emissions varies across sources. As discussed above this appears generally to be the case with utilities. When a firm plans to shut down one plant in the near future and to build a new one, leasing credits is a vastly superior alternative to the temporary installation of equipment in the old plant which would be useless when the plant was retired. The useful life of this temporary control equipment would be wastefully short.

We have also learned that ERC transactions have higher transactions costs than we previously understood. Regulators must validate every trade. When non-uniformly mixed pollutants are involved, the transactions costs associated with estimating the air quality effects are particularly high. Delegating responsibility for trade approval to lower levels of government may in principle speed up the approval process, but unless the bureaucrats in the lower level of government support the programme the gain may be negligible.

Emissions trading places more importance on the operating permits and emissions inventories than other approaches. To the extent those are deficient the potential for trades that protect air quality may be lost. Firms which have actual levels of emissions substantially below allowable emissions find themselves with a trading opportunity which, if exploited, could degrade air quality. The trading benchmark has to be defined carefully.

There can be little doubt that the emissions trading programme in the US has improved upon the command-and-control programme that preceded it. The documented cost savings are large and the flexibility provided has been important. Similarly emissions charges have achieved their own measure of success in Europe. To be sure the programmes are far from perfect, but the flaws should be kept in perspective. In no way should they overshadow the impressive accomplishments. Although economic incentive approaches lose their Utopian lustre upon closer inspection, they have none the less made a lasting contribution to environmental policy.

The role for economic incentive approaches should grow in the future if for no other reason than the fact that the international pollution problems which are currently commanding centre-stage fall within the domains where economic incentive policies have been most successful. Significantly many of the problems of the future, such as reducing tropospheric ozone, preventing stratospheric ozone depletion, moderating global warming, and increasing acid rain control, involve pollutants that can be treated as uniformly mixed, facilitating the use of economic incentives. In addition larger trading areas facilitate greater cost reductions than smaller trading areas. This also augers well for the use of emissions trading as part of the strategy to control many future pollution problems because the natural trading areas are all very large indeed. Acid rain, stratospheric ozone depletion, and greenhouse gases could (indeed should!) involve trading areas that transcend national boundaries. For greenhouse and ozone depletion gases, the trading areas should be global in scope. Finally, it seems clear that the pivotal role of carbon dioxide in global warming may require some fairly drastic changes in energy use, including changes in personal transportation, and ultimately land use patterns. Some form of charges could play an important role in facilitating this transformation.

We live in an age when the call for tighter environmental controls intensifies with each new discovery of yet another injury modern society is inflicting on the planet.

But resistance to additional controls is also growing with the recognition that compliance with each new set of controls is more expensive that the last. While economic incentive approaches to environmental control offer no panacea, they frequently do offer a practical way to achieve environmental goals more flexibly and at lower cost than more traditional regulatory approaches. That is a compelling virtue.

References

Anderson, F. R. et al. (1977), *Environmental Improvement Through Economic Incentives*, Baltimore, The Johns Hopkins University Press for Resources for the Future, Inc.

Atkinson, S. E. and Lewis, D. H. (1974), 'A Cost-Effectiveness Analysis of Alternative Air Quality Control Strategies', *Journal of Environmental Economics and Management*, 1, 237–50.

—— and Tietenberg, T. H. (1982), 'The Empirical Properties of Two Classes of Designs for Transferable Discharge Permit Markets', *Journal of Environmental Economics and Management*, 9, 101–21.

—— (1984), 'Approaches for Reaching Ambient Standards in Non-Attainment Areas: Financial Burden and Efficiency Considerations', *Land Economics*, 60, 148–59.

—— (1987), 'Economic Implications of Emission Trading Rules for Local and Regional Pollutants', *Canadian Journal of Economics*, 20, 370–86.

Baumol, W. J. and Oates, W. E. (1975), *The Theory of Environmental Policy*, Englewood Cliffs., N. J., Prentice Hall.

Bohm, P. and Russell, C. (1985), 'Comparative Analysis of Alternative Policy Instruments', in A. V. Kneese and J. L. Sweeney (eds.), *Handbook of Natural Resource and Energy Economics*, Vol. 1, 395–460, Amsterdam, North-Holland.

Bressers, H. T. A. (1988), 'A Comparison of the Effectiveness of Incentives and Directives: The Case of Dutch Water Quality Policy', *Policy Studies Review*, 7, 500–18.

Brown, G. M. Jr. and Johnson, R. W. (1984), 'Pollution Control by Effluent Charges: It Works in the Federal Republic of Germany, Why Not in the United States?', *Natural Resources Journal*, 24, 929–66.

de Lucia, R. J. (1974), *An Evaluation of Marketable Effluent Permit Systems*, Report No. EPA-600/5–74–030 to the US Environmental Protection Agency (September).

Dudek, D. J. and Palmisano, J. (1988), 'Emissions Trading: Why is this Thoroughbred Hobbled?', *Columbia Journal of Environmental Law*, 13, 217–56.

Feldman, S. L. and Raufer, R. K. (1987), *Emissions Trading and Acid Rain Implementing a Market Approach to Pollution Control*, Totowa, N. J., Rowman & Littlefield.

Hahn, R. W. (1984), 'Market Power and Transferable Property Rights', *Quarterly Journal of Economics*, 99, 753–65.

—— (1989), 'Economic Prescriptions for Environmental Problems: How the Patient Followed the Doctor's Orders', *The Journal of Economic Perspectives*, 3, 95–114.

—— and Noll, R. G. (1982), 'Designing a Market for Tradeable Emission Permits', in W. A. Magat (ed.), *Reform of Environmental Regulation*, Cambridge, Mass., Ballinger.

—— and Hester, G. L. (1989a), 'Where Did All the Markets Go? An Analysis of EPA's Emission Trading Program', *Yale Journal of Regulation*, 6, 109–53.

—— (1989b), 'Marketable Permits: Lessons from Theory and Practice', *Ecology Law Quarterly*, 16, 361–406.

Harrison, D., Jr. (1983), 'Case Study 1: The Regulation of Aircraft Noise', in Thomas C. Schelling (ed.), *Incentives for Environmental Protection*, Cambridge, Mass., MIT Press.

ICF Resources, Inc. (1989), 'Economic, Environmental, and Coal Market Impacts of SO_2 Emissions Trading Under Alternative Acid Rain Control Proposals', a report prepared for the Regulatory Innovations Staff, USEPA (March).

Krupnick, A. J. (1986), 'Costs of Alternative Policies for the Control of Nitrogen Dioxide in Baltimore', *Journal of Environmental Economics and Management*, 13, 189–97.

Liroff, R. A. (1980), *Air Pollution Offsets: Trading, Selling and Banking*, Washington, D.C., Conservation Foundation.

—— (1986), *Reforming Air Pollution Regulation: The Toil and Trouble of EPA's Bubble*, Washington D.C., Conservation Foundation.

Lyon, R. M. (1982), 'Auctions and Alternative Procedures for Allocating Pollution Rights', *Land Economics*, 58, 16–32.

McGartland, A. M. (1984), 'Marketable Permit Systems for Air Pollution Control: an Empirical Study', Ph.D. dissertation, University of Maryland.

Main, J. (1988), 'Here Comes the Big Cleanup', *Fortune*, 21 November, 102.

—— Maleug, David A. (1989), 'Emission Trading and the Incentive to Adopt New Pollution

Abatement Technology', *Journal of Environmental Economics and Management*, 16, 52–7.

Maloney, M. T. and Yandle, B. (1984), 'Estimation of the Cost of Air Pollution Control Regulation', *Journal of Environmental Economics and Management*, 11, 244–63.

Misiolek, W. S. and Elder, H. W. (1989), 'Exclusionary Manipulation of Markets for Pollution Rights', *Journal of Environmental Economics and Management*, 16, 156–66.

Montgomery, W. D. (1972), 'Markets in Licences and Efficient Pollution Control Programs', *Journal of Economic Theory*, 5, 395–418.

Oates, W. E., Portney, P. R., and McGartland, A. M. (1988), 'The Net Benefits of Incentive-Based Regulation: The Case of Environmental Standard Setting in the Real World', Resources for the Future Working Paper, December.

Opschoor, J. B. and Vos, H. B. (1989), *The Application of Economic Instruments for Environmental Protection in OECD Countries*, Paris, OECD.

Palmer, A. R., Mooz, W. E., Quinn, T. H., and Wolf, K. A. (1980), *Economic Implications of Regulating Chlorofluorocarbon Emissions from Nonaerosol Applications*, Report No. R–2524–EPA prepared for the US Environmental Protection Agency by the Rand Corporation, June.

Roach, F., Kolstad, C., Kneese, A. V., Tobin, R., and Williams, M. (1981), 'Alternative Air Quality Policy Options in the Four Corners Region', *Southwestern Review*, 1, 29–58.

Seskin, E. P., Anderson, R. J., Jr., and Reid, R. O. (1983), 'An Empirical Analysis of Economic Strategies for Controlling Air Pollution', *Journal of Environmental Economics and Management*, 10, 112–24.

Shapiro, M. and Warhit, E. (1983) 'Marketable Permits: The Case of Chlorofluorocarbons', *Natural Resource Journal*, 23, 577–91.

Spofford, W. O., Jr. (1984), 'Efficiency Properties of Alternative Source Control Policies for Meeting Ambient Air Quality Standards: An Empirical Application to the Lower Delaware Valley', Discussion paper D–118, Washington D.C., Resources for the Future, November.

Sprenger, R. U. (1989), 'Economic Incentives in Environmental Policies: The Case of West Germany', a paper presented at the Symposium on Economic Instruments in Environmental Protection Policies, Stockholm, Sweden (June).

Stahl, I., Bergman, L., and Mäler, K. G. (1988), 'An Experimental Game on Marketable Emission Permits for Hydro-carbons in the Gothenburg Area', Research Paper No. 6359, Stockholm School of Economics (December).

Stavins, R. N. (1989), 'Harnessing Market Forces to Protect the Environment', *Environment*, 31, 4–7, 28–35.

Terkla, D. (1984), 'The Efficiency Value of Effluent Tax Revenues', *Journal of Environmental Economics and Management*, 11, 107–23.

Tietenberg, T. H. (1980), 'Transferable Discharge Permits and the Control of Stationary Source Air Pollution: A Survey and Synthesis', *Land Economics*, 56, 391–416.

—— (1985), *Emissions Trading: An Exercise in Reforming Pollution Policy*, Washington, D.C., Resources for the Future.

—— (1988), *Environmental and Natural Resource Economics*, 2nd edn., Glenview, Illinois, Scott, Foresman and Company.

—— (1989a), 'Acid Rain Reduction Credits', *Challenge*, 32, 25–9.

—— (1989b), 'Marketable Permits in the U.S.: A Decade of Experience', in Karl W. Roskamp (ed.), *Public Finance and the Performance of Enterprises*, Detroit, MI, Wayne State University Press.

—— and Atkinson, S. E. (1989), 'Bilateral, Sequential Trading and the Cost-Effectiveness of the Bubble Policy', Colby College Working Paper (August).

Vos, H. B. (1989), 'The Application and Efficiency of Economic Instruments: Experiences in OECD Member Countries', a paper presented at the Symposium on Economic Instruments in Environmental Protection Policies, Stockholm, Sweden (June).

29

ECONOMIC PRESCRIPTIONS FOR ENVIRONMENTAL PROBLEMS: HOW THE PATIENT FOLLOWED THE DOCTOR'S ORDERS*

Robert W. Hahn

One of the dangers with ivory tower theorizing is that it is easy to lose sight of the actual set of problems which need to be solved, and the range of potential solutions. As one who frequently engages in this exercise, I can attest to this fact. In my view, this loss of sight has become increasingly evident in the theoretical structure underlying environmental economics, which often emphasizes elegance at the expense of realism.

In this paper, I will argue that both normative and positive theorizing could greatly benefit from a careful examination of the results of recent innovative approaches to environmental management. The particular set of policies examined here involves two tools which have received widespread support from the economics community: marketable permits and emission charges (Pigou, 1932; Dales, 1968; Kneese and Schultze, 1975). Both tools represent ways to induce businesses to search for lower cost methods of achieving environmental standards. They stand in stark contrast to the predominant "command-and-control" approach in which a regulator specifies the technology a firm must use to comply with regulations. Under highly restrictive conditions it can be shown that both of the economic approaches share the desirable feature that any gains in environmental quality will be obtained at the lowest possible cost (Baumol and Oates, 1975).

Until the 1960s, these tools only existed on blackboards and in academic journals, as products of the fertile imaginations of academics. However, some countries have recently begun to explore using these tools as part of a broader strategy for managing environmental problems.

This paper chronicles the experience with both marketable permits and emissions charges. It also provides a selective analysis of a variety of applications in Europe and the United States and shows how the actual use of these tools tends to depart from the role which economists have conceived for them.

* Originally published in *Journal of Economic Perspectives*, 1989, vol. 3, no. 2, pp. 95–114.

The Selection of Environmental Instruments

In thinking about the design and implementation of policies, it is generally assumed that policy makers can choose from a variety of "instruments" for achieving specified objectives. The environmental economics literature generally focuses on the selection of instruments that minimize the overall cost of achieving prescribed environmental objectives.

One instrument which has been shown to supply the appropriate incentives, at least in theory, is marketable permits. The implementation of marketable permits involves several steps. First, a target level of environmental quality is established. Next, this level of environmental quality is defined in terms of total allowable emissions. Permits are then allocated to firms, with each permit enabling the owner to emit a specified amount of pollution. Firms are allowed to trade these permits among themselves. Assuming firms minimize their total production costs, and the market for these permits is competitive, it can be shown that the overall cost of achieving the environmental standard will be minimized (Montgomery, 1972).

Marketable permits are generally thought of as a "quantity" instrument because they ration a fixed supply of a commodity, in this case pollution. The polar opposite of a quantity instrument is a "pricing" instrument, such as emissions charges. The idea underlying emissions charges is to charge polluters a fixed price for each unit of pollution. In this way, they are provided with an incentive to economize on the amount of pollution they produce. If all firms are charged the same price for pollution, then marginal costs of abatement are equated across firms, and this result implies that the resulting level of pollution is reached in a cost-minimizing way.

Economists have attempted to estimate the effectiveness of these approaches. Work by Plott (1983) and Hahn (1983) reveals that implementation of these ideas in a laboratory setting leads to marked increases in efficiency levels over traditional forms of regulation, such as setting standards for each individual source of pollution. The work based on simulations using actual costs and environmental data reveals a similar story. For example, in a review of several studies examining the potential for marketable permits, Tietenberg (1985, pp. 43–44) found that potential control costs could be reduced by more than 90 percent in some cases. Naturally, these results are subject to the usual cautions that a competitive market actually must exist for the results to hold true. Perhaps more importantly, the results assume that it is possible to easily monitor and enforce a system of permits or taxes. The subsequent analysis will suggest that the capacity to monitor and enforce can dramatically affect the choice of instruments.

Following the development of a normative theory of instrument choice, a handful of scholars began to explore reasons why environmental regulations are actually selected. This positive environmental literature tends to emphasize the potential winners and losers from environmental policies as a way of explaining the conditions under which we will observe such policies. For example, Buchanan and Tullock (1975) argue that the widespread use of source-specific standards rather than a fee can be explained by looking at the potential profitability of the affected industry under the two regimes. After presenting the various case studies, I will review some of the insights from positive theory and see how they square with the facts.

The formal results in the positive and normative theory of environmental economics are elegant. Unfortunately, they are not immediately applicable, since virtually none of the systems examined below exhibits the purity of the instruments which are the subject of theoretical inquiry. The presentation here highlights those instruments which show a marked resemblance to marketable permits or emission fees. Together, the two approaches to pollution control span a wide array of environmental problems, including toxic substances, air pollution, water pollution and land disposal.

Marketable Permits

In comparison with charges, marketable permits have not received widespread use. Indeed, there appear to be only four existing environmental applications; three of them in the United States. One involves the trading of emissions rights of various pollutants regulated under the Clean Air Act; a second involves trading of lead used in gasoline; a third addresses the control of water pollution on a river; and a fourth involves air pollution trading in Germany and will not be addressed here because of limited information (see Sprenger, 1986). These programs exhibit dramatic differences in performance, which can be traced back to the rules used to implement these approaches.

Emissions trading

By far the most significant and far-reaching marketable permit program in the United States is the emissions trading policy. Started over a decade ago, the policy attempts to provide greater flexibility to firms charged with controlling air pollutant emissions.[1] Because the program represents a radical departure in the approach to pollution regulation, it has come under close scrutiny by a variety of interest groups. Environmentalists have been particularly critical. These criticisms notwithstanding, the Environmental Protection Agency Administrator Lee Thomas (1986) characterized the program as "one of EPA's most impressive accomplishments."

Emissions trading has four distinct elements. Netting, the first program element, was introduced in 1974. Netting allows a firm which creates a new source of emissions in a plant to avoid the stringent emission limits which would normally apply by reducing emissions from another source in the plant. Thus, net emissions from the plant do not increase significantly. A firm using netting is only allowed to obtain the necessary emission credits from its own sources. This is called *internal trading*

because the transaction involves only one firm. Netting is subject to approval at the state level, not the federal.

Offsets, the second element of emissions trading, are used by new emission sources in "non-attainment areas." (A non-attainment area is a region which has not met a specified ambient standard.) The Clean Air Act specified that no new emission sources would be allowed in non-attainment areas after the original 1975 deadlines for meeting air quality standards passed. Concern that this prohibition would stifle economic growth in these areas prompted EPA to institute the offset rule. This rule specified that new sources would be allowed to locate in non-attainment areas, but only if they "offset" their new emissions by reducing emissions from existing sources by even larger amounts. The offsets could be obtained through internal trading, just as with netting. However, they could also be obtained from other firms directly, which is called *external trading*.

Bubbles, though apparently considered by EPA to be the centerpiece of emissions trading, were not allowed until 1979. The name derives from the placing of an imaginary bubble over a plant, with all emissions existing at a single point from the bubble. A bubble allows a firm to sum the emission limits from individual sources of a pollutant in a plant, and to adjust the levels of control applied to different sources as long as this aggregate limit is not exceeded. Bubbles apply to existing sources. The policy allows for both internal and external trades. Initially, every bubble had to be approved at the federal level as an amendment to a state's implementation plan. In 1981, EPA approved a "generic rule" for bubbles in New Jersey which allowed the state to give final approval for bubbles. Since then, several other states have followed suit.

Banking, the fourth element of emissions trading, was developed in conjunction with the bubble policy. Banking allows firms to save emission reductions above and beyond permit requirements for future use in

---

[1] Pollutants covered under the policy include volatile organic compounds, carbon monoxide, sulfur dioxide, particulates, and nitrogen oxides (Hahn and Hester, 1986).

Table 29.1 [*orig. table 1*]   Summary of emissions trading activity.

Activity	Estimated number of internal transactions	Estimated number of external transactions	Estimated cost savings (millions)	Environmental quality impact
Netting	5,000 to 12,000	None	$25 to $300 in Permitting costs: $500 to $12,000 in emission control costs	Insignificant in individual cases; Probably insignificant in aggregate
Offsets	1800	200	See text	Probably insignificant
Bubbles:				
Federally approved	40	2	$300	Insignificant
State approved	89	0	$135	Insignificant
Banking	<100	<20	Small	Insignificant

Source: Hahn and Hester (1986).

emissions trading. While EPA action was initially required to allow banking, the development of banking rules and the administration of banking programs has been left to the states.

The performance of emissions trading can be measured in several ways. A summary evaluation which assesses the impact of the program on abatement costs and environmental quality is provided in table 29.1. For each emissions trading activity, an estimate of cost savings, the environmental quality effect, and the number of trades is given. In each case, the estimates are for the entire life of the program. As can be seen from the table, the level of activity under various programs varies dramatically. More netting transactions have taken place than any other type, but all of these have necessarily been internal. The wide range placed on this estimate, 5000 to 12,000, reflects the uncertainty about the precise level of this activity. An estimated 2000 offset transactions have taken place, of which only 10 percent have been external. Fewer than 150 bubbles have been approved. Of these, almost twice as many have been approved by states under

generic rules than have been approved at the federal level, and only two are known to have involved external trades. For banking, the figures listed are for the number of times firms have withdrawn banked emission credits for sale or use. While no estimates of the exact numbers of such transactions can be made, upper bound estimates of 100 for internal trades and 20 for external trades indicate the fact that there has been relatively little activity in this area.

Cost savings for both netting and bubbles are substantial. Netting is estimated to have resulted in the most cost savings, with a total of between $525 million to over $12 billion from both permitting and emissions control cost savings.[2] By allowing new or modified sources to locate in areas that are highly polluted, offsets confer a major economic benefit on firms which use them. While the size of this economic benefit is not easily estimated, it is probably in the hundreds of millions of dollars. Federally approved bubbles have resulted in savings estimated at $300 million, while state bubbles have resulted in an estimated $135 million in cost savings. Average savings from federally

[2] The wide range of this estimate reflects the uncertainty which results from the fact that little information has been collected on netting.

approved bubbles are higher than those for state approved bubbles. Average savings from bubbles are higher than those from netting, which reflects the fact that bubble savings may be derived from several emissions sources in a single transaction, while netting usually involves cost savings at a single source. Finally, the cost savings from the use of banking cannot be estimated, but is necessarily small given the small number of banking transactions which have occurred.

The performance evaluation of emissions trading activities reveals a mixed bag of accomplishments and disappointments. The program has clearly afforded many firms flexibility in meeting emission limits, and this flexibility has resulted in significant aggregate cost savings – in the billions of dollars. However, these cost savings have been realized almost entirely from internal trading. They fall far short of the potential savings which could be realized if there were more external trading. While cost savings have been substantial, the program has led to little or no net change in the level of emissions.

Lead trading

Lead trading stands in stark contrast to the preceding marketable permit approaches. It comes by far the closest to an economist's ideal of a freely functioning market. The purpose of the lead trading program was to allow gasoline refiners greater flexibility during a period when the amount of lead in gasoline was being significantly reduced. (For a more detailed analysis of the performance of the lead trading program, see Hahn and Hester, 1987.)

Unlike many other programs, the lead trading program was scheduled to have a fixed life from the outset. Interrefinery trading of lead credits was permitted in 1982. Banking of lead credits was initiated in 1985. The trading program was terminated at the end of 1987. Initially, the period for trading was defined in terms of quarters. No

banking of credits was allowed. Three years after initiating the program limited banking was allowed, which allowed firms to carry over rights to subsequent quarters. Banking has been used extensively by firms since its initiation.

The program is notable for its lack of discrimination among different sources, such as new and old sources. It is also notable for its rules regarding the creation of credits. Lead credits are created on the basis of existing standards. A firm does not gain any extra credits for being a large producer of leaded gasoline in the past. Nor is it penalized for being a small producer. The creation of lead credits is based solely on current production levels and average lead content. For example if the standard were 1.1 grams per gallon, and a firm produces 100 gallons of gasoline, it would receive rights entitling it to produce or sell up to 110 (100 x 1.1) grams of lead. To the extent that current production levels are correlated with past production levels, the system acknowledges the existing distribution of property rights. However, this linkage is less explicit than those made in other trading programs.[3]

The success of the program is difficult to measure directly. It appears to have had very little impact on environmental quality. This is because the amount of lead in gasoline is routinely reported by refiners and is easily monitored. The effect the program has had on refinery costs is not readily available. In proposing the rule for banking of lead rights, EPA estimated that resulting savings to refiners would be approximately $228 million (U.S. EPA, 1985a). Since banking activity has been somewhat higher than anticipated by EPA, it is likely that actual cost savings will exceed this amount. No specific estimate of the actual cost savings resulting from lead trading have been made by EPA.

The level of trading activity has been high, far surpassing levels observed in other environmental markets. In 1985, over half of the refineries participated in trading. Approx-

---

[3] One of the reasons EPA set up the allocation rule in this way was to try to transfer some of the permit rents from producers to consumers. This will not always occur, however, and depends on the structure of the permits market as well as the underlying production functions.

imately 15 percent of the total lead rights used were traded. Approximately 35 percent of available lead rights were banked for future use or trading (U.S. EPA, 1985b, 1986). In comparison, volumes of emissions trading have averaged well below 1 percent of the potential emissions that could have been traded.

From the standpoint of creating a workable regulatory mechanism that induces cost savings, the lead market has to be viewed as a success. Refiners, though initially lukewarm about this alternative, have made good use of this program. It stands out amidst a stream of incentive-based programs as the "noble" exception in that it conforms most closely to the economists' notion of a smoothly functioning market.

Given the success of this market in promoting cost savings over a period in which lead was being reduced, it is important to understand why the market was successful. The lead market had two important features which distinguished it from other markets in environmental credits. The first was that the amount of lead in gasoline could be easily monitored with the existing regulatory apparatus. The second was that the program was implemented after agreement had been reached about basic environmental goals. In particular, there was already widespread agreement that lead was to be phased out of gasoline. This suggests that the success in lead trading may not be easily transferred to other applications in which monitoring is a problem, or environmental goals are poorly defined. Nonetheless, the fact that this market worked well provides ammunition for proponents of market-based incentives for environmental regulation.

Charges in Practice

Charge systems in three countries are examined. Examples are drawn from France, the Netherlands, and the United States. Particular systems were selected because they were thought to be significant either in their scope, their effect on revenues, or their impact on the cost effectiveness of environmental regulation. While the focus is on water effluent charges, a variety of systems are briefly mentioned at the end of this section which cover other applications.

Charges in France

The French have had a system of effluent charges on water pollutants in place since 1969 (Bower et al., 1981). The system is primarily designed to raise revenues which are then used to help maintain or improve water quality. Though the application of charges is widespread, they are generally set at low levels.[4] Moreover, charges are rarely based on actual performance. Rather, they are based on the expected level of discharge by various industries. There is no explicit connection between the charge paid by a given discharger and the subsidy received for reducing discharges (Bower et al., 1981, p. 126). However, charges are generally earmarked for use in promoting environmental quality in areas related to the specific charge. The basic mechanism by which these charges improve environmental quality is through judicious earmarking of the revenues for pollution abatement activities.

In evaluating the charge system, it is important to understand that it is a major, but by no means dominant, part of the French system for managing water quality. Indeed, in terms of total revenues, a sewage tax levied on households and commercial enterprises is larger in magnitude (Bower et al., 1981, p. 142). Moreover, the sewage tax is assessed on the basis of actual volumes of water used. Like most other charge systems, the charge system in France is based on a system of water quality permits, which places constraints on the type and quantity of effluent a firm may discharge. These permits are required for sources discharging more than some specified quantity (Bower et al., 1981, p. 130).

Charges now appear to be accepted as a way of doing business in France. They provide a significant source of revenues for

---

[4] Charges cover a wide variety of pollutants, including suspended solids, biological oxygen demand, chemical oxygen demand, and selected toxic chemicals.

water quality control. One of the keys to their initial success appears to have been the gradual introduction and raising of charges. Charges started at a very low level and were gradually raised to current levels (Bower et al., 1981, p. 22). Moreover, the pollutants on which charges are levied has expanded considerably since the initial inception of the charge program.[5]

Charges in the Netherlands

The Netherlands has had a system of effluent charges in place since 1969 (Brown and Bressers, 1986, p. 4). It is one of the oldest and best administered charge systems, and the charges placed on effluent streams are among the highest. In 1983, the effluent charge per person was $17 in the Netherlands, $6 in Germany, and about $2 in France (Brown and Bressers, 1986, p. 5). Because of the comparatively high level of charges found in the Netherlands, this is a logical place to examine whether charges are having a discernible effect on the level of pollution. Bressers (1983), using a multiple regression approach, argues that charges have made a significant difference for several pollutants. This evidence is also buttressed by surveys of industrial polluters and water board officials which indicate that charges had a significant impact on firm behavior (Brown and Bressers, 1986, pp. 12–13). This analysis is one of the few existing empirical investigations of the effect of effluent charges on resulting pollution.

The purpose of the charge system in the Netherlands is to raise revenue that will be used to finance projects that will improve water quality (Brown and Bressers, 1986, p. 4). Like its counterparts in France and Germany, the approach to managing water quality uses both permits and effluent charges for meeting ambient standards (Brown and Bressers, 1986, p. 2).[6] Permits tend to be uniform across similar discharges. The system is designed to ensure that water quality will remain the same or get better

(Brown and Bressers, 1986, p. 2). Charges are administered both on volume and concentration. Actual levels of discharge are monitored for larger polluters, while small polluters often pay fixed fees unrelated to actual discharge (Bressers, 1983, p. 10).

Charges have exhibited a slow but steady increase since their inception (Brown and Bressers, 1986, p. 5). This increase in charges has been correlated with declining levels of pollutants. Effluent discharge declined from 40 population equivalents in 1969 to 15.3 population equivalents in 1980, and it was projected to decline to 4.4 population equivalents in 1985 (Brown and Bressers, 1986, p. 10). Thus, over 15 years, this measure of pollution declined on the order of 90 percent.

As in Germany, there was initial opposition from industry to the use of charges. Brown and Bressers (1986, p. 4) also note opposition from environmentalists, who tend to distrust market-like mechanisms. Nonetheless, charges have enjoyed widespread acceptance in a variety of arenas in the Netherlands.

One final interesting feature of the charge system in the Netherlands relates to the differential treatment of new and old plants. In general, newer plants face more stringent regulation than older plants (Brown and Bressers, 1986, p. 10). As we shall see, this is also a dominant theme in American regulation.

Charges in the United States

The United States has a modest system of user charges levied by utilities that process wastewater, encouraged by federal environmental regulations issued by the Environmental Protection Agency. They are based on both volume and strength, and vary across utilities. In some cases, charges are based on actual discharges, and in others, as a rule of thumb, they are related to average behavior. In all cases, charges are added to the existing regulatory system

---

[5] For example, Brown (1984, p. 114) notes that charges for nitrogen and phosphorous were added in 1982.

[6] Emission and effluent standards apply to individual sources of pollution while ambient standards apply to regions such as a lake or an air basin.

which relies heavily on permits and standards.

Both industry and consumers are required to pay the charges. The primary purpose for the charges is to raise revenues to help meet the revenue requirements of the treatment plants, which are heavily subsidized by the federal government. The direct environmental and economic impact of these charges is apparently small (Boland, 1986, p. 12). They primarily serve as a mechanism to help defray the costs of the treatment plants. Thus, the charges used in the United States are similar in spirit to the German and French systems already described. However, their size appears to be smaller, and the application of the revenues is more limited.

Other fee-based systems and lessons

There are a variety of other fee-based systems which have not been included in this discussion. Brown (1984) did an analysis of incentive-based systems to control hazardous wastes in Europe and found that a number of countries had adopted systems, some of which had a marked economic effect. The general trend was to use either a tax on waste outputs or tax on feedstocks that are usually correlated with the level of waste produced. Companies and government officials were interviewed to ascertain the effects of these approaches. In line with economic theory, charges were found to induce firms to increase expenditures on achieving waste reduction through a variety of techniques including reprocessing of materials, treatment, and input and output substitution. Firms also devoted great attention to separating waste streams because prices for disposal often varied by the type of waste stream.

The United States has a diverse range of taxes imposed on hazardous waste streams. Several states have land disposal taxes in place. Charges exhibit a wide degree of variation across states. For example, in 1984, charges were $14/tonne in Wisconsin and $70.40/tonne in Minnesota (U.S. CBO, 1985,

p. 82). Charges for disposal at landfills also vary widely. The effect of these different charges is very difficult to estimate because of the difficulty in obtaining the necessary data on the quantity and quality of waste streams, as well as the economic variables.

The preceding analysis reveals that there are a wide array of fee-based systems in place designed to promote environmental quality. In a few cases, the fees were shown to have a marked effect on firm behavior; however, in the overwhelming majority of cases studied, the direct economic effect of fees appears to have been small. Several patterns repeat themselves through these examples.

First, the major motivation for implementing emission fees is to raise revenues, which are then usually earmarked for activities which promote environmental quality.[7] Second, most charges are not large enough to have a dramatic impact on the behavior of polluters. In fact, they are not designed to have such an effect. They are relatively low and not directly related to the behavior of individual firms and consumers. Third, there is a tendency for charges to increase faster than inflation over time. Presumably, starting out with a relatively low charge is a way of testing the political waters as well as determining whether the instrument will have the desired effects.

Implementing Market-based Environmental Programs

An examination of the charge and marketable permits schemes reveals that they are rarely, if ever, introduced in their textbook form. Virtually all environmental regulatory systems using charges and marketable permits rely on the existing permitting system. This result should not be terribly surprising. Most of these approaches were not implemented from scratch; rather, they were grafted onto regulatory systems in which permits and standards play a dominant role.

Perhaps as a result of these hybrid

---

[7] The actual application of fees is similar in spirit to the more familiar deposit-refund approaches that are used for collecting bottles and cans.

approaches, the level of cost savings resulting from implementing charges and marketable permits is generally far below their theoretical potential. Cost savings can be defined in terms of the savings which would result from meeting a prescribed environmental objective in a less costly manner. As noted, most of the charges to date have not had a major incentive effect. We can infer from this that polluters have not been induced to search for a lower cost mix of meeting environmental objectives as a result of the implementation of charge schemes. Thus, it seems unlikely that charges have performed terribly well on narrow efficiency grounds. The experience on marketable permits is similar. Hahn and Hester (1986) argue that cost savings for emissions trading fall far short of their theoretical potential. The only apparent exception to this observation is the lead trading program, which has enjoyed very high levels of trading activity.

The example of lead trading leads to another important observation; in general, different charge and marketable permit systems exhibit wide variation in their effect on economic efficiency. On the whole, there is more evidence for cost savings with marketable permits than with charges.

While the charge systems and marketable permit systems rarely perform well in terms of efficiency, it is important to recognize that their performance is broadly consistent with economic theory. This observation may appear to contradict what was said earlier about the departure of these systems from the economic ideal. However, it is really an altogether different observation. It suggests that the performance of the markets and charge systems can be understood in terms of basic economic theory. For example, where barriers to trading are low, more trading is likely to occur. Where charges are high and more directly related to individual actions, they are more likely to affect the behavior of firms or consumers.

The data from the examples given earlier can be used to begin to piece together some of the elements of a more coherent theory of instrument choice. For example, it is clear that distributional concerns play an important role in the acceptability of user charges. The revenue from such charges is usually earmarked for environmental activities related to those contributions. Thus, charges from a noise surcharge will be used to address noise pollution. Charges for water discharges will be used to construct treatment plants and subsidize industry in building equipment to abate water pollution. This pattern argues that different industries want to make sure that their contributions are used to address pollution problems for which they are likely to be held accountable. Thus, industry sees it as only fair that, as a whole, they get some benefit from making these contributions.

The "recycling" of revenues from charges points up the importance of the existing distribution of property rights. This is also true in the case of marketable permits. The "grandfathering" of rights to existing firms based on the current distribution of rights is an important focal point in many applications of limited markets in pollution rights (Rolph, 1983; Welch, 1983). All the marketable permit programs in the United States place great importance on the existing distribution of rights.

In short, all of the charge and marketable permit systems described earlier place great importance on the status quo. Charges, when introduced, tend to be phased in. Marketable permits, when introduced, usually are optional in the sense that existing firms can meet standards through trading of permits or by conventional means. In contrast, new or expanding firms are not always afforded the same options. For example, new firms must still purchase emission credits if they choose to locate in a non-attainment area, even if they have purchased state-of-the-art pollution control equipment and will pollute less than existing companies. This is an example of a "bias" against new sources. While not efficient from an economic viewpoint, this pattern is consistent with the political insight that new sources don't "vote" while existing sources do.

The review of marketable permits and charge systems has demonstrated that regulatory systems involving multiple instruments are the rule rather than the exception. The fundamental problem is to determine the most appropriate mix, with an eye to both economic and political realities.

In addition to selecting an appropriate mix of instruments, attention needs to be given to the effects of having different levels of government implement selected policies. It might seem, for example, that if the problem is local, then the logical choice for addressing the problem is the local regulatory body. However, this is not always true. Perhaps the problem may require a level of technical expertise that does not reside at the local level, in which case some higher level of government involvement may be required. What is clear from a review of implementing environmental policies is that the level of oversight can affect the implementation of policies. For example, Hahn and Hester (1986) note that a marked increase in bubble activity is associated with a decrease in federal oversight.

Because marketable permit approaches have been shown to have a demonstrable effect on cost savings without sacrificing environmental quality, this instrument can be expected to receive more widespread use. One factor which will stimulate the application of this mechanism is the higher marginal costs of abatement that will be faced as environmental standards are tightened. A second factor which will tend to stimulate the use of both charges and marketable permits is a "demonstration effect." Several countries have already implemented these mechanisms with some encouraging results. The experience gained in implementing these tools will stimulate their use in future applications. A third factor which will affect the use of both of these approaches is the technology of monitoring and enforcement. As monitoring costs go down, the use of mechanisms such as direct charges and marketable permits can be expected to increase. The combination of these factors leads to the prediction that greater use of these market-based environmental systems will be made in the future.

References

Barde, J., "Use of Economic Instruments for Environmental Protection: Discussion Paper," ENV/ECO/86.16, Organization for Economic Cooperation and Development, September 9, 1986.

Baumol, W. and Oates, W., *The Theory of Environmental Policy*. Englewood Cliffs, NJ: Prentice-Hall, 1985.

Becker, G., "A Theory of Competition Among Pressure Groups for Political Influence," *Quarterly Journal of Economics*, 1983, XCVII, 371–400.

Boland, J., "Economic Instruments for Environmental Protection in the United States," ENV/ECO/86.14, Organization for Economic Cooperation and Development, September 11, 1986.

Bower, B. et al., *Incentives in Water Quality Management: France and the Ruhr Area*. Washington, D.C.: Resources for the Future, 1981.

Bressers, J., "The Effectiveness of Dutch Water Quality Policy," Twente University of Technology, Netherlands, mimeo, 1983.

Brown, G., Jr., "Economic Instruments: Alternatives or Supplements to Regulations?" *Environment and Economics*, Issue Paper, Environment Directorate OECD, June 1984, 103–120.

Brown, G., Jr. and J. Bressers, "Evidence Supporting Effluent Charges," Twente University of Technology, The Netherlands, mimeo, September 1986.

Brown, G., Jr. and R. Johnson, "Pollution Control by Effluent Charges: It Works in the Federal Republic of Germany, Why Not in the U.S.," *Natural Resources Journal*, 1984, 24, 929–966.

Buchanan, J. and G. Tullock, "Polluters' Profits and Political Response: Direct Controls Versus Taxes," *American Economic Review*, 1975, 65, 139–147.

Campos, J., "Toward a Theory of Instrument Choice in the Regulation of Markets," California Institute of Technology, Pasadena, California, mimeo, January 26, 1987.

Coelho, P., "Polluters and Political Response: Direct Control Versus Taxes: Comment," *American Economic Review*, 1976, 66, 976–978.

Dales, J., *Pollution, Property and Prices*. Toronto: University Press, 1968.

David, M. and E. Joeres, "Is a Viable Implementation of TDPs Transferable?" In Joeres, E. and M. David, eds., *Buying a Better Environment: Cost-Effective Regulation Through Permit Trading*. Madison: University of Wisconsin Press, 1983, 233–248.

Dewees, D., "Instrument Choice in Environmental Policy," *Economic Inquiry*, 1983, *XXI*, 53–71.

Elmore, T. et al., "Trading Between Point and Nonpoint Sources: A Cost Effective Method for Improving Water Quality," paper presented at the 57th annual Conference/Exposition of the Water Pollution Control Federation, New Orleans, Louisiana, 1984.

Hahn, R., "Designing Markets in Transferable Property Rights: A Practitioner's Guide." In Joeres, E. and M. David, eds., *Buying a Better Environment: Cost Effective Regulation Through Permit Trading.* Madison: University of Wisconsin Press, 1983, 83–97.

Hahn, R., "Rules, Equality and Efficiency: An Evaluation of Two Regulatory Reforms," Working Paper 87-7, School of Urban and Public Affairs, Carnegie Mellon University, Pittsburgh, Pennsylvania, 1987.

Hahn, R. and G. Hester, "Where Did All the Markets Go?: An Analysis of EPA's Emission Trading Program," Working Paper 87-3, School of Urban and Public Affairs, Carnegie Mellon University, Pittsburgh, Pennsylvania, 1986. Forthcoming in the *Yale Journal on Regulation.*

Hahn, R. and G. Hester, "Marketable Permits: Lessons for Theory and Practice," *Ecology Law Quarterly*, forthcoming.

Hahn, R. and A. McGartland, "The Political Economy of Instrument Choice: An Examination of the U.S. Role in Implementing the Montreal Protocol," Working Paper 88-34, School of Urban and Public Affairs, Carnegie Mellon University, Pittsburgh, Pennsylvania, 1988.

Hahn, R. and Noll, R., "Designing a Market for Tradable Emissions Permits," In Magat, W. ed., *Reform of Environmental Regulation.* Cambridge, MA: Ballinger, 1982, 119–146.

Hahn, R. and Noll, R., "Barriers to Implementing Tradable Air Pollution Permits: Problems of Regulatory Interaction," *Yale Journal on Regulation*, 1983, *I*, 63–91.

Kashmanian, R. et al., "Beyond Categorical Limits: The Case for Pollution Reduction Through Trading," paper presented at the 59th Annual Water Pollution Control Federation Conference, Los Angeles, CA, October 6–9, 1986.

Kneese, A. and Schultze, C., *Pollution, Prices, and Public Policy.* Washington, D.C.: The Brookings Institution, 1975.

Montgomery, W. D., "Markets in Licenses and Efficient Pollution Control Programs," *Journal of Economic Theory*, 1972, *5*, 395–418.

Noll, R., "The Political Foundations of Regulatory Policy," *Zeitschrift fur die gesamte Staatswissenschaft*, 1983, *139*, 377–404.

Novotny, G., "Transferable Discharge Permits for Water Pollution Control In Wisconsin," Department of Natural Resources, Madison, Wisconsin, mimeo, December 1, 1986.

O'Neil, W., "The Regulation of Water Pollution Permit Trading under Conditions of Varying Streamflow and Temperature." In Joeres, E. and M. David, eds., *Buying a Better Environment: Cost-Effective Regulation Through Permit Trading.* Madison, Wisconsin: University of Wisconsin Press, 1983, 219–231.

Opschoor, J., "Economic Instruments for Environmental Protection in the Netherlands," ENV/ECO/86.15, Organization for Economic Cooperation and Development, August 1, 1986.

Panella, G., "Economic Instruments for Environmental Protection in Italy," ENV/ECO/86.11, Organization for Economic Cooperation and Development, September 2, 1986.

Patterson, D., Bureau of Water Resources Management, Wisconsin Department of Natural Resources, Madison, Wisconsin, telephone interview, April 2, 1987.

Pigou, A., *The Economics of Welfare*, fourth edition, London: Macmillan and Co., 1932.

Plott, C., "Externalities and Corrective Policies in Experimental Markets," *Economic Journal*, 1983, *93*, 106–127.

Rolph, E., "Government Allocation of Property Rights: Who Gets What?," *Journal of Policy Analysis and Management*, 1983, *3*, 45–61.

Sprenger, R. "Economic Instruments for Environmental Protection in Germany," Organization for Economic Cooperation and Development, October 7, 1986.

Thomas, L., memorandum attached to Draft Emissions Trading Policy Statement, Environmental Protection Agency, Washington, D.C., May 19, 1986.

Tietenberg, T., *Emissions Trading: An Exercise in Reforming Pollution Policy*, Washington, D.C.: Resources for the Future, 1985.

U.S. Congressional Budget Office, *Hazardous Waste Management: Recent Changes and Policy Alternatives*, Washington, D.C.: U.S. G.P.O., May 1985.

U.S. Environmental Protection Agency, "Costs and Benefits of Reducing Lead in Gasoline, Final Regulatory Impact Analysis," Office of Policy Analysis, February 1985a.

U.S. Environmental Protection Agency, "Quarterly Reports on Lead in Gasoline," Field Operations and Support Division, Office of Air and Radiation, July 16, 1985b.

U.S. Environmental Protection Agency, "Quarterly Reports on Lead in Gasoline," Field Operations and Support Division, Office of Air and Radiation, March 21, May 23, July 15, 1986.

Welch, W., "The Political Feasibility of Full Ownership Property Rights: The Cases of Pollution and Fisheries," *Policy Sciences*, 1983, *16*, 165–180.

Yohe, G., "Polluters' Profits and Political Response: Direct Control Versus Taxes: Comment," *American Economic Review*, 1976, *66*, 981–982.

30

SUSTAINABILITY: PRINCIPLES AND PRACTICE*

R. K. Turner

Principles: Introduction

A large and diverse literature has emerged in recent years concerned with the notion of sustainable development (SD)(1). Many definitions (often incompatible) have been suggested and debated, thereby exposing a range of approaches linked to different world views (Pearce et al. 1989; Pearce and Turner 1990).

From the ecocentric perspective, the extreme deep ecologists seem to come close to rejecting even the sustainable utilisation of nature's assets. According to some of their writings, the environment ought not to be conceived of as a collection of goods and services for human use at all (Rolston 1988). From the opposite technocentric perspective, other analysts argue that the concept of sustainability contributes little new to mainstream (largely neo-classical economic) approaches to intertemporal choice. Given this world view, the maintenance of a sustainable growth economy over the long run depends on the adequacy of investment expenditure. First and foremost, it is investment in physical and human capital (i.e. buildings, machines, etc., plus the stock of knowledge) that counts and only to a much lesser degree investment in natural capital (i.e. the stock of non-renewable and renewable resources provided by the bisophere and the opportunities for solar-power recycling) (Nordhaus 1992). A key assumption of this position is that there will continue to be a very high degree of substitutability between all forms of capital resources.

In a typology, outlined in a later subsection of this chapter, these two polar positions have been labelled respectively the 'very strong sustainability' (VSS) and the 'very weak sustainability' (VWS) positions (Turner 1992).(2) But defining sustainable development is not the only, and probably not the most important, problem. If the sustainability goal is accepted then a fundamental requirement is a set of sustainability principles that can give some concrete form to a sustainable development strategy. This strategy will necessarily have to encompass multiple and interrelated goals (reflecting the several dimensions of the sustainability concept) – social/cultural, economic, political, environmental and moral – and will have to deploy a package of enabling policy instruments.

The most publicised definition of sustainability is that of the World Commission on

* Originally published in Turner, R. K. (ed.) (1993) *Sustainable Environmental Economics and Management: Principles and Practice*, London: Belhaven.

Environment and Development (WCED) (the Brundtland Commission). The Commission defined SD as 'development that meets the needs of the present without compromising the ability of future generations to meet their own needs' (WCED 1987, 43). Both an equity dimension (intragenerational and intergenerational) and a social/psychological dimension (i.e. the term 'needs' is used rather than the economic term 'wants', which is tied into the concept of consumer sovereignty) are clearly highlighted by this definition. If a society accepts the desirability of the goal of SD then it must develop economically and socially in such a way that it minimises the effects of its activities, the costs of which are borne by future generations. In cases where the activities and significant effects are unavoidable, future generations must be compensated for any costs they incur.(3)

The Brundtland Commission also highlighted 'the essential needs of the world's poor, to which overriding priority should be given', and 'the idea of limitations imposed by the state of technology and social organisation on the environment's ability to meet present and future'. The central rationale for SD is therefore to increase people's standard of living (broadly defined) and, in particular, the well-being of the least advantaged people in societies, while at the same time avoiding uncompensated future costs.

So how does the environment fit this requirement? According to Norton (1992), while the environment is mentioned it is given a passive role in the Commission's analysis: it 'does not impose any non-negotiable limits on sustainable use, independent of limitations on the abilities of humans to control it'. Economic growth and environmental protection are therefore at least potentially compatible objectives. The Brundtland Commission and similar viewpoints are anchored to a greater or lesser extent to the technocentric world view and therefore advocate some version of a weak sustainability-type approach.

Advocates of the strong sustainability approach (linked to or at least influenced by elements of ecocentrism) would take issue (to varying degrees) with the assumption of almost infinite substitutability of resources and that projections of economic growth and levels of social well-being can be forecast without accounting for the *scale* of human activities and its implications for the health and integrity of ecosystems.(4) In the 1950s the Dutch economist Jan Tinbergen formulated what has become known as the 'Tinbergen rule'. Simply stated, the rule lays down that for every independent policy goal there must be a complimentary independent enabling policy instrument (Tinbergen 1952). One of the best-known advocates of the 'strong sustainability' viewpoint, Herman Daly, has introduced this 'targets and instruments' rule into the SD debate.

For Daly (1992) any environmental economic analysis of development is fatally flawed unless resource allocation and distribution objectives (which are supported by actual policy instruments) are augmented by a consideration of the 'scale' question.(5) 'Scale' is defined in a 'materials-balance sense'(6) in terms of the throughput of matter/energy in an economic system. It is the product of population times per-capita resource use. Because of the laws of thermodynamics, 'useful' resource inputs (matter/energy) are drawn into the economy and at a later stage reintroduced into the environment as relatively 'useless' waste products. The scale of economic activity should be related to the natural capacities of ecosystems to regenerate resource inputs for the economy and to assimilate (absorb and/or store without long-term damage) the waste flows from the economy. A desirable scale for economic activity would be one that does not erode the environmental carrying capacity over time. What is missing is the appropriate set of policy instruments to regulate the scale of economies. Severance taxes linked to natural capital depletion, and, in cases where uncertainty is great, assurance bonds tied to resource developments, have been suggested (Page 1977; Costanza and Perrings 1990; Costanza and Daly 1992).

Sustainable development, it is generally agreed, is therefore economic development that endures over the long run.(7) Economic development could be narrowly defined in traditional terms as real gross national product (GNP) per capita, or real consumption per capita. Alternatively, the traditional measures can be modified and extended to include a more comprehensive set of welfare indicators – education, health, quality of life, etc. Failure adequately to account for natural capital and the contribution it makes to economic welfare will also lead to misperceptions about how well an economy is really performing (Costanza and Daly 1992). A framework to reflect the use of natural resources at the national level is in the process of being agreed by the United Nations Statistical Office, but there is much debate as to the feasibility of the proposed alterations and other alternative amendments (Bartelmus et al. 1991; Bryant and Cook 1992; Daly and Cobb 1989; Nordhaus 1992).

SD now becomes fairly simply defined as, at least, non-declining consumption per capita, or per unit of GNP or some alternative agreed welfare indicator(s). This is how SD has come to be interpreted by a majority of economists addressing the issue (Pearce et al. 1990; Pezzey 1989; Mäler 1991a). Nevertheless, determining the necessary and sufficient conditions for achieving SD is an altogether more complicated task, as we have hinted in previous paragraphs. The so-called 'London School' of environmental economists, among others, has argued that a non-declining stock of natural capital over time is a necessary condition for sustainability, because of substitutability limits in production processes as well as other factors (Pearce et al. 1990; Pearce and Turner 1990). This is a strong sustainability (SS) position, with natural capital fulfilling the role of the fair/just compensatory bequest to future generations. To meet the requirements of SD the future must actually be compensated for any impairment of well-being caused by actions and activities engaged in by the current generation. The mechanism by which the current generation can ensure that the future is compensated, and is not therefore worse off, is through the transfer of capital bequests. Capital provides the opportunity and capability to generate well-being through the creation of material goods

and amenity and other services that give human life its meaning.(8)

The weak sustainability position (WS) does not single out the environment for special treatment, it is simply another form of capital (natural capital). Therefore, what is required under SD is the transfer of an aggregate capital stock no less than the one that exists now. WS is, as we pointed out earlier, based on a very strong principle of perfect substitutability between the different forms of capital.(9)

Some Salient Elements in the SD Debate

Intragenerational and intergenerational equity

SD is future-orientated in that it seeks to ensure that the future generations are at least as well off, on a welfare basis, as current generations. It is therefore in economic terms a matter of intergenerational equity and not just efficiency. The distribution of rights and assets across generations determines whether the efficient allocation of resources sustains welfare across human generations (Howarth and Norgaard 1992). The ethical argument is that future generations have the right to expect an inheritance sufficient to allow them the capacity to generate for themselves a level of welfare no less than that enjoyed by the current generation. What is required, then, is some sort of intergenerational social contract.

Specification of the sustainability inheritance asset portfolio

Victor (1991) has recently remarked that one of the contributions that economists have made to the SD debate has been the idea that the depletion of environmental resources (source and sink resources) in pursuit of economic growth is akin to living off capital rather than income. SD is then defined as the maximum development that can be achieved without running down the capital assets of the nation, which are its resource base. The base is interpreted widely to encompass man-made capital K_m, natural capital K_n, human capital K_h and moral

Table 30.1 [*orig. table 1*]   Sustainability rules and indicators.

	No critical natural capital	Critical natural capital
Very weak sustainability	$s/y - \delta k/y > 0$	Perfect substitution All K_n and K_m Growth economy
Weak sustainability	$s/y - \delta m/y - \delta n/y = WSI$ $WSI > 0$ $\lambda > h$ $n > Z$	$WSI > 0$ $\lambda > h$ $n > Z$ $\delta n^* \leq 0$
Strong sustainability	$\delta n \leq 0$ $WSI > 0$	$WSI > 0$ $\delta n \leq 0$ $\delta n^* \leq 0$ $\delta K_c \leq 0$
Very strong sustainability	Perfect Complementarity All K_n and K_m Stationary-state economy	$WSI > 0$ $\delta n \leq 0$ $\delta n^* \leq$ $n \leq 0$ $\delta K_c \leq 0$ $\delta K_e \leq 0$

Notes:

K	=	total capital assets
s	=	savings
δm	=	depreciation on man-made capital
δn	=	depreciation on natural capital
λ	=	technical change
h	=	rate of population growth
n^*	=	critical natural capital (no substitutes)
K_c	=	cultural capital
K_e	=	moral/ethical capital
Z	=	lower bound stock limit (determined by SMS) to ensure ecosystem stability

(ethical) and cultural capital K_c. Victor identifies four 'schools' of thought on the 'environmental as capital' issue: the mainstream neoclassical school; the London School (after Pearce, Barbier, Markandya and Turner); the post-Keynsian school; and the thermodynamic school (after Boulding, Georgescu-Roegen and Daly). Roughly speaking, this spectrum of views moves from a position we can label 'very weak sustainability' through to one we call 'very strong sustainability' (see also Klassen and Opschoor 1990). Table 30.1 formalises in simplistic fashion these various sustainability paradigms, which in practice are less clearly defined and are overlapping.

Very weak sustainability (Solow sustainability)

This VWS rule merely requires that the overall stock of capital assets ($K_m + K_n + K_h$) should remain constant over time. The rule is, however, consistent with any one asset being reduced as long as another capital asset is increased to compensate. This approach to sustainability is based on a Hicksian definition of income, the principle of constant consumption (buttressed by a Rawlsian maximin justice rule operating intergenerationally),(11) on production functions with complete substitution properties, and the Hartwick rule governing the reinvestment of resource rents (Common and Perrings 1992).

Thus, following Hicks, income is the

maximum real consumption expenditure that leaves society as well endowed at the end of a period as at the start. The definition, therefore, presupposes the deduction of expenditures to compensate for the depreciation or degradation of the total capital asset base that is the source of the income generations, i.e. conservation of the value of the asset base. Assuming a homogeneous capital stock (perfect substitution possibilities) the Hartwick rule states that consumption may be held constant in the face of exhaustible resources if and only if rents deriving from the intertemporally efficient use of those resources are reinvested in reproducible capital.

It is now possible to derive an intuitive weak sustainability measure or indicator (in value terms) for determining whether a country is on or off a sustainable development path (Pearce and Atkinson 1992). Thus a nation cannot be said to be sustainable if it fails to save enough to offset the depreciation of its capital assets. That is,

$$WSI > 0 \quad if \, S > \delta K \quad\quad\quad (1.1)$$

where WSI is a sustainability index, S is savings and δK is depreciation on capital. Dividing through by income (Y) we have

$$WSI > 0 \quad if \, (S/Y) > (\delta K/Y) \quad\quad (1.2)$$

or

$$WSI > 0 \quad if \, (S/Y) > [\delta m/Y + (\delta n/Y)] \quad (1.3)$$

where δm is depreciation on man-made capital, δn is depreciation on natural capital and K_n and K_m are substitutable.

Weak sustainability (modified Solow sustainability)

Perrings (1991) and Common and Perrings (1992) highlight the fact that the technological assumptions (substitution possibilities) of the weak sustainability approach violate scientific understanding of the evolution of thermodynamic systems, and ecological thinking about the complementarity of resources in system structure and the importance of diversity in system resilience.

The London School has also modified the VWS approach by introducing into the analysis an upper bound on the assimilative capacity assumption, as well as a lower bound on the level of K_n stocks necessary to support SD assumption, into the analysis (Barbier and Markandya 1989; Pearce and Turner 1990; Klassen and Opschoor 1990). The concept of critical natural capital (e.g., keystone species and keystone processes) has also been introduced to account for the non-substitutability of certain types of natural capital (K_n) – such as environmental support services – and man-made capital (K_m). Thus the requirement for the conservation of the value of the capital stock has been buttressed by constraints aimed at the preservation of some proportion and/or components of K_n stock in physical terms.

The implications of this modified Solow sustainability thinking seems to be the formulation of a sustainability constraint which will impose some degree of restriction on resource-using economic activities. The constraint will be required to maintain populations/resource stocks within bounds thought to be consistent with ecosystem stability and resilience. To maintain the instrumental value (benefits) humans obtain from healthy ecosystems, the concern is not preservation of specific attributes of the ecological community but rather the management of the system to meet human needs, support species and genetic diversity, and enable the system to adapt (resilience) to changing conditions.

A set of physical indicators will be required in order to monitor and measure biodiversity and ecosystem resilience. As yet there is no scientific consensus over how biodiversity should be measured. Measuring genetic diversity presents the least difficulty, but measuring species diversity is more problematic – measures of species richness, taxonomic richness, richness of genera or families have all to be investigated and none are without difficulties. Problems associated with measuring biodiversity at the community level are even greater but such measures would be very useful to policy-makers if the aim is conservation at the ecosystem level. Community classifica-

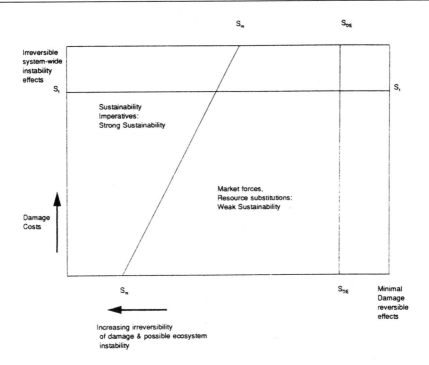

Figure 30.1 [*orig. figure 1.1*]   Safe minimum standards approach to sustainability.
Source: Adapted from B. Norton, Georgia Institute of Technology, quoted in Toman (1992).

tion schemes can be developed from the global level, through biogeographic provinces, down to regions within a country. Ecoregion classification is based primarily on the physical environment. A minimum set of 22 indicators (including species richness, species risk index, community diversity, etc.) has been proposed for wild and domesticated species (Reid et al. 1992).

For some commentators sustainability constraints of this type should be seen as expressions of the precautionary principle (O'Riordan 1992) and one that is akin to the safe minimum standards (SMS) concept (Bishop 1978). Toman (1992), quoting the work of Norton, has recently suggested that the SMS concept is a way of giving shape to the intergenerational social contract – see figure 30.1. Given irreversibility and uncertainty about the impact of economic activities on ecosystem performance, SMS posits a socially determined dividing line between moral sustainability imperatives

and the freeplay of resource trade-offs (e.g. S_mS_m in figure 30.1). To satisfy the intergenerational social contract, the current generation might rule out in advance (depending on the social opportunity costs involved) actions that could result in damage impacts beyond a certain threshold of cost and irreversibility. Social and not individual preference values will be part of the SMS setting process (Turner 1988a). Supporters of the technocentric paradigm might favour such a line as S_tS_t, while ecocentrics such as the deep ecologists might favour a line such as $S_{DE}S_{DE}$ (Pearce and Turner 1990).

Strong sustainability (ecological economics approach)

As we have seen, the weaker versions of sustainability are consistent with a declining level of environmental quality and natural resource availability as long as other forms of capital are substituted for K_n, or the imposition of SMS is judged to impose too high a social opportunity cost given the inevitable uncertainties involved in conservation benefits forecasting.

A number of analysts, from a variety of disciplines, have drawn attention to the 'missing elements' in the economic calculus that underlies the weak sustainability rules. Many ecosystem functions and services can be adequately valued in economic terms; others may not be amenable to meaningful monetary valuation. Critics of conventional economics have argued that the full contribution of component species and processes to the aggregate life-support service provided by ecosystems has not been captured in economic values (Ehrlich and Ehrlich 1992). Nor has the prior value of the aggregate ecosystem structure (life-support capacity) been taken into account in economic calculations; indeed, it is probably not fully measurable in value terms at all. There is the risk, therefore, that as environmental degradation occurs, some life-support processes and functions will be systematically eroded (because they are undervalued), increasing the vulnerability (reduced stability and resilience) of the ecosystem to further shocks and stress.

On this SS view, it is not sufficient to just protect the overall level of capital, rather K_n must also be protected, because at least some of K_n (critical K_n) is non-substitutable. Thus the SS rule requires that K_n be constant, and the rule would be monitored and measured via physical indicators. The case for this 'strong' view (linked to the precautionary principle) is based on the combination of a number of factors: presence of uncertainty about ecosystem functions and their total service value; presence of irreversibility in the context of some environmental resource degradation and/or loss; the loss aversion felt by many individuals when environmental degradation processes are at work; and the criticality (non-substitutability) of some components of K_n.

While akin to SMS, the SS rule is not the same since what is stressed in the latter approach is the combination of factors, irreversibility, uncertainty, etc., not their presence in isolation. Further, SMS says conserve unless the benefits forgone (social

opportunity costs) are very large. SS says, whatever the benefits forgone, K_n losses are unacceptable (i.e. constant 'aggregate' K_n, not constant K_n for each asset, with the exception of 'critical' components of K_n).

SS need not imply a steady-state, stationary economy, but rather changing economic resource allocations over time which are not sufficient to affect the overall ecosystem parameters significantly, i.e. beyond the point where the stability (resilience) of the system or key components of that system are threatened. A certain degree of 'decoupling' of the economy from the environment should, therefore, be possible through technical change and environmental restoration investment in a 'moderated' growth scenario.

Very strong sustainability (stationary state sustainability)

The VSS perspective concentrates on what Daly (1991; 1992) has termed the 'scale effect', i.e., the scale of human impact relative to global carrying capacity. For him the greenhouse effect, ozone layer depletion and acid rain all constitute evidence that we have already gone beyond a prudent 'plimsoll' line for the scale of the macroeconomy. The VSS approach reduces to a call for a steady-state economic system based on thermodynamic limits and the constraints they impose on the overall scale of the macroeconomy. The rate of matter and energy throughput in the economy should be minimised. The second law of thermodynamics implies that 100% recycling is impossible (even if it were socially desirable) and the limited influx of solar energy poses an additional constraint on the sustainable level of production in an economy (the solar influx potential is a matter of some dispute). Zero economic growth and zero population growth are required for a zero increase in the 'scale' of the macroeconomy. Supporters of the steady-state paradigm would, however, emphasise that 'development' is not precluded and that social preferences, community-regarding values and generalised obligations to future generations can all find full expression in the steady-state economy as it evolves (i.e. conservation of the moral capital (K_e) on which economic activity eventually depends (Hirsch 1976; Daly and Cobb 1989).

A systems and coevolutionary perspective for sustainability

The adoption of a systems perspective serves to re-emphasise the obvious but fundamental point that economic systems are underpinned by ecological systems and not vice versa. There is a dynamic interdependency between economy and ecosystem. The properties of biophysical systems are part of the constraints set which bound economic activity. The constraints set has its own internal dynamics which reacts to economic activity exploiting environmental assets (extraction, harvesting, waste disposal, non-consumptive uses). Feedbacks then occur which influence economic and social relationships. The evolution of the economy and the evolution of the constraints set are interdependent; 'coevolution' is thus a crucial concept (Common and Perrings 1992).

Norton and Ulanowicz (1992) advocate a hierarchical approach to natural systems (which assumes that smaller sub-systems change according to a faster dynamic than do larger encompassing systems) as a way of conceptualizing problems of scale in determining biodiversity policy. For them, the goal of sustaining biological diversity over several human generations can only be achieved if biodiversity policy is operated at the landscape level. The value of individual species, then, is mainly in their contribution to a larger dynamic, and significant financial expenditure may not always be justified to save ecologically marginal species. A central aim of policy should be to protect as many species as possible, but not all.

Ecosystem health (stability and resilience or creativity), interpreted in terms of an intuitive guide, is useful in that it helps focus attention on the larger systems in nature and away from the special interests of individuals and groups (Norton and Ulanowicz 1992). The full range of public and private instrumental and non-instrumental values all depend on protection of the processes that support the health of larger-scale eco-

logical systems. Thus when a wetland, for example, is disturbed or degraded, we need to look at the impacts of the disturbance on the larger level of the landscape. A successful policy will encourage a patchy landscape.

Environmental resource valuation

In particular, from the weak sustainability perspective there is an essential link between sustainable development and monetary valuations of the environment in terms of willingness to pay (WTP). The lack of meaningful monetary valuations of environmental assets would greatly circumscribe the weak sustainability case, based as it is on substitution possibilities between K_n and K_m.

According to conventional economic theory the value of environmental assets can be estimated by reference to the preferences for or against conservation of those assets, as displayed by individuals. Individuals possess a number of internal held values which result in objects being given various assigned values. In order to obtain as full an estimate of value as is feasible (known as 'total economic value') economists seek to quantify both use and non-use values. Monetary estimates of use values are relatively straightforward to obtain. Taking the example of a wetland ecosystem, humans derive *direct use values* in the form of outputs such as fish and needs, or service value such as recreation. The wetland ecosystem can also be utilised *indirectly* in the sense that the functions it can perform – floodwater storage and flood protection, effluent storage, storm buffering, etc. – are of benefit to humans. A combination of market prices and more indirect valuation methods can be used to estimate such use values (Turner and Jones 1991).

The valuation of some of the indirect use values are more difficult to derive, but many can be measured via the damage avoidance method (for example, flood damage avoided because of the wetland's capacity to retain excess water); and/or the substitute services (replacement cost) method (the costs of a sewage treatment works to substitute for the natural capacity of a large wetland to retain such effluent, up to some limit).

Individuals may also express a WTP for an option to use the environment some time in the future (option value), or display a willingness to pay to preserve the environment for their children and grandchildren (bequest value).

Non-use values are more complex and are not associated with any current or intended future use of an asset. Individuals may value just the very existence of certain species or whole ecosystems. In principle, many of the use and non-use values can be estimated by methods which infer WTP (or willingness to be compensated for a loss) such as the travel cost, hedonic price and contingent valuation methods.

Within economics, the debate about these monetary valuation techniques has basically been about technical issues. It has involved the theoretical validity and the reliability of the value estimates as the analyst moves from a consideration of use values in contexts familiar to individuals being studied to non-use values in contexts that individuals have little experience of. Questions about the form and type of information that survey respondents ought to be provided with and the type of aggregation processes required (i.e. to derive regional or national valuations from sample data) are typically debated. But outside economics, both the model of human behaviour that economics has assumed (rational self-interested maximisation), and the normative role of individualism (consumer sovereignty) have also been drawn into the debate and scrutinised critically.

Critics have argued that individuals can and do operate as both individual consumers (close to the conception of the preference- or want-dominated rational economic person assumed by conventional economics) and as citizens (a role that is influenced by 'ethical rationality' – duties, obligations, needs – and requires highly informed deliberation) (Sagoff 1988). Economics would play a less significant role in the context of citizens and social preference-dominated deliberations. However, we would not agree with Sagoff (1988) that economics plays almost no role (except for

cost-effectiveness assessments of predetermined standards/policies) in such deliberations and goal setting. Opportunity costs are ever present and therefore the choice of any particular goal carries an implicit cost (value) in relation to alternative goals. But the 'dual self-conception' of the individual as consumer and citizen is an important notion, one that has even attracted the attention of a minority of economists (Harsanyi 1955; Sen 1977; Margolis 1982; Lutz and Lux 1988). It has particular relevance for the contingent valuation method (Kahneman and Knetsch 1992). Overall, the axiom of individualism (consumer sovereignty) is weakened if the 'dual self-conception' thesis is both realistic (we believe it is) and significant in resource management contexts (again, we believe it is).

Sustainability, culture capital depletion and sustainable livelihoods

Coevolution is a local process, so local human subsystems are a significant starting point for the discussion of evolution in ecological economics (Berkes and Folke 1992). Traditional ecological knowledge may be important and therefore cultural diversity and biological diversity may go hand-in-hand as prerequisites to long-term societal survival. Diverse cultures encompass not just diverse environmental adaptation methods and processes, but also a diversity of world views (technocentrism and ecocentrism) that support these adaptations. The conservation of this rapidly diminishing pool of social experience in adaptability (cultural capital, K_c) may be as pressing a problem as maintenance of biodiversity since local cultures may be as important a reservoir of information as the genetic information contained in species currently threatened with extinction (Perrings et al. 1992).

The most convincing body of evidence for human self-organisational ability may be found in the literature of common property resources (Ostrom 1990). Institutions, coevolution and traditional ecological knowledge are all components of the common property dilemma. There exist a large number of self-regulating regimes governing access to resources in common property, and their scope in limiting the level of economic stress on particular ecological systems is clearly very wide (Bromley 1992).

Any sustainable strategy for the future will have to confront the question of how a vastly greater number of people can gain at least a basic livelihood in a manner which can be sustained; many will have to endure in environments which are fragile, marginal and vulnerable (Chambers and Conway 1992).

A working definition of sustainable livelihoods (SL) would be

a livelihood comprising the capabilities, assets (stores, resources, claims and access) and activities required for a means of living: a livelihood is sustainable which can cope with and recover from stress and shocks, maintain or enhance its capabilities and assets and provide sustainable livelihood opportunities for the next generation: and which contributes net benefits to other livelihoods at the local and global levels in the short and long term. (Chambers and Conway 1992, 27–8)

From the policy viewpoint the aim must be to promote SL security by means of vulnerability reduction. Both public and private action is required for vulnerability reduction – public action to reduce external stress and shocks through, for example, flood protection, prevention and insurance measures; and private action by households which add to their portfolio of assets and repertoire of responses so that they can respond more effectively in terms of loss limitation. Chambers and Conway (1992) offer a range of monetary value and physical indicators for SL monitoring and assessment of trends – for example, migration (and off-season opportunities), rights and access to resources, and net asset position of households.

Sustainability and ethics

For many commentators traditional ethical reasoning is faced with a number of challenges in the context of the sustainable development debate. Ecological economists would argue that the systems perspective demands an approach that privileges the requirements of the system above those of the individual. This will involve ethical judgements about the role and rights of present individual humans as against the system's survival and therefore the welfare of future generations. We have already argued that the poverty focus of sustainability highlights the issue of intra-generational fairness and equity.

So 'concern for others' is an important issue in the debate. Given that individuals are, to a greater or lesser extent, self-interested and greedy, sustainability analysts explore the extent to which such behaviour could be modified and how to achieve the modification (Turner 1988b; Pearce 1992). Some argue that a stewardship ethic (weak anthropocentrism; Norton 1987) is sufficient for sustainability, i.e. people should be less greedy because other people (including the world's poor and future generations) matter and greed imposes costs on these other people. Bioethicists would argue that people should also be less greedy because other living things matter and greed imposes costs on these other non-human species and things. This would be stewardship on behalf of the planet itself (Gaianism) in various forms up to 'deep ecology' (Naess 1973; Turner 1988a; Wallace and Norton 1992).

The degree of intervention in the functioning of the economic system deemed necessary and sufficient for sustainable development also varies across the spectrum of viewpoints. Supporters of the steady-state economy (extensive intervention) would argue that at the core of the market system is the problem of 'corrosive self-interest'. Self-interest is seen as corroding the very moral context of community that is presupposed by the market. The market depends on a community that shares such values as honesty, freedom, initiative, thrift and other virtues whose authority is diminished by the positivistic individualistic philosophy of value (consumer sovereignty) of conventional economics. If all value derives only from the satisfaction of individual wants then there is nothing left over on the basis of which self-interested, individualistic want satisfaction can be restrained (Daly and Cobb 1989).

Depletion of moral capital (K_e) may be more costly than the depletion of other components of the total capital stock (Hirsh 1976). The market does not accumulate moral capital, it depletes it. Consequently, the market depends on the wide system (community) to regenerate K_e, just as much as it depends on the ecosystem for K_n.

Individual wants (preferences) have to be distinguished from needs. For humanistic and institutional economists, individuals do not face choices over a flat plane of substitutable wants, but a hierarchy of needs. This hierarchy of needs reflects a hierarchy of values which cannot be completey reduced to a single dimension (Swaney 1987). Sustainability imperatives, therefore, represent high-order needs and values.

Sustainable Development: Operational Principles

The practical implications of SD have yet to be properly assessed, but some broad outlines are presented below. The shift from VWS through to VSS positions involves the progressive rejection of the axioms of consumer sovereignty and infinite substitution possibilities (in both utility and production functions) and their replacement with a new set of axioms. Norton (1992) has recently proposed five axioms of ecological management which are relevant to the stronger versions of sustainability:

(i)   'The Axiom of Dynamism' – nature is a set of processes in a continual state of flux, but larger systems change more slowly than smaller systems.
(ii) 'The Axiom of Relatedness' – all processes are interrelated.
(iii) 'The Axiom of Hierarchy' – systems exist within systems.
(iv) 'The Axiom of Creativity' – processes are the basis for all biologically based productivity.
(v)   'The Axiom of Differential Fragility' – ecological systems vary in their capacity to withstand stress and shock.

These axioms are then linked to a broad normative policy principle/objective, the maintenance of the health/integrity of ecosystems (Leopold 1949).

A number of rules (which fall some way short of a blueprint) for the sustainable utilisation of the natural capital stock can now be outlined (roughly ordered to fit the VWS to VSS progression):

(i)   Market and intervention failures related to resource pricing and property rights should be corrected.
(ii) The regenerative capacity of renewable natural capital (RNC) should be maintained – i.e. harvesting rates should not exceed regeneration rates – and excessive pollution which could threaten waste assimilation capacities and life-support systems should be avoided.
(iii) Technological changes should be steered via an indicative planning system such that switches from non-renewable natural capital (NRNC) to RNC are fostered; and efficiency-increasing technical progress should dominate throughput-increasing technology.
(iv) RNC should be exploited, but at a rate equal to the creation of RNC substitutes (including recycling).
(v)   The overall scale of economic activity must be limited so that it remains within the carrying capacity of the remaining natural capital. Given the uncertainties present, a precautionary approach should be adopted with a built-in safety margin.

Notes

1   A number of early issues and arguments encompassed by the sustainability debate were analysed in Turner (1988b).
2   In economic terms, the VSS position treats the economy and the environment as perfect complements; while the VWS position assumes almost perfect substitution between physical and human capital and natural capital. In fact, VWS assumes smooth production functions (i.e. functions describing the transformation process as raw material, labour and other inputs are transformed into outputs of goods and services and

non-productive outputs which often become waste flows) with perfect inputs substitution properties; and smooth welfare (utility) functions (i.e. functions describing economic well-being derived from the consumption of goods and services and ambient environmental conditions) in which all consumer wants are assumed to be substitutable.

3   SD therefore rejects the neo-classical 'potential Pareto criterion' (PPC) which underpins conventional economic cost–benefits analysis as applied to projects, policies or courses of action. The PPC sanctions activities (on efficiency grounds) if the gainers gain enough benefits hypothetically to compensate all the losers while still remaining better off than they were in the status-quo situation. An SD policy would mandate actual not hypothetical compensation for all losers (equity and efficiency trade-offs).

4   Following Leopold's (1949) 'land ethic' (much quoted by ecocentrically influenced writers), sustainable activities are activities that do not destabilise the large-scale, biotic and abiotic systems on which future generations will depend. 'Strong sustainability' (SS) resource management will therefore include a commitment (non-negotiable constraints or fixed standards) to protect the health and integrity of ecological systems (Norton 1992). The extensiveness of the commitment does, however, vary across the range of SS views and positions.

5   'Allocation' refers to the relative division of the resource flow among alternative product users. An efficient allocation (desirable) is conditioned by individual human preferences as weighted by the ability of the individual to pay (and therefore value). Broadly speaking, relative prices determined in a competitive market (or incentives to correct for market failures) are the relevant policy instrument for the allocative efficiency objective (target). 'Distribution' refers to the relative division of final goods and services (embodying resources) among current humans and future generations. Some relatively egalitarian distribution is considered fair or just and therefore socially acceptable. Taxes and welfare payments represent the relevant policy instrument.

6   The materials balance model is based on an appreciation of the first and second laws of thermodynamics which, put simplistically, say that matter and energy cannot be destroyed (only transformed) by human activity; and that there is a universal law of entropy which lays down that relatively ordered (low-entropy) and useful resource inputs are transformed by the economy (as a set of irreversible processes) into less well-ordered and less useful (high-entropy) non-product outputs which, if they are not recycled, re-enter the environment as waste flows (Ayres and Kneese 1989). The waste substances contaminate the environmental media into which they are introduced and can potentially cause pollution damage impacts. A non-zero level of contamination is impossible because 100% recyling is thermodynamically impossible. Pollution damage may or may not be economically significant, depending upon the effects on human welfare. Zero 'economic' pollution is probably technologically impossible and is in any case prohibitively costly (in clean-up or control cost terms) and therefore socially undesirable.

An optimal scale for the economy is one that is sustainable and at which humans have not yet sacrificed essential ecosystem functions and services that are at present worth more at the margin than the production benefits derived from further growth in the scale of resource use (Daly 1992).

7   Sustainability interpreted as non-declining utility of a representative member of society for millennia into the future (Pezzey 1992).

8   See, in particular, Page (1982) for an exposition of the 'justice as opportunity' thesis.

9   There is one important caveat that should be added because of the existence of non-renewable resources. The depletion of such resources must be accompanied by investment in substitute resources – for example, investment in renewable energy as a substitute for fossil fuels.

*   *   *

11   The philosopher John Rawls has used contractarian philosophy in order to formulate principles of justice chosen by rational and risk-averse individual representatives from contemporary society in an 'original position' (the negotiations) and generating from behind what he calls a 'veil of ignorance' (i.e. individuals are assumed not to know to which stratum of society they themselves belong). The contractarian approach is therefore based on actual or hypothetical negotiations which are said to be capable of yielding mutually agreeable principles of conduct, which are also binding upon all parties. One of the principles that Rawls derives is known as the

'differential principle' or maximin criterion. What it boils down to is the guarantee of an acceptable standard of living for the least well-off in contemporary society (Rawls 1972). Other writers have sought to place the maximin criterion in an intergenerational context (Page 1982; Norton 1989). In this context the rule becomes one of passing on over time an 'intact' resource base. The constant capital assets rule (VWS) or the constant natural capital assets rule (SS) would be relevant to this intergenerational equity case (Pearce et al. 1991, ch. 11).

References

Ayres, R. U. and Kneese, A. V. (1989). Externalities: Economics and thermodynamics. In Archibugi, F. and Nijkamp, P. (eds), *Economy and Ecology: Towards Sustainable Development*. Kluwer, Dordrecht.

Barbier, H. B. and Markandya, A. (1989). *The Conditions for Achieving Environmentally Sustainable Economic Development*, LEEC Paper 89-01, London Environmental Economics Centre, London.

Barret, S. (1991). The problems of global environmental protection. In Helm, D. (ed.), *Economic Policy towards the Environment*. Blackwell, Oxford.

Baumol, W. J. and Oates, W. E. (1988). *The Theory of Environmental Policy*, 2nd edn. Cambridge University Press, Cambridge.

Banelmus, P. Stahmer, C. and van Tongeren. J. (1991). Integrated environmental and economic accounting: framework for a SNA satellite system. *Review of Income and Wealth* 37, 111–48.

Beckerman, W. (1992). Economic growth and the environment: Whose growth? Whose environment? *World Development* 20, 481–96.

Berkes, F. and Folke, C. (1992). A systems perspective on the interrelations between nature, human-made and cultural capital. *Ecological Economics*, 5, 1–8.

Bishop, R. C. (1978). Economics of endangered species. *American Journal of Agricultural Economics* 60, 10–18.

Brandt Commission (1980). *North–South: A Programme for Survival*. Pan, London; and Brandt Commission (1983) *Common Crisis*. Pan, London.

Bromley, D. W. (1992). The commons, common property and environmental policy. *Environmental and Resource Economics* 2, 1–18.

Bryant, C. and Cook, P. (1992). Environmental issues and the national accounts. *Economic Trends* 469, 99–122.

Buckley, G. P. (ed.) (1989). *Biological Habitat Reconstruction*. Belhaven Press, London.

Chambers, R. and Conway, G. (1992). *Sustainable Rural Livelihoods: Practical Concepts for the 21st Century*, Discussion Paper 296. Institute of Development Studies, Sussex University, Sussex.

Common, M. and Perrings, C. (1992). Towards an ecological economics and sustainability. *Ecological Economics* 6, 7–34.

Costanza, R. and Daly, H. E. (1992). Natural capital and sustainable development. *Conservation Biology* 6, 37–46.

Costanza, R. and Perrings, C. (1990). A flexible assurance bonding system for improved environmental management. *Ecological Economics* 2, 57–76.

Daly, H. E. (1991). Towards an environmental macroeconomics. *Land Economics* 67, 255–59.

Daly, H. E. (1992). Allocation, distribution and scale: towards an economics that is efficient, just, and sustainable. *Ecological Economics* 6, 185–94.

Daly, H. E. and Cobb, J. B. (1989). *For the Common Good: Redirecting the Economy Towards Community, the Environment and a Sustainable Future*. Beacon Press, Boston.

Ehrlich, P. R. and Ehrlich, A. (1992). The value of biodiversity. *Ambio* 21, 219–26.

Frey, B. (1992). Pricing and regulation affect environmental ethics. *Environmental and Resource Economics* 2, 399–414.

Garrod, G. D. and Willis, K. G. (1991). *Some Empirical Estimates of Forest Amenity Value*, Countryside Change Working Paper 13. Countryside Change Unit, University of Newcastle upon Tyne.

Green, C. H. et al. (1990). The economic evaluation of environmental goods. *Project Appraisal* 5, 70–82.

Grubb, M. (1989). *The Greenhouse Effect: Negotiating Targets*. Royal Institute of International Affairs, London.

Harris, C. C. and Brown, G. (1992). Gain, loss and personal responsibility: The role of motivation in resource valuation decision-making. *Ecological Economics* 5, 73–92.

Harsanyi, J. C. (1955). Cardinal welfare, individualistic ethics and interpersonal comparisons of utility. *Journal of Political Economy* 61, 309–21.

Helm, D. (ed.) (1991). *Economic Policy towards the Environment*. Blackwell, Oxford.

Hirsch, F. (1976). *Social Limits to Growth*. Routledge, London.

Howarth, R. B. and Norgaard, R. B. (1992). Economics of sustainability or the sustainability of economics: Different paradigms.

Ecological Economics 4, 93–116.

Kahneman, D. and Knetch J. (1992). Valuing public goods: The purchase of moral satisfaction. *Journal of Environmental Economics and Management* 22, 57–70.

Klassen, G. K. and Opschoor, J. B. (1990). Economics of sustainability or the sustainability of economics: Different paradigms. *Ecological Economics* 4, 93–116.

Leopold, A. (1949). *Sand County Almanac.* Oxford University Press, Oxford.

Lutz, M A. and Lux, K. (1988). *Humanistic Economics: The New Challenge.* Bootstrap Press, New York.

Mäler, K. G. (1991a). National accounts and environmental resources. *Environmental and Resource Economics* 1, 1–17.

Mäler, K. G. (1991b). International environmental problems. In Helm, D. (ed.), *Economic Policy towards the Environment.* Blackwell, Oxford.

Margolis, H. (1982). *Selfishness, Altruism and Rationality: A Theory of Social Choice.* Cambridge University Press, Cambridge.

Maslow, A. (1970). *Motivation and Personality.* Harper and Row, New York.

Naess, A. (1973). The shallow and the deep, long range ecology movement: a summary. *Inquiry* 16, 95–100.

Nordhaus, W. (1992). Is growth sustainable? Reflections on the concept of sustainable economic growth. Paper presented to International Economic Association Meeting, Varenna, Italy, October.

Norton, B. G. (1987). *Why Preserve Natural Variety?* Princeton University Press, Princeton, NJ.

Norton, B. G. (1989). International equity and environmental decisions: A model using Rawls' veil of ignorance. *Ecological Economics* 1, 137–59.

Norton, B. G. (1992). Sustainability, human, welfare and ecosystem health. *Environmental Values* 1, 97–111.

Norton. B. G. and Ulanowicz, R. E. (1992). Scale and biodiversity policy: A hierarchical approach. *Ambio* 21, 244–9.

OECD (1992). *Environmental and Economics: A Survey of OECD Work.* OECD, Paris.

Opschoor, J. B. and Turner, R. K. (eds) (1993). *Economic Incentives and Environmental Policies: Principles and Practice.* Kluwer, Dordrecht, forthcoming.

Opschoor, J. B. and Vos, H. B. (1989). *The Application of Economic Instruments for Environmental Protection in OECD Countries.* OECD, Paris.

O'Riordan, T. (1992). *The Precaution Principle in Environmental Management.* CSERGE GEC Working Paper 92-02. CSERGE, UEA, Norwich and UCL, London.

Ostrom, E. (1990). *Governing the Commons: The Evolution of Institutions for Collective Action.* Cambridge University Press, Cambridge.

Page, T. (1977). *Conservation and Economic Efficiency.* Johns Hopkins University Press, Baltimore, MD.

Page, T. (1982). Intergenerational Justice as Opportunity. In Maclean, D. and Brown, P. (eds) *Energy and The Future.* Rowman and Littlefield, Totowa.

Parson, G. R. (1991). A note on choice of residential location in travel cost demand models. *Land Economics* 67, 360–4.

Pearce, D. W. (1992). Green economics. *Environmental Values* 1, 3–13.

Pearce, D. W. and Atkinson, G. D. (1992). *Are National Economics Sustainable? Measuring Sustainable Development.* CSERGE GEC Working Paper 92-11. CSERGE, UEA, Norwich and UCL, London.

Pearce, D. W. and Turner, R. K. (1990). *Economics of Natural Resources and the Environment.* Harvester Wheatsheaf, Hemel Hempstead.

Pearce, D. W., Markandya, A. and Barbier, E. B. (1989). *Blueprint for a Green Economy.* Earthscan, London.

Pearce, D. W., Markandya, A. and Barbier, E. B. (1990). *Sustainable Development.* Earthscan, London.

Pearce, D. W. et al. (1991). *Blueprint 2.* Earthscan, London.

Penn, J. (1990). Towards an ecologically-based society: A Rawlsian perspective. *Ecological Economics* 2, 225–42.

Perrings, C. (1991). The preservation of natural capital and environmental control. Paper presented at the Annual Conference of EARE, Stockholm.

Perrings, C. et al. (1992). The ecology and economics of biodiversity loss. *Ambio* 21, 201–11.

Pezzey, J. (1989). *Economic Analysis of Sustainable Growth and Sustainable Development.* Environmental Department: Working Paper No. 15. World Bank, Washington, DC.

Pezzey, J. (1992) Sustainability: An interdisciplinary guide. *Environmental Values* 1, 321–62.

Rawls, J. (1972). *A Theory of Justice.* Oxford University Press, Oxford.

Rolston III, H. (1988). *Environmental Ethics.* Temple University Press, Philadelphia.

Sagoff, M. (1988). Some problems with environmental economics. *Environmental Ethics* 10.

Sedjö, R. A. (1992). A global forestry initiative. *Resources* 109, 16–18.

Sen, A. K. (1977). Rational fools: A critique of the behavioral foundations of economic theory. *Philosophy and Public Affairs* 16, 317–44.

Smith, V. K. (1992). Non market valuation of environmental resources: An interpretative appraisal. Draft copy of unpublished paper.

Swaney, J. (1987). Elements of a neo-institutional environmental economics. *Journal of Environmental Issues* 21, 1739–79.

Tietenberg, T. H. (1991). Economic instruments for environmental regulation. In Helm, D. (ed.), *Economic Policy towards the Environment*. Blackwell, Oxford.

Tinbergen, J. (1952). *On the Theory of Economic Policy*. North Holland, Amsterdam.

Toman, M. A. (1992). The difficulty of defining sustainability. *Resources* 106, 3–6.

Turner, R. K. (1988a). Wetland conservation: Economics and ethics. In Collard, D. et al. (eds), *Economic Growth and Sustainable Environments*. Macmillan, London.

Turner, R. K. (ed.) (1988b). *Sustainable Environmental Management: Principles and Practice*. Belhaven Press, London.

Turner, R. K. (1992). *Speculations on Weak and Strong Sustainability*, CSERGE Working Paper, GEC 92-26. University of East Anglia, Norwich and University College, London.

Turner, R. K. and Jones, T. (eds) (1991). *Wetlands: Market and Intervention Failure*. Earthscan, London.

Turner, R. K., Bateman, I. and Pearce, D. W. (1992). United Kingdom. In Navrud, S. (ed.), *Valuing the Environment: The European Experience*. Scandinavian University Press, Oslo.

United Nations Development Programme (1992). *Human Development Report*. Oxford University Press, New York.

Victor, P. A. (1991). Indications of sustainable development: Some lessons from capital theory. *Ecological Economics* 4, 191–213.

Wallace, R. R. and Norton, B. G. (1992). Policy implications of Gaian theory. *Ecological Economics* 6, 103–18.

World Commission on Environment and Development (1987). *Our Common Future*. Oxford University Press, Oxford.

Willig, R. D. (1976). Consumer's surplus without apology. *American Economic Review* 66, 587–97.

Part VI

Political and Social Agendas

INTRODUCTION AND GUIDE TO FURTHER READING

31 Redclift, M. 1993: Sustainable development: needs, values, rights. *Environmental Values*, 2, 3–20.

This key paper, reproduced in full, explores the intellectual traditions within which the idea of sustainable development has emerged. It questions the authority of science and technology, and argues that more emphasis should be placed on cultural diversity.

32 Brundtland, G.H. 1994: The solution to a global crisis. *Environment*, 36, 10, 16–20.

This is an edited version of the Prime Minister of Norway's keynote address to the International Conference on Population and Development held in Cairo on 5 September 1994.

33 Koch, M. and Grubb, M. 1993: Agenda 21. In: Grubb, M., Koch, M., Thomson, K., Munson, A. and Sullivan, F. (eds) *The 'Earth Summit' Agreements: A Guide and Assessment. An Analysis of the Rio '92 UN Conference on Environment and Development*, London: Earthscan, pp. 97–158.

This is the introductory section of the authors' thorough review of the contents of Agenda 21.

34 Rees, J. 1991: Equity and environmental policy. *Geography*, 76, 4, 292–303.

This wide-ranging paper, reproduced in full, argues that the apparent consensus over the desirable nature of future environmental policy is in reality a sham.

35 Newby, H. 1991: One world, two cultures: sociology and the environment. *BSA Bulletin Network*, 50, 1–8.

36 Harrison, P. 1993: Executive Summary – The Third Revolution: Population, Environment and a Sustainable World. In: Harrison, P., *The Third Revolution: Population, Environment and a Sustainable World*, Harmondsworth: Penguin, pp. 323–30.

This is the entire executive summary of Paul Harrison's popular book, which was named Book of the Year 1992 by the *New Internationalist* and which also won a Population Institute Global Media Award.

Management of any kind implies that some person, or group of people, is managing some thing, or some other person. It thus involves particular social and political relationships; it is concerned with the maintenance of power and the allocation of resources. This is as true of environmental management as it is of any other kind of management. Indeed, the very idea of environmental management implies that one group of people is managing the environment on behalf of society at large. More often than not, such 'managers' are personnel in trans-national companies or other businesses, seeking to maximize their profits with little real regard for the environmental interests of the majority of the population. Invariably they end up by designing systems which allocate the main benefits to those already in positions of power; the poor and underprivileged rarely benefit from such environmental management.

However, it is also possible to conceive of societies managing their own environments as common property, for the benefit of the whole community rather than for the benefit of a small minority. Hence debates about the role of common property rights, at all scales from individual communities to the global context, are central to an interpretation of environmental management. Moreover, anyone involved in changing the environment can be seen as an environmental 'manager'. Thus farmers on small recently privatized plots of land in Estonia, multinational companies such as Shell involved in oil extraction in Nigeria, and nomadic herders in the Arabian peninsula, can all be seen as types of environmental manager.

Economic approaches to environmental management have invariably been formulated within a framework and language that reflect the dominant classes from which they have emerged. As Redclift (1993a, p. 108) has observed, 'Riding the wave of neo-liberal policies in all the advanced industrial societies, economics has grown towards, rather than away from, the world view of the powerful. This implies clothing oneself in the language of "objectivity", notably, a culturally grounded view of "management" as the necessary means to effect desirable changes.' It is only with some of the more radical deep ecological and Marxist approaches, what Turner (1993) describes as the very strong sustainability policies, that this world view is challenged. Redclift (1993a, pp. 108–9) goes on to note that 'The environment is actually the product of human culture (our discourse about nature) but it tends to be represented as being about science. In its attempt to maintain its authority to prescribe, economics has chosen to ignore the analysis of difference, in favour of an analysis based on similarity.'

Part VI therefore seeks to examine some of the ideological, political and social challenges that have been advanced against the idea of environmental management as a top-down, science-driven concept. In so doing, we can only touch on some of these critiques; the rise of radical, anarchist, and deep-ecological approaches to environmental issues can be explored further in the annotated list of suggested texts which follows.

Ideologies of Environmental Management

The concept of sustainable development has been one of the most successful and pervasive contributions of environmental economics. Yet, as already noted, it remains a highly ambiguous and controversial term. The first reading selected for Part VI therefore provides a powerful critique of the idea of sustainable development as developed in environmental economics. Michael Redclift argues that there are four fundamental flaws with the approach: first, that economists can only value the environment in monetary terms, and that these are very different from what it is really worth; second, he suggests that environmental economics' claim of value neutrality is flawed, since all values reflect particular social systems and cannot therefore be strictly neutral; third, he argues that these conventional approaches rest upon a very specific view of human nature and social relations, which is far from universal; and fourth, he notes that the assumption in much environmental economics that individuals are fully appraised of the situations in which they have to make choices, is rarely met. Redclift goes on to argue that science itself is part of the problem. As he suggests, 'It is clear that the view we take of the environment is closely bound up with the view we take of science. Increasingly, environmental problems are looked upon as scientific problems, amenable to scientific answers' (Redclift, 1993b p. 16). But as Redclift (1993b, p. 17) points out, 'the modes of inquiry in the natural sciences are themselves *social processes, into which crucial assumptions, choices, conventions and risks are necessarily built*'. According to such arguments, we need to change the very nature of society if we are really to grapple with the environmental problems, such as global warming, which are seen as leading to imminent global nemesis. Put at their simplest, these arguments suggest that science and modern economics have played a fundamental role in causing the unsustainable economic growth that has dominated the core capitalist countries of the world over the last century, and that solutions to these problems are therefore unlikely to be found within these modes of discourse and practice.

One of the strongest critiques of contemporary capitalist society has come from the feminist tradition, and the role of women is now widely seen as being crucial to the environmental debate. The chapter by Gro Harlem Brundtland, Prime Minister of Norway at the time the article was written in 1994, argues that it is essential to strengthen the role and status of women if we are to come to grips with key environmental agendas facing the global community. It is an edited version of the keynote speech she gave to the International Conference on Population and Development held in Cairo on 5 September 1995. As Chair of the World Commission on Environment and Development since 1983, Brundtland has played a major role in the formulation of international policy concerning the environment, and this chapter reflects the crucial importance she attributes to population pressure as a factor influencing the environment. Her central argument here is that in order to reduce population growth it is essential to empower women through legislative change, improved access to information, and a redirection of resources. In the longer term, she suggests that the most important solution for the world's environmental problems is to reduce population pressure by giving women greater access to equal rights. Although the chapter has little directly to say about the environment itself, it presents a direct challenge to male-dominated practices and discourses of environmental management. Brundtland also makes reference to the UN Conference on Environment and Development (UNCED) held in Rio de Janeiro in 1992, arguing that one of the failures of this confer-

ence was that it did not succeed with regard to population. Any understanding of environmental management must therefore address the global environmental political agenda, and the next two chapters therefore focus overtly on the politics of the environment.

The Political Context of the Environment

The UN Conference on Environment and Development, popularly known as the Rio Earth Summit, received considerable media hype across the world. However, many people, both before and after the conference, were highly critical of its agenda and of its conclusions. Whatever the eventual effects of the Rio Summit on the environment turn out to be though, it must itself be seen as an outcome of an unprecedented political process at a global scale which sought to address the links between the environment and development (Grubb et al., 1993).

Until the early 1970s, most environmental agendas were focused essentially at a local scale, and very little national, let alone international, attention was paid to the linkages between development and the environment. However, a growing realization that environmental issues spread across international boundaries, led to a call by Swedish politicians, concerned in particular by the connection between air pollution and the acidification of Scandinavian water bodies, for the UN to address such environmental concerns. The result was the 1972 UN Conference on the Human Environment held at Stockholm, which attracted representatives from some 113 countries and 400 intergovernmental and non-governmental organizations (McCormick, 1989). The formal outcome of the conference was twofold: the Stockholm Declaration on Human Environment, which established 26 key principles concerning the environment; and an Action Plan for the Human Environment which focused on a global assessment programme, environmental management activities and relevant supporting measures such as environmental education and training.

Over the subsequent decade a number of international initiatives were developed consequent on the Stockholm Conference's agendas. Notable among these was the complex set of negotiations on the Law of the Sea completed in 1982. However, the next major initiative was the creation of the World Commission on Environment and Development in 1983, which was charged with the task of re-examining the linkages between the environment and development and with formulating policies to deal with them. The resultant report, *Our Common Future* (WCED, 1987), known widely as the Brundtland Report after its chair, concentrated largely on the idea of sustainable development. In so doing, it advocated that the UN should prepare both a universal declaration on the environment and also initiate a formal programme on sustainable development. At around the same time, other important international bodies such as the International Union for the Conservation of Nature (IUCN) and the UN Environment Programme (UNEP) were also formulating strategies concerned with environmental conservation and environmental action programmes. Eventually, in December 1989, the UN General Assembly agreed to convene the United Nations Conference on Environment and Development (UNCED), and set in motion the negotiations which would lead to the agreements signed in Rio in 1992.

The Rio conference involved the signing of five main agreements: the Framework Convention on Climate Change; the Convention on Biological Diversity; Agenda 21; the Rio Declaration; and the Forest Principles (Grubb et al., 1993). The first two of these were legally binding conventions negotiated separately from the main UNCED processes. The remaining three general agreements were signed by nearly all of the participating countries, and it is on these that most popular attention has subsequently focused. The Rio Declaration outlines 27 principles of sustainable development, and was designed to flesh out the agreements embodied in the 1972 Stockholm Declaration, but concentrating more on implementation, blame and responsibility.

In contrast, Agenda 21, was designed as an action plan to implement sustainable development across the globe. As Grubb et al. (1993, p. 17) note, 'Many themes recur throughout Agenda 21. These include a "bottom-up" approach of putting emphasis upon people, communities and NGOs; the need for "open governance"; the importance of adequate information; the need for adequate cross-cutting institutions; and the complementarity between regulatory approaches and market mechanisms for addressing development and environmental needs.'

Agenda 21 has proved to be highly controversial, which is scarcely surprising given the debate surrounding the whole concept of sustainable development already outlined in the chapter by Redclift which introduces the selections chosen for Part VI. However, it is an important document, and its framework is discussed further in the chapter by Koch and Grubb, part of which is reprinted here. This establishes the style and status of the document, its structure, and the contents of the Preamble, which is seen as setting the whole tone for Agenda 21. Much of the criticism of UNCED, though, has been based on its overtly top-down, managerialist stance, which is seen by many as ignoring the poor and the underprivileged at a range of scales from the international to the local. As a team of writers for *The Ecologist* (1993, p. vi) has argued, 'as the limousines took the heads of state from their embassies to the massive conference centre on the outskirts of Rio, many of those going about their daily lives outside the conference hall – in effect, the many millions whose fate and livelihoods were under discussion – no doubt asked: what lies behind the razzamatazz? Were the world's political leaders attending the meeting out of a genuine concern for the "environment" or for "equity"? Or were the anonymous but high-ranking "representatives" of Gorbachev's "world community" simply repositioning themselves so as to minimize the changes that "constructing life in the 21st century" will inevitably entail? And if, as Maurice Strong, Secretary-General of UNCED, claimed, there was a "groundswell of support from grassroots for the objectives of the Earth Summit", why were the issues which have been central to the work of grassroots groups – in particular the right of local communities to determine their own future – excluded from the agenda?'

In her chapter, Judith Rees takes this discussion of environmental policy-making down to the practical scale of national politics, arguing that the pace, scale and complexity of environmental change has outstripped the ability of political decision-making processes to implement appropriate responses. Moreover, while acknowledging that intergenerational equity is an important concept in discussing sustainability, she reinforces the argument that the present-day distribution of wealth and resources is also crucial to the environmental debate. This applies at a range of scales from the international to the local. Thus in the international arena, the very different political interests of richer and poorer countries in managing environmental change must be taken into consideration when discussing the practical implications of fine-sounding UN rhetoric. Within individual countries, Rees also points out that taxes on energy use are regressive, with the poorest families shouldering a greater burden than the richest. She concludes that the apparent consensus over the future direction of environmental policy is in reality a sham, hiding fundamental disagreements over the sharing out of the cost burden of a really radical new approach to environmental management.

Social Reflections on Environmental Management

The final selection of papers in this volume develops further Rees's concern with social issues. Howard Newby returns once again to a challenge of the technological determinism which has dominated both the environmental debate and the policy formulation of environmental agendas. Rather than being exogenous to human society, Newby emphasizes that both science and technology change are closely influenced by socio-economic and political conditions. If we are to influence environmental change it

is therefore crucial that we understand sociologically the way in which these influences operate. However, Newby also argues that sociologists have in general failed to address environmental issues in a sufficiently thorough and rigorous manner. In part, he sees this as a result of the humanistic traditions of social sciences which have been diametrically opposed to the technological interests of the natural sciences (see also Unwin, 1992). Redclift and Woodgate (1994) have explored these ideas further, and they suggest that the lack of research by sociologists on environmental agendas can be traced back to the founding fathers of modern sociology who defined the natural environment negatively as that which was not social. In order to formulate an alternative practice to unsustainable development they advocate that a realist agenda be forged, in which sociology engages directly with the implications of the human relationship with nature.

A summary of Paul Harrison's popular book on the links between population, the environment and sustainability has been chosen both as a sociological example of populist writing on environmental issues, and also as a cogent summary of the socio-demographic factors influencing the environment. Paul Harrison was presented with a Global 500 award by the United Nations Environment Programme (UNEP) in 1988, and in his own personality can therefore be seen as representing a specific expression of the linkages between the increased global awareness of environmental agendas and their propagation in the written media. Harrison argues in unequivocal terms that the environmental crisis is the result of three things: the explosions in population, consumption and technology. Adopting a neo-Malthusian stance, he predicts global disaster unless urgent action is taken to reduce population levels.

Finally, the lack of attention paid to the interests of the mass of the world's population by the Rio Earth Summit, needs to be redressed by a consideration of the crucial role that the commons play in environmental management. It has been widely argued that the Earth Summit served only the interests of those in power. The authors of a report for *The Ecologist* (1993, p. 2) thus note that 'The demands from many grass-roots groups around the world are not for more "management" – a fashionable word at Rio – but for agrarian reform, local control over local resources, and power to veto developments and to run their own affairs. For them, the question is not *how* their environment should be managed – they have the experience of the past as their guide – but *who* will manage it and in *whose* interest.' They suggest that the fundamental environmental issue facing the future of human communities is whether or not they will be able to reclaim and defend their commons, so that they can benefit from the equity resulting from sharing a truly common future.

Such a rejection of the whole notion of 'environmental management' is far removed from the examples of technical control and environmental economics illustrated earlier in this collection of readings. However, it emphasizes the fundamental point that there are many different ways in which the relationship between people and the physical environment can be addressed. This book has sought to reflect both this diversity of approach, and also the crucial importance of a multidisciplinary framework in understanding and changing the human exploitation of the world in which we live.

References

Grubb, M., Koch, M., Thomson, K., Munson, A and Sullivan, F. 1993: *The 'Earth Summit' Agreements: a Guide and Assessment. An Analysis of the Rio '92 UN Conference on Environment and Development*. London: Earthscan Publications.

Lovelock, J. 1989: *The Ages of Gaia*, Oxford: Oxford University Press.

McCormick, J. 1989: *The Global Environmental Movement: Reclaiming Paradise*. London: Belhaven.

Redclift, M. 1993a: Environmental economics, policy consensus and political empowerment. In: Turner, R.K. (ed.) *Sustainable Environmental Economics and Management*. London: Belhaven, 106–19.

Redclift, M. 1993b: Sustainable development: needs, values, rights. *Environmental Values*, 2, 3–20.

Redclift, M. and Woodgate, G. 1994: Sociology and the environment: discordant discourse? In: Redclift, M. and Benton, T. (eds) *Social Theory and the Global Environment*. London: Routledge, 51–66.

The Ecologist 1993: *Whose Common Future? Reclaiming the Commons*. London: Earthscan.

Turner, R.K. 1993: Sustainability: principles and practice. In: Turner, R.K. (ed.) *Sustainable Environmental Economics and Management*. London: Belhaven, p. 3–36.

Unwin, T. 1992: *The Place of Geography*. Harlow: Longman.

WCED (World Commission on Environment and Development) 1987: *Our Common Future*. Oxford: Oxford University Press.

Suggested further readings

Adams, W.M. 1990: *Green Development: Environment and Sustainability in the Third World*. London: Routledge.

In this text, Adams discusses the problems of development and its environmental impact on the developing world. It is an important read for students and teachers of development and environmental studies who wish to examine the problems of striving for sustainable development.

Allen, T. and Thomas, A. (eds) 1992: *Poverty and Development in the 1990s*. Oxford: Oxford University Press.

This Open University text for a course on Third World Development introduces the major issues involved in understanding and analysing poverty and development. The book comprises three sections: introducing the problems of poverty; the historical context of poverty and its associated problems; and a consideration of development and the problems of implementation.

Angotti, T. 1993: *Metropolis 2000: Planning, Poverty and Politics*. London: Routledge.

This text offers an analysis of metropolitan development and planning. It provides many examples from all parts of the world, and under different economic and environmental conditions, including metropolitan development in the United States, the Former Soviet Union (FSU), and the 'dependent metropolis' of the developing world. A consideration of the problems of urban planning theory and practice in the metropolis and its communities is presented. Throughout the principle of 'integrated diversity' is emphasized, linking neighbourhood planning with a broader vision of a planned metropolis.

Blaikie, P.M. and Brookfield, H.C. 1987: *Land Degradation and Society*. London: Methuen.

This book clearly outlines the combined role of social, economic and environmental factors in influencing land degradation. Examples are taken from the Mediterranean and western Europe, pre-capitalist social formations, tropical rain forests, and socialist countries. It emphasizes the great complexity of the problems associated with environmental degradation and the need for social scientists and natural scientists to work together in pursuit of their solution.

Brandt, W. 1980: *North–South: A Programme for Survival*. The report of the Independent Commission on International Development Issues under the Chairmanship of Willy Brandt. London: Pan Books, 304 pp. Brandt, W. 1983: *Common Crisis North–South: Cooperation for World Recovery*. The Brandt Commission 1983. London: Pan Books.

These books are historically important, presenting the investigations and recommendations on the problems of inequality in the world and the failure of economic systems to tackle these issues. The authors describe the global crisis in trade, energy, and food supply, and concentrate on the overriding problem of how to provide the financial help, and ways of compensating for the decline in financial liquidity to reverse the decline in trade and to raise the overall world economy. The WCED followed these reports with its publication of *Our Common Future* (see below).

Ekins, P. 1992: *A New World Order: Grassroots Movements for Global Change*. London: Routledge.

In this book, Ekins analyses the environmental problems associated with war, insecurity, militarization, poverty and the denial of human rights. Examples are taken from several dozen countries, and the author argues that the objectives of the peace, social justice, feminist and green movements cannot really be separated. Solutions to these problems are presented at a grass-roots level.

Grubb, M., Koch, M., Thomson, K., Munson, A. and Sullivan, F. 1993: *The 'Earth Summit' Agreements: A Guide and Assessment. An Analysis of the Rio '92 UN Conference on Environment and Development*. London: Earthscan.

This is an extremely useful summary of the UN Conference on Environment and Development held at Rio de Janeiro in 1992. It provides an outline of the agreements leading up to the Rio conference, the outcomes of the conference, its key themes and lessons, and then detailed analyses of the UN Framework Convention on Climate Change, the UN Convention on Biological Diversity, Agenda 21, and Forest Principles.

Harrison, P. 1993: *The Third Revolution: Population, Environment and a Sustainable World*. Harmondsworth: Penguin.

This popular book on the relationships between population, environment and sustainability was named Book of the Year 1992 by the *New Internationalist*, and its central argument is that the environmental crisis is the result of a population and consumption explosion. The author suggests that if we are to survive this crisis we need to go through a 'Third Revolution', similar in its significance to the agricultural and industrial revolutions.

Johnson, R.J., Taylor, P.J. and Watts, M. 1995: *Geographies of Global Change*. Oxford: Blackwell Publishers.

This is a useful book exploring geoeconomic, geopolitical, geosocial, geocultural and geoenvironmental change. It considers the collapse of socialism, the reconfiguration of North Atlantic capitalism, the hypermobility of capital, the rise of ferocious nationalisms, global environmental change, the power of international media, and the social movements associated with population growth and international migration. The textbook provides a useful economic, political, social, cultural and ecological view of change at every geographical scale from the global to the local. It is an essential text for undergraduate and graduate students who are studying geography.

Moore Lappe, F. and Schurman, R. 1989: *Taking Population Seriously*. London: Earthscan.

This book provides a useful analysis of the reasons for population growth, discussing the need to understand the underlying social and economic causes of population growth to help implement effective population control.

Myers, N. and Simon, J.L. 1994: *Scarcity or Abundance?* London: W.W. Norton and Company.

This is an interesting book that summarizes a debate between Myers and Simon. Simon presents optimistic views that the environment has plenty and will support the improvement of the human condition, while Myers presents a view of ecological degradation with its potential catastrophic effects. The authors present a 'pre-debate' statement and a 'post-debate' statement in which their final arguments are a response to both the debate proceedings and their opponent's opening statement.

O'Riordan, T. 1981: *Environmentalism*. London: Pion-Methuen.

This is a seminal book from the early 1980s which argues that environmentalism is as much a state of being as a mode of conduct or set of policies. It provides a comprehensive account of the emergence of environmentalism and the search for solutions to the balance between policies advocating growth and non-growth. There is also a good section on environmental law.

Pepper, D. 1993: *Eco-socialism: from Deep Ecology to Social Justice*. London: Routledge.

In this book, Pepper provides a provocative anthropocentric analysis of the way forward for green politics and environmental movements, building on the works of Marx, Morris, Kropotkin and anarcho-syndicalism. It rejects biocentrism, simplistic limits to growth and overpopulation theses, and argues for a radical eco-socialism in which people must control their own lives and their relationships with the environment.

Redclift, M. 1984: *Development and the Environmental Crisis: Red or Green Alternatives*. London: Methuen.

This book is a useful review of the red and green alternatives to the effects of economic development on the environment.

Redclift, M. 1987: *Sustainable Development: Exploring the Contradictions*. London: Methuen.

This book argues that the development recommendations of the WCED report (WCED, 1987) need to be redirected to give greater emphasis to local (indigenous) knowledge and experience if effective political action is to be taken to minimize any environmental damage.

Redclift, M. and Benton, T. (eds) 1994: *Social Theory and the Global Environment*. London: Routledge.

This is an excellent collection of papers published in association with the UK's Economic and Social Research Council's Global Environmental Change Programme. It emphasizes the ways in which cultural, economic and political values are already involved in shaping the definitions of environmental problems for scientific analysis. It advocates the importance of social science for an understanding of the human costs of environmental degradation.

Sandbach, F. 1981: *Environment, Ideology and Policy*. Oxford: Blackwell.

This is a good and wide-ranging early survey of the role of ideology in influencing environmental policy.

Sarre, P. (ed.) 1991: *Environment, Population and Development*. London: Hodder and Stoughton.

This Open University text examines the environmental issues associated with population growth, and economic and technological development. It discusses population dynamics; agriculture, productivity and sustainability; urbanization, behaviour and social problems.

Sheehan, M.J. 1988: *Arms Control: Theory and Practice*. Oxford: Blackwell.

This is a useful analysis of the origins and development of arms control. It discusses the issues which underpin arms control, the problems of verifying treaties, and the political context in which arms control negotiations, both domestic and international, are considered.

Shumacher, F. 1973: *Small is Beautiful*. London: Blond and Briggs.

Small is Beautiful is an early and highly influential book challenging the doctrine of economic, technological and scientific specialization, and advocating instead the case for Intermediate Technology, based on small work units, utilizing local labour and resources. It suggests that human pursuit of profit and progress has resulted in gross economic inefficiency, environmental pollution and inhumane working conditions.

The Ecologist 1993: *Whose Common Future? Reclaiming the Commons*. London: Earthscan.

This stimulating book argues that environmental degradation stems from the enclosure of the commons and the dispossession of local communities. It draws its examples from contexts as diverse as medieval England and the streets of Bangkok and New York, to illustrate that communities depend on shared resources of the commons for their autonomy and identity. Critical of the UN Earth Summit in 1992, it suggests that rather than outside management, communities require control over their own resources.

Welford, R. 1995: *Environmental Strategy and Sustainable Development*. London: Routledge.

In this book, Welford outlines an interesting debate over environmental strategy in business and provides a radical business agenda for the future.

World Commission on Environment and Development (WCED) 1987: *Our Common Future*. Oxford: Oxford University Press.

This is the report of the World Commission on Environment and Development. It examines the critical environmental and developmental problems facing the world today. It contains many very useful tables and figures, being a good reference source for all college and university students, teachers, and policy-makers concerned with environmental and development issues.

Yearley, S. 1992: *The Green Case: A Sociology of Environmental Issues, Arguments and Politics*. London, Routledge.

This is an account of the basis of 'green' arguments and their social and political implications. It examines the reasons for the success of leading campaign groups (such as Greenpeace), and analyses the developments in green politics and green consumerism.

Young, S.C. 1993: *The Politics of the Environment*. Manchester: Baseline Books.

This is a short, readable, book assessing environmental problems and their economic, social and political contexts. It focuses on Britain and gives particular attention to the emergence of green protest and the activities of green parties.

31

SUSTAINABLE DEVELOPMENT: NEEDS, VALUES, RIGHTS*

Michael Redclift

[handwritten: * All of these (following) chapters will be useful for me to look at so]

[handwritten margin note: Political]

Sustainable development remains a confused topic. Like motherhood, and God, it is difficult not to approve of it. At the same time, the idea of sustainable development is fraught with contradictions (Redclift 1987). This has not prevented 'sustainability' from being invoked in support of numerous political and social agendas. This paper argues that sustainable development has gained currency precisely because of the way it can be used to support these various agendas.

The idea of sustainability is derived from science, but at the same time highlights the limitations of science. It is used to carry moral, human, imperatives, but at the same time acquires legitimacy from identifying biospheric 'imperatives' beyond human societies. This paper argues that married to the idea of 'development', sustainability represents the high water mark of the Modernist tradition. At the same time, the emphasis on cultural diversity, which some writers view as the underpinning of sustainability, is a clear expression of Post-Modernism. The strength of the idea of sustainable development, then, lies in its ambiguity, and its range.

Sustainable development can be viewed from different perspectives. The concept has its origin in the need to assert the primacy of living within ecological limits, without forfeiting the idea of 'progress'. In this the idea of sustainable development reflects ambivalent social goals. It is socially constructed, and yet the concept expresses a 'realist' position on the world. It is used to suggest that however we 'construct' the environment, living within limits is a challenge to human ingenuity that we cannot avoid. Sustainable development suggests the need to engage with the world, but our vision of the world is itself a social construction, and so, inevitably, a relativist one.

Antecedents: The Etymology of 'Sustainable'

The word 'sustainable' is derived from the Latin *sus tenere*, meaning to uphold. It has been used in English since 1290, but the etymology of 'sustainable' carries interesting, and important, implications for the way the word is used. As de Vries (1989) reminds us, 'sustaining' can mean "supporting a desired state of some kind" or, conversely, "enduring an undesired state". The verb 'to sustain' carries a passive connotation, while the adjective 'sustainable' is used in an active sense. As we shall argue below, the juxtaposition of both

* Originally published in *Environmental Values*, 1993, vol. 2, pp. 3–20.

normative/active, and positive/passive meanings, has enabled the idea of sustainability to be employed in a variety of contradictory ways.

The following propositions serve to illustrate some of these competing, but powerful, interpretations:

1   The first proposition is that sustainable development has *proved useful as a concept, precisely because it combines the idea of prescriptive action, with that of enduring, defendable properties, located in scientific principles.* As we shall see the reference to 'scientific' principles, as the basis for rational human action, is central to the environmental message. At the same time, of course, Green thinking takes issue with much reductionist science, regarding it as part of the problem to be addressed.

2   The second proposition is that the idea of sustainable development *is born of intellectual necessity, as much as political necessity. It emerges, in fact, from problems generated by Modernism itself,* including our faith in science. As David Cooper argues, the modern concept of The Environment ". . . is symptomatic not only of a predilection for a scientific perspective, but of the situation of today's intellectuals" (Cooper and Palmer 1992: 171). It follows from this proposition that the idea of sustainability reflects unease about the human condition; we use nature and the environment to mirror the discontents of human societies.

We will return to these propositions later. For the present, it is worth considering the historical legacy from which the idea of sustainable development has grown. Although we increasingly refer to other cultures, and other epochs, in defence of the idea of sustainability, it should properly be seen as the outcome of a quite specific set of events, beginning with the idea of progress, and associated with the Enlightenment in Western Europe. Our willingness to authenticate sustainability by reference to societies which possess no such concept, is

both historically and intellectually revealing. As we shall see, it carries serious implications for 'global' strategies of development, which ensure the continued economic hegemony of the northern, industrialized countries. The 'globalization' of environmental problems, and suggested solutions, today reflects a continuing history of colonial and post-colonial thought associated with the (Northern) idea of 'development'. The specific history of 'development' and 'sustainability' as *complementary* ideas is a product of this Modernist discourse.

Sustainability and Development: The Discourse Surrounding Needs

In his classic study of the environmental idea, Clarence Glacken argues that "the association of the idea of progress with the environmental limitations of the earth" was, necessarily, a post-Enlightenment development (Glacken 1973: 654). The thinkers of the European Enlightenment, such as Condorcet, Godwin and Malthus, had all developed a primitive concept of 'carrying capacity', but they did not explore the implications of environmental changes that were driven by human behaviour. In the late eighteenth century even the most radical thinkers assumed a stable physical environment, as the backdrop for human progress. The earth could be cultivated "like a garden" but Enlightenment thinking did not consider that "an environment deteriorating as a result of long human settlement, might offer hard choices in the future . . ." (Glacken 1973: 654).

Later, Glacken makes his point even more forcefully, pointing to elements in the modern equation which pre-nineteenth century thinkers ignored. He writes that:

With the eighteenth century there ends in Western civilization an epoch in the history of man's relationship to nature. What follows is of an entirely different order, influenced by the *theory of evolution, specialization in the attainment of knowledge, acceleration in the transformation of nature* . . . (Glacken 1973: 705) (emphasis added)

These three elements – evolutionary theory, scientific specialization, and economic development on an unprecedented scale, throughout the nineteenth century – define the context in which sustainability was to become important. They also define the context for Modernism and, I would argue, the human (or 'inner') limits placed on the development of productive forces.

There has been considerable debate about what defines Modernism. For our purposes, Modernism is the view that ideas grounded in (essentially) Western philosophy and science can serve as the basis for social criticism and understanding. In the opinion of Post-Modernist writers like Lyotard, Modernism's mistake was to have recourse to "the grand narratives of legitimation" which are no longer credible (Lyotard 1984: 23). In terms of our discussion of sustainability, Modernism represents an attempt to deal with the problems of nature through reference to 'natural laws': to 'external limits' imposed on human societies. These laws were formulated throughout the eighteenth century, but it was not until the nineteenth century that scientific disciplines, and the successful application of science through technology, assumed the authority it has today. In a sense, then, Modernism sought to pit human ingenuity against the 'external limits', in the Promethean spirit. Some thinkers, including Engels, were dimly aware of the costs attached to the advance of science, principally in terms of the difficulties in managing nature (Engels 1970). However, managing nature was quite a different thing from understanding it, and the Modernist position placed human beings above the environment from which they had sprung.

There are two key elements in the Modernist perspective on the environment. The first is based on certain ideals of reason and freedom which are associated with the idea of 'progress'. Evolutionary theory, scientific specialization and economic development provided the context for this essentially optimistic perspective.

The second element concerns the way in which the basis for legitimation, in the new intellectual discourse, as well as the basis for criticism, was provided by science. Science provided the means to transform nature, and at the same time provided our critique of nature. As belief in the 'progress' of science has met more obstacles, so our ability to invoke 'science' in defence of human actions appears more problematic. Ultimately we are faced with questions that are both philosophical and political: does the modernist sense of progress entail destruction of the environment? Does our ability to elaborate notions of sustainability, without endorsing the idea of progress and development, mean jettisoning the modernist discourse altogether?

According to Sklair (1970) the nineteenth century represented a watershed in thinking about progress, because it was the period in which science became institutionalized, in which material progress was linked to new ways of thinking:

By the middle of the nineteenth century those who wrote about society were in a position to elaborate a theory concerning the relations between scientific and social progress from an entirely different . . . point of view. (Sklair 1970: 33)

The centrality of science to society has made it almost impossible to consider the idea of progress without thinking of its critique, which is by no means confined to Post-Modernist writing. Almost forty years ago Claude Lévi-Strauss, in his celebrated tract on *Race and History* (1958) pointed out that historicist conceptions of progress are ultimately flawed: ". . . progress . . . is neither continuous nor inevitable; its course consists of leaps and bounds . . . These . . . are not always in the same direction" (Lévi-Strauss 1958: 21). Later he goes on to write that advancing humanity is like a gambler throwing dice, and each throw giving a different score. "What he wins on one [throw], he is always liable to lose on another, and it is only occasionally that history is 'cumulative', that is to say, that the scores add up to a lucky combination" (Lévi-Strauss 1958: 23).

Notwithstanding the 'progressive' nature of science and technology, however, and the material improvements to which they helped give rise, 'development' still has to confront a deep-rooted difficulty. If human progress can only be achieved at the expense of destroying the environment, and ultimately the resources on which development depends, then a theory of development lacks legitimacy. Fraser and Nicholson (1990) note that Modernism ". . . narrates a story about the whole of human history which purports to guarantee that the pragmatics of the modern sciences and of modern political processes – the norms and ends which govern these practices – are themselves legitimate" (1990: 22).

For ideas to retain their power, they must be legitimated. Sustainable development is one such idea, which seeks to legitimate its own propositions by recourse to what are assumed to be universal values. By incorporating the concept of 'sustainability' within the account of 'development', the discourse surrounding the environment is often used to strengthen, rather than weaken, the basic supposition about progress. Development is read as synonymous with progress, and made more palatable because it is linked with 'natural' limits, expressed in the concept of sustainability. The essential discourse surrounding nature, and what are assumed to be natural laws, is viewed not as part of a broader socially constructed view of 'progress', but as part of an essentially non-human logic, located in biological systems.

Sustainable development then becomes a methodology, as well as a normative goal, a model for planning, a strategy involving purposeful management of the environment. Like other models derived from nature, 'sustainability' acquires legitimacy from its biological origins. This also takes on ethical importance. Some of the approaches which recognise these political parameters, it must be added, are extremely useful within specific contexts. Michael Jacobs, for example, makes good use of the idea of 'sustainability planning' in his case for more recognition, from the political Left, of the role of market interventions in environmental management (Jacobs 1990). As we shall argue later, sustainable development

seeks to define an 'ends/means' structure, based on a hierarchy of needs. As de Vries puts it ". . . planning for sustainable development assumes that a blueprint for Utopia can, *and should* be made . . . a recipe for how to travel towards the end of the road" (de Vries 1989: 8). Other writers, notably Merchant (1980) find an echo in today's 'managerial ecology' of earlier, seventeenth century approaches, which also ". . . subjected nature to rational analysis for long-term planning. . . " (Merchant 1980: 252).

The Allocative Principles Behind Sustainable Development

If the idea of sustainable development is a product of Modernism, it also answers to the problems of Modernism, in a variety of ways. First, and here the comparisons with Marxism are particularly interesting, sustainable development *invokes the concept of 'need', in the context of 'development', to meet problems of resource allocation in time and space*. The discussion of needs illustrates the essentialist discourse surrounding sustainability.

Problems of allocation in time, between 'now' and 'later', between present and future generations, are central to the discourse surrounding sustainability. Intergenerational equity, in the way used by economists, is a concern to register the preferences and choices of future generations, as yet unborn. It is used to extend Neo-Classical theory, which would otherwise fail to fully reflect future choices over the way the environment is valued. The fact that intergenerational equity considerations play such a large part in environmental economics, reflects the constraints under which the Neo-Classical paradigm is employed as a tool of economic policy. Societies recognize questions of 'intergenerational equity' in a variety of ways, and these concerns are reflected in a variety of social science disciplines, notably in anthropology, jurisprudence and philosophy. Only economics has sought to incorporate the choices of future generations into its methodological corpus, making environmental questions more amenable to economic treatment.

Sustainable development also answers to problems of allocation in space: allocation of resources between 'here' and 'elsewhere'. These are the problems of intragenerational equity, especially between different societies and between North and South, which are often ignored in the discussion. Instead of addressing these issues, attention has focused on the future costs of development to our own societies, as if the satisfaction of our future needs is the principal bone of contention, rather than the way we currently satisfy our needs at other people's expense.

The importance, indeed primacy, of inter-generational equity is vividly illustrated by the currency given to the definition of sustainable development used by the Brundtland Commission: ". . . development that meets the needs of the present without compromising the ability of future genera-tions to meet their own needs" (WCED 1987). This is an enticing definition, in many ways, but it begs at least as many questions as it purports to answer. This definition carries the clear implication that 'needs' can be divorced from the development process itself, that they are not part of development, but can be arrived at independently.

The experience of the last two centuries in the industrialized world hardly bears out this proposition. Some would argue that needs can be viewed from a relativist posi-tion; that is, they are essentially historically determined. How needs are defined then depends on who is doing the defining. The knowledge we have of needs changes over time, and is linked to our ability to satisfy them. Each society 'defines' needs in its own way, and has evolved quite complex mech-anisms to reassure its members that their needs are being met. Others, including many within the contemporary Green movement, would argue that we need to discriminate between these mechanisms, as the 'satisfiers' of needs, and the needs themselves, which are thought to be ahistorical.

Doyal and Gough (1991) point out that 'needs' are different things to different people, and distinguish different ideological positions on human needs. Thus, to Neo-Classical economists needs are preferences; to the New Right needs are dangerous; to Marxists needs are historical; to anthropolo-gists needs are group specific; to radical democrats needs are discursive; and to phenomenologists they are socially constructed (Doyal and Gough 1991). Again a reflection of Lévi-Strauss is quite illumin-ating in this instance. He writes that:

> . . . cultures . . . appear to us to be in more active *development* when moving in the same direction as our own, and stationary when they are following another line . . . (Lévi-Strauss 1958: 25)

In other words, cultures only appear to be stationary because their ways of meeting their own needs are unfamiliar to us, often without meaning, and 'cannot be measured in terms of the criteria we employ' (Lévi-Strauss 1958: 23).

Sustainable Development and the Problem of Legitimation

It was argued above that a distinguishing feature of Modernism is that it purports to legitimate its own discursive practices. The discourse surrounding sustainable develop-ment is therefore a metadiscourse, in which the claims to provide insights can only be evaluated in terms of the discourse itself. From this point of view, the important point about sustainable development is that *the choice of a biological concept (sustainability) leaves open the possibility that it can be treated both as a model and as a point of legitimation.* The natural world is used, in other words, both as a model for systems based on human intervention and, ultimately, a constraint on human development. Sustainability appears to provide a point of reference outside the confines of human experience, which can also serve to guide human choices. We are dealing, then, with both the naturalization of social behaviour, and the legitimacy conferred on that behaviour by reference to natural laws. But, if sustainable develop-ment appears to provide a reference outside the confines of human experience, which is relevant to that experience, how seriously should we take this comparison? To what extent is the choice of biological systems merely one of metaphor, and to what extent is the natural environment treated as a point of reference? To answer these questions involves a short digression into the biological significance of 'sustainability'.

The concept of sustainability in ecology is an important one. Within plant ecology 'sustainable' is related to the successional changes in plant communities which might serve as a model for the management of forests and rangeland. The key idea is that environmental management can benefit from referring to natural succession, from utilizing the knowledge we have acquired about natural, ecological systems. The principle of 'sustainable yields' has become well established in certain fields of environmental management, particularly in fisheries management and forestry.

In ecological terms the most mature natural systems are those in which energy shifts away from production towards the maintenance of the system itself. The best example are tropical forests, which exhibit all the features of a 'climax system', including enormous natural diversity, and in which the survival of the species is guaranteed by the complexity of system interactions. *provided human is zero*

From an ecological point of view, agricultural systems are always modifications, in which natural ecosystems have been interfered with by human beings, usually to enhance their productivity at the expense of their sustainability. To an ecologist, then, agricultural systems should demonstrate the capacity to renew themselves, to regenerate, in the face of disturbances in their natural evolution.

The ability of agricultural systems to withstand disturbances, stresses and shocks, is the principal characteristic of sustainable agricultural systems, according to Conway (1985). It was also the guiding principle which lay behind the first World Conservation Strategy, published in 1980, that of 'sustainable utilization' (WCS 1980). In the hands of writers like Odum (1971) the ecologists' interest in 'evolutionary adjustment', the idea that ecosystems only evolve

successfully when they are protected from rapid changes, served as a guide to the way that power was exercised in human society. Ecology provided ideas about the way systems work, including systems subject to human intervention, at the same time highlighting the point beyond which such systems are no longer 'sustainable'.

The use of a concept drawn from ecology, rather than one of the social sciences, can be regarded as conveying several ambiguities:

1   It is often not clear when biological systems are being used as a metaphor, and when as a referent.
2   Sustainable systems occurring in nature were used as a model for environmental and resource management, without reference to the differences introduced by human needs and choices.
3   *Incorporating the ecological idea of sustainability represents a way of viewing the shortcomings, or contradictions, of 'development'. These shortcomings are usually seen as a 'malfunction' of the system and, as such, one which can be addressed by human intervention.* For example, the Brundtland Commission reported that ". . . as a system approaches ecological limits, inequalities (in access to resources) sharpen. . ." (WCED 1987: 49). The implication is that distribution problems, of intragenerational equity, are made worse by the failure to adhere to sound ecological principles, which should inform global policy as well as local environmental management.

The confusion surrounding our inability to behave according to biological injunctions, represented here by the absence of sustainability, is central to the appeal of sustainable development. It eases the passage from 'scientific' uncertainty to political prescription. It provides a moral force, which we have seen as essential to Modernist discourse, that seeks to engage our emotions as well as our minds. As we shall argue later, this tendency to provide a normative foundation for allegedly 'scientific' injunctions, reaches its clearest expression in relation to global climate change. The call to take measures to avert 'global warming', and the suggestion that we have already reached unsustainable levels in our dependence on hydrocarbons, acquire moral, as well as scientific authority, from the profligacy of the current development model. Whereas the dilemma posed by the environmental lobby in the 1960s and 1970s was that the scarcity of resources posed 'limits to growth', in the 1990s the principal threat to our survival is identified in the 'externalities' (notably global warming and ozone depletion) of the growth model itself.

In the next section we examine the theory of value that underpins much of the discourse on sustainable development. If the concept of sustainable development represents an attempt to bring together ethical injunctions and scientific authority, the importance of the concept can be seen most clearly in the way it has come to inform economic policy making. Environmental economics, we shall argue, proceeds by setting aside most of the problems associated with science as a social process, and the social authority of scientific knowledge. Recourse is made instead to nature, and the laws of 'natural capital', providing increased legitimacy for the otherwise contested concept of 'development'.

Economic Values and Sustainable Development: Paradigm Regained

Environmental economics has sought to extend Neo-Classical theory, by encompassing the environment, and attaching monetary values to losses in natural capital. The revisionist case, from within economics, begins with the observation that changes in wealth are only recorded in the national income figures *when they pass through the market*. This leaves changes in the stock of environmental capital outside the basis of calculation. There are two aspects of the 'conventional' approach to environmental assets that are criticised:

1   Conventional economic accounting frequently regards the destruction of

resources as a contribution to wealth. For example, the destruction of tropical forests is recorded as an increase in Gross Domestic Product (GDP) in the national accounts.

2   At the same time, the cost of making good any environmental damage is recorded as a positive contribution to GDP. The costs of reducing pollution, for example, and of measures to prevent pollution, are registered as contributions to economic growth.

However, these two propositions lead to a paradox. If the costs of environmental redress are counted as a contribution to economic wealth creation, then logically pollution itself should be recorded as a cost against economic growth. We know, after all, at the intuitive level, that pollution (and other forms of resource degradation and depletion) actually reduces the value of the environment. Economics, in its conventional form, seems unable to recognize this fact.

Following our intuitive logic, nevertheless, carries problems. If we count pollution as a 'cost' against natural capital, we must also argue that developed countries, which pay a considerable amount for environmental abatement, are 'poorer' as a consequence, although pollution abatement is clearly a benefit.

They are paying, in effect, to maintain the quality of their natural capital stock. Presumably countries which are unable to pay the price of environmental abatement, or choose not to do so (including most developing countries), are therefore 'richer' than they would have been had they paid for the costs of abatement. This appears to lack logic, and to run counter to the experience of the real world, in which resource management in developing countries is increasingly unsustainable.

There is another problem, too. If we measure changes in environmental quality, such as air pollution, land degradation and the loss of species, how do we do it? There are two obvious answers which economists give to this question. First, we can measure the reduction in environmental quality by the cost of restoration. Second, we can calculate how much consumers are prepared to spend to maintain environmental quality (contingent valuation). In practice there are problems with both answers. Some environments, like that of the African Sahel for example, cannot be restored to their former quality, even if we knew precisely what it was. In addition, contingent valuation, for its part, is a very inadequate tool for measuring the value of environments to groups of people, as we shall see later.

It is clear, then, that problems with the definition, methodology and techniques of environmental economics, strike at the heart of the debate surrounding sustainable development. It is all the more important, then, that we are clear about the territory on which we seek to argue. Environmental economists like Pearce have advanced thinking within economics in a number of important ways, all of which retain importance for other social sciences such as sociology.

(a) They recognize that changes in natural capital stocks involve both costs and benefits. Unlike most conventional economics, environmental values are not determined by income flows alone, but by the stock of (natural) capital. Wetlands, for example, are converted, through drainage, for agricultural use. Similarly, the open seas are used as sinks for the disposal of wastes (Pearce et al. 1989: 5).

(b) There is also a recognition that 'non-use' values are important, as well as 'use-values'. Environmental values include 'existence values' which can be arrived at notionally, and a (theoretical) price attached to them.

(c) Environmental economics also recognizes uncertainty and irreversibility as principles in resource conservation, and as reasons for conserving the stock of natural capital, even if it is not 'optimal' (Pearce et al. 1989: 7).

(d) It is also recognized that 'optimality' refers only to economic uses and efficiency, and effectively excludes any social goals for conserving resources. (This does not prevent Pearce from encountering a problem in 'non-

efficiency' goals, in that they might "be better served by converting natural capital into man-made capital. . ." [Pearce 1989: 7].)

The difficulty in fully incorporating social goals within the analysis of optimal resource utilization is, paradoxically, demonstrated by the principle which is used to defend it. Pearce declares that *we know natural capital is valuable because people are willing to pay to preserve it . . .*

A simple conceptual basis for estimating a benefit is to find out what people are willing to pay to secure it. Thus, if we have an environmental asset and there is the possibility of increasing its size, a measure of the economic value of the increase in size will be the sums that people are willing to pay to ensure that the asset is obtained. *Whether there is an actual market in the asset or not is not of great relevance. We can still find out what people would pay if only there were a market* (Pearce 1989: 8).

It is clear from the foregoing discussion that environmental economists like Pearce have proved able to push back the boundaries of the Neo-Classical paradigm, and to accommodate environmental concerns in their analysis. However, this accommodation has come at a price. Essentially, the analysis has widened the bounds of consumer choice, enabling the individual's preferences to be expressed. In the next section we will examine the limitations of this approach. For the moment we need only register the fact that environmental economics leaves the Neo-Classical paradigm intact. Market values, or imputed market values, can be used to provide a fuller account of natural capital, and the benefits of sustainability. In seeking sustainable development, Pearce notes that "... *what constitutes development, and the time horizon to be adopted, are both ethically and practically determined*" (Pearce 1989: 3). This observation should lead us to consider not only the political context in which decisions are taken about the environment, but also the circumstances under which environmental economics is used to help facilitate decisions. If 'development' is subject to value judgements, and lies outside the compass of objective science, why is environmental economics not subject to the same value judgements? Is development to be subject to value judgements, but not the paradigm within which it is understood . . . ?

The Limitations of Environmental Economics: Paradigm Lost

We have argued that environmental economics has succeeded in enlarging the Neo-Classical paradigm, with important consequences for the way in which the economic values of environmental impacts are calculated. It now remains to examine the assumptions of the paradigm itself.

The first problem with the paradigm is that it fails to recognize that monetary values are *always* exchange values, not use values. When Pearce refers to 'use benefits' and 'use values' he is referring to exploitation values. Use values do not attract monetary values because they exist outside the framework of market pricing. Environmental economists will argue that this is no impediment to using monetary values for them, and that the way we arrive at these prices is a matter of methodological refinement. Economics has developed techniques to impute such values, in the form of shadow pricing, and contingent valuation. There are no barriers to attaching prices on environmental goods and services, it will be objected – merely misplaced ideological objections.

This is to miss the point. *Economists cannot value what the environment is worth; merely its value in monetary terms.* As Oscar Wilde maintained, it is possible to find a price for everything and the value of nothing. The point is that monetary valuations do not capture the worth of the environment to different groups of people. Giving increased value to environmental assets is not simply a question of attaching larger figures to assets in the course of cost-benefit analysis. As Elson and Redclift (1992) demonstrate, this means attaching *cardinal valuations* through monetary measures, such as prices and taxes, when *ordinal valuations* (more/less valuable) may be more appropriate, and useful. We can undertake

valuation by establishing the thresholds that operate for real people in the real world, rather than through monetization.

Let us use women's labour in the forest communities of the developing world as an example. Men and women value the environment differently because of the different use they make of it. The value women attach to the environment is usually invisible to others because the use they make of it is not subject to market values. Nevertheless women's activities, such as collecting firewood, gathering plants and fetching water, for both use and exchange, are vital for the sustainability of poor rural households. Many of the environmental goods that women collect, and households use, are 'free goods' in nature but vitally important for survival. Elson and Redclift (1992) note that one tribal community in Andra Pradesh could identify one hundred and sixty nine different items of consumption, drawn from forest and bush land. Environmental accounting is ill-equipped to measure the real value of the environment to women, when these use values are part of direct household provisioning.

The second problem with the paradigm is that it claims 'value neutrality', when environmental economics itself expresses the preferences and biases of the society in which it was developed. Values are a reflection of specific social systems, and express degrees of commitment to a specific social order, the order which espouses them. The values we place on nature, not surprisingly, reflect our priorities, not *the value of nature itself*. Nature is a mirror to our system of values, and in seeking monetary values for environmental goods and services we are merely 'naturalizing' our concern with market values.

Environmental economics provides a good illustration of the way we seek to construct the environment socially, through the mechanism of monetary valuation. Progress within the discipline aims to extend the paradigm, rather than to place it within its political and social context. Development projects, for example, such as large dams or irrigation schemes, are said to have 'environmental consequences', which environmental economics is well-placed to address.

This is to ignore the fact that development projects are socially created and socially implemented. They already internalize a view of nature, in their methodology and practices. They also seek to acquire legitimacy for the idea of projects – another instance of the way they are socially constructed. Development projects have already internalized a view of nature *from which environmental consequences themselves spring*. (In the same way ecological projects have internalized a position on society; but this is a still more difficult nut to crack!) What is at issue, then, is the appropriateness of environmental economics, which does not recognize its own relativism, in evaluating projects which are themselves an expression of specific values and interests in the social order.

There is a third area in which the Neo-Classical model can be faulted. It is that this model fails to recognise that *conventional economic analysis rests on a particular view of human nature and social relations*. It is important to establish the elements in this model of social relations which underlie the methodology of monetary valuation.

First, environmental economics sees social interaction as instrumental. That is, it is designed to maximize the individual's utility. As Hodgson points out, within environmental economics the tastes and preferences of individuals are considered a given (Hodgson 1992: 40).

Second, and related to this, environmental economics does not see social interaction as constituting value in its own right, because of the intrinsic value of human beings. Social interaction reveals the person as an 'object', surrounded by other 'objects', rather than a 'subject person' (in Max-Neef's phrase) able and willing to behave in ways that do not correspond to short-term economic advantage (Ekins and Max-Neef 1992). It is this failure to recognize human behaviour as culturally determined, and capable of a very wide range of variability, which cannot be easily married with the reductionism of economics.

The 'rational, individual calculator'

beloved by economists sits uneasily in cultures other than those which helped to develop the paradigm in the first place. (And none too easily in many areas of behaviour within the developed countries.) This individual is supposed to make choices, expressed as market preferences, within a 'neutral' cultural context. Social and economic processes are never 'rational' in a universal sense: rationality is always culturally grounded. The calculations of individuals are not the same as individual calculations.

The calculations of individuals can be best understood as the outcome of social processes, peculiar (and unique) to every society. Concepts like the 'willingness to pay' concept, used by environmental economists, presuppose a set of cultural and ideological assumptions. Returning to the example we gave from Pearce earlier, although economists might look upon the North Sea as a 'waste sink resource', fishing communities in the area would view it otherwise, as would holidaymakers, or artists, or any individual or group of individuals.

The problem for modern environmental economics is compounded by a fourth set of issues, which concern the degree to which the 'individual, rational calculator' is fully appraised of the situation in which he is being asked to make choices. As Gleick puts it:

Modern economics relies heavily on the efficient market theory. Knowledge is assumed to flow freely from place to place. The people making important decisions are supposed to have access to more or less the same body of information . . . (Gleick 1988: 181)

These objections to the paradigm on which environmental economics is founded, suggest that environmental economics has a certain technical competency, in attaching monetary values to environmental benefits and losses, but that this competence, important though it is, should not be confused with an adequate basis for environmental valuation. Indeed, we need to look at environmental economics within a wider context, in which we consider it as a product of society itself.

Before considering where this leaves our discussion of sustainable development, we should examine the wider policy context from another perspective, which builds on the points above. We need to look at the environment within the context of the way science itself is socially constructed.

The Environment and the Social Construction of Science

It is clear that the view we take of the environment is closely bound up with the view we take of science. Increasingly, environmental problems are looked upon as scientific problems, amenable to scientific 'answers'. An example is the current set of policy prescriptions surrounding global environmental changes, particularly global warming. Since global warming is a 'scientific' problem, it is assumed that it must have a scientific solution. The 'greenhouse effect' is viewed as carrying social and economic implications, but scarcely as an 'effect', in that the human behaviour which underlies global warming is rarely considered. More attention is paid to ways of mitigating the effects of global warming, than to its causes in human behaviour and choices, the underlying social commitments which make up our daily lives.

Part of the problem with this approach is that the modes of inquiry in the natural sciences are themselves *social processes, into which crucial assumptions, choices, conventions and risks, are necessarily built.* Once we regard science as outside ourselves it becomes impossible to take responsibility for its consequences. And so it is with global warming: relegated to the sphere of 'consequences' we are able to avoid the environmental implications of our own behaviour, and that of our societies.

This process of disengagement from the consequences of our behaviour is well established in a variety of ways. The second World Conservation Strategy document *Caring for the Earth* (IUCN 1991), provides a useful example, in diagrammatic form, of the model through which we manage our

resources (IUCN 1991: 76). The model portrays the way in which economic development, driven by fossil fuel resources, has taken society along a path which ignores the limits imposed by renewable solar energy (in all its forms). Instead our development model has channelled the material wealth which fossil hydrocarbons have helped to create, towards the creation of capital goods, themselves dependent on further fossil fuel exploitation. The model thus ensures a continued and spiralling demand for scarce and ultimately finite resources, which are fast contributing to global nemesis, by posing the ultimate, unsustainable, problem for the economies which consume them. Global warming, the loss of biodiversity, the problems associated with the 'ozone hole', and other global environmental changes, represent the ultimate 'externality', and point to problems in the growth model itself.

As *Caring for the Earth* points out, by concentrating investment on surplus value in order to maximize the accumulation of industrial capital, we have tended to neglect natural capital (environmental goods and services). Instead we have focused our support on the development of human capital, within a small intellectual elite, working within spheres of knowledge which are closely allied to new technologies and often wasteful resource uses.

This illustrates the lack of congruity between technological and scientific knowledge, and the social implications of using that knowledge in specific ways. Wherever we look – nuclear power, toxic wastes, pesticides, air pollution, water quality – we see examples of our failure to grasp the social implications of the scientific knowledge we possess, and the costs which are passed on to the environment. We know that environmental science cannot make political choices about the consequences of technology for the environment. At the same time, environmental policy is nothing more than the formulation of one set of social and political choices, governing environmental uses, over another set of choices. It is hardly surprising that the discussion and practice of sustainable development is intimately linked to the social authority of our science and technology. In the North this authority is increasingly contested, especially by environmental groups and interested citizens. In the South it is frequently ignored, notably by development institutions, whose model of 'development' acknowledges no social authority but that of science. As we have argued, that is why development in the South is ultimately not socially and politically sustainable.

Where does this leave our discussion of sustainable development? It soon becomes clear that we cannot achieve more ecologically sustainable development without ensuring that it is also socially sustainable. We need to recognize, in fact, that our definition of what is ecologically sustainable answers to human purposes and needs as well as ecological parameters.

By the same token, we cannot achieve more socially sustainable development in a way that effectively excludes ecological factors from consideration. If the model for better environmental policy merely 'adds on' environmental considerations to existing models it is not equipped to provide a long-term view. The strong sense of 'sustainable development' emphasizes the sustainability of the interrelationship between biological, economic and social systems, rather than that of the component parts. Each system involves elements – social 'needs', levels of production, biodiversity – which are subject to modification. It follows that any environmental social science that does not seek to rethink the development agenda is ill-equipped to address environmental problems.

Conclusion

We have argued that much of the writing on sustainable development takes its message from the natural sciences. In the past this has been a message of hope, as people have lived longer, and consumed more goods, especially in the North. Sustainable development, in this tradition, is about seeking consensus and agreement, in the belief that we can manage the contradictions of development better. Sustainable development, then, represents a renewal of Modernism.

A more critical perspective regards science as part of the problem, as well as the solution. It takes issue with the inevitability of economic growth, and its consequences for the aptly-named 'hydrocarbon society'. It argues that the limits placed on development, are not merely limits in the resources available to us, as was believed in the 1970s. The limits today are 'external limits' too, represented most vividly by the challenge of global warming. The more critical perspective suggests that environmental management, as a strategy to cope with the externalities of the development model, is found wanting. Modern economics has played a major role in the unsustainable development that characterizes North and South. For the pursuit of growth, and neglect of its ecological consequences, has its roots in the classical paradigm which informed both market economies and state socialist ones.

If we are to meet the problems presented by imminent global nemesis, we need to go beyond the assertion that such problems are themselves socially-constructed. We need to embrace a realist position, while recognizing and acknowledging the relativism of our values and our policy instruments. The challenge is to develop a 'third view', which enables us to assume responsibility for our actions, while exploring the need to change our underlying social commitments. We need to develop a broader and deeper foundation for the formulation of a realist policy agenda, and one which, unlike environmental economics, does not exclude 'interests' from its calculus. Sustainable development answers to problems initiated in the north, with our 'global' development model. We have also understood sustainable development within a cultural context of our own – *our* view of nature, *our* problems with science and technology, *our* confidence in the benefits of economic growth. But sustainable development has become a 'global' project, and our capacity to find solutions is seriously reduced by our inability to recognize we are the prisoners of our history. The global project is being developed in ignorance of the intellectual history which contributed to global environmental problems in the first place and made us poorly equipped to deal with them. It is time to redraw the frontiers of knowledge and belief, and to recognize that they both have a part to play in avoiding global nemesis.

References

Conway, G. 1985. "Agro-ecosystem analysis". *Agricultural Administration* 20: 31–55.

Cooper, D.E. and Palmer, J.A. 1992. *The Environment in Question*. London, Routledge.

de Vries, H.J.M. 1989. *Sustainable Development*. Groningen, Netherlands.

Doyal, L. and Gough, I. 1991. *A Theory of Human Need*. London, Macmillan.

Ekins, P. and Max-Neef, M. (eds.) 1992. *Real-life Economics*. London, Routledge.

Elson, D. and Redclift, N. 1992. *Gender and Sustainable Development*. London, Overseas Development Agency.

Engels, F. 1970. "Introduction to the dialectics of nature". In *Selected Works*, edited by K. Marx and F. Engels (one volume). London, Lawrence and Wishart.

Fraser, N. and Nicholson, L.J. 1990. "Social criticism without philosophy: an encounter between Feminism and Postmodernism". In *Feminism/Postmodernism*, edited by Linda J. Nicholson. New York and London, Routledge.

Glacken, C. 1973. *Traces on the Rhodian Shore*. University of California Press.

Gleick, J. 1988. *Chaos: Making a New Science*. London, Heinemann.

Hodgson, G. 1992. "Rationality and the influence of institutions". In *Real-life Economics*, edited by P. Ekins and M. Max-Neef, pp. 40–8. London, Routledge.

IUCN (International Union for Conservation of Nature) 1991. *Caring for the Earth*. Gland, Switzerland, IUCN/UNEP/WWF.

Jacobs, M. 1990. *The Green Economy*. London, Pluto Press.

Lévi-Strauss, C. 1958. *Race and History*. UNESCO, Paris.

Lyotard, J. 1984. *The Postmodern Condition*. University of Minnesota Press.

Merchant, C. 1980. *The Death of Nature*. New York, Harper and Row.

Odum, H.E. 1971. *Environment, Power and Society*. New York, John Wiley.

Pearce, D., Barbier, E. and Markandya, A. 1989. *Sustainable Development: Economics and Environment in the Third World*. London, Earthscan.

Redclift, M.R. 1987. *Sustainable Development: Exploring the Contradictions*. London, Methuen.

Sklair, L. 1970. *The Theory of Progress*. London, Routledge and Kegan Paul.

WCED (World Commission for Environment and Development) 1987. *Our Common Future*. Oxford University Press.

WCS 1980. *World Conservation Strategy*. Gland, Switzerland, IUCN/UNEP/WWF.

32

THE SOLUTION TO A GLOBAL CRISIS*

Gro Harlem Brundtland

We are gathered here to answer a moral call to action. Solidarity with present and future generations has its price. If we do not pay it in full, we will be faced with global bankruptcy.

This conference is about the future of democracy, how to widen and strengthen its forces and scope. Unless we empower our people, educate them, care for their health, and allow them to enter an economic life rich in opportunity on an equal basis, poverty will persist, ignorance will be pandemic, and the people's needs will suffocate under their numbers. The items and issues of this conference are therefore not merely items and issues, but building blocks in our global democracy.

It is entirely proper to address the future of civilization here in a cradle of civilization. We owe a great debt to President Mubarak and the people of Egypt for inviting us to the banks of the Nile, where the relationship between people and resources is so visible and where the contrast between perma- nence and change is so evident. We are also indebted to Nafis Sadik and her devoted staff, who have prepared the conference with intensive care and inspiration.

Ten years of experience as a physician and 20 as a politician have taught me that improved life conditions, a greater range of choices, access to unbiased information, and true international solidarity are the signs of human progress.

We now possess a rich library of analyses of relationships between population growth, poverty, the status of women, and consumption patterns, as well as of policies that work and policies that do not, and of the environmental degradation that is acceler- ating at this very moment.

We are here not to repent but to make a pledge. We pledge to change policies. When we adopt our plan of action, we sign a promise – a promise to allocate more resources next year than we did this year to health care systems, education, family plan- ning, and the struggle against AIDS. We promise to make men and women equal before the law by rectifying disparities and promoting women's needs more actively than men's until we can safely say that equality is reached.

We need to use our combined resources more efficiently through a reformed and better coordinated United Nations system. This is essential to counteract the crisis threatening international cooperation today.

In many countries where population growth is higher than economic growth, the

* Originally published in *Environment*, 1994, vol. 36, no. 10, pp. 17–20.

problems are exacerbated each year. The costs of future social needs are soaring. The punity for inaction will be severe and a legacy that future generations do not deserve.

The benefits of policy change are so great that we cannot afford not to change policy. We must measure the benefits of successful population policies in savings – in public expenditure on infrastructure, health and social services, housing, sewage treatment, and education. Egyptian calculations show that for every pound invested in family planning today, 30 pounds are saved in future expenditures on food subsidies, education, water, sewage, housing and health.

Ninety-five percent of population increase takes place in developing countries – in those communities least equipped to bear the burden of rising numbers. They are the ecologically fragile areas where current numbers already reflect an appalling disequilibrium between people and Earth's resources.

The preponderance of young people in many of our societies means that there will be an absolute increase in the population figures for many years ahead, regardless of the strategy we adopt here in Cairo. But the outcome of the Cairo conference may significantly determine whether global population can be stabilized early enough and at a level for humankind and the global environment to survive.

The final program of action must embody irreversible commitments toward strengthening the role and status of women. We must all be prepared to be held accountable. That is how democracy works.

The plan of action must promise access to education and basic reproductive health services, including family planning as a universal right for all.

Women will not become more empowered merely because we want them to be, but through legislative changes, increased information, and redirection of resources. It would be fatal to overlook the urgency of this issue.

For too long, women have had restricted access to equal rights. It cannot be repeated often enough that there are few investments that bring greater rewards for the population as a whole than the investment in women. But still they are being patronized and discriminated against in terms of access to education, productive assets, credit, income and services, inclusion in decision-making, and equal treatment in working conditions and pay. For too many women in too many countries, development has been only an illusion.

Women's education is the single most important path to higher productivity, lower infant mortality, and lower fertility. The economic returns on investment in women's education are generally comparable to those of men, but the social returns in terms of health and fertility by far exceed what we gain from the education of men. The girl who receives her diploma will have fewer babies than her sister who does not. Let us pledge to watch over the numbers of girls enrolling in school. Let us monitor also the numbers of girls that complete their education and question disparities between the two.

It is encouraging that there is already so much common ground between us. I am pleased by the emerging consensus that everyone should have access to the whole range of family planning services at an affordable price. Sometimes, however, religion is a major obstacle. This happens when family planning is made a moral issue. Morality cannot be only a question of controlling sexuality and protecting unborn life. Morality is also a question of giving individuals the opportunity of choice, of suppressing coercion of all kinds, and of abolishing incrimination of individual tragedy. Morality becomes hypocrisy when it means accepting the suffering and death of women in connection with unwanted pregnancies and illegal abortions and accepting that unwanted children live in misery.

None of us can ignore that abortions occur. Where they are illegal or heavily restricted, the life and health of the pregnant woman are often at risk. The decriminalization of abortion is, therefore, necessary to protect the life and health of women.

Traditional religious and cultural obstacles can be overcome by economic and social development, which focuses on the enhancement of human resources. Buddhist Thailand, Moslem Indonesia, and Catholic Italy demonstrate that relatively sharp reductions in fertility can be achieved in an amazingly short time.

It is encouraging that the International Conference on Population and Development [ICPD] will expand the focus of family planning programs to include concern for sexually transmitted diseases and emphasize the processes of pregnancy, delivery, and abortion. It is tragic that it had to take a disaster like the HIV/AIDS pandemic to open our eyes to the importance of combatting sexually transmitted diseases. It is also tragic that so many women died from pregnancies before we realized that the traditional mother-and-child health programs – effective in saving the lives of so many children – have done little to save the lives of women.

In planning for the future, therefore, it seems sensible to combine health concerns that deal with human sexuality under the heading "reproductive health care." I have tried in vain to understand how that term can possibly be read as promoting abortions or endorsing abortion as a means of family planning. Rarely, if ever, have so many misrepresentations been used to imply meaning that was never there in the first place.

The total number of abortions in Norway stayed the same after abortion was legalized, while illegal abortions sank to zero. The same is true of other countries; namely, the law has an impact on the decision-making process and on the safety of abortion – but not on the total number of abortions. Our abortion rate is one of the lowest in the world.

Unsafe abortion is a major public health problem in most corners of the globe. We know full well that wealthy people often manage to pay their way to safe abortion regardless of the law.

A conference of this status and importance should not accept attempts to distort facts or neglect the agony of millions of women risking their lives and health. I simply refuse to believe that the stalemate reached over this crucial question will be allowed to block a serious and productive outcome of the Cairo conference, based on full consensus and adopted in good faith.

Reproductive health services not only deal with problems that have been neglected, they also cater to clients who have previously been overlooked. Young people and single persons have received too little help, and continue to do so, because family planning clinics seldom meet their needs. Fear of promoting promiscuity is often said to be the reason for restricting family planning services to married couples. But we know that lack of education and services does not deter adolescents and unmarried persons from sexual activity. On the contrary, there is increasing evidence from many countries, including my own, that sex education promotes responsible sexual behavior, including abstinence. Lack of reproductive health services makes sexual activity more risky for both sexes, but particularly for women.

As young people stand at the threshold of adulthood, their emerging sexuality is too often met with suspicion or plainly ignored. At this vulnerable time in life, adolescents need both guidance and independence, they need education as well as the opportunity to explore life for themselves. This requires tact and a delicately balanced approach from their parents and society at large. It is my sincere hope that ICPD will contribute to increased understanding and greater commitment to the reproductive health needs of young people, including the provision of confidential health services.

Visions are needed to bring about change. But we must also realize our visions through the allocation of resources. The price tag for the program that we are to adopt here has been estimated to be between $17 billion and $20 billion per year.

The real challenge begins when ICPD is over, when it is time to translate the new approach and objectives into implementation of programs. Norway will continue to participate in a dialogue with our bilateral and multilateral partners. We are pleased to see that important donors, such as the United States and Japan, are now increasing their support to population issues. Other countries should follow suit. Hopefully, Norway will soon be joined by other donor countries fulfilling the target of allocating at least 4 percent of official development assistance [ODA] to population programs.

It is also important that governments devote 20 percent of their expenditures to the social sector and that 20 percent of ODA is allocated toward eradicating poverty.

To meet the cost requirements of this plan of action, however, another long-standing target needs to be fulfilled – the 0.7 percent of gross domestic product for ODA. The so-called donor fatigue – attributed to the general budgetary problems of the industrialized world – will certainly not facilitate this challenge. The budgetary priorities and allocations are debated by national governments every year. And the 1-percent-and-above allocation to ODA, which Norway has been able to defend over the last 15 years or so, does not materialize without serious political work. Our work would be greatly facilitated if other donor countries were to begin approaching the target of 0.7 percent, and – important both to Norway and to the whole donor community – if this conference proves by its outcome that we are truly committed to creating a new, real solidarity with the world's poor and underprivileged.

Population growth is one of the most serious obstacles to world prosperity and sustainable development. We may soon be facing new famine, mass migration, destabilization, and even armed struggle as peoples compete for ever more scarce land and water resources.

In the more developed countries, today's generations may be able to delay their confrontation with the imminent environmental crisis, but their children will be facing the ultimate collapse of vital resource bases.

To achieve a sustainable balance between the number of people and the amount of natural resources that can be consumed, both the peoples of the industrialized North and the privileged in the developing South have a special obligation to reduce their ecological impact.

Changes are needed in both the North and the South, and these changes need to occur within the frameworks of democracy. Only when people have the right to take part in the shaping of society by participating in democratic political processes will changes be politically sustainable. Only then can we fulfill the hopes and aspirations of generations yet unborn.

I take this opportunity to summon and challenge this conference to step up to its responsibility toward coming generations. We did not succeed in Rio with regard to population. Cairo must be successful – for Earth's sake.

33

AGENDA 21*

Matthias Koch and Michael Grubb

The Nature of Agenda 21

Agenda 21 is intended to set out an international programme of action for achieving sustainable development in the 21st Century. It seeks to be comprehensive in its scope, and to make recommendations on the measures which should be taken to integrate environment and development concerns. To this end, it provides a broad review of issues pertaining to sustainable development, including statements on the basis for action, objectives, recommended activities, and means of implementation. These are based on experience and analysis of the issues, combined with the interests of different parties, including interests brought forward by both developing and industrialised countries.

Style and status of the document

Agenda 21 is an immense document, but it is still far from clear quite what has been created. It is not a legal agreement; governments are not required to follow each recommendation, paragraph by paragraph and line by line. This, and the fact that it was prepared through a relatively broad participatory process, allowed a different flavour to emerge from that which is typically found in international conventions. Compared with these, Agenda 21 contains far more about the nature of the problems, aims, possible approaches and desirable policies, often expressed in rather general and non-committal ways.

On casual reading, many parts of the Agenda 21 text appear either as statements of the obvious, or as a simplistic policy 'wish list'. Yet many things are 'obvious' once stated, and the elements of Agenda 21 frequently point to ways in which policy could perhaps readily be improved. A negotiated compendium of 'good' policy guidance still has some political weight, and thus in many respects, governments negotiated the text with all the ponderous care, sharp sensitivities and microscopic detail of a legal text.

Agenda 21 is perhaps best seen as a collection of agreed and negotiated wisdoms as to the nature of the problems and relevant principles of the desirable and feasible paths towards solutions, taking into account national and other interests. It stands as a grand testimony and guide to collected national insights and interests pertaining to sustainable development, against which government and other actions can and will be compared.

Agenda 21 covers several topics which are covered by the UNCED Conventions or

* Originally published in Grubb, M., Koch, M., Thomson, K., Munson, A. and Sullivan, F. *The 'Earth Summit' Agreements: a Guide and Assessment. An analysis of the Rio '92 UN Conference on Environment and Development*, London: Earthscan, pp. 97–158.

other existing conventions. As legal instruments, clearly the Conventions take precedence, and it was intended that the Agenda 21 chapters should support the Conventions, by outlining broader approaches and helping to set a framework of ideas for implementation. In some cases this was achieved, and the Agenda 21 text goes beyond that of the corresponding convention in its scope. However, many of the tensions during negotiations on the Conventions emerged again in the debate on Agenda 21, and in such cases language from the corresponding Convention was usually adopted. With its many cross-references to other meetings, statements and agreements, Agenda 21 also serves as a useful marker for the range of activities which have already been undertaken, or which are planned, including many of the most significant outputs from events held in the run-up to Rio.

With a mandate to address almost every issue pertaining to sustainable development, with a global scope, the process leading to Agenda 21 was inevitably of Byzantine complexity. In general, the texts which served as the basis of negotiations were presented by the UNCED Secretariat. Sometimes the Secretariat largely wrote the first drafts on the basis of many and varied inputs; in other cases, whole sections of text were lifted from documents prepared for UNCED by governments or non-governmental groups, that either volunteered contributions or were commissioned to do so. Various interested governments and non-governmental groups then proposed additional or alternative texts as negotiations proceeded. Most of the final text was then honed down during negotiations at the final PrepCom, and the most difficult sections were sent for final negotiation at Rio.

Merely to have completed such an undertaking within 22 months from the first PrepCom is not a trivial achievement. In addition to the sheer breadth of subjects, Agenda 21 also faced the difficulties raised by trying to address the concerns and problems of very disparate countries, ranging from the least to the most developed (although the main focus was upon developing countries). For example, it is very difficult to propose human health measures that can apply both to areas where even basic health services have not been developed and to those where services are established but the health problems are very different. In bringing together such disparate interests and attitudes, the process of developing Agenda 21, and UNCED itself, also served as an immense global educational exercise for those involved.

The main purpose of this chapter is simply to summarise the main content of Agenda 21, and to note key points of interpretation or dispute in each of its chapters. The concluding section then notes a number of underlying themes which recur throughout Agenda 21. These include the 'bottom-up' approach of putting the emphasis upon people, communities and NGOs; the need for 'open governance'; the importance of adequate information; the need for adequate cross-cutting institutions; and the complementarity between regulatory approaches and market mechanisms for addressing development and environmental needs. In general, Agenda 21 really does seek to integrate environment and development, and succeeds in highlighting the linkages between many different specific issues. However, inevitably some issues are absent or only weakly addressed so that despite its very broad scope Agenda 21 cannot quite stand as a comprehensive document.

Despite its limitations, Agenda 21 represents a remarkable and unique endeavour. Whether or not it is judged successful will depend very much upon its translation into national policies, action plans, legislation and guidelines at all levels of society, and the extent to which it influences activities within the UN system and other international bodies. Agenda 21 created the UN Commission on Sustainable Development to 'oversee its implementation', but this cannot 'enforce' the agreement in any legal sense. The real impact of Agenda 21 will depend upon the extent to which national governments and all the various groups discussed in the document, from local councils and trade unions to scientific groups, business and industry, absorb and pursue the recommendations therein, influenced also by the continued efforts of environment and development groups.

Structure of the document

Agenda 21 is a report of over 500 pages, comprising 40 chapters, organised into four sections which address the major areas of political action, i.e.

- social and economic development (Chapters 1–8);
- natural resources, fragile ecosystems and related human activities, byproducts of industrial production (Chapters 9–22);
- major groups (Chapters 23–32); and,
- means of implementation (Chapters 33–40).

Of these, the 13 chapters of the second section occupy almost half the total volume of Agenda 21. Each chapter focuses on a distinct issue, eg. poverty, freshwater, indigenous communities, legal instruments. Various programme areas in each chapter concentrate on particular problems of concern. The programme areas are generally structured as follows:

- *The Basis for action* addresses the background of concern, discusses efforts yet to be undertaken and problems which still persist.
- The *Objectives* section summarises the main issues to be addressed, and outlines the general themes to be implemented. Usually these do not set targets, but for some areas, including health and freshwater, specific goals were adopted.
- The *Activities* section outlines a variety of specific measures that the relevant bodies 'should' undertake[1]
- The *Means of implementation* section concludes by outlining the infor-

mation, resources and institutional requirements for implementing the programme area. This is usually subdivided into the following aspects:

i) financial and cost evaluation: the costs for each programme area as estimated by the UNCED secretariat;

ii) scientific and technological means: the need for further research and pilot projects, transfer of technology and sharing of expertise, as well as seminars and conferences to be undertaken;

iii) human resource development: generally expresses concern about the lack of skilled personnel particularly in developing countries and addresses the strengthening of training opportunities and education programmes;

iv) capacity-building: the need for institutional and technical capacities especially in developing countries, including aspects such as policies and administration, organisational structures, institutional and legal mechanisms, international cooperation and national coordination.

This structure is not always adopted, and sometimes overlaps or inconsistencies seem to have been unavoidable.

Financial issues permeated all the negotiations, being considered the crucial factor for the implementation of Agenda 21. Initially it was hoped that the funding mechanisms, including detailed proposals for the financial requirements of each programme area, could be negotiated, but this proved impossible. Instead, the estimates provided by the UNCED Secretariat were noted at the end of each programme area, together with a strong *caveat* indicating that these were 'indicative and order of magnitude estimates only and have not been reviewed by Governments'.

---

[1] The phrasing used to introduce these activities generally takes the form:
'Governments at the appropriate level/ Governments/States, with the support of/with the assistance of/in cooperation with . . . relevant United Nations bodies/international and regional organisations/intergovernmental and non-governmental organisations/private sector, [sometimes adding: where appropriate/in accordance with...] should...' The reason for different forms of phrasing is frequently unclear, but the general message is that the bodies addressed 'should' undertake the programme activities then set out.

The Preamble

The Preamble to Agenda 21 differs from the form adopted for legal documents; it serves more as a conventional introduction, allowing for general observations concerning expectations for the document, and *caveats* concerning it. The opening remarks outline the fundamental perceptions and goal of the exercise:

Humanity stands at a defining moment in history. We are confronted with a perpetuation of disparities between and within nations, a worsening of poverty, hunger, ill health and illiteracy, and the continuing deterioration of the ecosystems on which we depend for our well-being. However, integration of environment and development concerns and greater attention to them will lead to the fulfilment of basic needs, improved living standards for all, better protected and managed ecosystems and a safer, more prosperous future. No nation can achieve this on its own; but together we can – in a global partnership for sustainable development.

The Preamble reflects the complexity of addressing both environmental and developmental issues as a single overall objective. There is concern on the one hand about the devastating social and economic situation in many developing countries and on the other hand about the worsening perspective of the global environment.

The Preamble cites the 1989 UN enabling resolution (44/228) on the need to integrate environment and development concerns. To this end 'Agenda 21 addresses the pressing problems of today and also aims at preparing the world for the challenges of the next century. It reflects a global consensus and political commitment at the highest level on development and environment cooperation'.

The responsibilities and modalities are clearly established: 'Its successful implementation is first and foremost the responsibility of Governments. National strategies, plans, policies and processes are crucial in achieving this'. These shall be 'supported and supplemented' by international agencies, primarily the UN, and other international, regional and sub-regional organisations; and the 'broadest public participation and the active involvement of the non-governmental organizations and other groups should also be encouraged'.

The negotiations emphasised the deep concerns of developing countries about their economic situation in general, and the present net outflow of financial resources from the South to the North. They fear that their gloomy development perspective might even further be threatened by international environmental agreements, unless the international economic situation changes fundamentally. Consequently:

The developmental and environmental objectives of Agenda 21 will require a substantial flow of new and additional financial resources to developing countries, in order to cover the incremental costs for the actions they have to undertake to deal with global environmental problems and to accelerate sustainable development.

The 'new and additional' phrasing reflects the concern of the South that scarce development aid might only be redirected from existing overseas aid budgets to international environment funds. The cost estimations of the UNCED secretariat are noted as 'an indicative order of magnitude assessment, further examination and refinement is needed'.

The Rio Declaration is taken as providing guiding principles for Agenda 21: 'Agenda 21 . . . will be carried out . . . in full respect of all the principles contained in the Rio Declaration on Environment and Development'. In fact, this sentence was inserted primarily as a compromise designed to avoid disputes in Agenda 21 over references to the rights of 'people under occupation', but it has far broader implications.

Two additional *caveats* received special mention. The situation of East European and CIS countries ('economies in transition') is recognised in terms of 'particular circumstances', for which 'special attention' is needed. Also, the insistence of several oil-exporting Arab countries led to the addition that 'environmentally sound' always means 'environmentally safe and sound' – phrasing long understood as reflecting caution towards nuclear power. Both these elements of the Preamble were inserted at a very late stage, as a way of avoiding repetition of the corresponding contentious phrases throughout Agenda 21.

Thus, except in its brevity, the Preamble sets the tone for Agenda 21 itself: an ungainly compromise, with specific *caveats* for special concerns and interests; but one which nevertheless usefully defines the context, and clears the political path, for agreement on the subsequent 500 pages.

34

EQUITY AND ENVIRONMENTAL POLICY*

Judith Rees

The Environmental Challenge

Environmental policy makers face a massive and unprecedented challenge. It is now widely accepted that current and past generations have eaten into the Earth's environmental resources to such an extent that the well-being of future generations could be threatened. Intergenerational equity demands a major change in the way we utilise environmental systems. As the Government's White Paper *This Common Inheritance* put it: "we must not sacrifice our future well-being for short-term gains, nor pile up environmental debts which will burden our children" (Department of the Environment, 1990). However, such well-intentioned words cannot disguise the fact that the pace, scale and complexity of environmental change have far outrun the capacity of the political system to implement effective responses.

Today the political decision-making process appears to be far slower than the process of environmental change. To halt or markedly slow down the rate of environmental degradation there is widespread consensus that *three* very closely related alterations need to be made in our approach to environmental policy-making.

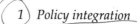 1 Policy integration

To date the environment has been treated as a separate policy arena, divorced from such mainstream political tasks as economic management, infrastructural development or social planning. This is most obvious at the global scale, where trade policies, international financial flows, patterns of direct overseas investment, aid and economic development strategies have rarely been seen as having much to do with the environment. It has now become fashionable for Governments to embrace the concept of sustainable development and to endorse the notion that environmental and economic policies are inextricably intertwined. The Commission of the European Community, for example, in its Fourth Environmental Action Programme 1987, argued that environmental considerations must be seen "as an essential part of economic and social development". However, it has proved far easier to talk about policy integration than it has been to achieve it. One important difficulty is that those making policy frequently are not those making the key economic decisions. As Chris Patten (then Secretary of State for the Environment) once put it "the most important parts of environmental

* Originally published in *Geography*, 1991, vol. 76, no. 4, pp. 292–303.

policy are handled elsewhere – the levers aren't in my office". In reality, many of the levers aren't in any government office, but in the complex processes which create social values, and in the institutions and structures which make up the global political economy.

2) Anticipatory or preventative policy

Typically environmental policy has been reactive, with attempts being made to control the undesirable outcomes of decisions made elsewhere in the economy. At worst policy has degenerated into fire-fighting or damage limitation exercises, concentrating on whichever specific environmental problems happen to have achieved political salience at different points in time. This 'react and regulate' approach is known to be costly; it inevitably follows, rather than prevents, pollution damage, and policy makers, regulators and the regulated constantly find themselves chasing last year's hare. In the meantime new hares are busily entering the race as scientific knowledge improves, a different potential threat is identified and an awakened public awareness creates demands for further action. For at least fifteen years the defects of reactive policy have been identified and calls made to improve anticipatory planning mechanisms. However once again, it has proved far easier to talk about such mechanisms than it is to actually find ones which are effective, implementable and politically acceptable.

3) Global policy

Although environmentalists have preached for over a century about the interconnectedness of all the Earth's environmental systems, it was not really until the 1980s that we all became accustomed to the idea that physical interdependencies require global mitigation strategies. International environmental conventions are now commonplace but as yet the organisations and mechanisms available to ensure real co-operative action can only be regarded as rudimentary. There have been numerous calls for the creation of new international institutions; Gro Brundtland, for example, has forcibly argued that reform is urgently needed to establish stronger institutions "even with supranational authority for decision-making, monitoring and to ensure compliance" (EADI, 1990). Most Governments have no difficulty in endorsing the view that global policies are needed to control the pace of environmental change; indeed most refuse to take unilateral actions which would affect the relative competitiveness of their businesses on world markets. However, few are prepared to follow the Brundtland scenario and "put in hock our sovereignty" (Nicholas Ridley, 1989) to any international enforcement agency. Somewhat ironically the need for stronger institutions comes at a time when the credibility of interventionist strategies has crumbled, capitalist market systems are seen to have triumphed over the command economy, and when more and more ethnic groups are demanding the freedom to decide their own future (Brookfield, 1990).

The Illusion of Consensus

At a broad conceptual level, the debate over environmental change has been remarkable for the degree of consensus it has produced. The concepts of sustainable development, precautionary action, global co-operation and the stewardship ethic are widely employed in the rhetoric of international environmental politics. However, the degree of consensus is an illusion. The concepts on which it is apparently based are imprecise; there is little agreement about what they mean and still less about how they can be translated into implementable management strategies. The areas of disagreement are numerous but one above all is crucial, namely the question of who should pay the costs of the required policy shift. At present the greatest single barrier inhibiting the implementation of controls capable of tackling environmental change is the SEMPY (somebody else must pay) syndrome. The real environmental challenge will be to achieve consensus over what constitutes an equitable allocation of environmental protection costs between social groups, sectors within national economies and, above all, between nations.

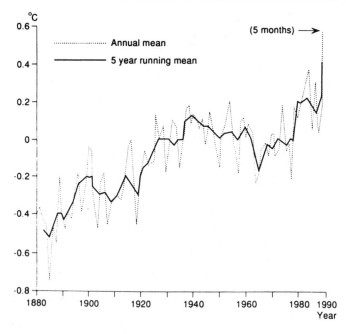

Figure 34.1 [*orig. figure 1*]   Global surface temperature rise 1880–1990.
Source: Pearce (1990).

One of the beguiling features about the notion of sustainable development, as popularised by the Brundtland Commission (1987), was that it appeared to suggest that we do not need to sacrifice economic growth and real incomes in order to safeguard our environmental future. In other words it contained the comforting message, that we can have our cake and eat it, although the flavour would have to change. The gloomy conclusions of the 1960s Limits to Growth debate were rejected and in its place there was the idea that wealth creation, increasing real incomes and technological change would be the vehicles for environmental improvement and the long-term sustainability of the planet. Moreover, since poverty was regarded as the key to ecosystem degradation in Third World Countries, these nations were to be offered a share in the good life. In other words sustainable development appeared to suggest that it was possible for everyone to gain by shifting the path of economic growth to minimise environmental change.

However, as soon as we start to consider what mechanisms to use actually to achieve our environmental aims then it is clear that there must be losers. Importantly different control strategies will produce different sets of losers and this makes it infinitely more difficult to achieve agreement on the way forward.

I will attempt to put some flesh on these very general arguments by considering the strategies for dealing with global warming.

Global Warming

Following the publication of the evidence from the Intergovernmental Panel on Climate Change (IPCC), it is generally accepted that the warming threat is real; temperatures are rising and will continue to do so (figure 34.1).

By the year 2030 average global surface temperatures are expected to have increased by between 1.5°C and 4.5°C if present emission trends of all greenhouse gases continue (figure 34.2). There is also little dispute that the top end of this temperature rise range would involve major socio-economic consequences as climatic zones shift and sea levels rise. Agreement on the

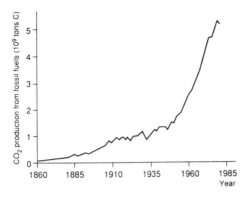

Figure 34.2 [*orig. figure 2*]   History of world fossil fuel energy consumption, as reflected in carbon emissions. *Source*: Darmstadter (1986).

threats and the urgency of the situation has, however, produced pitifully little action. For example in the Government's 1990 White Paper on the Environment, global warming is described as an unprecedented challenge to mankind, but the proposals to "improve guidance to motorists on how to save fuel" and "encourage energy labelling" for consumer durables hardly constitute an effective response. As Andrew Warren (Director of the Association for the Conservation of Energy) has put it – the Government appears to have swallowed a mountain to produce a mouse!

This lack of substantive action is not surprising; the Government is only too aware that it is pointless to incur the costs of control if other nations will not do the same. Moreover, scientists have not yet been able to say *whether* and *when* global warming will have any significant impact on the British economy itself. It is at least conceivable that climatic zone shifts would be beneficial; increasing water prices and reduced agricultural output in the Eastern Counties could be more than compensated by lower heating bills and the growth of tourism. As Howard Newby, Chairman of the Economic and Social Research Council, has recently argued, the relationship between science and policy making has shifted:

Traditionally this relationship has been

based on the belief that science provides decision-makers with objective *hard* facts on which to base their *soft* value-ridden policies. But now we find scientists delivering *soft* uncertain facts to politicians and policy-makers who face hard decisions. (Newby, 1990)

As was argued earlier, environmental policy-makers are used to reacting to proven pollution damage; they are now being asked to anticipate risk, but with no real knowledge of the costs and benefits involved. Moreover, to have prevented a potential problem brings no observable benefits to those having to bear the costs involved.

In 1992 it is hoped that the chances of real action being taken will be significantly improved by the signing of a climatic protocol at the Environment and Development Conference to be held in Brazil. The aim is to reduce one policy-making uncertainty by getting all nations to agree to limit their emissions of greenhouse gases. The chances that this agreement will prove meaningful are extremely *slim*. There is no meeting of the minds over which nations should bear the greatest control burdens.

In general the advanced countries are talking about stabilising their emissions of CO_2 at 1990 levels by the years 2000 or 2005, although some, most notably Germany, Australia, France and New Zealand are prepared to actually reduce their emissions by 20–30 per cent by the same dates. There are, however, two vital exceptions – both the United States and the USSR are still not in favour of establishing emission control targets: together they account for over 40 per cent of global CO_2. While stabilisation or 20 per cent reductions at least represent some improvement over past trends, it is clear that Northern nations want to minimise the pain of any policy shift. They are relying on technological innovation to achieve any emission targets without the need for major lifestyle changes, without any diminution in rates of economic growth, without any net adverse effects on their trade balances, and importantly without any significant *net* costs over the longer term to the economy as a whole. The targets set are not ambitious

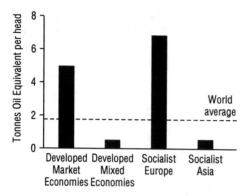

Figure 34.3 [*orig. figure 3*]   Per capita energy consumption, 1987. *Source*: Rees (1990).

when compared to the *existing* trends in energy efficiency. Japan's GNP, for example, grew by 46 per cent between 1973 and 1985 with virtually no increase in energy use. Recent studies have even shown that the United States could stabilise its emission levels with little or no additional cost in *aggregate* (Boyle, 1990). This has considerably weakened the US case that it could not afford the previously projected $4,000 billion bill for CO_2 control and exposes the real concerns for the powerful lobby groups – car owners, fossil fuel producers and the electricity utilities.

For Third World nations stabilisation

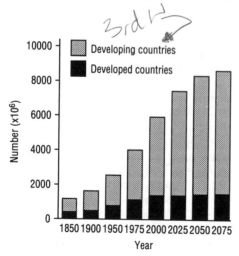

Figure 34.4 [*orig. figure 4*]   Global population growth. *Source*: Crosson (1986).

cannot possibly be achieved at zero net cost. Energy consumption is still exceptionally low per capita (figure 34.3) and given the inevitably large rise in population levels (figure 34.4) attempts to cut the rate of increase in CO_2 output would imply major reductions in the per capita use of fossil fuels. With justification Third World countries question the equity of stabilising the situation at an arbitrary date so perpetuating, indeed exacerbating, past inequalities in energy use and thus in greenhouse gas emissions. Today 20 per cent of the world's population consumes over 70 per cent of commercial energy. LDCs understandably argue that the costs of controlling global atmospheric change must be borne by those who caused the problems in the first place, who have profited by using up the absorptive capacity of the environment (figure 34.5) and who are *able* to pay the costs involved. Support for such arguments about equity comes from the fossil fuel producers and multinational trading companies. Their forecast output growth is critically dependent on expanded LDC markets as developed country projected energy consumption growth is at best sluggish.

However, it is known that any suggestion that LDCs should be allowed to continue on their past path of CO_2 production will negate any possible benefits from developed country stabilisation. The fastest growing component of global CO_2 emissions is the LDCs, and in particular, China, which already contributes 10 per cent of world output, is increasing its emissions by some 5 per cent per annum and is relying on quadrupling its coal production by the year 2000 to meet projected energy needs.

Even minimal attention to equity would suggest that (1) Northern nations must cut their emissions by more than is necessary to allow some increase from LDCs; (2) the North should bear the costs of developing non-fossil fuel energy sources; (3) the North must freely transfer the technology needed to reduce emissions to the South. Guestimates place the costs involved for LDCs in shifting from fossil fuels and adopting other carbon-reducing technologies at some $5–6,000 billion, over the next

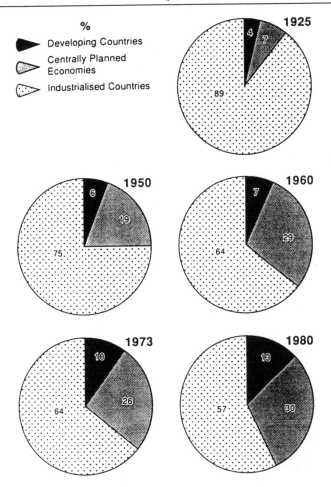

%
◆ Developing Countries
◁ Centrally Planned
   Economies
◁ Industrialised Countries

Figure 34.5 [*orig. figure 5*]   World energy consumption by developed and developing countries. *Source*: Darmstadter (1986).

30 years, but this sum does not include any potential development losses to countries, such as Colombia, who will lose opportunities to trade in fossil fuels.

Common sense dictates that LDCs will not adhere to any stabilisation agreements unless significant financial help is forthcoming. Some advanced nations clearly recognise this and a new fund has been established, administered by the World Bank with the support of the United Nations (Land, 1991). However, its size at $1.5 billion is clearly inadequate, particularly as it has to cover marine pollution, energy conservation, reafforestation, CFC controls and the preservation of the planet's diversity of plant and animal species. A minimal contribution to the flow of funds from North to South is also being made by 'Debt-for-Nature' swops, whereby conservation agencies, such as the World Wildlife Fund, buy up (at discounted rates) a country's debt to a particular financial institution, who had long since been resigned to the loss of the loan. The 'released' money is then pledged to conservation, most commonly protection of rain forests. However, such efforts merely touch the tip of the debt iceberg; Brazil alone owes some $120 billion and numerous LDCs have to commit at least 20 per cent of their export earnings to *interest* payments alone (Economist Atlas, 1989). Debt levels mean that there is a net flow of income from the South to North, which clearly reduces still further LDC ability to bear global warming control costs.

Equity and Market Controls

The target setting approach to the curbing of greenhouse gases is merely a version of traditional 'react and regulate' environmental policy. Assuming that agreed targets are actually set, numerous problems remain of which four are of particular importance:

1   Individual countries still have the problem of how to achieve them.
2   Standards will be 'bargained' and are unlikely to have much relationship to environmental quality goals. It is already known that stabilisation of emissions will not stabilise their concentration in the atmosphere. At 1990 output levels the supply of greenhouse gases exceeds the rate of destruction.
3   Merely setting a target does nothing to ensure that it is complied with; it will also be necessary to monitor each country's emissions and to punish those caught cheating. Unless there is some agency charged with enforcement, the so-called 'free loader problem' will inevitably occur (a country will seek to benefit economically by ignoring the target and letting others reduce their emissions).
4   Governments will decide which sectors of the economy will bear the control costs; the result is unlikely to be the most efficient, least cost solution.

These well known problems with all 'command and control' approaches to pollution control have helped focus attention on the potential role of economic incentives or market based control strategies, such as carbon taxes and tradeable permits. Following the publication of the 'Pearce Report' *Blueprint for a Green Economy* (Pearce, Markandya and Barbier, 1989) there has been a mass of literature on the theoretical advantages of incentive systems and at

least 50 modelling exercises which try to predict the tax levels needed to affect various reductions in CO_2 emissions. While in no way disputing that market mechanisms could produce a more *efficient* means of achieving emission targets, the resulting distributions of environmental capacity and emission control costs need be neither *equitable* nor politically acceptable.

All pricing solutions are based on the assumption that *willingness* to pay is a measure of the utility of environmental goods and services. However, any system which essentially allows £1 = 1 vote, must bias the allocation of environmental capacity towards the desires of the relatively wealthy and away from the basic needs of the poor. In reality ability to pay is crucial.

Carbon taxes

At a national level, it is now well known that any tax on energy use is regressive (figure 34.6). If the imposition of a carbon tax increased energy prices by just 15 per cent then the poorest families would cut their energy consumption by almost 10 per cent, but would still pay 2–3 per cent of their total income in tax. This would be a considerable burden since for the poorest 20 per cent of families fuel and power already takes 11.5 per cent of their total income, and this increases to 16.1 per cent for low income pensioners. On the other hand wealthy households would barely cut their consumption at all (1.2 per cent) and yet would only pay out 0.3 per cent of their income in the new tax. Unfortunately most studies have concluded that tax rates well in excess of 15 per cent would actually be required to have much effect on CO_2 emission levels. In Britain, for example, a 100 per cent tax seems to be necessary just to keep CO_2 emissions constant; while work for the European Commission suggests that a massive 421 per cent tax would be needed to allow CO_2 to be reduced by 20 per cent (Boyle, 1990). Such price increases are clearly untenable not only because of their effect upon inflation but also because the vision of thousands of old age pensioners dying of hypothermia would do little for any Government's election chances.

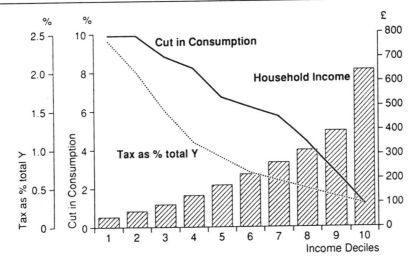

Figure 34.6 [*orig. figure 6*]   Distributional consequences of carbon tax causing a 15 per cent real price rise for domestic fuel and power (UK). *Source*: Drawn from tabular material produced in Pearson and Smith (1990).

The classic economist's solution to this problem is that the income needs of the poor should be met through the social security or pension system. Even at a national level this solution is not popular with the Treasury. Of course, if carbon taxes were used internationally, there is no global social security system and, as we have already seen, advanced countries have been loath to contemplate a wealth transfer to the South. There are also numerous practical problems inherent in the use of pollution taxes at an international scale. Most obviously difficulties are caused by variations in exchange rates and by the need to ensure that the real tax levels are maintained in the face of inflation. In addition a bureaucracy would have to be created to administer the tax system and to allocate the resulting tax revenue. Such problems have led to the suggestion that another form of market mechanism – the tradeable permit – could be a more acceptable and practicable alternative.

Tradeable permits

Conceptually the idea behind tradeable permits is a simple one and practical experience with their use has been gained, largely in the United States, in the allocation of water rights and, more recently, of rights to emit pollutants into controlled airsheds. In theory, rights to use the available capacity of the environment would be bought and sold in the market so ensuring that environmental capacity would go to those users who valued it most highly. Additionally, industries would have strong incentives to develop pollution control technology, since any unused disposal rights could be sold. In other words the market would establish the cheapest, most efficient way of meeting specific environmental standards (Helm, 1990).

However, theoretical simplicity belies the practical problems. In the global warming case, tradeable permits would ideally be established (figure 34.7) by calculating the capacity of the global system to absorb CO_2 and other warming gases; this capacity would then be translated into units of allowable annual emissions. The units would be allocated between individual countries and trading could then begin. At present, our knowledge of the complex inter-reactions between elements in the global environmental system is highly imperfect; therefore there are enormous problems in actually establishing CO_2 limits compatible with the 'no global warming' constraint. However, if we assume that some meaningful capacity

Table 34.1 [*orig. table 1*]   Tradeable permits and CO_2 emissions from fossil fuels.

Country	1980	1985	Permit Allocation (Grandfathering) %	Permit Allocation (Population) %
United States	1,249	1,186	26	5
Europe	853	780	17.3	10.2
United Kingdom			*(3)*	*(1)*
Japan	243	244	5.4	2.5
Soviet Union	871	958	21.3	5.8
China	395	508	11.3	21.8
Other Developing Countries	658	819	18.2	52.8

Source: Mintzer (1988).

limits can be established and accepted by all nations, the difficulties involved in implementing tradeable permits are by no means over.

One critical question which has to be addressed is what criteria are to be used to allocate the emission units between countries? Where permit trading has already been implemented within individual countries, 'grandfathering' is the normal allocative criterion. This means that *existing* polluters are given the available rights. In reality such an arrangement is necessary to 'buy' the agreement of the established polluters to join the trading system. Pearce and Helm (1989) suggest that if tradeable permits are used to combat global warming, countries would receive an allocation based on 1990 emission levels less 20 per cent to take account of the fact that current emissions are known to be too high to prevent the continued accumulation of greenhouse gases. As table 34.1 clearly shows, this system would give developed countries the giant share of the available permit units. Inevitably countries of the Third World will not regard this allocation as equitable, particulary as the rights to use the environment in the future will be given to those very countries that have in the past used up the capacity of the global system to absorb CO_2 and other warming gases.

To be fair Pearce and Helm (1989) do acknowledge that 'grandfather' allocations "don't offer a lot for third world states". They suggest that an alternative allocative criterion could be population, which has the advantage that poor countries unable to use their emission allocations could earn much needed revenue by selling them on the market. This suggestion has obvious equity

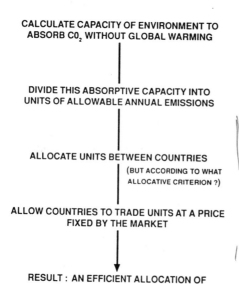

TRADEABLE PERMITS

CALCULATE CAPACITY OF ENVIRONMENT TO ABSORB CO_2 WITHOUT GLOBAL WARMING

DIVIDE THIS ABSORPTIVE CAPACITY INTO UNITS OF ALLOWABLE ANNUAL EMISSIONS

ALLOCATE UNITS BETWEEN COUNTRIES
(BUT ACCORDING TO WHAT ALLOCATIVE CRITERION ?)

ALLOW COUNTRIES TO TRADE UNITS AT A PRICE FIXED BY THE MARKET

RESULT : AN EFFICIENT ALLOCATION OF ENVIRONMENTAL CAPACITY

Figure 34.7 [*orig. figure 7*]   Tradeable permits.

advantages, but is extremely unlikely to prove acceptable to the developed world. Countries, such as the United States and the Soviet Union, would certainly oppose any system which reduced their emission allocations from over 20 per cent of the total to less than 6 per cent (table 34.1 column 4). Both countries would suffer extreme economic dislocation as they struggled to buy back enough units to allow their industries to continue in production. Given the state of the Soviet economy and the fact that the United States already has a massive trade deficit the repercussions of a population-based tradeable emission permit system would be enormous, and these repercussions would be felt throughout the global trading system.

Even in the unlikely event that agreement could be reached on an equitable initial allocation of emission units, there is still no guarantee that permit trading would actually produce an efficient use of available environmental capacity. It is well known that efficiency is the outcome of trade only under very specific conditions (perfectly competitive free markets, economically rational production, consumption and trading behaviour, perfect information about trading opportunities and so forth). Such conditions of perfection do not exist in reality and it is impossible to divorce the potential trade in emission rights from the question of trading relations in the world economy as a whole. The widespread existence of protectionism, imperfection in world capital markets, the existence of imperfectly competitive transnational companies, the generally poor prices achieved by Third World primary product producers and the clear technological dependence of LDCs, will all produce inefficiencies in an emissions trading system. Moreover, there will still be the need to create an international agency capable of monitoring each country's emission behaviour. Inevitably there will be enormous incentives for nations to emit greenhouse gases in excess of their entitlements. Third World nations, for example, may well be tempted to sell emission units to obtain immediate access to capital, but be unable to afford to buy back permits to meet future development needs; in such circumstances the temptation to 'cheat' would be great.

Conclusions

An attempt has been made in this paper to show that the apparent consensus over the desirable nature of future environmental policy is in reality a sham. Agreement over the desirability of sustainable development has provided a facade, behind which remains hidden the ultimate barrier to a truly new approach to global environmental problems, namely disagreement over the sharing of the cost burden. Northern Governments, are by and large, adopting a 'business as usual' stance; sustainability will seemingly be achieved through technological change which will allow gradual and relatively minor changes to occur in the developmental growth path with minimal economic pain. Governments of the South, on the other hand, can see little immediate advantage to them in controlling greenhouse gas emissions. Technology certainly does not afford the same panacea since it will largely have to be purchased from the North. LDCs are therefore unlikely to accede in any significant way to emissions controls unless the controls are accompanied by a massive redistribution of real wealth and welfare. Such a redistribution at present seems highly unlikely since it involves confronting the institutions within the global political economy which have acted to create and perpetuate inequalities. As the Brundtland Commission made clear, sustainability and equity are two sides of the same coin; unless the latter is confronted many countries will simply not have the resources to meet the requirements of environmental sustainability.

References

Boyle, S. (1990) "Lessons from the past, what people do when energy costs more", *New Scientist*, 3 November, p. 38.

Brookfield, H. (1990) "Environmental sustainability with development: what prospects for a research agenda?", in *EADI (European Association of Development Research and*

Training Institutes) Newsletter No. 2, Dec., Geneva: EADI.

Brundtland Commission (1987) [World Commission on Environment and Development], *Our Common Future*, WCED, London: Oxford University Press.

Brundtland, Gro Harlem (1990) "A global agenda for change", in *EADI (European Association of Development Research and Training Institutes) Newsletter* No. 2, Dec., Geneva: EADI.

Commission of the European Communities (1987) Fourth Environmental Action Programme (1987–92), *Official Journal* C328, 7 Dec., Brussels.

Crosson, P. (1986) "Agricultural development looking to the future", in Clark, W. and Munn, R. (eds.) *Sustainable Development of the Biosphere*, Cambridge: International Institute for Applied Systems Analysis and Cambridge University Press.

Darmstadter, J. (1986) "Energy patterns in retrospect and prospect", in Clark, W. and Munn, R. (eds.) *Sustainable Development of the Biosphere*, Cambridge: International Institute for Applied Systems Analysis, Laxenburg and Cambridge University Press.

Department of the Environment/Welsh Office (1990) *This Common Inheritance*, London: HMSO.

Economist Books (1989) *The Economist Atlas (The Shape of the World Today)*, Economist Books/Hutchinson Business Books.

Helm, D. (1990) "Who should pay for global warming", *New Scientist*, 3 Nov., pp. 36–9.

Land, T. (1991) "$1.5bn fund promotes Third World pollution control", *Lloyds List*, 18 Feb.

Mintzer, I. (1988) "Our changing atmosphere; energy policies, air pollution and global warming", in WMO, *The Changing Atmosphere*, Geneva: WMO.

Newby, H. (1990) "Opening the door to social scientists", *New Scientist*, 8 September, p. 26.

Pearce, D. and Helm, D. (1989) "Assessment: economic policy towards the environment", *Oxford Review of Economic Policy*, 6(1), pp. 1–15.

Pearce, D., Markandya, A. and Barbier, E.B. (1989) *Blueprint for a Green Economy*, London: Earthscan Publications.

Pearce, F., (1990) *Turning Up the Heat*, London: Bodley Head.

Pearson, M. and Smith, S. (1990) *Taxation and Environmental Policy: Some Initial Evidence*, IFS Commentary No. 19, Institute for Fiscal Studies.

Rees, J. (1990) *Natural Resources: Allocation, Economics and Policy*, Second Edition, London: Routledge.

Ridley, N. (1989) [Then Secretary of State for the Environment] quoted in *ENDS (Environmental Data Service)* Report 172, May.

35

ONE WORLD, TWO CULTURES: SOCIOLOGY AND THE ENVIRONMENT*

Howard Newby

In this lecture I want to assess the contribution of sociology to the study of the environment. In doing so I shall be making two underlying claims, one of which will be entirely congenial to the majority of sociologists; the other, I suspect, less so.

The first claim, which I shall go on to elaborate in a moment, is that environmental change, as it is currently defined, is as much a social science as a natural science issue. Although this need scarcely be argued within the social science research community, as far as the public at large is concerned the study of the environment continues to be perceived predominantly as a matter on which only natural scientists can form authoritative judgements, while the influence of the social sciences is somewhat more muted. I thus want to take up a little of my time in reminding ourselves why the social sciences are central to, and not marginal to, the study of the environment, even though I acknowledge that I am speaking beyond the heads of this particular audience to a rather wider public.

My second claim concerns sociology specifically. I will argue that the contribution of sociology to the public debate on environmental change, and even to debates within the social sciences, has hitherto been negligible and quite out of proportion with the burgeoning interest not only in geography and economics but in anthropology, political science and international relations. Compared with the intellectual challenge which environmental issues present and the intellectual capacity which the discipline of sociology, not least in this country, has to hand, the slender contribution of sociologists to the study of the environment has been, to put it mildly, disappointing. I shall argue, however, that the discipline of sociology is inhibited, not merely by particular conjunctural factors, but by a deeply-rooted set of theoretical and conceptual difficulties which must be overcome if sociology is successfully to rise to the challenge which I believe to be there. To do so will require a degree of intellectual renewal which will have to accompany the inevitable turnover in the research community over the next decade. Needless to say, this second part of my argument is very much addressed to the 'private troubles' of the sociology profession

* Originally published in *BSA Bulletin Network*, 1991, no. 50, pp. 1–8.

and not directly to the 'public issues' which comprise the contemporary debate on the environment.

Environmental Concern and Public Debate

Let me begin, however, by briefly sketching in a few reminders of how environmental issues have been treated in recent public debate. There is little need for me to underline how the environment has re-emerged as an issue which has captured the public imagination in recent years. Such concern is, of course, not new and has proceeded in a somewhat cyclical fashion since at least the middle of the nineteenth century. Similarly, the term 'environment' covers a wide range of issues relating not only to the natural resources of the geosphere, biosphere and atmosphere but to rather more aesthetic and cultural matters often summarised by the term 'amenity' (see Lowe and Goyder, 1983; Newby 1990). Thus, each up-turn in public concern for the environment has been accompanied by both a change in emphasis and an addition to the catalogue of issues which constitute environmental politics. In Britain in the last few years, for example, the focus has moved away from comparatively parochial wildlife and countryside issues to problems on a *global* scale, particularly climate change. In part, this has been underscored by advances in scientific observation, but it has also been influenced by a renewed public awareness of scarcity. There is now a greater emphasis on the finite nature of the earth's resources and a growing suspicion that we may, in ways which remain not yet fully understood, be irreversibly tampering with the habitability of our planet.

These concerns amount to much more than an irrational atavism. The public is broadly aware of the fact that environmental degradation, whether at the global, national or local level, is a result of *human* intervention in natural systems and, in particular, our current patterns of economic development and social organisation which place a burden on the earth's resources which are unsustainable in the long run. We should

not, therefore, be too surprised that environmental issues are once more in the forefront of political debate, for in the broadest sense they are deeply political, raising concerns about the expansion of individual choice and the satisfaction of social needs, about individual freedom versus a planned allocation of resources, about distributional justice and the defence of private property rights, and about the impact of science and technology on society. Beneath the concern for 'the environment' there is, therefore, a much deeper conflict involving fundamental issues about the kind of society we wish to create for the future.

Public debates about the environment thus tend to proceed at two different levels. One is essentially symbolic: in this case 'the environment' needs to be placed within inverted commas. For here we find environmental issues being used to express deeply-rooted anxieties about the nature of human progress. 'The environment' is a readily available idiom in which anxieties can be articulated. To some degree this has been recognised by sociological observers, and it is an issue to which I shall return below. However, there is a second, material level which relates to the degradation of natural resources, and what has become known as sustainable development, which relates to our systems of production, consumption and exchange. This has been given less attention by the sociological profession – yet it is our very social and economic arrangements which determine whether or not our use of natural resources can be sustained in the long run.

All of this suggests that environmental change can never be simply a scientific and technological issue. Advances in the natural sciences will enable us to establish the parameters of environmental change, but they will describe the symptoms and not explain the causes. The causes lie in *human* societies and their systems of economic development. Stated in this fashion, such observations seem self-evident, almost trite; yet they are at some variance to the principles which have guided the public policy debate on the environment in this country in recent years and wholly contrary to the policies which have, so far, underpinned the disposition of the UK research effort.

It is not too difficult to identify the event which has set the terms of the recent policy debate. In September 1988 Margaret Thatcher, in a speech to the Royal Society, drew attention to the looming threats to the global environment, particularly climate change, and stated her government's intention of taking an international lead in the investigation of, and response to, the greenhouse effect. The UK possessed an international lead in many aspects of climate modelling and so Britain's acknowledged scientific expertise in this area was to be used to spearhead an international effort to secure future sustainability. Before politicians and policy-makers meddled, she observed, what was needed was "good science to establish cause and effect".

By the first half of 1989 the environment was established as an important agenda item at the summit meetings of the world's political leaders. This process culminated in the creation of IPCC – the Inter-Governmental Panel on Climate Change – as a forum for the world's leading scientific experts to assess the evidence on global warming and advise the political leaders of the world's most industrialised nations on the likely nature and extent of climate change. Over the next twelve months an impressively organised international scientific establishment not only undertook this task but, in doing so, defined the nature of the problem in such a way that, even now, it is easy to equate global environmental change with climate change and to regard both as essentially scientific problems amenable to technological solutions.

It is instructive to dwell briefly on the organisation of the IPCC and the approach taken in its reports, published last autumn. The panel divided itself into three working groups, each of which produced its own report. The first working group dealt with 'processes'. It summarised and interpreted the available evidence on the extent and speed of global climate change, as well as assessing the basic science of the greenhouse effect. The report confirmed that the greenhouse effect is real, that by 2020 mean global temperatures will have risen by around 2°C, with higher levels over large land masses, that sea levels will rise by about 20cm by 2030, and that within the next century there will be a major outflow of ice from the Western Antarctic ice sheet due directly to greenhouse warming. By general consent, the group's report was easily the most cogent and attracted the greatest (though not total) consensus. For these reasons (and not just because its chairman happened to be British) it attracted the greatest public attention in this country.

The second working group was concerned with 'impacts'. It focussed on food production and was composed almost entirely of biological scientists and agronomists. It made a valiant attempt to assess the likely impacts of global warming on agriculture, but was hampered by the existing incapacity of climate models to predict climate change at a regional level with any degree of certainty. It made some half-hearted attempts to assess the impact in certain critical regions (for example, the major grain-growing areas and the Mediterranean) but these looked distinctly frayed around the edges. Crucially, it recognised that the impact of climate change would depend on the capacity of farmers to adapt their agricultural regimes to the new circumstances, but this, the group recognised, was essentially a socio-economic matter and, possessing no social scientists to speak of, it could only urge further research on the issue and offer a few throw-away lines about population growth.

The third report was concerned with 'responses', principally *policy* responses but also the prospects of what has become known as 'clean technologies'. This group struggled to achieve any consensus at all and failed to produce a report of any cogency or substance, having speedily become enmeshed in the discussion of different policy options, which were inevitably tinged with political considerations. It was, for this reason, widely regarded as the least 'scientific' of the reports.

In two important respects the IPCC exercise reflected the prevailing policy-maker's view of how the issue of global environmental change is to be tackled. The first concerns the organisation of knowledge into 'processes', 'impacts' and 'responses'. This is not just a taxonomy but is used as an *aetiology*: for this purpose change begins *exogenously* of human society, and socio-economic issues only become relevant in relation to impacts and responses. This categorisation is all the more worrying since there are signs, in science policy circles, of them becoming reified in ways which will influence future investigation. Thus, the social sciences, though not ignored, are assigned a very particular role in the scheme of things. It is science which changes, or, in this case, saves the world; social scientists come along afterwards to study its effects. Built into the heart of IPCC is a technological determinism which will now be extremely difficult to dislodge.

My second comment follows on from this. As the reports moved from the 'science' to the 'policy' (often a euphemism for politics), so the authorative character of the reports broke down. This is a factor which I shall return to below. For the time being we may merely note that, if you scratch the surface of the scientific and policy-making debates on the environment, you will often find old-style technological determinism lurking beneath it. This adopts 'technical fix' solutions to the environmental problem ('clean technologies', etc.), mediated only by somewhat abstract expressions of concern for population growth or the Third World. Issues like the latter are deemed to be 'political' problems and thus beyond the pale of 'science'. An essentially social science kind of expertise, which analyses, evaluates and communicates about *options* for policy and technological change is thereby omitted from consideration.

I should emphasise that I am taking the IPCC deliberations as only one example, albeit a very high-profile example, of this style of thinking, but it is illustrative of a very pervasive approach to the study of environmental change, whether global or local. The irony is that IPCC was preceded by a rather different style of enquiry into global environmental change, one which was not afflicted by technological deter-

minism, but which IPCC almost totally ignored. I am referring, of course, to the report of the World Commission on Environment and Development, better known as the Brundtland Report, and published as *Our Common Future* in 1987. It was this report which promoted the concept of *sustainable development* as the crucial policy goal which would tackle the problem of resource depletion and other aspects of environmental change on a global basis. Following the false start of the *Limits to Growth* debate of the 1970s, the Brundtland Report placed on a much more secure intellectual footing the proposition that existing patterns of economic development are simply not sustainable in the long term. The report emphasised development and not simply growth – that is, it contextualised economic growth (per capita GDP) in order to place it within a set of social and environmental objectives. This led inevitably to questions of choice, utility and equity, raising a series of tricky issues relating to national sovereignty, international trade, Third World development and what was to become known as 'environmental security'. Moreover, some of the Brundtland Report's conclusions represented, at the very least, a challenge to the developed world's short-term economic interests, but it argued that this was the necessary price to pay for global sustainability.

The Brundtland Report, then, offers an entirely contrary standpoint to IPCC. It begins with the notion of sustainable development and subsumes science and technology to this overriding policy goal. Such a normative approach does not, of course, corrupt disinterested enquiry, whether natural scientific or social scientific, but it does attempt to harness the output of this enquiry to the pursuit of a stated policy objective. The British response to the Brundtland Report was, to put it mildly, muted. Fortunately, the Report's approach was rescued by the work of David Pearce and his colleagues (Pearce et al, 1989) who demonstrated that sustainable development was not inextricably linked to Scandinavian social democracy and that this approach, no less than the work of IPCC, could be translated directly into practical policy measures. It is doubtful, however, whether the various members of IPCC were aware of the Brundtland Commission's approach or, if they were, regarded it as within their frame of reference. As a result, IPCC and Brundtland remain as a testimony to how the study of one world remains divided by two cultures – to the detriment of both.

Two Cultures: Part One

These two quite divergent episodes also demonstrate how two cultures of a rather different kind encounter environmental problems. As I have written elsewhere, one of the more fascinating aspects of recent debates about the global environment is the way in which it reflects a new kind of relationship between research and policy (or, if you prefer, 'science' and 'politics'). In the past, to use the distinction made by Jerry Ravetz (1990), the relationship was predicated on the belief that science provided decision-makers with objective, 'hard' facts on which to base their 'soft', value-ridden policies (see also, Rayner, 1989). This was certainly the presumption which underlay the establishment of IPCC. But the study of the environment is but one area where we now find scientists delivering only 'soft', uncertain 'facts' to decision-makers facing 'hard' decisions. Scientists want to discuss precautionary environmental principles, rather than rainfall predictions in the south of England, and politicians demand to know the location of sites for the 'safe' disposal of toxic waste, while scientists can only state that nothing is ever risk-free.

To social scientists, long accustomed to dealing with the subtle distinction between objectivity and value-neutrality, this experience is a familiar one. Social scientists have often tended to bring policy-makers *news* about social problems, rather than solutions; global environmental change is placing natural scientists in a somewhat analogous situation, since the 'solutions' lie no more in their sphere of influence than they do in that of social scientists. As Ravetz has pointed out:

Typically we find that facts are uncertain, values in dispute, stakes high and decisions urgent; and the framing of the problem involves politics and values as much as science . . . Indeed, in dealing with such complex problems the scientific experts are, in their own way, amateurs. They bring essential skills and information, but their contributions are to a dialogue of exploration and consensus, rather than to a rigid demonstration which conclusively proves something. The critical assessment of their information, in the context of its use, is a task for all participants, and not merely those with narrowly defined, technical expertise. (Ravetz, 1990)

There are many natural scientists who find this role uncomfortable, since it disrupts the established taken-for-granted relationship between science and politics. It also presents a problem for politicians in search of scientific legitimacy for their decisions: an appeal to the scientific 'facts' is a handy device to shut down the much more messy debate necessary to manage uncertainty as well as reconcile conflicting interests. But it is ultimately dangerous where the problem is intractable to scientific investigation and technical fixes *alone*, as a catalogue of recent cases from nuclear energy policy to food hygiene has shown. As far as environmental change is concerned, as we move away from the expression of abstract goals – even those with a socio-economic content like 'sustainable development' – to the actual means of achieving them – regulation, taxation, tradeable permits, etc. – so the costs become clearer to the public at large and conflicts of interest will inevitably emerge. How else is political debate maintained over something – the destruction of the environment – to which we are *all* opposed? As the distributional consequences of these costs are made manifest to the public, so *its* attention will concentrate on the *uncertainties* of the science (such as climate predictions) in order to persuade politicians to remove the *certain* burden of the increased costs. Scientists will then be left to rue the apparent 'irrationality' of the public in refusing to accept the best scientific evidence, while politicians will be looking for scapegoats for their public unpopularity.

Familiar? Social scientists learned their lesson the hard way in the failure to link social science with social engineering in the 1960s. When it comes to scapegoating, the allegations do not have to be true, merely plausible – ask any urban planner. More recently, and involving the natural sciences, you have to look no further than the many and various food scares as an example of this chain of events. Solutions to environmental problems are rarely amenable to technical fixes alone, no more than they can be handled by an equal and opposite 'social fix'. It is the *interplay* between the technical and the human which will hold the key. Public consent to changes in policy, let alone life-styles, will be essential. Hence, the centrality of social science enquiry to the study of the environment.

Two Cultures: Part Two

These comments are directly relevant to a consideration of sociology's distinctive role in the study of the environment. I need hardly stress the extent to which virtually all of the previous discussion has raised issues which are entirely relevant to the discipline of sociology. And there are, of course, many which I have not had time to raise, both theoretical and empirical. Why, then, has sociology remained so silent? Even within the social sciences, the running has been made by geographers (represented in this country by Tim O'Riordan) and by economists (such as David Pearce). Sociologists have, on the whole, allowed the environmental debate to proceed in a manner which has enabled the contribution of a sociological perspective to be ignored. Where, for example, has been the cogent sociological critique of neo-classical environmental economics to set alongside the Pearce Report?

Now, I recognise that I exaggerate for effect and that at one level I do not care *who* presents a sociological perspective on the environment, as long as it is made. I am, therefore, happy to acknowledge the contribution of writers as diverse as Ted Benton, Phillip Lowe, Michael Redclift and Brian Wynne, some of whom might well, with some justification, resist the label of sociologist. But, in comparison with the scale of the issues involved and the importance granted to them by the public at large, the contribution of sociology has, as I alleged earlier, been negligible. Furthermore, I doubt whether more than a few of our students are being asked to consider environmental issues, that even fewer of them (and us!) have read the Brundtland Report – a document by no means immune to sociologically-informed criticism – than the natural scientists on whom I was so hard earlier, and that where such a treatment is offered it is in the outer reaches of a third-year undergraduate option where it can be safely marginalised alongside the sociology of development, or some such.

This really will not do. A consideration of the environment, and particularly of the concept of sustainable development, raises precisely the same kind of fundamental questions about the organisation of society that faced the early founders of the discipline when they encountered the rise of industrialism, capitalism and liberal democracy. But our syllabuses and textbooks remain silent and the research cupboard is almost bare. Why? Why has there been such a collective failure of the sociological imagination of this order of magnitude? Common observation suggests that it cannot solely be the personal values of individual sociologists, even though many, quite rightly, retain a healthy scepticism of some of the rhetoric in which environmental arguments are often couched. Nor, I believe, can it be accounted for by the fact that we are an ageing research community, reluctant perhaps to re-tool our intellectual capital and add to our sense of mid-life crisis.

The reasons, I believe, lie much deeper than this and have more to do with the intellectual inheritance of sociology and how the discipline defines what is distinctively 'sociological'. Let me approach this question somewhat obliquely. One aspect of the study of the environment to which sociologists have contributed quite extensively is the study of environmental values and environmental organisations. The environment, in other words, has been approached via the study of environmental*ism* – as an ideology or as a social movement (see, for example, Lowe and Goyder; Redclift). As far as it goes, much of this work has been useful and important, not least in de-mystifying much of the rhetoric of environmentalism and demonstrating how, in certain cases, it has proved to be little more than a vocabulary which cloaks certain sectional interests and presents them as the common good. Bill Williams and Ray Pahl had identified NIMBY more than thirty years before it was labelled as such and entered public discourse.

There is also no denying that the study of the environmental movement represents an interesting sociological problem. For, having stripped away the NIMBYs, one is still left with a substantial body of adherents who seek to gain no personal benefit for themselves or their families and who possess, moreover, no intrinsic political base. As I have argued elsewhere, their demands are more akin to the demand for citizenship rights, albeit increasingly on a global scale and in a situation – "the tragedy of the commons" – where citizenship rights and duties stand in urgent need of re-definition. Orthodox explanations of social movements have some difficulty in handling this. Hence the tendency towards sociological de-mystification is taken a stage further. 'The environment' is interpreted not materially, but culturally, as a set of symbols which furnish, in the contemporary world, the predominant vocabulary of discontent. Environmental change is analysed, not for what it is, but for what it symbolises: environmentalism, in other words, is reduced to a moral panic. From this it is but a short step to explaining environmental*ism* (and, by extension, environmental change itself) *away*. It is 'merely' the expression of some 'deeper' form of social discontent or existential pessimism.

As far as conventional sociology is concerned, this is the end of the problem. Here, to paraphrase David Hume, the socio-logical mind comes to rest. But the sociological mind has entirely missed the point that 'the environment' is not *only* a set of symbols, powerful though these are; it also has a material aspect – sustainable development again – which cannot be either ignored or explained away in this fashion. This, in turn, bears upon a problem which is as old as social science itself: the place of humankind in nature and the place of nature in ourselves (Benton, 1983).

When, as sociologists, we consider the relationship of the human to the natural, we are immediately weighed down with the cultural baggage of our own discipline. The very character of sociology in the nineteenth century emerged out of attempts to delineate the 'social' from 'human nature', and to assert the dominance of the cultural (nurture) over the evolutionary (nature). In the twentieth century we are only too aware of the political misuse of biological determinism and its profoundly authoritarian connotations. Put together, these pre-dispositions have provided a powerful set of factors which have led sociologists to fight shy of re-examining and re-conceptualising the link between society and nature. The very *raison d'être* of sociology has rested upon identifying and demarcating a disciplinary paradigm quite distinct from, and irreducible to, the natural and the biological. And, even to pick at this problem runs the risk of being tainted with an abhorrent political philosophy.

This, I believe, is the fundamental reason why sociology has found it difficult to come to terms with the study of environmental change. Nor is this limited to the environmental issue: the same considerations have applied to the study of gender, of health (witness the current intellectual onslaught from neuro-biology and molecular biology) and of demography. As Benton has argued:

The idea of human nature as an empty and infinitely flexible vessel into which may be poured any cultural content we choose is, like its philosophical partner, the empiricist 'tabula rasa' conception of the human mind . . . quite intellectually indefensible . . . There is reason to suppose that important and persistent areas of explanatory weakness in the human sciences can be linked precisely to this thoroughgoing anti-naturalism. (1983, p. 8)

Other social sciences have proved to be less inhibited by the need to re-conceptualise the relationship between the human and the natural. The problem of human agency in relation to the environment is well recognised in the literature of, for example, geographers. Here there has been a willingness to recognise that environmental management, and conflicts over the environment, are *both* about the way people dominate each other *and* the way they seek to dominate nature. Michael Redclift, in a thoughtful summary of this debate, has stated the issue succinctly. "The 'Green' agenda", he writes, "is not simply about the environment *outside human control;* it is about the implications for social relations of bringing the environment within human control." (Redclift, 1990, p. 6; emphasis in the original).

I find this aphorism very appealing. It returns us to some of the questions I raised earlier in this presentation about the environmental debate also being a debate about the kind of society we wish to create for the future. This lifts the potential contribution of sociology above the provision of descriptive data on 'social attitudes' or studies of the diffusion of clean technologies – important though these can be for policy-makers and the public at large – to genuinely challenging analyses which have always been at the core of the sociological tradition. In this respect our pressing environmental problems pose the same possibilities for no less a sociological enquiry than did the rise of industrialism, capitalism and liberal democracy from the late eighteenth century onwards. In this sense, the sociology of the environment ought not to be regarded as some esoteric, sub-disciplinary specialism, but rather something which defines the sociological enterprise as a pervasive and contextualising theme.

Conclusions

It is impossible for me in this lecture to begin the task of theoretical and conceptual reconstruction which is required, nor have I any desire to do so, for prescription would be inappropriate. However, I would urge the sociology profession to address the issue of environmental change more seriously and more systematically than hitherto, not just because of its supreme practical importance, but because it offers the kind of intellectual challenge to sociology quite on a par with the more familiar, albeit unresolved, issues of class or gender. When one contemplates the relative silence of sociology in this regard, one recognises that the notion of two cultures cuts both ways.

I am aware of the irony in what I am arguing. Having berated the natural science community for its failure to regard the environment as a social, economic and political issue, having accused it of a naive technological determinism and a failure to come to terms with the diversity of knowledge creation which the issue demands, the discipline of sociology seems to me to be afflicted by the obverse of the same problems. Are we not, equally and oppositely, committed to the social fix, suspicious and even dismissive of technology, and unwilling to, even apprehensive of, exploring the interface between the human and the natural?

Whatever the origins of the existing debate on the environment, especially the global environment, it cannot be dismissed as *just* a political fad. Exploring the limits and substance of sustainable development will demand the skills of sociologists; sociologists, I believe, have a responsibility to offer them. But this need not to be taken on as a glum public duty: the ramifications of environmental change offer exciting possibilities of intellectual renewal which should not be lightly cast aside. Perhaps the BSA, as a first practical step, could attempt to kick-start the debate within the discipline by setting up its own panel of enquiry, with a remit to report to a future annual conference? The Association will, quite rightly, wish to be no more prescriptive than I have been, but it can act as a focus and as a catalyst. The case for social science still needs to be made to a wider audience, even if *we* all accept it. A report as cogent as John Houghton's for IPCC would help here. Such an exercise might also stimulate debate within sociology itself about our future horizons. I do not wish to exaggerate the importance of the environment in relation to other pressing issues, but on grounds of timeliness and public interest alone, I hope the BSA will consider this proposal seriously. We may produce the proverbial camel, but at least it will be an environmentally-friendly camel and not one which continues to spit sociology in the eye!

More seriously, if we are convinced of the importance of the social sciences in general, and sociology in particular, to the study of the environment, it is essential we seize the opportunities that are offered and communicate our perspective in a manner which is intelligible to other interested parties. Sociological techniques and insights are an *integral* part of finding solutions to our pressing environmental problems. It is vital for all our futures that we lose no opportunity to acquire the appropriate knowledge about ourselves and our relationship to our planet.

References

Benton, E., "Biological ideas and the cultural crisis". Paper presented to *The Royal Institute of Philosophy*, Spring, 1983.

The Brundtland Report, "Our Common Future", Oxford University Press, 1987.

Department of the Environment, *Sustaining Our Common Future*, 1989.

P. Lowe and J. Goyder, *Environmental Groups in Politics*, Allen and Unwin, 1983.

H. Newby, "Ecology, amenity and society: social science and environmental change", *Town Planning Review*, Vol. 60, 1990, No. 1, pp. 3–13.

D. Pearce, D. Barbrer and A Markandya, *Blueprints for a Green Economy*, Earthscan, London, 1989.

J. Ravetz, "Knowledge in an uncertain world", *New Scientist*, Sept 22, 1990, p. 18.

S. Rayner, "Human choice and the global environment: prospects for the new science and the politics". Paper presented to American Association for the Advancement of Science, New Orleans, February, 1990.

M. Redclift, "Sustainable development through popular participation: a framework for analysis". Paper to UNRISD workshop on *Sustainable Development Through Popular Participation in Resource Management*.

36

THE THIRD REVOLUTION: POPULATION, ENVIRONMENT AND A SUSTAINABLE WORLD – EXECUTIVE SUMMARY*

P. Harrison

In the last two decades environmental problems have risen from regional to planetary in scale and scope. The 1970s fears of spreading deserts and shrinking forests were joined in the 1980s by red tides, acid rain, the ozone hole and the threat of global warming.

The same two decades saw human populations soar from 3.7 billion in 1970 to 5.5 billion in 1992. At the same time the consumer revolution spread from the West to East Asia, and to a growing middle class in other countries.

The conjunction is no coincidence: for the environmental crisis is the outcome of the population and consumption explosion. Technology is crucial too. For each unit we consume, we are still using too many resources, and emitting too much pollution.

Yet we may be on the threshold of deeper crisis. If present trends go on unchecked, the consequences are incalculable. For in the 1990s we are growing faster than ever, adding 97 million extra people – almost two United Kingdoms – a year. The decade will add a billion people – the equivalent of four Americas.

World population is expected to grow to 8.5 billion in 2025 and could pass 10 billion by 2050. It may rise to 11.5 billion before levelling off. Consumption per person will at least double. The impact of our present population and consumption is already too high, but by 2050 it could increase by four times.

The growing environmental crisis is so deep that we need the broadest analysis and strategy to cope with it. Yet most theories in the field stress single causes – population, overconsumption, technology, inequality – and deny the importance of others. One-sided analysis leads to one-sided measures. We have to see population as one factor among many. All interact, and all must be dealt with.

They need to be linked

Factors – Direct and Indirect

Three factors work directly on the environ-

* Originally published in Harrison, P. *The Third Revolution: Population, Environment and a Sustainable World*. Harmondsworth: Penguin, 1993, pp. 323–30.

ment: *population* – the number of people; *consumption* – the amount each person consumes; *technology*, which decides how much space and resources are used, and how much waste is produced, to meet consumption needs. These three are *never* found apart. Environmental impact is the result of all three multiplied together:

$$I = P \times C \times T$$

The three factors exist everywhere. Excess consumption in the rich countries bears the main blame for damage to global commons. The average person in a developed country emits roughly twenty times more water and climate pollutants than their counterpart in the South. Hence the 57.5 million population grow in the North expected during the 1990s will pollute the globe more than the expected extra 911 million Southerners.

But population growth even at modest consumption levels leads to deforestation. And the global consumption balance is changing as both incomes and populations grow in the South. By the year 2000, developing countries will account for 60 per cent of fertilizer use, and by 2025, for 44 per cent of carbon dioxide output from fossil fuels.

Many other factors affect the environment indirectly. They include factors stressed by the left, like *poverty* and *inequality*; by the right, like the degree of *democracy*, *market freedom*, or *property rights*; and by feminists, such as *women's rights*. A comprehensive analysis and strategy must include all of these.

By comparing the annual rate of change in our three direct elements, we can find out their relative importance. The population impact can be expressed as a percentage of the overall environmental impact. This varies widely from one field to another.

The population impact is lower where technology is changing fast, such as output of CFCs, or where population is growing slowly, as in Western countries. But with basic needs in developing countries, such as expansion of arable and irrigated land, and livestock numbers, it runs at 69–72 per cent.

Major Areas of Impact

Humans use the environment in three ways: as living space, as a *source of resources* and as a *sink for wastes*.

We need *space* for farms and cities. As we expand, wild habitats and the species that depend on them contract. In just fifteen years from 1973 forests in developing countries shrank by 1.45 million square kilometres – six times the area of the UK. The needs of population growth accounted for 79 per cent of the loss. If we grow to 11.5 billion, we will need at least an extra 12.6 million square kilometres of land that is currently wildlife habitat – double the area of all nature reserves today.

Resources have been the focus of Malthusian concerns. We are not close to a crisis of *non-renewable resources* such as minerals or fossil fuels. Reserves of these have expanded over time. However, future demands will be vast. A world of 11.5 billion people, consuming at today's US levels, would need a new energy source equal to all the world's oil reserves every seven years.

By contrast, *renewable resources* are already showing signs of stress. The annual fish catch in 1988 already exceeded the sustainable yield of the oceans. Between 1978 and 1989 food production fell behind population growth in sixty-nine out of 102 developing countries, *Global* food production kept ahead of population growth till 1979, but since 1985 it has fallen behind.

Some thirty-nine developing countries are already using *water* at a rate that causes regional problems. Without *soil* conservation, the developing world could lose 18 per cent of its potential rainfed cropland and 29 per cent of potential food production.

An even greater threat stems from our output of *wastes*. Increased use of *fertilizers* pollutes waterways and coasts. Population growth accounts directly or indirectly for 72 per cent of the increase in fertilizer use in developing countries.

We are probably well short of the *resource-carrying capacity* of the earth. But for many waste gases we have already passed the *waste-carrying capacity* – the earth's absorption limits. The ozone hole shows we are beyond it for CFCs, though population growth was a weak contributor here. Industrialized areas have passed the limits on gases that create acid rain. Population growth accounted for 25 per cent of the upward pressure on these in Western countries. It accounts for about 40 per cent of the global increase in carbon dioxide.

The Dynamics of Change

Humans do not remain passive in the face of environmental problems. They adapt. They change technologies, fertility levels and consumption patterns in response.

And they change their ways of managing the environment. For one resource after another we have begun by *gathering* open-access resources. As our numbers and consumption grow we move on to *mining* or *pillaging*. The ensuing *crisis* is usually followed by a transition to more *sustainable management*.

Our management of *waste sinks* also evolves. We begin by scattering our solid, liquid and gaseous wastes, move on to *dumping*, then to *spoil* and *crisis*. The move to sustainable management has yet to happen for most types of waste.

At certain times the combined pressure of number, consumption and technologies leads to more widespread crisis. At such times human culture and social organization may undergo major *revolutions*.

Environmental crisis may be a key agent in biological evolution and has driven some of the greatest leaps in human civilization. When early hunter-gatherers exhausted supplies of wild game and crops, their response was to cultivate crops and domesticate animals. This was the first, *agricultural revolution*.

This enabled populations to grow faster. But deforestation progressed as more and more land was cleared. Western Europe plunged into an acute wood shortage, which forced the move to fossil fuels and minerals and spurred the *industrial revolution*.

Each revolution solved one set of problems – but led on to others. Use of fossil fuels and minerals freed us from dependence on the land but increased output of pollutants and greenhouse gases.

The first two crises were *resource crises*. But the present crisis is different. It is primarily a *crisis of pollution and degradation*, by nature much harder to deal with. Problems of this type build slowly and invisibly. The science and technology to understand and deal with them may yet have to be developed. Often the institutions needed to take effective action do not exist, especially at regional and global levels. Responses are often delayed and sometimes blocked for longer periods. Adaptation may move too slowly to avert severe damage. Often environmental problems have to reach these crisis proportions before action is taken. Indeed, damage and crisis are the stimuli that drive adaptation.

Pulling Out All the Stops – a Strategy for Change

The global scale of our impact today means that the human race is playing with high explosives. Environmental change is not always smooth. It can pass critical thresholds where massive shifts occur. In the case of changes in climate and ocean circulation the shifts could be catastrophic. We don't know exactly where the trigger points lie: we only know that we cannot risk passing them. Survival demands that we play safe.

The scale and speed of adaptation required over the next half century are greater than any the human race has faced before. We must pass through a *Third Revolution*, just as momentous as the first two. It has already begun in small ways. Eventually it will affect all aspects of our lives, cultures and societies. The end result will be to reduce our impact on the environment to a sustainable level.

Somehow we must abolish poverty and achieve social and economic development for the world's majority. Yet we must do so without endangering the chances of future generations and other species. We cannot pick and choose which elements to work on. We must work on population, consumption and technology and on all the factors that influence them.

The most promising solutions to start with are those that involve the lowest costs

Knock on effect

and promise the greatest combination of benefits.

Least tractable will be the level of individual consumption. The world's 1.1 billion absolute poor must *increase* their consumption. The rest of us will not readily reduce it. Even in rich countries consumption continues to grow. Only a radical change in values and culture can slow it down voluntarily – especially if concern for Gaia achieves religious status on a wide scale. There is at least a chance that the present crisis will evoke such a value shift.

The most promising approach for the present is to reduce the impact of consumption levels by working on technology and population.

Technology change is crucial. It will succeed best where it involves low costs and high returns – as with energy conservation, soil and water conservation or agroforestry. But it cannot do the job alone. Energy efficiency grew in most parts of the world over the 1973–87 period – yet because of population and consumption growth, world energy use grew by 20 per cent.

Priority for Human Resource Development

Action on population may be the most promising avenue of all. It offers not just slower growth of numbers and lower eventual totals. Fewer, better timed births also promise lower infant mortality, improved mother and child health and better nutrition and education. Strong evidence suggests economic benefits too. In the 1980s the 50 per cent of countries with slower population growth saw average incomes grow 2.5 per cent a year faster than the 50 per cent with faster population growth.

This was not so much because faster economic growth leads to slower population growth. Population growth in 1965–80 explained 23 per cent of the variation in economic growth in developing countries in the 1980s. But economic growth in the earlier period explained only 7 per cent of the variation in population growth in the 1980s.

Change in reproductive behaviour can move as fast as the speediest shift in technology. In Thailand and Kerala fertility dropped from 6.5 to 2.3 children per woman in only twenty years.

But it is crucial to get the tactics right. *We are not talking about coercion or about low-quality programmes to throw contraceptives at the problem.* These approaches do not work: they abuse women's rights, spread distrust and actually slow down the spread of family planning. 'Population control' is impossible without shooting people.

The fastest results are achieved through a broad combination of measures which work together to multiply the speed of change. Foremost is the *extension of women's rights* to property, credit, jobs, equal pay and equal power. Equalizing *female education and literacy* with male, while raising both, empowers women to assert their rights. It also leads to better child health and nutrition and lower birth rates.

We also need *improved maternal and child health* to reduce infant mortality and give parents confidence that nearly all their children will survive.

Fertility will decline slowly of itself when these elements are present. But it will decline rapidly only with *universal access to good-quality family planning.* Services should be close to home and easily affordable. There should be a wide choice of methods, with good counselling and good medical back-up in the case of problems.

There is a very strong case for making this broad package of *human resource development* the central priority for aid to developing countries in the 1990s. It would improve the quality of life for women and children – and men too. It would lay the groundwork for faster economic growth. And it would slow population growth and environmental damage.

At the same time we must work on all those indirect factors that keep population growth and environmental damage high. There must be *an end to absolute poverty*, and *improved distribution* of land and other assets. We must work to make the *international economic order* more just. We must work to spread and perfect *democracy* and *local control* over the environment. We must *free markets* from unnecessary controls, while making them responsive to environmental and social costs.

Making a Difference

With serious efforts we could bring the population total for 2050 down to the low UN projection of 8 billion. After that world population could fall. The annual cost would be equivalent to about three days of military spending.

But the 1980s saw weak efforts. Human resources rated low in aid budgets. Government spending was hit by debt and adjustment. If these trends continue, we could see a world population of 12.5 billion in 2050, possibly rising to 21 billion in 2100.

The difference by 2050 would be 4.5 billion – one whole earth of 1980.

A difference of 4.5 billion people could determine whether we cross the critical thresholds – or stay on the safe side of them. If one day we face absolute ceilings on output of pollutants like carbon dioxide or chemical fertilizers, then lower populations could enjoy higher 'rations' per person. The individual 'ration' in a world of 8 billion people would be 56 per cent higher than in one of 12.5 billion.

Time is of the essence. Hamlet did not kill Claudius till his hand was forced – and lost his own life as a result of delay. If we wait till our hand is forced, it may be too late. We must act without delay.

INDEX